Lecture Notes in Biomathematics

Vol. 1: P. Waltman, Deterministic Threshold Models in the Theory of Epidemics. V, 101 pages. 1974. DM 16,–

Vol. 2: Mathematical Problems in Biology, Victoria Conference 1973. Edited by P. van den Driessche. VI, 280 pages. 1974. DM 28,–

Vol. 3: D. Ludwig, Stochastic Population Theories. VI, 108 pages. 1974. DM 18,–

Vol. 4: Physics and Mathematics of the Nervous Systems. Edited by M. Conrad, W. Güttinger, and M. Dal Cin. XI, 584 pages. 1974. DM 45,–

Lecture Notes in Biomathematics

Managing Editor: S. Levin

4

Physics and Mathematics of the Nervous System

Proceedings of a Summer School organized by the International Centre for Theoretical Physics, Trieste, and the Institute for Information Sciences, University of Tübingen, held at Trieste, August 21–31, 1973

Edited by
M. Conrad, W. Güttinger, and M. Dal Cin
Institute for Information Sciences, University of Tübingen

Springer-Verlag
Berlin · Heidelberg · New York 1974

Michael Conrad
Werner Güttinger
Mario Dal Cin
Institut für Informationsverarbeitung
Universität Tübingen
74 Tübingen/Köstlinstraße 6
BRD

Library of Congress Cataloging in Publication Data

Summer School on the Physics and Mathematics of the Nervous System, Trieste, 1973.
The physics and mathematics of the nervous system.
(Lecture notes in biomathematics ; v. 4)
1. Nervous system. 2. Biological physics.
3. Biomathematics. I. Conrad, Michael, ed.
II. Güttinger, W., ed. III. Dal Cin, M., 1940- ed.
IV. International Centre for Theoretical Physics.
V. Tübingen. Universität. Institut für Informationsverarbeitung. VI. Title. VII. Series.
[DNLM: 1. Biophysics--Congresses, 2. Information theory--Congresses. 3. Mathematics--Congresses.
4. Neurophysiology--Congresses. W1 LE334 b. 4/ WL102 S955n 1973]
QP355.2.S85 1973 599' .01'88 74-23218

AMS Subject Classifications (1970): 68-XX, 68-02, 68A10, 68A15, 68A20, 68A25, 68A30, 68A35, 68A40, 68A45, 92-XX, 92-02, 92A05, 92A25, 93-XX, 93-02, 93A99, 93B99, 93C10, 93C15, 93C20, 93C25, 93 C30, 93C40, 93C45, 93C55, 93C60, 93D99, 94-XX, 94-02, 94A05, 94A10, 94A15, 94A20, 94A25, 94A30, 94A55

ISBN-13: 978-3-540-07014-6 e-ISBN-13: 978-3-642-80885-2
DOI: 10.1007/ 978-3-642-80885-2

Preface

This volume is the record and product of the Summer School on the Physics and Mathematics of the Nervous System, held at the International Centre for Theoretical Physics in Trieste from August 21-31, 1973, and jointly organized by the Institute for Information Sciences, University of Tübingen and by the Centre.

The school served to bring biologists, physicists and mathematicians together to exchange ideas about the nervous system and brain, and also to introduce young scientists to the field. The program, attended by more than a hundred scientists, was interdisciplinary both in character and participation.

The primary support for the school was provided by the Volkswagen Foundation of West Germany. We are particularly indebted to Drs. G. Gambke, M.-L Zarnitz, and H. Penschuck of the Foundation for their interest in and help with the project. The school also received major support from the International Centre for Theoretical Physics in Trieste and its sponsoring agencies, including the use of its excellent facilities. We are deeply indebted to Professor A. Salam for his kind cooperation and also to Professor P. Budini, Dr. A.M. Hamende, and to the many members of the Centre staff whose hospitality and efficiency contributed so much to the success of the school. We are pleased to acknowledge the generous aid and cooperation of the University of Tübingen and would like to thank its President, A. Theis, for his personal engagement and for his opening address. We would also like to thank Professor John Taylor for his efforts throughout the organizational process and during the school.

Finally we would like to express our debt to Mrs. K. Schwertfeger and Mrs. M. Jochim for their help in typing a difficult manuscript, to Mrs. D. Conrad for her invaluable help throughout, and also to Mrs. A. Peters of Springer Verlag for her patient cooperation.

M. Conrad

W. Güttinger

M. Dal Cin

ORGANIZING COMMITTEE

Werner Güttinger (Co-Chairman)
Abdus Salam (Co-Chairman)
Michael Conrad
Mario Dal Cin
John G. Taylor

SESSION CHAIRMEN AND WORK GROUP LEADERS

H. Barlow
L. Bianchi
V. Braitenberg
H. Bremermann
M. Conrad
M. Dal Cin
J. George
W. Güttinger
H. Hahn
C. Legéndy
R. Livingston
H.H. Pattee
E. Pfaffelhuber
W. Precht
A. Salam
J.G. Taylor

SPONSORING ORGANIZATIONS

Volkswagen Foundation
International Centre for Theoretical Physics
University of Tübingen

CONTENTS

PART VIII Dynamical and Chemical Systems

Introduction

The Summer School on the Physics and Mathematics of the Nervous System had the double task of acculturating newcomers to the problems of neuroscience and crossbreeding ideas from different disciplines at a fairly sophisticated level. To achieve this, it was divided into lectures and workgroup sessions, with the lectures beginning with the role of physics and mathematics in biology and then turning (not always in strict temporal sequence) to the molecular biophysics of the nervous system and brain, the biophysics of nerve cells, sensory perception, network approaches to the brain, and artificial intelligence. The workgroups dealt with relevant mathematical or physical ideas (from automata, information, and dynamical systems theory) and with discussion built around certain lecture topics (especially in the workgroup on neural nets).

The flow of ideas in the book reflects that basic organization. Part I consists of two papers dealing with the mathematical and physical problems surrounding the description of biological systems, one from a physicist's and the other from a biologist's point of view. The first of these, by W. Güttinger, describes the qualitative, topological approach to complex biological phenomena and the second, from R.B. Livingston, discusses the unique complexities of biological systems and the kind of mathematics and physics which must be developed to deal with these complexities.

The book then turns, in Part II, to the nervous system and brain at the molecular level. This part starts with E. Neumann's model of the chemical control of electrical potential changes in neuronal membranes, moves on to a paper, by M. Conrad, on molecular information processing and its relation to learning and memory, and concludes with a close analysis, by H.H. Pattee, of the physical nature of switching processes and its implications for the brain and the computer.

The papers in Part III come from the group of lectures on cellular biophysics and sensory physiology. P. Fatt begins by describing transmitter induced changes in the postsynaptic cell and together with G. Falk gives a discussion of presynaptic processes and limitations to single photon sensitivity in vision. H. Barlow then reviews the story of visual perception and visual acuity, which actually leads up to his discussion of redundancy and perception in Part VII. J. Roederer closes out the section with a description of the auditory system and some notes on the perception of musical sound.

The papers in Part IV represent the group of lectures on neural networks, cerebellum, and cerebrum. J. Taylor describes the network approach and the basic facts about the brain relevant to this approach, W. Precht guides us through the essential facts and theories about the cerebellum, C. Legéndy puts forth a theory of early learning, and V. Braitenberg outlines a new interpretation of the structure of the cerebral cortex. The section ends with J. Taylor's summary of the workgroup on neural nets (since the topics discussed in this group are most relevant to the papers in this part and also Part III).

The contributions in Part V deal with artificial intelligence and the functional approach to biological systems. H. Bremermann begins by describing and developing new ideas of complexity and their fundamental importance for artificial and natural systems and also for the physical limits of computation. B. Ulrich then describes a model neuron based on the Josephson effect with potential applications for the construction of artificial neural nets and O. Rössler describes a new approach to the brain based on the functional requirements of natural behavior. B.D. Josephson concludes the section by describing his artificial intelligence/psychology approach, overviewing its potential impact on brain science, and making some asides on language and quantum mechanics.

The papers in Part VI come from the workgroup on molecular and modifiable automata. The main theme of this group was the relation between structure and function in automata and biological systems, the mechanism and formal description of transformation of function, and the relation between transformation of function and learning. W. Merzenich begins the section with an introduction to finite automata in the context of tessellation structures and also develops the idea of computation spaces. R. Vollmar takes us into the world of Turing computability and describes the properties of Turing machines with modifiable tape structures. O. Rössler shows how to simulate automata with chemical reaction networks and M. Conrad formalizes the molecular folding processes which underlie gradual transformation of function and describes their connection to the concept of programmability and the nervous system. E. Dilger then directs the discussion to automata whose reliable function is based on structural and dynamical redundancy and shows how adaptable components allow for remarkable reliability properties. M. Dal Cin concludes the section by weaving the relation between reliability and adaptability into the formal structure of tolerance geometry and developing a scheme for describing the performance of modifiable automata and their learning processes in terms of this geometry.

The next set of papers (Part VII) comes from the workgroup on information theory and learning in biology. H. Barlow begins by suggesting some new ideas about the role of environmental redundancy in perception, E. Pfaffelhuber formalizes notions about redundancy and perception into information theoretical terms, and R. Heim describes an approach to information theory capable of putting these ideas on an algorithmic basis. E. Pfaffelhuber closes out the section with an overview of the program and summary of the workgroup discussion.

The final section (Part VIII) contains contributions from the workgroup on dynamical and chemical systems. J. George starts with an overview of dynamical systems theory, especially bifurcation theory, and its potential applications. G. Meyer gives a tutorial discussion on the basic principles of biochemical kinetics and H. Hahn provides a mathematical justification of the pseudo-steady-state hypothesis in biochemistry, with comments on the relation between the hypothesis and catastrophe theory. O. Rössler concludes the section with an excursion into the strange world of exotic enzyme kinetics (including exotica analogous to the nerve impulse and conditioning). The book ends with John Taylor's concluding impressions of the school.

The above papers describe the nervous system and related problems from diverse points of view as well as diverse points of view designed to describe the nervous system. This is a reflex of the Trieste Summer School, with its interdisciplinary spirit, with its invitation to new approaches, and its recurrent switching between mathematics, biology, and physics.

The Editors

PART I

Physics and Mathematics in the Biological Context

CATASTROPHE GEOMETRY IN PHYSICS AND BIOLOGY*

W. Güttinger
Institute for Information Sciences,
University of Tübingen, Tübingen,
Federal Republic of Germany**

and

Department of Physics and Astronomy,
University of Wyoming, Laramie, Wyoming, USA

1. Introduction

In contrast to mathematics, the biological and human sciences are interested in life and the course it takes. If, up to now, these sciences are so little mathematized and remain inexact, this is because our immediate intuitive and qualitative comprehension of biological, social and economic facts is entirely sufficient to cope with the necessities of life. Even in the physical sciences an intuitive, phenomenological approach to complex problems generally is often superior to, and more effective than, precise quantitative methods and, indeed, is always at the root of new fundamental conceptions.

Relying on a qualitative understanding of processes that have proved themselves too complex for rational quantitative analysis is a basic human instinct. Whenever we run into difficulties or are overwhelmed by a mass of data and contradictions -- as in biology and economics today -- it provides us with conceptual guidance to single out the most significant phenomena and to simplify matters to the point where we can come to terms with them. This may be a matter of common sense but, then, finding out just how this structural approach to complex phenomena comes about would add a new dimension to man's knowledge.

Attempting a qualitative description of, and gaining global geometric insight into complex systems dates back to H. Poincaré toward the end of the nineteenth century [1]. Out of it topology was born. It soon followed quite entangled paths in the hands of increasingly abstract mathematicians. Thus it took a hundred years until R. Thom returned to topology's intuitive geometric roots, making it into a fundamental and universally valid basis for a qualitative, structural understanding of complex phenomena

---

*Work supported in part by the German Science Foundation

**Author's present address

in the animate and inanimate world [2]. Truly enough, ever since Pythagorean times the Gods have made geometry. But it took Thom's genius to recognize that growth and change of forms must be attributed to a genericity hypothesis that, in every circumstance, nature realizes that morphology which locally is the simplest and the least fragile compatible with given local initial data. The ensuing concept of structural stability (insensitivity of a system against slight perturbations) leaves but a few possible modes of sudden qualitative changes of structure and behavior -- Thom's elementary catastrophes --, at least as long as nature behaves reasonably smooth. This discovery has, of course, far-reaching consequences, in particular for those inherently qualitative sciences which, justifiably, rely more upon intuition than upon precise quantitative mathematical description, notably biology and the human and behavioral sciences, economics and all that.

However, the observed accordance between empiric morphologies and topology -- this preestablished harmony between nature and mathematics -- has both a geometric and a physical origin. Catastrophe theory by itself can only explain the local geometry and not the forces that are shaping it. On the other hand, looking at local forces alone cannot explain the geometry unless structure formation can be attributed to local minima of entropy production. This presents one of the most challenging problems at this time, viz., to unify catastrophe geometry with the thermodynamics of dissipative structures [3].

The reader may very well pause at this point and ask what on earth we are talking about and why, for that matter, bother with catastrophes in biology. Let me try to answer this question by giving an example from motivation analysis. Rational behavior is a rather rare activity of our nervous system, especially when we are angry and frightened at the same time and then jump from attack to flight and vice versa. Attack and flight are two extreme forms of aggressive behavior caused by the two conflicting drives rage and fear. Rage-only causes attack, fear-only causes flight, neither causes neutral behavior, but both competing drives mixed together cause either attack or flight, depending on chance if equally strong. The catastrophes are the sudden qualitative changes of behavior. It is precisely this story of drives and behavior, causes and effect, K. Lorenz's dogs and N. Tinbergen's seagulls are telling us [4], e.g., the dog in Fig. 1a that faces a rival. It is a topological theorem, due to Thom and elucidated by E.C. Zeeman [5], that this elementary behavior catastrophe has to be visualized as the surface of an overhanging cliff whose middle sheet represents the unstable situation and whose two edges, when projected onto the control plane made up by the two drives, form a cusp inside of which the two drives are in conflict (Fig. 1b).

But what if, as time goes on, a third competing drive -- say, love, w -- comes into play? Then the single cusp eventually bifurcates into three cusps enclosing a pocket with three conflicting regimes as a new organizing center around the point O in Fig. 1c which marks the beginning of a terrace inside the cliff.

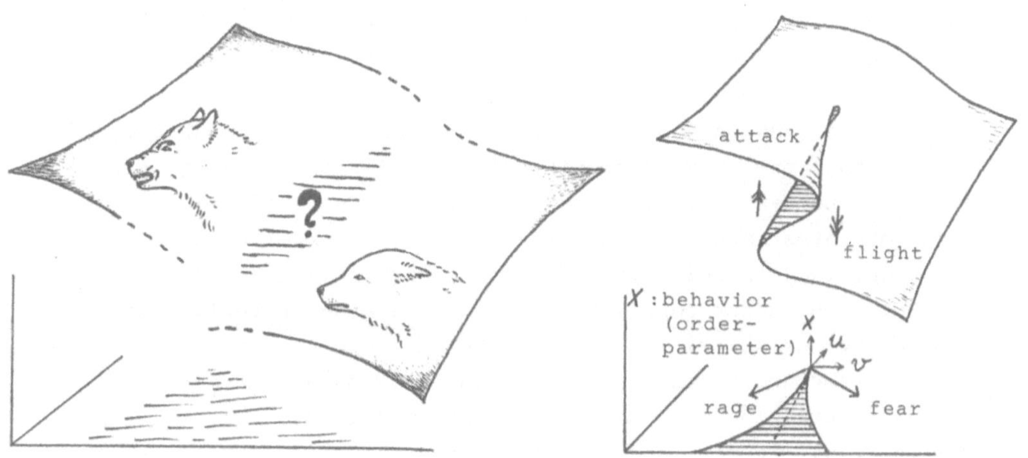

Figure 1a

Figure 1b

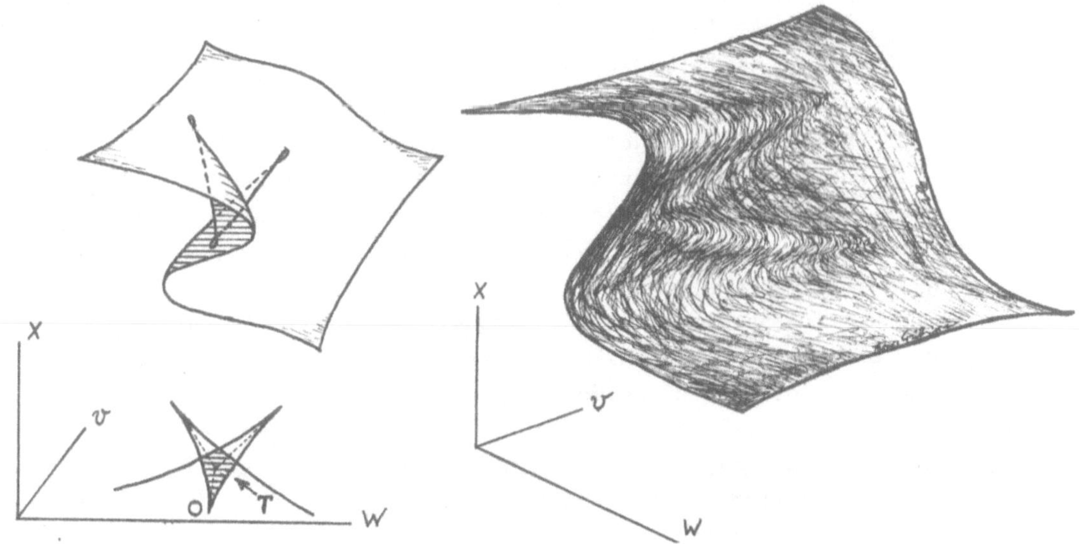

Figure 1c

As another example, these graphs gain immediate physical significance as phase transition surfaces (properly amputated according to the 2nd law of thermodynamics) when Fig. 1b is viewed as van der Waals' diagram for liquid-gas transition [6,11] and T in Fig. 1c is recognized as tricritical point in ternary fluid mixtures, metamagnets, etc. [7,11].

Throughout nature we observe such fascinatingly similar critical phenomena: A laminar flow suddenly turns turbulent, rain turns suddenly into snow crystals, plasmas exhibit shock waves and nerve cells suddenly fire. A gradual relaxation after an economic squeeze causes an inflationary burst, slow changes in environment cause sudden evolutionary or cultural breakdowns, and nations suddenly go to war. Their common characteristics is that one or more significant behavior variables or order parameters (in Landau's terminology) x,y,... undergo sudden large, discontinuous or catastrophic changes if their slow, competing but continuous driving controls u,v,... cross a bifurcation set (the cusps in Figs. 1b,c) and enter conflicting regimes causing instabilities in the behavior variables. The latter measure the degree and kind of ordering or structure which is built up (or destroyed) near the bifurcation set. Examples of behavior variables (x in Fig. 1b) are aggressiveness, particle density and magnetization. Typical control variables (u,v,...) include rage and fear, pressure, temperature, and magnetic field in ferromagnets. One or more controls (v in Fig. 1b) drive the system from one "phase" to another. Other control variables (u in Fig. 1b) drive the "orthogonal" changes, i.e., toward or away from the cusp vertices, and thus act as splitting factors because they separate the behavior pattern into its extremes.

Common to all these catastrophe phenomena (besides their qualitative similarity) is their universality expressed by the fact that the details of the system undergoing sudden transitions are almost irrelevant. While this empirical evidence is at the root of Landau's phenomenological theory of phase transitions [8] and thus hints at thermodynamic principles, already the dimensional analysis underlying familiar scaling laws [9] points to a qualitatively invariant description of the phenomena under consideration. This means that we have to disregard algebraic structures for a while (we cannot add two phases to yield a third one), and confine ourselves to precisely those two concepts which alone appear sufficient to allow qualitative conclusions, viz. order structures and topology. It amounts to saying that the qualitative laws of nature are written in the language of thermodynamics and geometry. This I shall now explain.

2. Dynamical Systems

To explore the topological features of an autonomous dynamical system we describe its temporal evolution by the solutions of the first order differential system

$$dx_i/dt = X_i(x_j,\mu) \quad \text{or} \quad \dot{x} = X(x,\mu) \; . \tag{1}$$

The state variables $x_i(t)$, depending on time t and forming a vector $x \in R^n$, may represent concentrations of interacting chemical substances, particle positions, etc. The X_i, depending on parameters μ describing external factors, define a vector field X in R^n which, by the existence theorem, determines a unique flow ϕ in R^n. Given a set of initial data, there is a unique trajectory (parametrized by t) or solution curve $x = \phi(t,\mu)$ through each point of R^n, i.e., a curve to which the vector field is everywhere tangent, X giving the speed of the motion at x [10].

The linear harmonic oscillator $z + \gamma z + z = 0$ provides the simplest example when written as a first order system $x = y$, $y = -x-\gamma y$ in position-velocity space R^2 of points $x = z$, $y = z$. Its trajectories around the stationary point O, $x = y = 0$, which is reached asymptotically for $t \to \pm\infty$, are shown in Fig. 2.

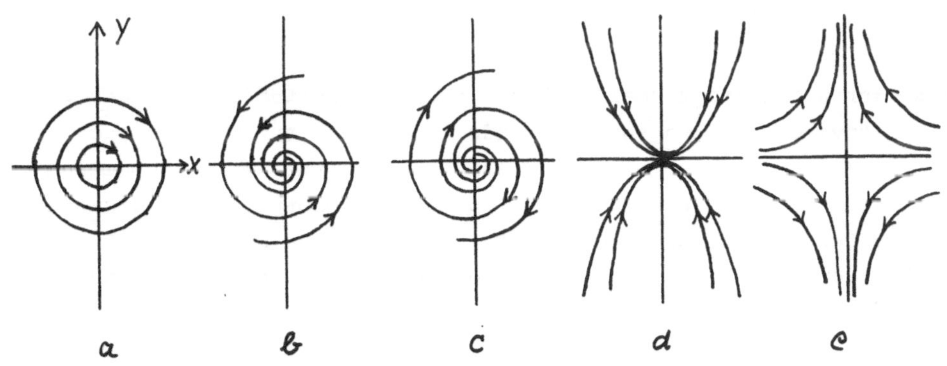

Figure 2

For $\gamma = 0$ the orbits are concentric circles around the origin O which is a center (Fig. 2a). For $\gamma > 0$, O is an attractor with everything outside spiraling toward it (Fig. 2b), while for $\gamma < 0$, O is a repellor (Fig. 2c). In the aperiodic case (Fig. 2d) O is a stable node. O becomes a saddle point (Fig. 2e) if the last term z in the oscillator equation is replaced by $-z$. New phenomena are springing up when nonlinearities come into play, as in van der Pol's self-sustained relaxation oscillator obtained by replacing the constant damping γ by an amplitude-dependent one, $\gamma = 3z^2-\mu$, μ real, to yield $\ddot{z} + (3z^2-\mu)\dot{z} + z = 0$ or the equivalent system $\dot{x} = -x^3 + \mu x - y$, $\dot{y} = x$. For $\mu < 0$, the stationary point O, $x = y = 0$, is a stable attractor as in Fig. 2d. If μ crosses the value $\mu_c = 0$ and becomes positive, O changes into a repellor and there appears a periodic, self-sustained oscillation or limit cycle Γ (Fig. 3a) which itself is a stable attractor: Everything inside spirals out toward the closed orbit Γ and everything outside spirals in. We call $R^2 - O$ the basin of attraction of Γ. It is no accident that a plane u = const. through the cliff of Figs. 1b,8 gives the dotted $\mathfrak{z}$-shaped curve in Fig. 3a.

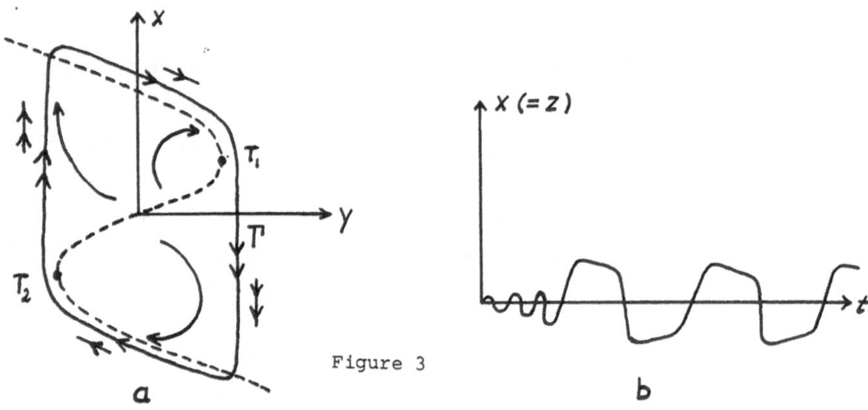

Figure 3

Indeed, for $\mu > 0$ and z small (when $\gamma < 0$) the system absorbs energy from its environment and the amplitude increases, but when it becomes too large ($\gamma > 0$) energy will be dissipated and the amplitude decreases again. Then the game starts anew (Fig. 3b). Absorption and dissipation of energy tend to balance each other throughout the motion ($\oint_\Gamma dE = 0$) and, whatever the initial state of the system, it tends to the periodic oscillation Γ. Along nearly horizontal sections of the trajectories the motion is very slow (single arrows) whereas the representative phase points move very fast along nearly vertical sections of the orbits (double arrows): slow and fast motions are transversal except for the thresholds T_1, T_2 where they undergo sudden transitions (catastrophes). Identifying the hysteresis phenomenon underlying van der Pol's relaxation oscillator with the behavior of transistors and ferromagnets [11], circadian clocks [12], the heartbeat [13] and other oscillatory systems nowadays is a familiar exercise.

A singular or stationary point $x_o = x_o(\mu)$ of the flow described by Equ. (1) is defined by $X(x_o,\mu) = 0$. The behavior of a flow near, and the stability properties of, a singular point are determined by the location of the eigenvalues $\lambda = \lambda(\mu)$ of the Jacobian matrix $J = (\partial X_i/\partial x_k)_{x_o} = J(x_o)$ in the complex λ-plane. This is because in the linear approximation to Equ. (1) the solutions vary in time as $\exp(\lambda t)$. The point x_o is an attractor (stable equilibrium) of the system if $Re\lambda(\mu) < 0$ for all λ. Suppose now that, as μ increases beyond a critical value μ_c, a pair of complex eigenvalues cross the imaginary axis. Then, for $\mu > \mu_c$, the point x_o is no longer attracting but usually changes into a repellor and there appears a new, isolated attracting periodic solution Γ_μ of (1) (limit cycle) close to x_o (Fig. 4) whose amplitude is proportional to $\sqrt{\mu-\mu_c}$ in first approximation.

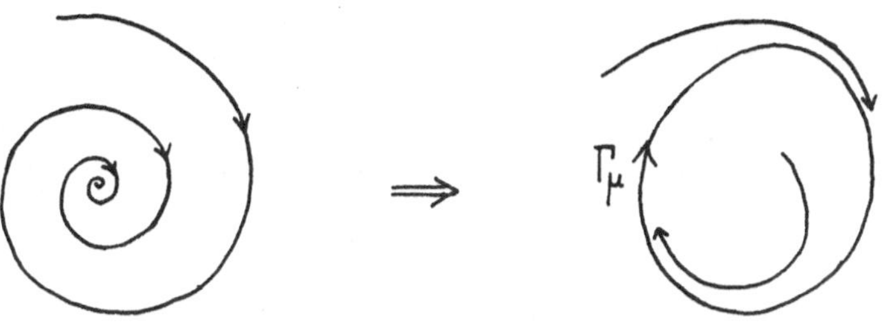

Figure 4

This phenomenon, where out of a stable structure a qualitatively new structure emerges, is called (Hopf-) bifurcation. The limit cycle Γ_μ may lose its stability as μ increases further past another critical value μ_c' where the periodic oscillation may bifurcate into one of longer period or into a nonperiodic, "turbulent" one. Bifurcation phenomena are typical of nonlinear dissipative systems and have been observed in various chemical reactions [14], hydrodynamics (turbulence) [15], etc. but the theory is still in its infancy [16] with the exception of gradient systems [2]. The onset of bifurcation may be understood geometrically by visualizing singular points x_o as loci of intersections of hyperplanes defined by the equation system $X(x_o,\mu) = 0$. For example, in $R^2(x_1,x_2)$ a singular point x_o is given by the intersection of the two solution curves C_1, C_2 of the equations $X_1(x_1,x_2,\mu) = 0$, $X_2(x_1,x_2,\mu)$

= 0, respectively. For a given μ, C_1 and C_2 may intersect transversally (i.e., "without contact") as in Fig. 5a (x_o non-degenerate: $\det J \neq 0$) or with a common tangent (i.e., "with contact") as in Figs. 5b,c (x_o degenerate: $\det J = 0$). Changing μ results in a deformation of the curves, say, $C_2 \to C_2' = C_2 + \delta C_2$, which in turn gives rise to new intersections near x_o and new types of singularities (if $\det J = 0$) as shown in Figs. 5d,e, and, consequently, to new topological situations and modes of behavior. We remark that the distinction between intersections with and without contact underlies Ehrenfest's original classification of phase transitions.

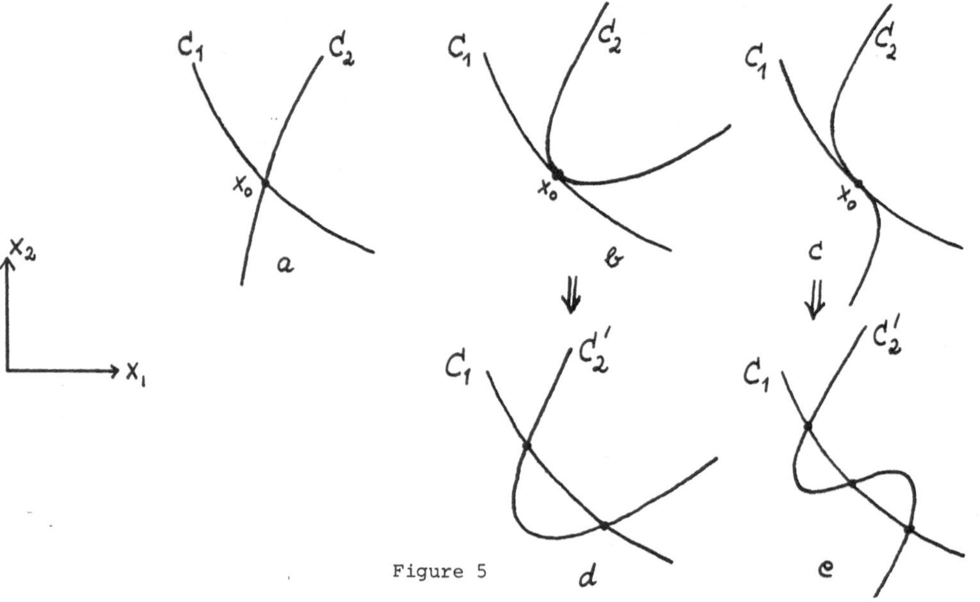

Figure 5

After all that has been said above, physicists who are used anyway to believing that the world is made up of oscillators will have no problems in accepting the following generalizations, though their differential-topological formulation is difficult. We define a dynamical system (M,X) as a smooth vector field X on a smooth (and compact) manifold $M \subset R^n$. Smooth means sufficiently differentiable. We call "X-orbit" a curve in M to which X is everywhere tangent. X determines a unique flow ϕ on M (i.e. a smooth map ϕ: $R \times M \to M$ such that $\forall t \in R$, ϕ: $t \times M \to M$ is a diffeomorphism). The basic geometric units making up the system are (i) a set of fixed points: attractors (sinks), repellors (sources) and saddle points, (ii) a set of periodic orbits or limit cycles (attractors and repellors) [adding saddle-type "horseshoes", the system is called a

Smale flow if the number of units is finite]. Besides, there may be centers which, however, we shall soon eliminate. We might further expect that almost all of the space M may be partitioned into the basins of attractors (towards which almost all orbits eventually flow) and that this geometric structure characterizes entirely the qualitative behavior of the system.

The aim of this construction is not to study individual trajectories (or solutions to Equ. (1)) but to describe qualitatively the family of all the orbits as a whole from a global geometric point of view, and to find out if and how the system undergoes drastic qualitative changes of shape and form under perturbations.

3. Structural Stability

Many systems and structures encountered in nature enjoy an inherent stability property: they preserve their quality under slight distortions. Otherwise we could hardly think about or describe them, and today's experiment would not reproduce yesterday's result. We give the scheme (M,X) some content by this hypothesis of structural stability. We call the dynamical system (M,X) structurally stable if, for a sufficiently small perturbation δX of the vector field X, the perturbed system $(M, X+\delta X)$ is, roughly speaking, topologically equivalent to the unperturbed system. Thus the system (1), $\dot{x} = X$, is structurally stable if there exists a continuous one-to-one map that carries each trajectory of the perturbed system (with $|\delta X| < \varepsilon \quad |\partial \delta X| < \varepsilon$)

$$\dot{x} = X + \delta X \tag{2}$$

onto a trajectory of the unperturbed one.

By tilting all arrows (X) in the flow of Fig. 2a of the undamped harmonic oscillator slightly inwards (by δX), all the circular orbits change their quality basically and become infinite spirals towards the center, as in the damped oscillator (Fig. 2b). Thus, the undamped oscillator is structurally unstable, and so are all conservative Hamiltonian systems: adding a small damping term results in energy dissipation, the trajectories spiral down to energy minima and the dissipative flow is no longer equivalent to the conservative flow. On the other hand, dissipative systems such as the damped oscillator (Fig. 2b) and van der Pol's oscillator (Fig. 3) are structurally stable (for $\gamma > 0$ and $\mu > 0$, respectively): tilting the vectors by δX turns spirals into spirals. Considering a system (1) controlled by parameters μ, we may interpret a small change in μ as a perturbation δX. Then, according to what has been said in the context of Fig. 5, the system may suddenly change in quality, attractors might bifurcate or coalesce if singular points degenerate and μ crosses a critical set. The study of such changes is bifurcation and catastrophe theory.

4. Catastrophe Theory

To characterize structurally stable systems we must consider the whole set $\mathbb{D}$ of vector fields on M, i.e., a space $\mathbb{D}$ of dynamical systems, and choose an equivalence relation and a topology on the set $\mathbb{D}$. We may call X, $X' \in \mathbb{D}$ equivalent if a homeomorphism of M exists throwing X-orbits on X'-orbits, and choose a C^0- (or C^1-) topology on $\mathbb{D}$, i.e., define X,X' ($X' = X+\delta X$) to be close if the vectors (and their derivatives) are close. Then define $X \in \mathbb{D}$ to be structurally stable if it has a neighborhood of equivalents in $\mathbb{D}$. Besides classifying structurally stable systems, the outstanding question is to know if they are dense, that is, if any system can be approximated by structurally stable ones, so that almost all systems are structurally stable and one can ignore the unstable residue. This difficult problem gave rise to deep mathematical studies [10, 18] showing for example, that Smale flows (Sec. 2) are structurally stable and dense in the C^0-topology. Suppose we had developed a good stability theory and could decompose $\mathbb{D}$ into an open-dense set $\mathbb{S} \subset \mathbb{D}$ of structurally stable (stable, for short) systems and the complementary bifurcation set $\Sigma = \mathbb{D} - \mathbb{S}$ of unstable systems, $\mathbb{D} = \mathbb{S} \cup \Sigma$ (Fig. 6).

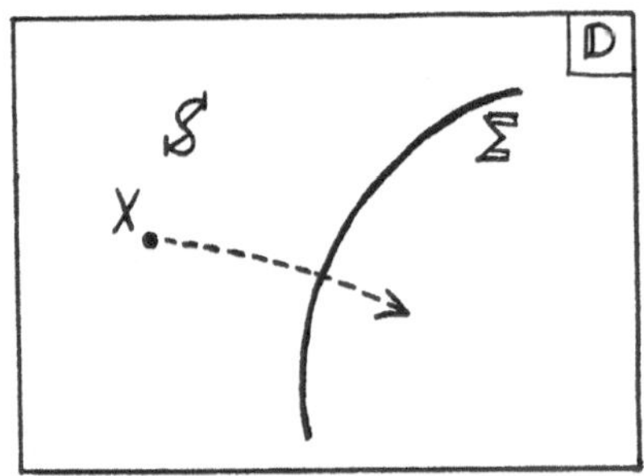

Figure 6

Consider then a system (X in Fig. 6) controlled by a parameter. The system's change can be represented by an arc in $\mathbb{D}$. If the arc crosses Σ, the system changes its quality. The crossing points are called catastrophes.

Naturally enough, the geometric structure of the bifurcation set of vector fields may be very wild, and the only category for which the program has thus far proved feasible is that of a local gradient dynamics $V: M \to R$, $X = -\text{grad } V$. Its singularities make up those of the map V, and perturbing the system is related to the problem of stability of differentiable maps between two manifolds (cf., however [17]). Notwithstanding its limitations, gradient systems offer an exceptionally rich and beautiful approach to qualitative dynamics.

Let the vector field $X = X(x,\mu)$ be derived from a real-valued, differentiable potential $V = V_\mu(x)$, depending on a parameter set μ,

$$X(x,\mu) = -\underset{x}{\text{grad}}\, V_\mu(x), \tag{3}$$

so that Equ. (1) becomes the gradient system

$$\dot{x} = -\underset{x}{\text{grad}}\, V_\mu(x) \quad \text{or} \quad \dot{x}_i = -\partial V_\mu(x)/\partial x_i \quad . \tag{4}$$

The singular points x_o (attractors, repellors, saddles) of the flow ($X(x_o) = 0$) are then just the local extrema (minima, maxima, inflexion points) of the potential V, grad $V(x_o) = 0$, i.e., its stationary or equilibrium points. We call x_o structurally stable (or a non-degenerate quadratic point of the map $x \to \nabla V$) if the Hessian $H(x_o) := |\partial^2 V/\partial x_i \partial x_k|_{x=x_o} \neq 0$ (i.e., if $\det J \neq 0$, cf. Sec. 2). Around such a point, V is a non-degenerate quadratic form and perturbing V by δV ($\partial^\nu \delta V/\partial x_i^\nu$ small for all ν and i) does not change its shape. For example, the potential $V = x^2$ ($x \in R^1$) is invariant against small perturbations $\delta V = ax + b$ near its stationary point $x_o = 0$: $V + \delta V$ has the same form as V because $H(0) = V''(0) \neq 0$ (Fig. 7a). If, however, $H(x_o) = 0$, i.e. if x_o is not structurally stable but is a degenerate or inflexion point, then a perturbation of V by δV results in a number (assumed finite) of topologically and physically different types of potentials $V + \delta V$, called universal unfoldings of V (or of x_o). For example, a perturbation of $V = V_o = x^3/3$ ($x \in R^1$) around its stationary point $x_o = 0$ by $\delta V = ax^2 + bx + c$ gives $V_b := V + \delta V = x^3/3 + bx$ (since the quadratic and constant terms can be removed by translations). This perturbed potential $V_b(x)$ is quite different from $V_o = V_{b=0} = V$ in shape for $b < 0$ (where it has two extrema) and $b>0$ (where it has none): Fig.7b. The point $x_o=0$ ($H(0)=V''(0)=0$)

is structurally unstable and so is the form of $V = V_o$, in that a small change of the parameter b in V_b around b = 0 changes the shape of V_b completely. On the other hand, changing b a bit within the domains $b > 0$ or $b < 0$ leaves the form of V_b qualitatively unaffected. For $b > 0$ or $b < 0$, $V_b(x)$ remains geometrically the same, i.e., V_b is structurally stable under small parameter variations in these domains. We call V_b the "universal unfolding" of $V = V_o$ and b = 0 a "catastrophe point" of the family of potentials $V_b(x)$ that separates two stable regions $b \lessgtr 0$. V_b determines a fold curve. The singularity x=0, or V_o, acts as an "organizing center".

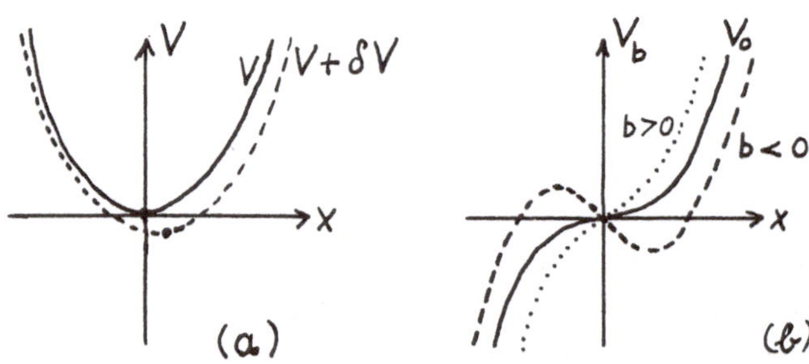

Figure 7

Similarly, the potential $V_{oo} = x^4/4$ has a degenerate minimum at $x = x_o = 0$. The perturbed potential or universal unfolding of V_{oo} is given (near x = 0) by

$$V_{uv}(x) = x^4/4 + ux^2/2 + vx \tag{5}$$

with two "unfolding parameters" u,v (cubic and constant terms can be removed by translations, and higher powers of x are negligible in a neighborhood of x = 0). The stationary points x_o of $V_{uv}(x)$ are given by the roots of the equation

$$S: \quad \text{grad}V = dV_{uv}(x)/dx = x^3 + ux + v = 0. \qquad (6)$$

The set $S(V_{uv}): x = x_o = x(u,v)$ of these equilibrium points form a surface S in (x,u,v)-space $R^3 = R^1 \times C$, $C = R^2$, showing the beginning (at $x=u=v=0$) of an overhanging cliff made up of two folds (Figs. 8a, 1b), which is best modeled by a rubber sheet. For fixed $(u,v) \in C$ (control space) the points $x = x_o$ $(=x_1, x_2, x_3)$ on S give the local minima (regimes) and maxima of V_{uv} at (u,v), corresponding respectively to the upper or lower and the middle sheet of the surface S. The various forms of $V_{uv}(x)$ when u,v vary in C are shown in Fig. 8b. The cliff of Fig. 8 (and the double cliff of Fig. 1c) may be modeled by rubber sheets. The analogy of (6) to the Van der Pol case is obvious.

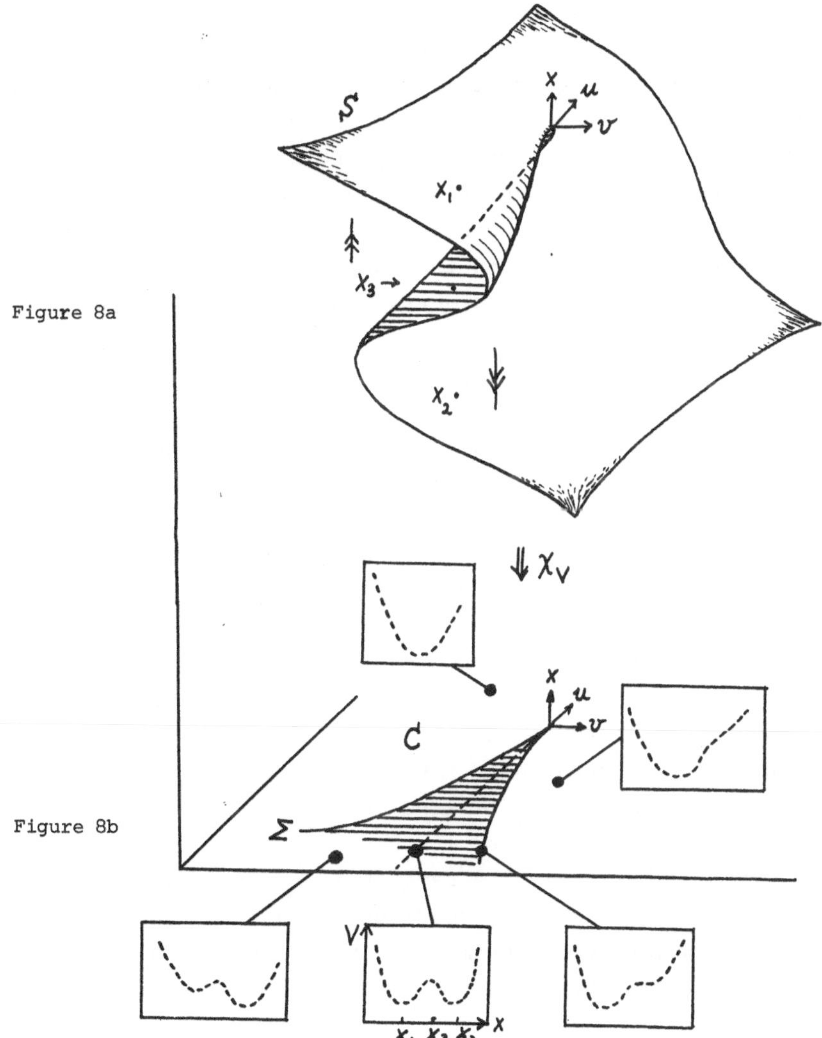

Figure 8a

Figure 8b

The cusp (bifurcation set) Σ in Fig. 8b is the (vertical) projection onto the plane C of the two edges of the cliff S of Fig. 8a (where the tangents to S become vertical: $V''_{uv}(x) = 0$), and thus follows by eliminating x from Equ. (6) and

$$d^2V_{uv}(x)/dx^2 = 3x^2 + u = 0 \tag{7}$$

to yield the cusp equation

$$\Sigma: \quad 4u^3 + 27v^2 = 0. \tag{8}$$

Σ is the set of (u,v)-values for which two stationary points (a minimum and a maximum) of V_{uv} coalesce. Imagine V_{uv} to describe locally a geographic landscape varying in shape by forces u,v which change the number, position and relative heights of valleys. Suppose that the stable states accessible to the system are given by the non-degenerate quadratic minima of V_{uv} (the attractors reached by the trajectory of a material point) and corresponding to the upper and lower sheets of the cliff, while the maximum represents the unstable state and corresponds to the cliff's middle sheet. Then, crossing the cusp Σ from inside ($4u^3 + 27v^2 > 0$) to outside ($4u^3 + 27v^2 < 0$) or vice versa annihilates an attractor or generates a new one, and the state must either flow towards a different attractor or decide between two possible minima where to stay or go on to, until one of them disappears when the other branch of the cusp is crossed. If the speed of the flow is large compared with the speed of change of control--which in general is the case and can be formally achieved by multiplying the l.h.s. of Equ. (4) by a small parameter ε--then the system will appear to jump into a qualitatively new state when Σ is crossed. These jumps, from the cliff's upper sheet over its edge down to the lower sheet, or up at the other side, are the "catastrophes" and the cusp Σ is the catastrophe set. The two examples given in Sec. 1 in the context of Fig. 1b illustrate precisely this "cusp catastrophe". Why is it so fundamental?

We have embedded the potential $V = V_{oo}(x)$ (a germ) into the two-parameter family of potentials $V_{uv}(x)$ (the unfolding of V_{oo}). While V_{oo} is a very fragile object (because its singular point is degenerate) the forms of V_{uv} remain basically the same inside or outside the cusp, respectively: $V_{uv}(x)$ is stable under small perturbations not crossing Σ. The cusp is a singularity of the map $\chi = \chi_V$ of the surface S onto the plane C, $\chi: S \to C$, induced by the projection $R^3 = R^1 \times C \to C$. Intuitively, bending S (or V_{uv}) smoothly without creasing the rubber sheet (by a diffeomorphism h) into a surface S', and deforming C smoothly into another plane surface C' by a change of coordinates (diffeomorphism k) does not disturb the qualities of the cusp catastrophe: the ensuing map χ' of S' onto C', $\chi':S'\to C'$, remains equivalent to χ (i.e., equal up to smooth transformations). Consider then the space $\mathbb{V}$ of all 2-parameter families of potentials $V = V(x;u,v)$ and their surfaces S_V of stationary points defined by $dV/dx = 0$. Let us call $V \in \mathbb{V}$ structurally stable or generic if it has a neighborhood of equivalents, i.e., if χ_V is equivalent to $\chi_{V'}$ for all $V' = V + \delta V \in \mathbb{V}$ close to V. Then, and this is the content of the Thom-Whitney theorem, the only singularities (defined by $dV/dx = d^2V/dx^2 = 0$) of the projection $S_V \to C$ are folds and cusps. This means that the most complicated behavior that can happen locally is the cusp catastrophe of Fig. 8 and that any $V(x;u,v)$ (u,v playing the role of μ in Equs. (3), (4)) is locally equivalent to $V_{uv}(x)$, with the cusp singularity characterizing the breakdown of stability. While this is a geometric fact, its physical meaning is obvious: if a system with order parameter x is slowly driven from one phase to another by a control v, and an orthogonal drive u sets in to split the quality of the phases (different states of order or symmetry, etc.) and if a phase may persist for a while with the transition to the other delayed, the cusp catastrophe is intuitively the simplest model and, as we have seen, the least fragile.

Extending these considerations to potentials $V_\mu(x)$ with n behavior variables or order parameters $x \in R^n$, and k control parameters $\mu \in R^k$, and classifying the elementary modes of sudden qualitative changes of behavior and structure -- the elementary catastrophes -- which are locally possible in structurally stable (gradient) systems, leads to the following adaptation of Thom's theorem (generalized by Mather, Siersma and others [18]):

<u>Theorem</u>. Let R^n be the space of states x (behavior variables, order parameters etc., with coordinates x_i ($i=1,2,\ldots,n$) denoted by x,y,z,...) and let $C = R^k$ ($k \leq 6$) be the control space of parameters μ (with coordinates u,v,w,...). Let $\mathbb{V}$ be the space of generic potentials $V = V_\mu(x)$, i.e. of smooth (C^∞) maps $V: R^n \times C \to R$. Then there exists an open-dense subset of $\mathbb{V}$ such that for almost every $V \in \mathbb{V}$: (i) the surface S of stationary points of V, defined by $\operatorname{grad}_x V = 0$, forms a smooth manifold $S \subset R^n \times C$; (ii) any singularity ($\operatorname{grad}_x V = 0$, $H = |\partial^2 V/\partial x_i \partial x_k| = 0$) of the map χ_V: $S \to C$ is equivalent to one of the elementary catastrophes supposed to be at $x = \mu = 0$ and listed below; and (iii) χ_V is stable under small

perturbations of V.

The dimension k of C determines the number (m) of basically different catastrophes: k(m) = 1(1), 2(2), 3(5), 4(7), 5(11), 6(14),... . It is important that the number n of state variables x_i does not enter into the classification (at least for $k < 7$). Thus, when $k \leq 6$ and there are $n = 10^3$ or more state variables x (concentrations, spins, etc.) in a system, its local behavior is still governed by $m \leq 14$ elementary, irreducible modes. They all exhibit striking sudden changes or catastrophes (though smoothed out by noise) when μ crosses the image in C of the set of singularities of χ_V, the bifurcation or catastrophe set $\Sigma \subset C$, determined by the map χ_V of S onto the parameter (μ-) space C, viz., by eliminating x from grad $V_\mu = 0$, $|\partial^2 V_\mu(x)/\partial x_i \partial x_k| = 0$.

List of elementary catastrophes:

Name	k (dim C)	n' (corang R^n)	Universal Unfolding $V_\mu(x)$	Fig.
Fold	1	1	$x^3/3 + bx$	9
Cusp	2	1	$x^4/4 + ux^2/2 + vx$	8
Swallowtail	3	1	$x^5/5 + ux^3/3 + vx^2/2 + wx$	10
Butterfly	4	1	$x^6/6 + tx^4/4 + ux^3/3$ $+ vx^2/2 + wx$	11 & 1c
Hyperbolic Umbilic	3	2	$x^3 + y^3 + wxy + ux + vy$	12a
Elliptic Umbilic	3	2	$x^3 - 3xy^2 + w(x^2+ y^2)$ $+ ux + vy$	12b
Parabolic Umbilic	4	2	$x^2y + y^4 + wx^2 + ty^2$ $+ ux + vy$	12c
Higher Order catastrophes:				
Wigwam	5	1	$x^7/7 + px^5/5 + tx^4/4$ +...(as in Butterfly)	13a
Star	6	1	$x^8/8 + qx^6/6 +\dots$ (as in Wigwam)	13b
Double Cusp	7	2	$x^4 + y^4 +\dots$	14
....	..	..	......	

What the above theorem tells a physicist is this: In any gradient-like physical situation where continuously changing controls $\mu \in C$ cause discontinuous changes (possibly smeared out by noise) in a state space R^n of any dimension, one can choose local coordinates (in R^n and $C = R^k$) such that the modes of change which can occur topologically are one of the above list. Thus if, for example, we have a system with n order parameters $x \in R^n$, controlled by, say 4 parameters $\mu = (u,v,w,t) \in R^4$, then its behavior is locally governed by the first 7 catastrophes of the list, and of the n states but n' = 2 (x,y) remain significant order parameters. The challenge lies in singling them out in observed phenomena. Lower-order catastrophes are trivially contained in higer-order ones, and however complicated V may be, they are always equivalent to the polynomials (the universal unfoldings) in the list.

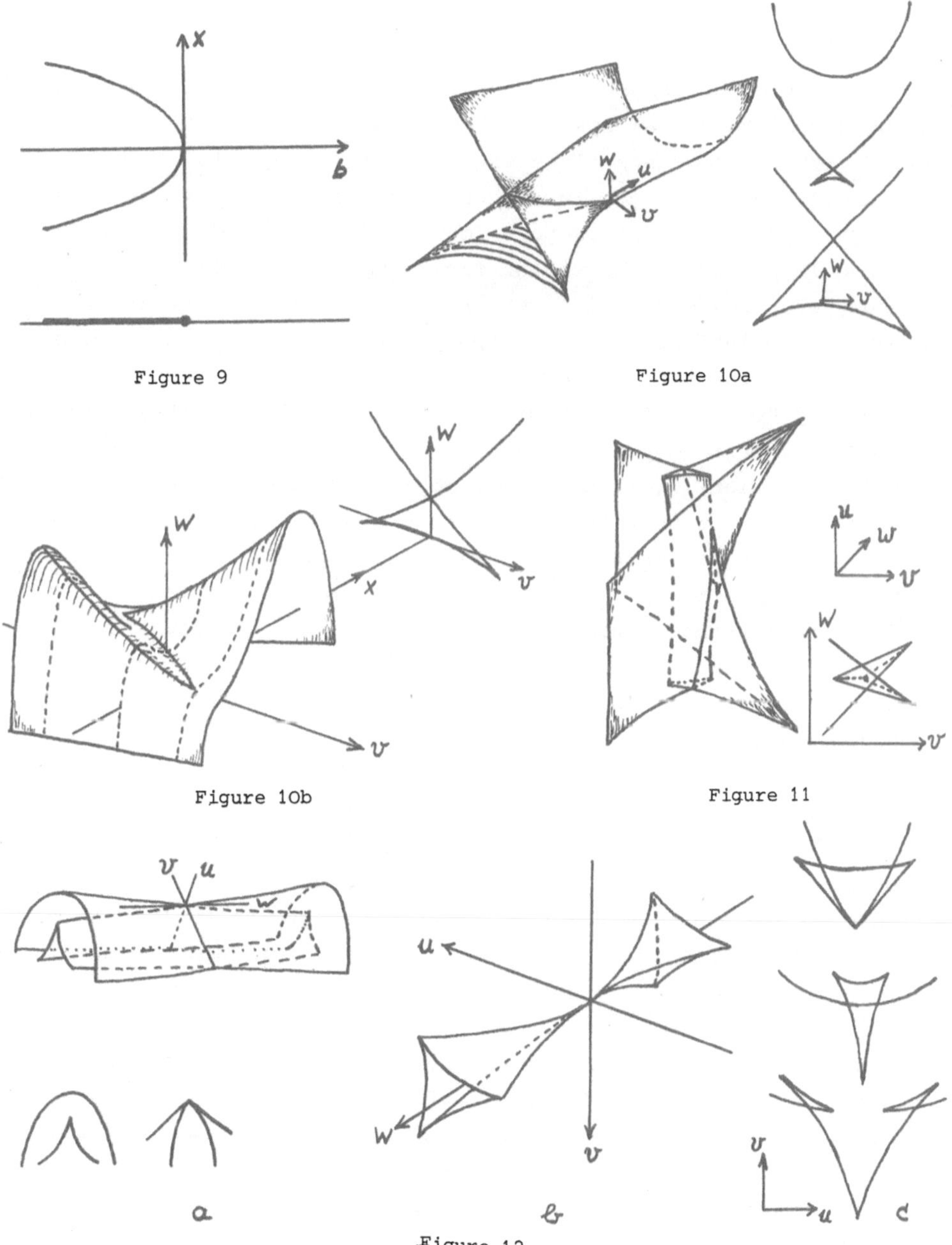

Figure 9

Figure 10a

Figure 10b

Figure 11

Figure 12

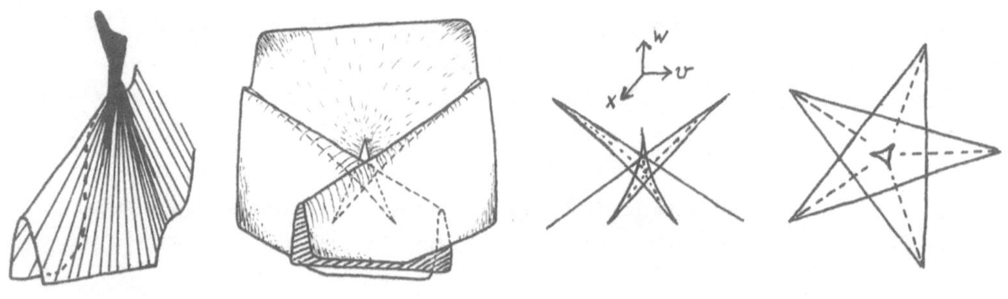

Figure 13a Figure 13b Figure 14

The restriction to gradient systems can be relaxed somewhat by considering any differential system on R^n, parametrized by μ, such that $V_\mu(x)$, V_μ: $R^n \to \mathbb{R}$, decreases along trajectories. V may be an energy, cost or Lyapunov function. But, in any case, M must have only isolated fixed points corresponding to attractors (minima of V), repellors (maxima of V) and saddle points of the vector field, so that almost every trajectory eventually reaches a minimum. Recurrences are obviously excluded. Some physical principles must be superimposed on the geometry even at this stage, because the systems considered are basically static. With a terminology borrowed from phase transition phenomena, these principles include: (i) a state may always select the lowest available minimum (Maxwell convention), (ii) the state may remain at a minimum until this disappears (delay or hysteresis convention) or a threshold between another one becomes low enough or, (iii) a state may choose a certain minimum as soon as it appears (saturation convention), etc.

When the internal dynamics of a system is not of gradient type, the theory of bifurcation becomes very difficult, and is virtually unexplored. But one may expect that the gradient-like situation keeps some validity though generalized catastrophes may spring up. For example, in a Morse-Smale system, made up of finite sets of fixed points and closed orbits (limit cycles) -- as, e.g. in the van der Pol case (Sec. 2) -- bifurcation is governed by the elementary catastrophes. However, extending the theory to cope with the creation, coalescence and bifurcation of limit cycles and high-dimensional attractors, passing to infinite-dimensional spaces and providing a dynamical setting for catastrophes in physical terms remain major challenges for further research. In pursuing this program one may then associate with each $\mu \in C$ a vector field $X_\mu(x)$ on a state space M and defined by $\dot{x} = X_\mu(x)$, $x \in M$. A vector field on M×C (and tangent to the fiber M) defines a metabolic field G on C, G generating a map of C into the space $\mathbb{D}$ of vector fields. If μ is on $\Sigma \subset C$, G_μ is a field of $\mathbf{\Sigma}$ (the complement of $\mathbf{S}$ in Fig. 6) and the local catastrophe set Σ in C appears as the counter-image, $\Sigma = G^{-1}(J)$, of a universal catastrophe set J in $\mathcal{D}$ if the map is transversal to J. A domain in C, e.g. space-time, with a metabolic field on it defines a morphogenetic field. When G is not transversal to Σ, the induced morphology G^{-1} becomes quite complicated and little can be said about the ensuing generalized catastrophes. On this background Thom has discussed a wealth of structures in the animate and inanimate world from the qualitative point of view, covering an enormous spectrum of material from developmental biology to linguistics to the structure of galaxies [2].

There remains a gap between the general theory of structurally stable models Thom puts forth and practical schemes for applying the theory of classification and deformation of singularities in the sciences. It is for this reason that the theory often looks more like an art than a science.

5. Catastrophes in Physics and Biology

When one accepts the idea that, because of their universality, critical phenomena in nature have a common topological basis, they must be governed by unfolding singularities in a structurally stable way, i.e. by the irreducible elementary catastrophes. Their physical origin may, roughly speaking, be traced to nonlinearities due to heterogeneities in otherwise homogeneous media (semiconductors, crystals, membranes, etc.). And it is in fact nonlinear dynamics where catastrophe geometry promises to provide both a general framework and a method. In a physical system with order parameters or behavior variables $x \in M$, driven by competing controls $\mu \in C$, one assumes tentatively a local gradient dynamics and chooses a "potential function" $V = V_o(x)$ having the origin (representing, in general, highest disorder or symmetry) as a degenerate singularity. This function acts as an "organizing center" and may represent, when unfolded into $V_\mu(x)$, a free energy , Lyapunov functions such as excess entropy or information gain etc. In biological systems, M may, for example, represent the state space of cells and C the space of morphogenes, with a subsequent map onto space-time revealing morphogenesis as images of catastrophes caused, e.g. by competing concentration gradients.

Having found the elementary geometric laws governing a structurally stable catastrophic world, let us try to understand them in physical terms. To begin with, we start from the linear oscillator's next best structurally stable version, the forced pendulum's anharmonic Duffing approximation, whose displacement z(t) satisfies the equation $\ddot{z}+\gamma\dot{z}+\omega_o^2 z+\alpha z^3=F\sin\omega t$. To lowest order, neglecting γ for the moment, its iterative solution is $z(t)=A\sin\omega t+S(A,F)\sin 3\omega t$, where $3\alpha A^3/4-(\omega^2-\omega_o^2)A-F=0$

and $S=const.A^3/[A^3+4A\omega_o^2/3\alpha-4F/3\alpha]$. Identifying $A\to x$, $(\omega^2-\omega_o^2)\to u$, $F\to -v$, the oscillator's nonlinear resonance response x to the drives u,v is seen to be described by the overhanging cliff of the cusp catastrophe (Figs. 1b,8). A plane u=const through the cliff's surface shows the familiar resonance curve and hysteresis cycle, and the domain inside Σ corresponds to in/out-of-phase behavior. The poles of S determine the same surface, which is indicative of the role played by catastrophes in S-matrix theory [19]. Next to the shadow of our dog's mind (the area inside the cusp of Figs. 1a,1b where-u-v=rage and -u+v=fear) comes, of course, the transition of a ferroelectric near the Curie point from the non-pyroelectric to the pyroelectric phase. We describe it by the unfolding of the thermodynamic potential $V=\Phi(D)$ around its singularity D=O (D=x=polarization). With the simplest crystalline symmetry assumption and confining ourselves to a 2nd order transition (k=2), i.e., $V\to V_{uv}$, one finds Equ. (6) with the identification $u=T-T_c$, v=E (E=electric field), and the behavior near the critical temperature T_c is therefore governed by the cusp catastrophe in accordance with the Ginzburg-Landau theory [8,11].

Proceeding with oscillator models, we observe that subharmonic oscillations start to bifurcate from harmonic ones as we go on to higher approximations, or to higher-order nonlinearities. They gain particular importance in entrainment and synchronization phenomena that abound in biological systems [24]. A relatively simple example is obtained by combining the van der Pol oscillator (Sec. 2) with the above anharmonic one into the equation

$$\ddot{z} + \gamma(z^2-\beta)\dot{z} + z + \alpha z^3 = f(t) \tag{9}$$

where f is the external stimulus. Assuming $f=F\sin\omega t$, determining the periodic solution z(t) and examining its stability in the (F=w, ω=v)-plane exhibits a swallowtail-like situation (Fig. 10b) with the region below the cusps' overlaps corresponding to harmonic entrainment, while beats occur beyond that domain. We hope to classify the--admittedly rather messy--entrainment phenomena in terms of catastrophes in a forthcoming paper [20]. They are difficult to analyze even topologically because of the occurrence of Smale's "horseshoes".

Not satisfied to learn catastrophe physics in terms of a single oscillator, let us consider many, and indeed a continuum of oscillators. That's where field theory starts, classical or quantized, and the trouble begins--also with the catastrophes, because their geometry becomes now a part of a much more comprehensive theory which is as yet almost unexplored. If we consider a partial differential equation, whose characteristics satisfy Equ. (4), it is clear that the solutions of the latter, and therefore the catastrophes, characterize the propagation of singularities of the former [17]. When these principles are suitably extended and bifurcation theory [16] finally joins catastrophe theory [2], a practicable scheme for handling singularities and modeling continuum phenomena in the sciences will ultimately emerge. But even at the present stage, recognizing the cuspoids of the corang-1-singularities as loci of caustics and projections of ruled surfaces leaves no one surprised at finding magnetohydrodynamic [21] or even chemical shock waves to be the result of cusp geometries [20]--fusion, and fission for that matter, being commonly accepted catastrophes anyway [25]. The single-mode laser [22] is, perhaps, the best known example of a continuous set of nonlinear oscillators of the above type, exhibiting sudden transitions in power output if the input exceeds threshold. From the microscopic theory with its enormous number of degrees of freedom, Haken derived for the electric field strength E of the laser light the equation $E = a\nabla^2 E + bE + cE^3 + \dot{F}$ which, when the space-dependent part $\nabla^2 E$ and the fluctuation force F are neglected, reduces (via $E = -\mathrm{grad}V(E)$) to the (properly amputated) cusp catastrophe for the onset of coherent radiation. This is no longer surprising when one views the onset of laser radiation as a phase transition. Nor should, then, the chemists among you be surprised on discovering that the very same laser equation--with E replaced by chemical concentration variables Fe^{++}/Fe^{--}, etc.--is at the root of Winfree's scroll wave models [23].

Phase transitions--in ferroelectrics, antiferromagnets (metamagents), ^3He-^4He solutions, multicomponent-fluid mixtures, chemical reactions, particle reactions, and so on--are, as you will believe me by now, a conceptually simple theoretical (though not simple experimental) laboratory for catastrophe geometry, and, in fact, provide its most direct physical realization. It is, of course not so that one should look at Fig. 1b as representing the liquid-gas transition, which rather is given by the amputated swallowtail (Fig. 10b) of the Gibbs free energy G(P,T) [11]. Instead, we have to start from the microscopic dynamics and--this is the clue offered by our theorem in Sec. 4--single out (via the grand canonical partition function, by averaging, by the Bogoliubov-Uhlenbeck contraction, etc.) of the underlying R^n those few (n' = corang R^n) macroscopic order parameters that are ultimately the most relevant observable physical ones [11]. This results in the following <u>classification hypothesis</u>: One-order-parameter transitions are dominated by the cuspoid geometry of corang 1, two-order-parameter transitions by the umbilic geometry of corang 2. Then, under the assumption of structurally stable scaling, a Landau-type mean-field theory (zeroth approximation in the system dimension (d)) will describe the critical behavior qualitatively correctly. After all, although its quantitative predictions are not too good, they are not that bad, either: you cannot expect more from topology alone. Confining ourselves for the moment to the cuspoid geometry with one order parameter $\eta = x$, we describe a multi-component system by the unfolding $V_\mu = \Phi := \sum_{r=0}^{2N} a_\nu x^\nu$ ($\mu = \{a_\nu\} \in C = R^k$) of a free energy V_0 with 2N-th order singularity at $x = 0$, where the a_ν are related to the physical "field"variables (P,T, chemical potentials, etc.) by a diffeomorphism [11]. With the "coexistence curves" (conflict strata or coexisting minima of equal depth) indicated by dotted lines in Figs. 1c, 10, 11, 13b, one immediately discovers the tricritical and higher order critical points in fluid mixtures, etc. where three or more phases become simultaneously identical. I don't know if they have all been detected experimentally by now. At least the critical point of the star pocket (Fig. 13b) seems not yet to be on the experimentalist's desk--but it must be there, sooner or later, otherwise the Gods must have already made a serious mistake when they created Pythagoras.

Things become even more exciting when we look for "corang-2-singularities", i.e. for systems with two coupled order parameters x,y. Ferro-antiferromagnetic, ferro-piezoelectric and crystalline-superfluid systems provide familiar examples, along with two-mode lasers and BCS pairings [26, 11]. In these cases we start with the partition function $Z = Tr\{exp[-\beta(H_0 + V_\mu(x,y))]\}$ and describe the system as before by the free energy Φ given by the (averaged) universal unfolding $V_\mu(x,y)$ corresponding to the umbilics or double cusps (with k=3,4 or k=7 type critical points, respectively) given in Sec. 4. Deferring a detailed discussion, we show in Figs. 12, 14 and 15 a few topological situations which also are typical of binary mixtures, MHD and biological systems as well. A consideration of bifurcation phenomena in dynamical equations and of the role caustics play in critical phenomena may be found in [11].

Let me finally touch on a field I have already been bothering with for two decades and which I see now sailing rapidly toward an edge of the overhanging cliff. I mean particle physics and quantum field theory in terms of structural stability concepts. Because of the close analogy between critical behavior in phase transitions (Widom scaling) and deep inelastic particle scattering (Bjorken scaling), critical point dominance in particle physics may be expected to follow rules similar to those developed above [19,27]. After all, scaling is but a special sort of stable mapping, and symmetry breaking is an obvious part of it.

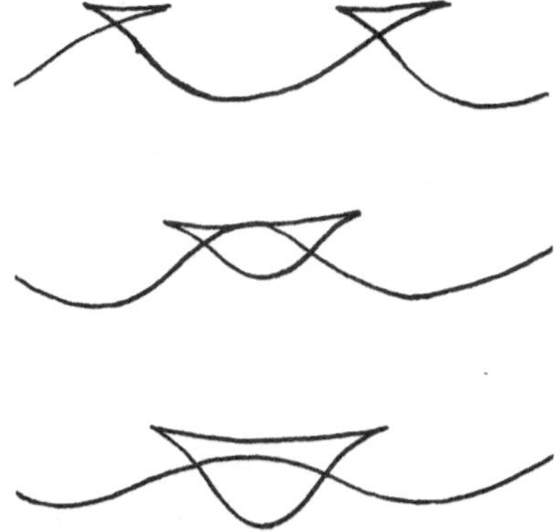

Fig. 15. Higher-order phase transition

Leaving the story of particle catastrophes for a more physically-minded audience, let me just add the following remarks. What we urgently have to do is to build fluctuations into the above frame and get at the thermodynamic roots of catastrophe geometry by considering the unfoldings $V_\mu(x)$ as an excess entropy or information gain $\delta^2 s$ [3] and, as the following discussion will show, extend the considerations to spaces with infinitely many dimensions.

While addressing the physicists, I feel that I may be boring the majority of the biologists among you. By the first day of this conference I already realized that some biologists--especially molecular ones--are nowadays a bit unsure of Thom's ideas, probably because he asked them recently [28] to "finally start thinking". Of course, but just extracting that little piece of theorem (Sec. 4) from Thom's writings and recognizing it as a first small step into an enormously wide field, is a task for an expert in topology, not molecular biology. On the other hand, biologists (like other scientists) are quick to recognize in the newcomers a lack of familiarity with the subtle complexities of their field and to dismiss them as naive, but often fail to understand the necessity of mathematical models because their own category of thought is basically unmathematical and, indeed, qualitative. Thus we are back to what I said at the beginning of this lecture--and what is my adaptation of Thom's philosophy--namely, that the qualitative laws of nature are written in the language of geometry. Biologists cannot be blamed for the sins of mathematicians who have entirely forgotten this fact, but they can help to revive the old geometric virtues. Their immediate task should be to look at biological phenomena the same global way Landau looked at phase transitions when creating his mean field theory of complicated dynamical situations, and adding geometric concepts the way I have just described. Let me, then, sketch in terms of a few examples how the beginning of such a program--anticipated by Thom with fascinating imagination--could be realized.

Modeling the evolution of biological macromolecules in terms of catastrophes is still in its infancy although, according to Eigen and others [29], the temporal evolution of such systems may be described in terms of the kinetic equation (1) with concentrations $x \in M$ of chemicals, and a set C of competing parameters μ representing monomers, etc. Thus, evolution and selection may quite well be describable in terms of catastrophes, though it is certainly not an easy task, even on the basis of the gradient system (4), to single out the relevant macroscopic order parameters. What one has to do, of course, is to supplement Equs. (1) or (4) by differential equations for the parameters μ, thought of as functions of the x, as seen, e.g., in the neuron example described later. I have learned lately that J. Tyson [30] is pursuing this program and I would, after all, not be too much surprised if selection, hypercycles and all that turn out to be phase-transition-like catastrophes. Formally analogous considerations are the main features of recent studies of the reaction-diffusion equation following from (1) by making x space-dependent:

$$\dot{x} = \nabla^2 x + X(x), \tag{10}$$

where X is a nonlinear function of the variables $x = x(t,\vec{r})$. Bifurcations and catastrophes caused by nonlinearities in $X(x)$ are physically due to autocatalytic and crosscatalytic reactions. Equ. (10) is a generalization of Lienard-type oscillators like (9) with the z,f made $\vec{r}$-dependent. As we have already seen, it governs laser transitions as well as scroll waves, and you will also believe that Equ. (10) describes mitosis, etc. [31]. That same equation is at the root of Lefever's, Nicolis', and Howard and Kopell's work [31,14], to name a few, but as we have seen before, the propagation of catastrophe singularities through the solutions of partial differential equations is a delicate matter.

This manifests itself already in topological models of neural behavior relating to the production, control and propagation of nerve impulses. If we denote by $I(x,t)$ a neuron's input current (related to the PSP) and by $\psi(x,t)$ the membrane potential at space point x and time t, then the following partial differential equation describes the neuron's behavior fairly reasonably [32]:

$$a\psi_{xx} + b\psi_{tt} + N'(\psi)\psi_t + c\psi_{xxt} + dN(\psi) = I + I_t \qquad (11)$$

Here $N(\psi)$ is the neuron's nonlinear current-voltage characteristics, $N'(\psi)$ its derivative, and since it is a cubic function, Equ. (11) is indeed of the same type as the equations we have found above in quite different contexts (laser, chemistry, etc.) Thus it is hardly surprising that neurons exhibit catastrophic phenomena: jumps, bursts and all that. What one discovers, e.g., when analyzing solutions of (11), are resonance phenomena (catastrophe concatenations) giving rise to rhythmic discharges as shown in Fig. 16, where the energy E stored in a neuron is plotted against the PSP. Things become almost trivial when in (11) we pass to a neural element and arrive at an equation identical with Equ. (9) with $z = \psi(t)$ and $f(t) = I(t) + \dot{I}(t)$ (the space coordinate drops out when passing to that limit) modeled by the circuit of Fig. 17. Writing it as a system of the type

$$\begin{aligned} \dot{x} &= -x^3 + ax + bu + cv \\ \dot{u} &= dx + eu \end{aligned} \qquad (12)$$

where $x = \psi(t)$, and identifying x with the membrane potential (or Na activation), u with the recovery variable (or H concentration) and v with the stimulus (PSP), Equ. (12) is obviously represented by the cusp catastrophe of Figs. 1b,8. This model (similar to FitzHugh's BvP model [33]) is of course far simpler and, through Equ. (11) more comprehensive than the one E.C. Zeeman recently proposed [13], but I refer the reader to [32] for an elaboration of simple neuron models.

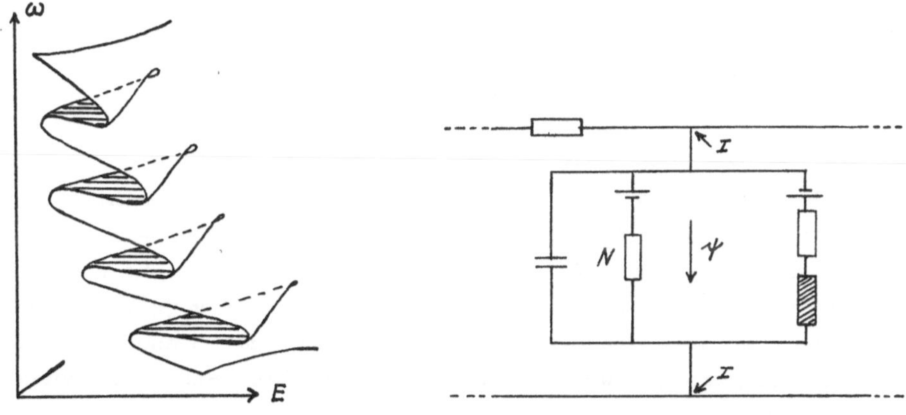

Figure 16

Figure 17

A set of diffusion-reaction equations of the type (10) are well-known from Prigogine's [3], Turing's, and Gierer-Meinhardt's [34] models of pattern formation due to competing concentration gradients μ. Describing slime molds and brains [35] may well follow similar lines. It is at this point where the map of M×C onto space-time I talked about at the beginning of this section comes into play, and the morphogenetic field of Sec. 4.

Identifying morphogenesis, and developmental biology in particular, with fundamental geometric structures is Thom's basic idea. One can indeed easily identify the phase transition cuspoids and umbilics in Figs. 10-15 with developing phases in the embryo (Fig. 18, [36]) and infer from the existence of an eventually unfolding organizing center (V_o) in the former case the existence of such a center in the latter one. This center may be the result of a local minimum of entropy production, but making such intuitions explicit is the great burden Thom has put upon us.

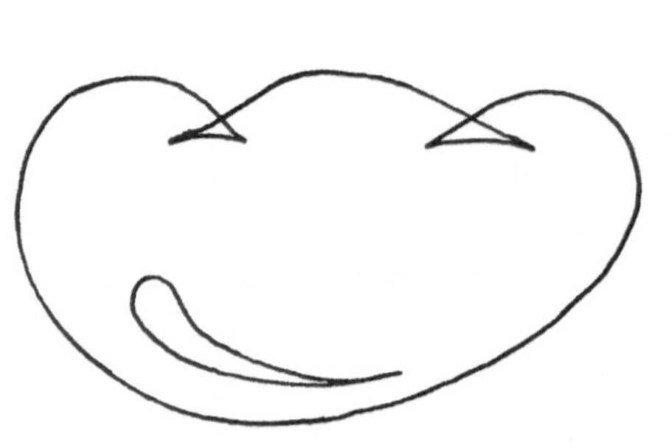

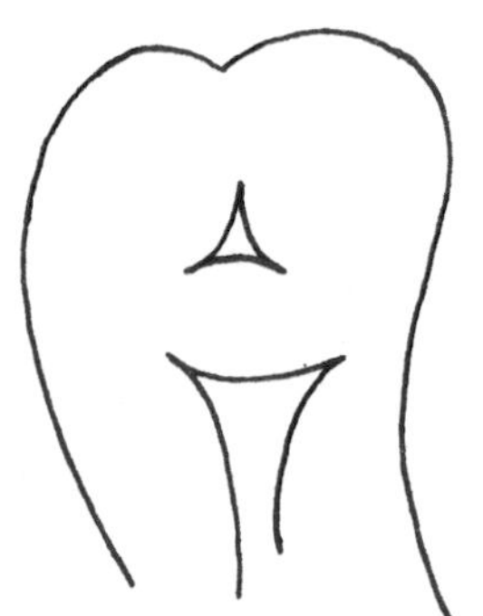

Amphibian gastrulation

Neural tube

Figure 18

6. Conclusion

We started by saying that the qualitative laws of nature are written in the language of geometry. This provides us with something to hold on to in our search for a natural philosophy. Though all of Thom's colleagues admire his beautiful and comprehensive view of the universe, few, if any, embrace the totality of his ideas, and only very few, if any, claim to understand them. But he has undoubtedly opened the door to the next great era of awakening of human intellect--that of understanding the qualitative content of complex phenomena.

Acknowledgments

I thank René Thom for persuading me that his statements are too natural to need proving, Horst Eikemeier for convincing me that this is so, and Tom Güttinger for modeling and drawing the catastrophes.

References

1. Poincaré, H.: Sur les courbes définies par une équation differentielle, Oevres Completes, Vol. I, Paris, 1881.

2. Thom, R.: Topological models in biology. In: Towards a Theoretical Biology, vol. 1, ed. by C.H. Waddington. Edinburgh University Press, Edinburgh, 1970.

   Thom, R.: Stabilité Structurelle et Morphogenèse, Benjamin, New York, 1972.

3. Glansdorff, P. and Prigogine, I.: Thermodynamics of Structure, Stability and Fluctuations, New York, 1971.

   Schlögl, F.: On stability of steady states, Z. Phys. 243, 303 (1971).

4. Lorenz, K.: On Aggression, New York, 1966.

   Tinbergen, N.: The Study of Instinct, Oxford, 1969.

5. Zeeman, E.C.: The geometry of catastrophe, Times Lit. Suppl., Dec. 10, 1971.

6. Fowler, D.H.: The Riemann-Hugoniot catastrophe and van der Waal's equation. In: Towards a Theoretical Biology, vol. 4, ed. by C.H. Waddington. Edinburgh University Press, 1972.

   Thom, R.: Phase transitions as catastrophes. In: Statistical Mechanics, ed. by S.A. Rice, K.F. Freed, and J.C. Light. Chicago, 1972.

   Shulman, L.S. and Revzen, M.: Phase transitions as catastrophes, Collect. Phen. 1, 43 (1972).

7. Griffiths, R.B.: Proposal for notation at tricritical points, Phys. Rev. B7, 545 (1973).

   Widom, B. and Griffiths, R.B.: Multicomponent-fluid tricritical points, Phys. Rev. A8, 2173 (1973).

   Wegner, F.J. and Riedel, E.K.: Phys. Rev. Lett. 29, 349 (1972).

   Chang, T.S., Hankey, A. and Stanley, H.E.: Phys. Rev. B8, 346 (1973).

   Shulman, L.S.: Tricritical points and type-three phase transitions, Phys. Rev. B7, 1960 (1973).

8. Landau, L.D.: In: Collected Papers of L.D. Landau, ed. by D. ter Haar. New York, 1965.

9. Wilson, K.: Phys. Rev. D3, 1818 (1971).

   Kadanoff, L.P.: Critical behavior, universality and scaling. In: Proceedings of the International School of Physics "Enrico Fermi", ed. by M. Green. New York, 1973.

10. Markus, L.: Lectures in Differentiable Dynamics, Amer. Math. Soc. Reg. Conference Series 3 (1971).

Peixoto, M. (Ed): Qualitative Theory of Differential Equations and Structural Stability, New York, 1967.

Smale, S.: Differentiable dynamical systems, Bull. Amer. Math. Soc. 73, 747 (1967).

Peixoto, M. (Ed.): Salvador Symposium on Dynamical Systems, New York, 1973.

11. Güttinger, W. and Eikemeier, H.: Catastrophe theory of critical phenomena, University of Tübingen preprint 1974 (to be published).

12. Winfree, A.T.: Time and timelessness in biological clocks. In: Temporal Aspects of Therapeutics, ed. by J. Urquardt and F.E. Yates. New York, 1973.

13. Zeeman, E.C.: Differential equations for the heartbeat and nerve impulse. In: Towards a Theoretical Biology, vol. 4, ed. by C.H. Waddington. Edinburgh University Press, 1972.

(Further references to Zeeman's work can be found in this volume.)

14. Kopell, N. and Howard, L.N.: Plane wave solutions to reaction-diffusion equations, Stud. Appl. Math. LII, 291 (1973).

15. Ruelle, D. and Takens, F.: On the nature of turbulence, Comm. Math. Phys. 20, 167 (1971).

16. Sattinger, D.H.: Topics in Stability and Bifurcation Theory, Lect. Notes in Mathematics, Vol. 309, Berlin, 1973.

17. Guckenheimer, J.: Catastrophes and partial differential equations, Ann. Inst. Fourier 23, Fasc. 2, 31 (1973); Bifurcations and Catastrophes. In: Salvador Symposium on Dynamical Systems, cf. Ref. [10].

Jänich, K.: Caustics and catastrophes, Univ. of Warwick preprint, 1973 (to be published in Math. Ann. 1974).

18. Arnold, V.I.: Lectures on bifurcations in versal families, Russian Math. Surv. 27, 5 (1972).

Siersma, D.: Singularities of C^∞ functions of right-codimension smaller or equal than eight, Indag. Math. 25, 31 (1973), and Amsterdam thesis 1974.

Mather, J.: Right equivalence, unpublished manuscript, Princeton University, 1970. T. Bröcker, Univ. of Regensburg Lecture Notes,1973.

Wasserman, G.: Stability of unfoldings, University of Regensburg thesis, 1973. T. Poston, R. Woodcock, Lect. N. Math. 373, 1973.

19. Güttinger, W. and Eikemeier, H.: Catastrophes in particle physics (to appear 1974).

20. Güttinger, W. (Ed.): Wyoming Symposium on Catastrophe Theory, 1973 (to be published), and University of Tübingen lecture notes 1973/ 1974.

21. See, e.g., Bazer, J. and Fleischman, O.: Propagation of weak hydromagnetic discontinuities, Phys. Fluids 2, 366 (1959).

22. Haken, G.: Introduction to synergetics. In: Synergetics, ed. by H. Haken. Stuttgart, 1973.

23. Winfree, A.T.: Scroll-shaped waves of chemical activity in three dimensions, Science 181, 937 (1973).

24. Pavlidis, T.: Biological Oscillators, New York, 1973.

25. Süssmann, G. et al.: First estimates of the nuclear viscosity constant. In: Physics and Chemistry of Fission 1973, to be published by IAEA, Vienna, 1974.

26. See, e.g., Liu, K.S. and Fisher, M.E., J. Low Temp. Phys. 10, 655 (1973) and Ref. 22.

Imry, Y., Scalapino, D.J., Gunther, L.: Phase transitions in systems with coupled order parameters (unpublished conference report).

27. Jaffe, A. and Glimm, J.: Critical point dominance in quantum field models (to be published).

Schroer, B.: Report at the 5th Brazilian Symposium on Theoretical Physics, Jan. 1974 (to be published).

28. Thom, R.: A global dynamical scheme for vertebrate embryology, AAAS Philadelphia meeting 1971 (to be published).

29. Eigen, M.: Mclecular selforganization and the early stages of evolution, Quart. Rev. Biophys. 4, 149 (1971).

Prigogine, I., Nicolis, G., Babloyantz, A.: Thermodynamics of evolution, Phys. Today 25, Nov./Dec. 1972.

Hahn, H.: Geometrical aspects of the pseudo steady state hypothesis in enzyme reactions, this volume.

30. Tyson, J.J.: Competition, Selection and Evolution, unpublished Göttingen manuscript, 1974.

31. Lefever, R. and Nicolis, G.: Chemical instabilities and sustained oscillations, J. theoret. Biol. 30, 267 (1971).

Nicolis, G.: Mathematical problems in biology. In: Lecture Notes in Mathematics, Vol. 322, Berlin, 1973.

Kauffman, S.: Measuring the mitotic oscillator, Bull. Math. Biophys. 1974 (in press).

32. Güttinger, W. and Hahn, H.: Physical principles of nonlinear neuronal information processing, Int. J. Neuroscience, vol. 6, no. 1, 53 (1973), 3, 61, 67 (1972).

33. FitzHugh, R.: Mathematical models of excitation and propagation in nerve. In: Biological Engineering, ed. by H.P. Schwan, New York, 1969.

Rinzel, J. and Keller, J.B.: Traveling wave solutions of a nerve conduction equation, Biophys. J. 13, 1313 (1973).

34. Turing, A.: The chemical basis of morphogenesis, Phil. Trans. Roy. Soc. B 237, 32 (1952).

Crick, F.: The scale of pattern formation, Symp. Soc. exp. Biol. 25, 429 (1971).

Gierer, A. and Meinhardt, H.: A theory of pattern formation, Kybernetik 12, 30 (1972) and private communications.

35. Cohen, M.H. and Robertson, A.: Cell migration and the control of development. In: Statistical Mechanics, ed. by S. Rice et al. [Ref. 6]

Cowan, J.D.: Stochastic models of neuroelectric activity. In: Statistical Mechanics, ed. by S. Rice [Ref. 6]

Von der Malsburg, Ch.: unpublished Göttingen manuscript 1973.

Legéndy, Ch., Tübingen, private communications 1973.

36. Balinsky, B.I.: Introduction to Embryology, Philadelphia, 1965 and Refs. [2,28].

Zeeman, E.C.: Reports at the Göttingen Symposium on Catastrophe Theory, Sept. 1973, and at the AAAS meeting, San Francisco, 1974 (unpublished).

SOME LIMITATIONS AFFECTING PHYSICS AND MATHEMATICS AS APPLIED TO BIOLOGY AND ESPECIALLY TO THE NERVOUS SYSTEM

R.B. Livingston
Department of Neuroscience, University of California San Diego, La Jolla, California

1. Introduction

Contributions by the Physical Sciences

The great triumphs of physics, including all the theoretical and practical contributions that have so enriched and empowered western civilization, have derived from the asking of two questions:

(i) What happens during interactions among a few bodies?

(ii) What happens during interactions among a large number of bodies?

In addressing the first question, physical scientists have attempted to render as complete an account as possible of all events and relationships occurring. At the atomic level there still remain apparently irreducible indeterminacies and unpredictable quantum perturbations (of Heisenberg and Bohr). The number of bodies for which reasonable predictions can be made is surprisingly small, both because of the practical burdens of computation and the more devastatingly cumulative progressions of uncertainties.

Examination of the two-body problem has been accomplished in many instances. Examination of the three-body problem often gets out of predictive control even over short time courses, and the possibility of working with higher degrees of complexity is seldom entertained.

Analysis of the second question, that concerning interactions among a large number of bodies, has yielded substantial results. In dealing with this question, physical scientists have had to be content with more limited specifications of what happens -- here nothing more explicit can be established than the probabilities of certain events occurring. For the practical management of problems encompassed by this question, it is necessary that the bodies be quite homogeneous. Virtually the only generally useful achievements relating to the treatment of large numbers of components have depended on maintaining strict homogeneity.

Most of modern technology has been based on insights derived from investigation of these two questions, and man's imagination and cultural heritage has been even more enduringly enriched thereby.

2. Some Problems Posed by Biology

The Problem of Heterogeneity

Organic structures and all living systems depend on the symmetrical valences of the carbon atom which tend to form long chain compounds with oxygen, hydrogen, nitrogen, and a relatively small number of other atoms. Biological componentry obviously involves the many-body problem, but it lies in a domain that has heretofore not been triumphantly addressed by traditional physical and mathematical techniques -- the domain of extreme heterogeneity. Biology, therefore, poses a third root question,

(iii) what happens during interactions among a large number of heterogeneous bodies?

Until recently, this question has been considered too ambiguous and difficult -- and therefore perhaps too uninteresting from the points of view of the physical and mathematical disciplines. Perhaps the most familiar example of the potentialities of traces of inhomogeneity is that of the transistor. Trace contaminants in a germanium crystal radically alter that crystal's potentialities for conducting electricity from a simple resistance to amplification and other radically different effects.

Similarly, minute contaminants in glass convert a glass electrode from being relatively non-selective to being differentially highly selective for hydrogen, potassium, sodium and other biologically interesting ions. George Eisenman has suggested that the effects of such trace contaminants in glass may provide a physical model for possible mechanisms for ionic discrimination by living membranes. He reasons that the phosphates, abundant in membranes, have four oxygen atoms which are arranged around the phosphorus atom in the same symmetrical way that the four oxygens are arranged around silicon in glass. The discrimination in membranes could possibly be provided by a shift of minute "contaminants" within the lattice of the membrane phosphates. Another possibility, favored by E. Neumann, is that short peptide chains with acidic groups embedded in lipid bilayers could perform the functions of ionic discrimination and control. The lesson is the same, namely that heterogeneity provides remarkable alterations in the operation of many-body structures.

Many of the most important discoveries in solid state physics have been empirically rather than theoretically derived. It was possible to work in the reverse direction, to construct theoretical interpretations based on more fundamental physical principles in order to account for the experimental findings. Even now, twenty years later, predictions, even in amorphous materials, are practically impossible. Unexpected functional qualities are continually being revealed as dependent on modest increases in heterogeneity.

In no branch of physics, including biophysics, does the structural complexity that is being successfully approached begin to reflect the level of structural heterogeneity that is commonplace in living cells. Yet, the examples of heterogeneity that have thus far been encountered have whetted the appetites of physicists and mathematicians -- along with biologists -- to seek a more profound conceptual and experimental grasp of living systems.

It seems to be clear that the theoretical and practical advantages of studying heterogeneity are far greater than anything thus far undertaken by the physical sciences.

The Finite Class Problem

It has been authoritatively calculated by Walter Elsasser that the number of potential atomic structural configurations of even the simplest living cell is an immense number when compared to the maximum potential number of such cells. Thus, organisms belong to finite classes of objects -- so limited in fact that the membership would be exhausted before the limits of present physical theory could be thoroughly tested on them. In the past, physics has always been able to depend, implicitly at least, on infinite classes of objects for experimental analysis. Not so with biology.

The Electron Orbitting Problem

At present, the physical sciences are limited to quite simple notions concerning the detailed structures of living things. Energies observed in living systems lie far below the energies necessary for atomic (nuclear) effects. Even close interactions appear to implicate only the outermost electronic orbits characteristic of chemical reactions. Large molecular structures permit a very wide range sharing of electronic orbits and energies. This, in turn, yields unexpected and far-reaching electro-elastic effects. It is obvious that biological heterogeneity is not haphazard, and that success in evolution has depended upon some minimum degree and extent of orderliness. But it is necessary to know precisely what orderliness and exactly where, in order to understand biological systems.

The energy it is necessary to impart into a system in order to locate (even approximately) the electrons within that system is large, and it increases exponentially as the complexity of the system is increased. It is not necessary to determine the "positions" of all of the outer orbitting electrons. But the ones that we are most interested in -- the ones that operate as the most crucial controls -- may have generally wider ranges, consistent with cross-coupling between interdependent complexities. Many significant biological control mechanisms evidently depend on particular electronic locations and movements. Increasing the precision of their location n times imparts energy to the system according to n^2.

Therefore, uncertainties relating to the electronic orbit, crucial to our comprehension of individual parts, become spread over larger realms of the system and definitely limit the extent of the system that can be analyzed in this way. A single cell vastly exceeds this range of achievable analysis -- this range of tolerable uncertainty, according to Elsasser.

Limitations of the Cartesian Assumption

A commonplace assumption underlying all disciplines of science is the Cartesian notion that an entity -- no matter how complex -- can be understood in its entirety by taking it apart and analyzing the interaction of its parts. From seriatim analysis of all such interactions, Descartes assumed, the behavior of the entirety can be comprehended. This assumption is true only for mechanisms of strictly limited complexity. This limit lies far below the level of complexity of single cells much less of nervous systems.

3. The Way Out

It is clear that the large number of parts and the immense heterogeneity of atomic structural organization of living things remove them from grasp by means of those approaches of physics and mathematics which have been so elegant and powerful when addressed to simpler systems. Biology is therefore rather remote from understanding in terms of systems-as-a-whole through the analysis of the few-body (atomic reconstruction) question (i), and is, from the outset, practically completely excluded from benefiting from the homogeneous many-body question (ii), because of the intrinsic heterogeneity of biological systems. In order to proceed along traditional physical experimental lines, we are obliged to experiment with critically deformed and isolated parts of the whole.

Although this tradition of physical experimentation obviously has been extremely profitable for the analysis of parts of living systems, and still has a very long way to go along the same lines, we must recognize its intrinsic limitations for the practical and conceptual understanding of living systems as entireties. We need to recognize these limitations in order to encourage ourselves to undertake alternative patterns of scientific inquiry that will enable us to progress beyond the limits to which traditional, physical and mathematical concepts and methods of approach are obligatorily confined. While many scientists are following proven fruitful but confined methods, some should be encouraged to pioneer in finding new ways of approaching these difficulties. There is a need for new physical and mathematical developments expressly for biological systems.

4. The Traditional Hierarchy of Scientific Disciplines

How does it come about that biology has traditionally been supposed to be "explained" in terms of physical theories which have never been put to test on systems of equal complexity? The traditional assumption of the explanatory hierarchy of the sciences was forcefully put forward in 1848 by four talented young students of Johannes Müller: Helmholtz, du Bois Reymond, Ludwig and von Brücke. They sought to rescue biology from vitalism and other conceptual traps of that period. They were determined that it would be possible to account for living processes as mechanisms, strictly and solely defined according to concepts originating from physics, chemistry and mathematics. We are under the same spell one hundred and twenty-five years later. Physical concepts and techniques have come a long way since 1848 -- and with them the prospects for their further advancement. Yet, a meaningful definition, at an atomic structural level, of any organism, or even of a single cell, is quite beyond achievement on both theoretical and practical grounds.

For physics and mathematics to contribute most effectively to biology in general and to neurosciences in particular, will require more than the application of what has been so effective in addressing questions (i) and (ii). There is need for exploration of question (iii) by new approaches at the same time as more traditional approaches are being furthered. It is not an "either/or" proposition. Traditional pursuits should be continued and intensified. And new mathematical and physical techniques are essential for the examination of complex configurations as unified, dynamic whole entities -- as systems. The main obstacle is to quit thinking about truly complex systems as if they were complex at the level of clocks.

Suppose there were clocks with as many gears as there are feed-back loops in a single cell and that these clocks belonged to a finite class such that by no means could you examine all of the gears (because to test some of the gears, you must destroy one clock). Then, out of 10,000 gears, you could examine up to 5,000 but no more! How would you establish the theory of operation of such a clock? Or model it?

In systems of readily definable complexity (and single cells far exceed this) (a) only an insufficient amount of atomic structural analysis can be accomplished before the class is eliminated, (b) heterogeneity and (c) electron orbitting and (d) transactional computations become too problematical to solve.

5. Toward a "Calculus of Potentiality"

When Newton in his efforts to work out planetary motions, and Leibnitz in his efforts to work out a practical means to prevent deep mines from flooding, developed the calculus of probability they made facile the solution of otherwise cumbersome and difficult mathematical procedures. But they knowingly gave up some precision that was available through more primitive mathematical treatments.

It has been suggested by Peter Drucker, among others, that we now need a "calculus of potentiality". This would allow us to expedite analyses presently too involved and cumbersome -- and it would accommodate a loss of precision and certainty to which we are now accustomed but which we could learn to do without in order to accommodate more difficult concepts and calculations.

It would be a more biological mathematics -- one more suited to the problems of evolution, homeostasis, embryology and almost all of the problems of neuronal integration. We would know how to predict less precisely but some predictions of limits and variance would become possible that are not now possible. We would probably be able to come as close as feasible to predicting accurately for truly complex systems.

Even moderately complex metabolic systems and nerve nets involve many components which appear to be severally reversible, irreversible, closed, open, forced, facultative, stable, metastable, and unstable, within the same system. They involve multiple, mutually interdependent, dynamic components in simultaneous action, constituting understructures that possess swirls and eddys of feed-back loops and altogether without any apparent single central controlling focus. The very word we have been using -- interaction -- is no longer adequate. A more meaningful word is transaction.

The examples given are among the simplest existing among biological systems. The requirements for biology of the future are more general and more demanding than these examples indicate. There exist layers upon layers, and orders upon orders, and orders of magnitude of layers and orders in the real world -- system interdependencies out of which qualitatively different regularities emerge, for which we may have legitimate curiosity and need to know (e.g. the regulation of respiration, temperature control, and still more complex transactions relating to consciousness, perception, learning, memory, judgment, creativity etc.).

6. The Hierarchy of Emergent Biological Capabilities

The hierarchy of qualitative changes that have emerged through biological evolution -- and may be seen to emerge independently from widely different biological systems -- cannot be modelled, much less predicted, by any amount of devotion to only the lower levels. This generalization appears to be equally cogent at the atomic, molecular, cellular, organismic and social levels.

One can work downward from more complex levels to discover explanations that derive from lower levels -- or at least that follow laws that regulate lower levels. The examination of more complex levels is so entrancing for its luxury of details that the general principles we are seeking to understand tend to become lost to sight. Yet what may be eminently satisfying generalizations at one level of complexity may have little or no bearing at the next level of greater complexity.

For example, the assumptions underlying the straightforward theories of Linderstrøm-Lang, which deal with proteins in dilute salt solutions, are obviously inadequate for dealing with proteins in cells.

Successively more complicated systems need to be analyzed as systems in their own right, as systems, at each different level, from atomic physics to social psychology. An additional goal, of overriding importance, is to search for those general features or principles which may distinguish each of the qualitatively different emergent levels. There are probably discernable even more general principles (operating vertically throughout successively higher levels of complexity) which need to be elucidated in order to account for the obvious but still completely unexplored general tendency for the recurrence of regularities at every level of evoluting complexity.

7. Successive Stages of Scientific Discovery

When Dewey and Bentley published their book Knowing and the Known (Beacon, Boston, 1949), they described the stages in the development of scientific explanations: self-action, interaction and transaction. In the stage of explanation by self-action, things are viewed as acting under their own powers. This is the case in Aristotelian substance in which things possess being which is eternal in action according to their nature. Galileo's development of insight into inertia characterizes comprehension of interaction systems. Hobbes, Descartes, Newton took up this line of thinking. The laws of Newton are classic examples of interactions: "the effects of simple forces acting on unalterable bodies". Space and time are omitted because they constitute a fixed framework within which events take place.

Transaction implies a more comprehensive view point. Space and time are integral parts of all events. The concept of immutable substance gives way to properties of fields; what were thought of as unalterable particles become recognized as mutable and dynamic loci of energy.

Biology, and particularly the neurosciences, are at cross-roads, basing their explanations partly on interactions and partly on transactions. Of course, the more comprehensive explanation does not destroy the usefulness of interactions for purposes of everyday discussions, anymore than relativity and quantum mechanics destroy the utility of the simpler Newtonian explanations. Yet there has emerged in the scientific community, a more sophisticated conception -- one that has practical importance -- of the physical universe. If physics had remained at the Newtonian level of discourse there would be much in the universe left unexplored, unexplained and unexploitable. Modern biology is reaching out for more transactional explanations. Ecology speaks of evolution of the habitat simultaneously with the organism. Animals are in dynamic transaction with their environment.

Information necessary for the adult organism is contained in the environment as well as in the gene structure (e.g. the environmental feature of gravitational force). The gene is reckoned to have different actions by virtue of its neighbors; the effective environment for the gene may include many other genes, or all genes in the same chromosome and within other chromosomes as well.

The generation of neuronal patterns is becoming better explained by reference to principles of extracellular ligands, specific cellular adhesions, exploratory zones of membranes exhibited by neuronal growth cones etc. The full explanation will surely rise above considerations of mere interactions among cells in free culture or the sorting out phenomena of different cell types in tissue culture.

Biology deals even more obviously than physics and mathematics with processes that transcend interactional mechanisms. On this account, biology, physics and mathematics will emerge as mutually donor and recipient disciplines in the development of many new, unexpected and enriching intellectual and technical achievements in relation to exploration of the nervous system. It can be a grand and exhilerating shared adventure.

PART II

Molecule and Brain

AN INTEGRAL PHYSICO-CHEMICAL MODEL FOR BIOEXCITABILITY

Eberhard Neumann
Max-Planck-Institut für Biophysikalische Chemie
D - 34 Göttingen-Nikolausberg , Postfach 968

Contents

1. Introduction

Bioelectrical excitability is a universal property of all higher organisms. Although numerous experimental data have been accumulated in the various disciplines working on bioelectricity, the mechanism of nerve excitability and synaptic transmission, however, is still an unsolved problem. Many mechanistic interpretations of nerve behaviour and the various molecular and mathematical models generally cover only a part of the known facts, are thus selective and of only limited value. According to Agin, there is no need for further (physically) unspecific mathematical models (such as for instance the Hodgkin-Huxley scheme for squid giant axons). "What is needed is a quantitative theory based on elementary physicochemical assumptions and which is detailed enough to produce calculated membrane behavior identical with that observed experimentally. At the present time such a theory does not exist." (Agin, 1972).

The first proposal of a physically specific mechanism for the control of bioelectricity is due to Nachmansohn (Nachmansohn 1955, 1971). The chemical hypothesis of Nachmansohn remained, however, on a qualitative level and until recently there was no detailed link between the numerous biochemical data on excitability and the various electrophysiological observations. A first attempt at integrating some basic data of biochemical and pharmaco-electrophysiological studies has been initiated by the late Aharon Katchalsky and a first report has been recently given (Neumann et al., 1973).

The present account is a further step towards a specific quantitative physico-chemical model for the control of ion flows during excitation. After an introductory survey on fundamental electrophysiological and biochemical observations our integral model is developed and some basic electrophysiological excitation parameters are formulated in terms of specific membrane processes.

2. Biomembrane Electrochemistry

It is well-known that bioelectricity and nerve excitability are electrically manifested in stationary membrane potentials and the various forms of transient potential changes such as, e.g., the action potential (Hodgkin, 1964; Tasaki, 1968). Although these electrical properties reflect membrane processes, bioelectricity and excitability are intrinsically coupled to the metabolic activity of the excitable cells. Now, inherent to all living cells is a high degree of coupling between energy and material flows. But already on the subcellular level of membranes intensive chemo-diffusional flow coupling occurs and apparently time-independent properties reflect balance between active and passive flows of cell components.

The specific function of the excitable membrane requires the maintenance of nonequilibrium states (Katchalsky, 1967). The intrinsic nonequilibrium nature of membrane excitability is most obviously reflected in the unsymmetric ion distributions across excitable membranes. These nonequilibrium distributions are metabolically mediated and maintained by "active transport" (e.g., Na^+/K^+ exchange-pump). A fundamental requirement for such an active chemo-diffusional flow coupling is spatial anisotropy of the coupling medium (Prigogine, 1968), including the membrane structure (Katchalsky, 1967).

Anisotropy is apparently a characteristic property of biological membranes. Furthermore, cellular membranes including nerve membranes may be physico-chemically described in terms of a layer structure. For instance it was found that the internal layer of the axonal membrane contains proteins required for excitability. Proteolytic action on this layer causes irreversible loss of this property (Tasaki and Singer, 1966).

The membrane components of excitable membranes include ionic constituents: fixed charges (some of them may serve to facilitate locally perm-selective ion diffusion), mono- and divalent metal ions, especially Ca^{2+} -ions, and in particular ionic side groups of membrane proteins directly involved in the control of electric properties. Local differences in the distribution of fixed charges may create membrane regions with different permeability characteristics, and (externally induced) ionic currents may amplify either the accumulation or the depletion of ions within the membrane. Structural inhomogeneity of excitable membranes may thus lead to the observed nonlinear dependencies between certain physical parameters, e.g. between current intensity and potential (Cole, 1968). A well-known example of this nonlinearity is the current rectification in resting stationary states of excitable membranes. Correspondent to current rectification, there is a straightforward dependence of the membrane potential on the logarithm of ion activities only in limited concentration ranges (Hodgkin and Keynes, 1955; Tasaki and Takenada, 1964). Thus, various chemical and physical membrane parameters show the intrinsic structural and functional anisotropy of excitable membranes. This recognition reveals the approximate nature of any classical equilibrium and nonequilibrium approach to electrochemical membrane parameters on the basis of linearity and involving the assumption of a homogeneous isotropic membrane. Since, however, details of the membrane anisotropy are not known, any exact quantitative nonequilibrium analysis of membrane processes and ion exchange currents across excitable membranes faces great difficulties for the time being.

3. Stationary Membrane Potentials

It is only in the frame of a number of partially unrealistic, simplifying assumptions that we may approximately describe the electric-chemical behaviour of excitable membranes, within a limited range of physico-chemical state variables (Agin, 1967). Explicitly, we use the formalism of classical irreversible thermodynamics restricted to linearity between flows and driving forces. The application of the Nernst-Planck equation to the ion flows across excitable membranes represents such a linear approximation, widely used to calculate stationary membrane potentials.

There is experimental evidence that large contributions to the resting stationary membrane potential, $\Delta\psi_r$, are attributable to ion selectivities. Indeed, it appears that excitable membranes have developed dynamic structures which are characterized by permselectivity for certain ion types and ion radii. This property may be described by "Nernst" terms which, however, are strictly valid only for electrochemical equilibria. Nevertheless, the Nernst approximation for the resting stationary potential is useful for a number of practical cases.

For the condition of stationarity, i.e. at zero net membrane current (I_m=O), we may write

$$\Delta\psi_r \simeq (\Delta\psi)_{I_m=0} = \frac{RT}{F} \sum_i \frac{t_i}{Z_i} \ln \frac{a_i^{(o)}}{a_i^{(i)}} \qquad (1)$$

In Eq. (1), t_i is the transference number representing the fraction of membrane current carried by ion type i (the fraction t_i may vary between O and 1); RT is the (molar) thermal energy, F is the Faraday; Z_i is the valency and a_i is the thermodynamic activity of ion i, (o) outside and (i) inside the excitable cell.

Although stationary states of excitable membranes always reflect balances between active transport processes and passive "leakage" fluxes, we may safely neglect flow contributions associated with active transport as long as the time scale considered does not exceed the millisecond range.

It should be realized that the measured potential difference (measured for instance with calomel electrodes in connection with salt bridges) cannot be used to calculate the exact value for the average electric field, $\bar{E}$, across the permeation barrier of thickness d.

$$\bar{E} = -\frac{1}{d}\Delta\psi' \tag{2}$$

The electrical potential difference $\Delta\psi'$ associated with the permeation barrier results from different sources. Although these sources are mutually coupled we may formally separate the various contributions. Due to the presence of fixed charges at membrane surfaces (Segal, 1968) there are Donnan potentials $\Delta\psi_D^{(o)}$ and $\Delta\psi_D^{(i)}$ from inside (i) and outside (o) interfaces between membrane and environment. Furthermore, there are interdiffusion potentials $\Delta\psi_U$ arising from differences in ionic mobilities within ion exchange domains of intramembraneous fixed charges (Tasaki, 1968). We may sum up these terms to $\Delta\psi_{DU} = \Delta\psi_D^{(o)} + \Delta\psi_D^{(i)} + \Delta\psi_U$. Thus,

$$\Delta\psi' = \Delta\psi_N + \Delta\psi_{DU} \tag{3}$$

Compared to $\Delta\psi_{DU}$ the "Nernst" contributions to $\Delta\psi$ appear relatively large. Inserting Eqs. (1) and (3) in Eq. (2) we obtain for the resting state:

$$(\bar{E})_{I_m=0} = -\frac{1}{d}\left(\Delta\psi_{DU} + \frac{RT}{F}\sum_i \frac{t_i}{Z_i}\ln\frac{a_i^{(o)}}{a_i^{(i)}}\right) \tag{4}$$

Among the various ions known to contribute to $\Delta\psi$ (in vivo) are the metal ions Na^+, K^+, Ca^{2+}, protons and Cl^- ions. In the resting state the value of the stationary membrane potential, $\Delta\psi_r$, is different for different cells and tissues. Distribution and density of fixed charges, ion gradients and the extent to which they contribute to $\Delta\psi_r$ differ for different excitable membranes.

Variations of the natural ionic environment lead also to alterations in the Donan and interdiffusion terms. Since ions are an integral part of the membrane structure, for instance as counterions of fixed charges, any change in ion type and salt concentration but also variations in temperature and pressure may cause changes in the lipid phase (Traeuble and Eibl, 1974) and in the conformation of membrane proteins. As a consequence of such structural changes membrane "fluidity" and permeability properties may be considerably altered. It is known that, for instance, an increase in the external Ca^{2+} ion concentration increases the electrical resistance of excitable membranes (Cf., e.g. Cole, 1968).

4. Transient Changes of Membrane Potentials

A central problem in bioelectricity is the question: what is the mechanism of the various transient changes of membrane parameters associated with nerve activity? Before discussing a specific physico-chemical model for the control of electrical activity it appears necessary to recall a few fundamental electrophysiological and chemical observations on excitable membranes.

The various transient expressions of nerve activity such as depolarizations and hyperpolarizations of the resting stationary potential level reflect changes of dielectric-capacitive nature, interdiffusional ion redistributions, and "activations" or "inactivations" of different gradients or changes of the extent with which these gradients contribute to the measured parameters. At constant pressure and temperature, the ion gradient contributions to $\Delta\psi$ may be modelled as variations of the t_i parameters; see Eq. (1).

4.1 Threshold behaviour.

Transient changes in $\Delta\psi$ may be caused by various physical and chemical perturbations originating from adjacent membrane regions, from adjacent excitable cells, or from external stimuli. Phenomenologically we differentiate sub- and suprathreshold responses of the excitable membrane to stimulation. Subthreshold changes are, for instance, potential changes which attenuate with time and distance from the site of perturbation; this phenomenon has been called electrotonic spread. If, however, the intensity of depolarizing stimulation exceeds a certain threshold value, a potential change is triggered that does not attenuate but propagates as such along the entire excitable membrane. This suprathreshold response is called "regenerative" and has been christened action potential or nerve impulse.

4.2 Stimulus characteristics

In order to evoke an action potential the intensity of the stimulation has to reach the threshold with a certain minimum velocity (Cole, 1968). If the external stimulus for instance is a current the intensity, I, of which gradually increases with time, t, the minimum slope condition for the action potential may be written

$$dI/dt \geq (dI/dt)^{min} \qquad (5)$$

The observation of a minimum slope condition for the generation of action potentials is of crucial importance for any mechanistic approach to excitability.

On the other hand, when rectangular "current pulses" are applied, there is an (absolute) minimum intensity, I_{th}, called rheobase, and a minimum time interval, Δt, of stimulus duration. It is found that the product of suprathreshold intensity and minimum duration is approximately constant. Thus, for $I \geq I_{th}$,

$$I \cdot \Delta t \approx \text{constant}, \tag{6}$$

describing the well-known strength-duration curve. Whereas the strength-duration product does not depend on temperature, the temperature coefficient of the rheobase (dI_{th}/dT related to a temperature increase of 10°C) Q_{10} is generally about 2 (see, e.g.,Cole, 1968).

It is recalled that the value of I_{th} is history dependent. The threshold changes as a function of previous stimulus intensity and duration. Hyperpolarizing and depolarizing subthreshold prepulses also change the threshold intensity of the stimulating current. See Section 3(d).

It appears that, in general, the induction of the action potential requires the reduction of the intrinsic membrane potential $\Delta\psi_r$ below a certain threshold value $\Delta\psi_{th}$. This potential decrease $\Delta(\Delta\psi) = \Delta\psi_r - \Delta\psi_{th}$ is usually about 15 to 20 mV and has to occur in the form of an impulse, $\int \Delta(\Delta\psi)dt$, in which a minimum condition between membrane potential and time must be fulfilled.

The condition for the initiation of an action potential may be written

$$\Delta\psi_r - \int \left(\frac{d\ \Delta\psi}{dt} \right)^{min} dt = \Delta\psi_{th} \tag{7}$$

or in the more general form:

$$\Delta\psi_r - \frac{1}{\Delta t} \int \Delta(\Delta\psi)\ dt \leq \Delta\psi_{th}. \tag{8}$$

The potential change $\Delta(\Delta\psi)$ is equivalent to a change in the intrinsic field, ΔE, across the membrane of the thickness d. Thus $\Delta E = -\Delta(\Delta\psi)/d$ and with Ohm's law we have $d \cdot \Delta E = -R_m \cdot I$, where R_m is the membrane resistance.

Including Eq. (1) we may write

$$\int \Delta(\Delta\psi)dt = -d \int \Delta E\ dt = \int R_m \cdot I\ dt = \frac{RT}{Z_i F} \int \Delta \ln \frac{a_i^{(o)}}{a_i^{(i)}}\ dt. \tag{9}$$

With Eqs. (8) and (9) we see how the intrinsic membrane potential may be changed: by an (external) field (voltage or current) pulse or by an ion pulse involving those ions that determine the membrane potential. Such an ion pulse may be produced, if the concentration of K^+ ions on the outside of excitable membranes is sufficiently increased within a sufficiently short time interval (impulse condition). Similarly, a pH change can cause action potentials (Tasaki, Singer, and Takenaka, 1965).

As seen in Eq. (1) the membrane potential is dependent on temperature. Therefore temperature (and also pressure) changes alter the membrane potential in excitable membranes. Moreover, action potentials can be evoked by thermal and mechanical shocks (see, e.g., Julian and Goldman, 1962).

It is furthermore remarked that under natural conditions electrical depolarization produces an action potential only if the excitable membrane is kept above a certain (negative) level of polarization; in squid giant axons the minimum membrane potential is about -30 to -40 mV. This directionality of membrane polarization and the requirement of field reduction for regenerative excitation are a further manifestation of the intrinsic anisotropy of excitable membranes.

4.3 Propagation of local activity

In many excitable cells $\Delta\psi_r$ is about -60 mV to -90 mV, where the potential of the cell interior is negative. Assuming an average membrane thickness d = 100 Å, we calculate with Eq. (2) an average field intensity of approximately 60 to 90 kV/cm. (Cf., however, Carnay and Tasaki, 1971). The field vector in the resting stationary state is directed from the outside to the inside across the membrane of the excitable cell. (It is this electric field which partially compensates the chemical potential gradient of the K^+ ions.) Any perturbation which is able to change locally the intrinsic membrane potential, i.e. the membrane field, will also affect adjacent parts of the membrane or even adjacent cells. If the field change remains below threshold or does not fulfill the minimum slope condition there will be only subthreshold attenuation of this field change. A nerve impulse in its rising phase may, however, reduce the membrane field in adjacent parts to such an extent and within the required minimum time interval, that suprathreshold responses are triggered. It is recalled that electric fields represent long-range forces (decaying with distance).

4.4 Refractory phases

Important observations suggestive for the nature of bioelectricity are the refractory phenomena. If a nerve fibre is stimulated a second time shortly after a first impulse was induced, there is either an impulse that propagates much slower or only a subthreshold change. Immediately after stimulation a fibre is absolutely refractory; no further impulse can be evoked. This period is followed by a relatively refractory phase during which the threshold potential is more positive and only a slowly propagated impulse can be induced. The various forms of history dependent behaviour are also called accommodation or adaptation of the fibre to preceding manipulations (see, e.g., Tasaki, 1968).

4.5 Time constants

If rectangular current pulses of fixed duration but of variable amplitude and amplitude directions are locally applied to an excitable membrane, a series of local responses are obtained. Fig. 1 shows schematically the time course of these local responses upon hyper- and depolarizing stimulations: subthreshold potential changes and action potential (Katz, 1966).

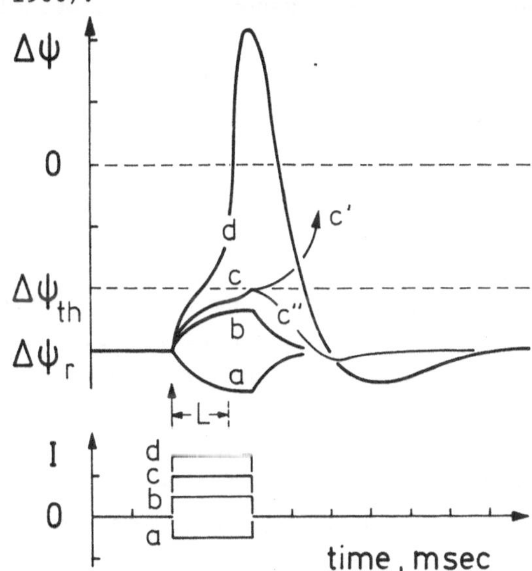

Fig. 1. Sub- and suprathreshold responses of the membrane potential, $\Delta\psi$, to stimulating rectangular current pulses of fixed duration but variable intensity, I. (a) inward current pulse causing hyperpolarization, (b) subthreshold outward current ($I < I_{th}$) causing depolarization, (c) threshold outward current (I_{th}) causing either an action potential (path c') or return to resting stationary potential $\Delta\psi_r$ (path c"), (d) suprathreshold outward pulse ($I > I_{th}$) causing action potential. L, latency phase of the action potential; $\Delta\psi_{th}$, threshold potential.

It appears that the threshold level reflects a kind of instability. When the stimulus is switched off, fluctuations either result in the action potential (path c → c') or in a return to the resting level (path c → c"). If for phase c the stimulus of the same intensity would last longer, an action potential would develop immediately from the hump visible in the late phase of stimulation.

It has been found that large sections of the time course of subthreshold responses and the first phase of the action potential (also called latency; see Fig. 1) can be described with linear differential equations of first order. Furthermore, the time constants τ_m associated with these sections are nearly equal.

The actual values for τ_m are different for different excitable membranes and are dependent on temperature. In squid giant axons the value of τ_m at about 20° C is about 1 msec; the corresponding value for lobster giant axons is about 3 msec. The temperature coefficient ($d\tau_m^{-1}/dT$ related to a temperature increase of 10°C) Q_{10} is about 2 (Cole, 1968).

As outlined by Cole, a time constant in the order of milliseconds can hardly be modelled by simple electrodiffusion: an ion redistribution time of 1 msec requires the assumption of extremely low ion mobilities (about 10^{-8} cm sec^{-1}/Vcm^{-1}). However, even in "sticky" ion exchange membranes ionic mobilities are not below 10^{-6} cm $sec^{-1}/V\ cm^{-1}$. Furthermore, the temperature coefficients of simple electrodiffusion are only about 1.2 to 1.5.

Thus, magnitude and temperature coefficient of τ_m suggest that electrodiffusion is not rate-limiting for a large part of the ion redistributions following perturbations of the membrane field.

This conclusion is supported by various other observations. The time course of the action potential can be formally associated with a series of time constants, all of which have temperature coefficients of about 2 to 3 (see, e.g., Cole, 1968). The various phases of the action potential are prolonged with decreasing temperature. In the framework of the classical Hodgkin-Huxley phenomenology prolonged action potentials should correspond to larger ion movements. However, it has been found recently that in contrast to this prediction, the amount of ions actually transported during excitation decreases with decreasing temperature (Landowne, 1973).

It thus appears that ion movements (caused by perturbations of stationary membrane states) are largely rate-limited by membrane processes. Such processes may comprise phase changes of lipid domains or conformational changes of membrane proteins. Changes of these types usually are cooperative and temperature dependencies are particularly pronounced within the cooperative transition ranges.

Prolonged potential changes and smaller ion transport at decreased temperatures may be readily modelled, if at lower temperature configurational rearrangements of membrane components involve smaller fractions of the membrane than at higher temperature.

There is a formal resemblance between the time constant τ of a phase change or a chemical equilibrium and the RC term of an electrical circuit with resistance R and capacity C (Oster, Perelson and Katchalsky, 1973). If a membrane process is associated with the thermodynamic affinity A and a rate $J_r = d\xi/dt$, where ξ is the fractional advancement of this process, we may formally define a reaction resistance $R_r = (\partial A/\partial J_r)$ and an average reaction capacity $\bar{C}_r = -(\partial A/\partial \xi)^{-1}$. The time constant is then given by

$$\tau = R_r \cdot C_r . \qquad (10)$$

The time course of electrical parameters may thus be formally modelled in terms of reaction time constants for membrane processes of the type discussed above. For the chemical part of the observed subthreshold change in the membrane potential

$$\Delta\psi = I_m R_m \, (1 - \exp[-t/\tau_m]) \qquad (11)$$

caused by the rectangular current pulse of the intensity I_m across the membrane resistance R_m, we have $\tau_m = R_r C_r$.

As already mentioned, the time constants for subthreshold changes and for the first rising phase of the action potential are almost the same. Furthermore, the so-called "Na^+-ion activation-inactivation curves (in voltage clamp experiments; see Fig. 2) for both sub- and suprathreshold conditions are similar in shape, although much different in amplitude" (Plonsey, 1969). These data strongly suggest that sub- and suprathreshold responses are essentially based on the same control mechanism; it is then most likely the molecular organization of the control system that accounts for the various types of response.

5. Proteins Involved in Excitation

Evidence is accumulating that proteins and protein reactions are involved in transient changes of electrical parameters during excitation. As briefly mentioned the action of proteases finally leads to inexcitability. Many membrane parameters such as ionic permeabilities are pH-dependent. In many examples this dependency is associated with a pK value of about 5.5 suggestive for the participation of carboxylate groups of proteins.

Sulfhydryl reagents and oxidizing agents interfere with the excitation mechanism, *e.g.* prolonging the duration and finally blocking the

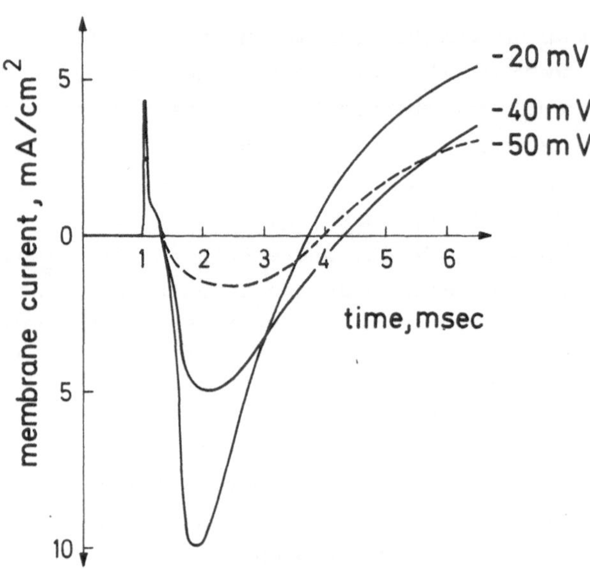

Fig. 2. Time course of membrane currents during voltage clamp experiments of internally perfused giant axons of squid Loligo pealii; resting potential $\Delta\psi_r \simeq -65$ mV. Dashed line: response to a clamp of -50 mV, i.e. a depolarization of 15 mV that, under unclamped conditions, does not evoke an action potential. The solid lines represent suprathreshold membrane currents. (Scheme of data by Narahashi et al., 1969).

action potential (Huneeus-Cox et al., 1966). The membrane of squid giant axons stain for SH groups provided the fibres had been stimulated (Robertson, 1970). These findings strongly suggest the participation of protein specific red-ox reactions in the excitation process.

Of particular interest appear the effects of ultraviolet radiation on nodes of Ranvier. The spectral radiation sensitivity of the rheobase and of the fast transient inward current (normally due to Na^+ ions) in voltage clamp is very similar to the ultraviolet absorbance spectrum for proteins (von Muralt and Stämpfli 1953; Fox 1972). The results of Fox also indicate that the fast transient inward component of the action current is associated with only a small fraction of the node membrane.

The conclusions on locally limited excitation sites are supported by the results obtained with certain nerve poisons. Extremely low concentrations of tetrodotoxin reduce and finally abolish the fast inward component of voltage clamp currents (for review see Evans, 1972). The block action of this toxin is pH dependent and is associated with a pK value of about 5.3.

It thus appears that the ionic gateways responsible for the transient inward current involve protein organizations comprising only a small membrane fraction.

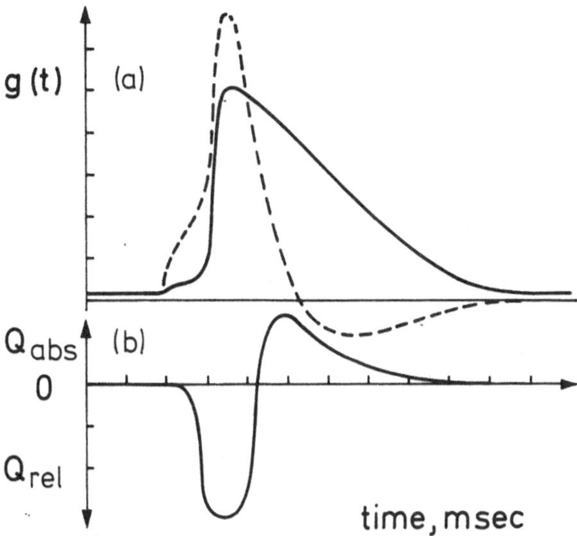

Fig. 3. Schematic representation of (a) conductivity change, g(t), accompanying the action potential (dashed line) and (b) heat exchange cycle. Q_{rel}, heat released during rising phase and Q_{abs}, heat absorbed during falling phase of the action potential.

6. Impedance and Heat Changes

The impedance change accompanying the action potential is one of the basic observations in electrophysiological studies on excitability (Cole, 1968). It is generally assumed that the impedance change reflects changes in ionic conductivities (ionic permeabilities). Fig. 3 shows schematically that the conductance first rises steeply and in a second phase decays gradually towards the stationary level. In contrast to the pronounced resistance change, the membrane capacity apparently does not change during excitation (Cole, 1970). This result, too, suggests that only a small fraction of the membrane is involved in the rather drastic permeability changes during excitation.

The action potential is accompanied by relatively large heat changes (Abbot, Hill and Howarth, 1958; Howarth, Keynes and Ritchie, 1968); see Fig. 3. These heat changes may be thermodynamically modelled in terms of a cyclic variation of membrane states. The complex chain of molecular events during a spike may be simplified by the sequence of state changes $A \rightarrow B \rightarrow A$, where A represents the resting stationary state and B symbolizes the transiently excited state of higher ionic permeability.

The heat changes occur under practically isothermal-isobaric conditions. If we now associate the Gibbs free energy change, ΔG, with the (overall) excitation process A $\rightleftharpoons$ B, we may write

$$\Delta G = \Delta H - T\Delta S, \qquad (12)$$

where ΔH is the reaction enthalpy (as heat exchangeable with the environment) and ΔS the reaction entropy.

Since the action potential most likely involves only a small fraction of the excitable membrane, the measured heat changes Q_{rel} and Q_{abs} appear to be very large. Since on the other hand the mutual transition A→B→A "readily" occurs, the value of ΔG (= $\Delta G_{A \to B}$ = $-\Delta G_{B \to A}$) cannot be very large. In order to compensate a large ΔH (here $\simeq$ Q), there must be a large value for ΔS; see Eq. (12). This means that the entropy change associated with the membrane permeability change during excitation is also very large.

It is, in principle, not possible to deduce from heat changes the nature of the processes involved. However, large configurational changes - equivalent to a large overall ΔS - in biological systems frequently arise from conformational changes of macromolecules or macromolecular organizations such as membranes or from chemical reactions. In certain polyelectrolytic systems such changes even involve metastable states and irreversible transitions of domain structures (Neumann, 1973).

The large absolute values of Q (and the irreversible contribution ΔQ, see Neumann, 1974) suggest structural changes and/or chemical reactions to be associated with the action potential. Furthermore, there are observations indicating the occurrence of metastable states and nonequilibrium transitions in excitable membranes at least for certain perfusion conditions (Tasaki, 1968).

7. The Cholinergic System and Excitability

Among the oldest known macromolecules associated with excitable membranes are some proteins of the cholinergic system. The cholinergic apparatus comprises acetylcholine (AcCh), the synthesis enzyme choline-O-acetyltransferase (Ch-T), acetylcholine-esterase (AcCh-E), acetylcholine-receptor (AcCh-R) and a storage site (S) for AcCh. Since details of this system are outlined elsewhere (Neumann and Nachmansohn, 1974), only a few aspects essential for our integral model of excitability are discussed here.

7.1 Localization of the AcCh system

Chemical analysis has revealed that the concentration of the cholinergic system is very different for various excitable cells. For instance, squid giant axons contain much less AcCh-E (and Ch-T) than axons of lobster walking legs (see e.g. Brzin el at., 1965); motor nerves are generally richer in cholinergic enzymes than sensory fibres (Gruber and Zenker, 1973).

The search for cholinergic enzymes in nerves was greatly stimulated by the observation that AcCh is released from isolated axons of various excitable cells, provided inhibitors of AcCh-E such as physostigmine (eserine) are present. This release is appreciably increased upon electric stimulation (Calabro, 1933; Lissak, 1939) or when axons are exposed to higher external K^+ ion concentrations (Dettbarn and Rosenberg, 1966). Since larger increases of external K^+ concentration depolarize excitable membranes, reduction of membrane potential appears to be a prerequisite for AcCh liberation from the storage site.

Evidence for the presence of axonal AcCh storage sites and receptors is still mainly indirect. However, direct evidence continuously accumulates for extrajunctional AcCh receptors (Porter et al., 1973); binding studies with α- bungarotoxin (a nerve poison with a high affinity to AcCh-R) indicate the presence of receptor-like proteins in axonal membrane fragments (Denburg et al., 1972).

There are still discrepancies as to the presence and localization of the cholinergic system. However, the differences in the interpretations of chemico-analytical data and the results of histochemical light- and electron microscopy investigations are gradually being resolved; there appears progressive confirmation for the early chemical data of an ubiquitous cholinergic system. For instance, AcCh-E reaction products can be made visible in the excitable membranes of more and more nerves formerly called non-cholinergic (see, e.g., Koelle, 1971). Catalysis products of AcCh-E are demonstrated in pre- and postsynaptic parts of excitable membranes (see, e.g., Barrnett, 1962; Koelle, 1971; Lewis and Shute, 1966). In a recent study stain for the α-bungarotoxin-receptor complex is visible also in presynaptic parts of axonal membranes (see Fig. 1 in Porter et al. 1973). These findings suggest the presence of the cholinergic system in both junctional membranes and thus render morphological support for the results of previous studies on the pre- and postsynaptic actions of AcCh and inhibitors and activators of the cholinergic system (Masland and Wigton, 1940; Riker et al., 1959).

7.2 The barrier problem

Histochemical, biophysical and biochemical, and particularly, pharmacological studies on nerve tissue face the great difficulty of an enormous morphological and chemical complexity. It is now recognized that due to various structural features not all types of nerve tissue are equally suited for certain investigations.

The excitable membranes are generally not easily accessible. The great majority of nerve membranes are covered with protective tissue layers of myelin, of Schwann- or glia cells. These protective layers insulating the excitable membrane frequently comprise structural and chemical barriers that impede the access of test compounds to the excitable membrane. In particular, the lipid-rich myelin sheaths are impervious to many quaternary ammonium compounds such as AcCh and curare.

In some examples such as the frog neuro-muscular junction, externally applied AcCh or the receptor inhibitor curare have relatively easy access to the synaptic gap whereas the excitable membranes of the motor nerve and the muscle fibre appear to be largely inaccessible. On the other hand, the cholinergic system of neuromuscular junctions of lobsters are protected against external action of these compounds whereas the axons of the walking legs of lobster react to AcCh and curare (cf., e.g., Dettbarn, 1967).

Penetration barriers also comprise absorption of test compounds within the protective layers. Furthermore, chemical barriers in the form of hydrolytic enzymes frequently cause decomposition of test compounds before they can reach the nerve membrane. For instance, phosphoryl phosphatases in the Schwann cell layer of squid giant axons cause hydrolysis of organophosphates such as the AcCh-esterase inhibitor diisopropylfluorophosphate (DFP), and impulse conduction is blocked only at very high concentrations of DFP (Hoskin et al., 1966).

A very serious source of error in concentration estimates and in interpretations of pharmacological data resides in procedures that involve homogenization of lipid-rich nerve tissue (see, e.g, Nachmansohn, 1969). For instance, homogenization liberates traces of inhibitors (previously applied) which despite intensive washing still adhered to the tissue. Even when the excitable membrane was not reached by the inhibitor, membrane components react with the inhibitor during homogenization (Hoskin et al., 1969). Thus block of enzyme activity observed after homogenization is not necessarily an indicator for block during electrical activity (cf., e.g., Katz, 1966, p. 90).

As demonstrated in radiotracer studies, failure to interfere with bioelectricity is often concomitant with the failure of test compounds to reach the excitable membrane. Compounds like AcCh or d-tubocurarine (curare) act on squid giant axons only after (enzymatic) reduction of structural barriers (Rosenberg and Hoskin, 1963). Diffusion barriers even after partial reduction are often the reason for longer incubation times and higher concentration of test compounds as compared to less protected membrane sites. In this context it should be mentioned that the enzyme choline-O-acetyltransferase, sometimes considered to be a more specific indicator of the cholinergic system, is frequently difficult to identify in tissue and is in vitro extremely unstable (Nachmansohn, 1963).

In the light of barrier and homogenization problems it appears obvious that any statements on the absence of the cholinergic system or on the failure of blocking compounds to interfere with excitability are only useful if they are based on evidence that the test compound had actually reached the nerve membrane.

7.3 Electrogenic aspects of the AcCh system

Particularly suggestive for the bioelectric function of the cholinergic system in axons are electrical changes resulting from eserine and curare application to nodes of Ranvier, where permeability barriers are less pronounced (Dettbarn, 1960). Similar to the responses of certain neuro-muscular junctions, eserine prolongs potential changes also at nodes; curare first reduces the amplitude of the nodal action potential and then also decreases the intensity of subthreshold potential changes in a similar way as known for frog junctions. See Fig. 4.

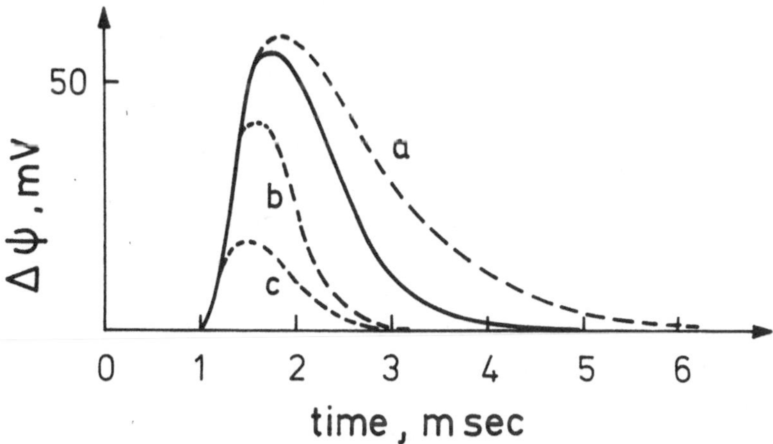

Fig. 4. Action potentials measured at a node of Ranvier of a single fiber of frog Sciatic nerve, T = 23° C; solid line: reference; dashed lines: (a) in the presence of 5×10^{-6} M eserine at pH 7.7 (10^{-3} M eserine blocks the action potential within 30 sec); (b) and (c): gradual decrease of the amplitude with time in the presence of 10^{-3} M curare. The inhibitor solution has been applied externally on top of the Schwann cell layers covering the excitable membrane at the nodes. (Scheme from data by Dettbarn, 1960).

(a)

$$CH_3-\overset{\oplus}{N}(CH_3)_2-CH_2-CH_2-O-C(=O)-CH_3$$

(b)

$$CH_3-\overset{\oplus}{N}(CH_3)_2-CH_2-CH_2-O-C(=O)-C_6H_4-NH-C_4H_9$$

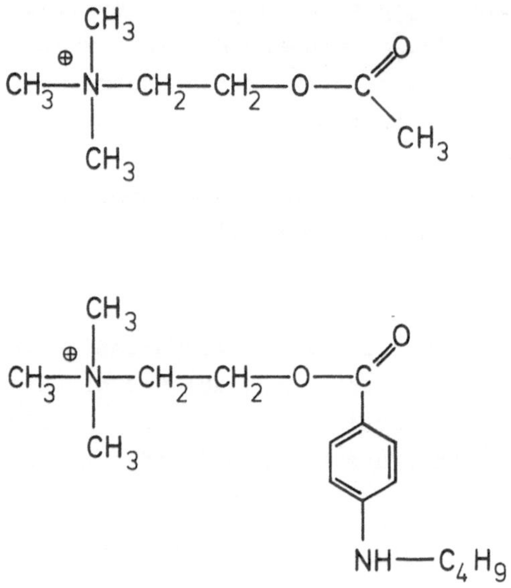

Fig. 5. Chemical structure of the acetylcholine ion (a) and of the tetracaine ion (b). Note that the structural difference is restricted to the acid residue: (a) CH_3 and (b) the amino-benzoic acid residue rendering tetracaine lipid-permeable.

In all nerves the generation of action potentials is readily blocked by (easily permeating) local anesthetics such as procaine or tetracaine. Due to structural and certain functional resemblance to AcCh which is particularly pronounced for tetracaine (see Fig. 5), these compounds may be considered as analogs of AcCh. In a recent study it was convincingly demonstrated how (by chemical substitution at the ester group) AcCh is successively transformed from a receptor activator (reaching junctional parts only) to the receptor inhibitor tetracaine reaching readily junctional and axonal parts of excitable membranes (Bartels, 1965). Local anesthetics are also readily absorbed in lipid bilayer domains of biomembranes (see for review Seeman, 1972).

External application of AcCh without esterase inhibitors faces not only diffusion barriers but esterase activity increases the local proton concentration (Dettbarn, 1967); the resulting changes in pH may contribute to potential changes.

Only very few nerve preparations appear to be suited to demonstrate a direct electrogenic action of externally applied AcCh. Diffusion barriers and differences in local concentrations of the cholinergic system may be the reason that the impulse condition for the generation of action potentials cannot be fulfilled everywhere (see

Section 4.2). Some neuroblastoma cells produce subthreshold potential changes and action potentials upon electrical stimulation as well as upon AcCh application (Harris and Dennis, 1970; Nelson et al., 1971; Hamprecht, 1974).

There are thus, without any doubt, many pharmacological and chemical similarities between synaptic and axonal parts of excitable membranes as far as the cholinergic system is concerned. On the other hand there are various differences. But it seems that these differences can be accounted for by structural and chemical factors.

As to this problem two extreme positions of interpretation may be distinguished. On the one side more emphasis is put to the differences between axonal and junctional membranes. An extreme view considers the responses of axons to AcCh and structural analogs as a pharmacological curiosity (Ritchie, 1963); and, in general, the cholinergic nature of excitable membrane is not recognized and acknowledged. However, the presence of the cholinergic system in axons and the various similarities to synaptic behaviour suggest the same basic mechanism for the cholinergic system in axonal and synaptic parts of excitable membranes.

Since there is until now no direct experimental evidence for AcCh to cross the synaptic gap, the action of AcCh may be alternatively assumed to be restricted to the interior of the excitable membrane of junctions and axons. This assumption is based on the fact that no trace of AcCh is detectable outside the nerve unless AcCh-esterase inhibitors are present. According to this alternative hypothesis intramembraneous AcCh combines with the receptor and causes permeability changes mediated by conformational changes of the AcCh-receptor.

This is the basic postulate of the chemical theory of bioelectricity, attributing the primary events of all forms of excitability in biological organisms to the cholinergic system: in axonal conduction, for subthreshold changes (electronic spread) in axons and pre- and postsynaptic parts of excitable membranes.

In the framework of the chemical model the various types of responses, excitatory and inhibitory synaptic properties are associated with structural and chemical modifications of the same basic mechanism involving the cholinergic system. Participation of neuroeffectors like the catecholamines or GABA and other additional reactions within the synapse possibly give rise to the various forms of depolarizing and hyperpolarizing potential changes in postsynaptic parts of excitable membranes. The question of coupling between pre- and postsynaptic events during signal transmission cannot be answered for the time being.

It is, however, suggestive to incorporate in transmission models the transient increase of the K^+ ion concentration in the synaptic gap after a presynaptic impulse.

7.4 Control function of AcCh.

It is recalled that transient potential changes such as the action potential result from permeability changes caused by proper stimulation. However, a (normally proper) stimulus does not cause an action potential, if (among others) certain inhibitory analogs of the cholinergic agents such as, *e.g.*, tetracaine are present. Tetracaine also reduces the amplitudes of subthreshold potential changes; in the presence of local anesthetics, for instance procaine, mechanical compression does not evoke action potentials (Julian and Goldman, 1962). It thus appears that the (electrical and mechanical) stimulus does *not directly* effect sub- and suprathreshold permeability changes, suggesting preceding events involving the cholinergic system.

If the AcCh-esterase is inhibited or the amount of this enzyme is reduced by protease action, subthreshold potential changes and (postsynaptic) current flows as well as the action potential are prolonged (see, *e.g.*, Takeuchi and Takeuchi, 1972; Dettbarn, 1960). Thus AcCh-E activity appears to play an essential role in terminating the transient permeability changes. The extremely high turnover number of this enzyme (about 1.4×10^4 AcCh molecules per sec, i.e., a turnover time of 70 μsec) is compatible with a rapid removal of AcCh (Nachmansohn, 1959).

In summary, the various studies using activators and inhibitors of the cholinergic system indicate that both initiation and termination of the permeability changes during nerve activity are (active) processes associated with AcCh. It seems possible, however, to decouple the cholinergic control system from the ionic permeation sites or gateways. Reduction of the external Ca^{2+} ion concentration appears to cause such a decoupling; the result is an increase in potential fluctuations or even random, *i.e.* uncontrolled firing of action potentials (see, *e.g.*, Cole, 1968).

8. The Integral Model

In the previous sections some basic electrophysiological observations and biochemical data are discussed that any adequate model for bioelectricity has to integrate. It is stressed that among the features excitability models have to reproduce are the threshold behaviour, the various similarities of sub- and suprathreshold responses, stimulus characteristics and the various forms of conditioning and history-dependent behaviour.

In the present account we explore some previously introduced concepts for the control of electrical membrane properties by the cholinergic system (Neumann et al., 1973). Among these fundamental concepts are the notion of a basic excitation unit, the assumption of an AcCh storage site particularly sensitive to the electric field of the excitable membrane, and the idea of a continuous sequential translocation of AcCh through the cholinergic proteins (AcCh-cycle). Finally we proceed toward a formulation of various excitation parameters in terms of nonequilibrium thermodynamics.

8.1 Key processes

In order to account for the various interdependencies between electrical and chemical parameters it is necessary to distinguish between a minimum number of single reactions associated with excitation.

A possible formulation of some of these processes in terms of chemical reactions has been previously given (Neumann et al., 1973). The reaction scheme is briefly summarized.

(1) Supply of AcCh to the membrane storage site (S), following synthesis (formally from the hydrolysis products choline and acetate).

$$S + AcCh = S_1(AcCh) \tag{13}$$

For the uptake reaction two assumptions are made: (i) the degree of AcCh association with the binding configuration S_1 increases with increasing membrane potential (cell interior negative), (ii) the uptake rate is limited by the confomational transition from state S to S_1, and is slow as compared to the following translocation steps; see also Sect. 8.5 (c).

Vesicular storage of AcCh as substantiated by Whittaker and coworkers (see, *e.g.*, Whittaker, 1973) is considered as additional storage for membrane sites of high AcCh turnover, for instance at synapses.

(2) Release of AcCh from the storage form S_1(AcCh), for instance by depolarizing stimulation.

$$S_1(AcCh) = S_2 + AcCh \tag{14}$$

Whereas S_1(AcCh) is stabilized at large (negative) membrane fields, S_2 is more stable at small intensities of the membrane field. The field-dependent conformational changes of S are assumed to gate the path of AcCh to the AcCh-receptors.

The assumptions for the dynamic behaviour of the storage translocation sequence

$$S + AcCh \underset{k_{-1}}{\overset{k_1}{=}} S_1(AcCh) \underset{k_{-2}}{\overset{k_2}{=}} S_2 + AcCh \qquad (15)$$

may be summarized as follows: the rate constant, k_2, for the release step is larger than the rate constant, k_1, for the uptake, and also $k_2 >> k_{-2}$. (See also Sect. 8.5)

(3) Translocation of released AcCh to the AcCh-receptor (R) and association with the Ca^{2+}-binding conformation $R_1(Ca^{2+})$. This association is assumed to induce a conformational change to R_2 that, in turn, releases Ca^{2+} ions.

$$R_1(Ca^{2+}) + AcCh = R_2(AcCh) + Ca^{2+} \qquad (16)$$

(4) Release of Ca^{2+}-ions is assumed to change structure and organization of gateway components, G. The structural change from a closed configuration, G_1, to an open state, G_2, increases the permeability for passive ion fluxes.

(5) AcCh hydrolysis. Translocation of AcCh from $R_2(AcCh)$ to the AcCh-esterase, E, involving a conformational transition from E_1 to E_2.

$$R_2(AcCh) + E_1 = E_2(Ch^+, Ac^-, H^+) + R_2 \qquad (17)$$

The hydrolysis reaction causes the termination of the permeability change by re-uptake of Ca^{2+},

$$R_2 + Ca^{2+} = R_1(Ca^{2+}) \qquad (18)$$

concomitant with the relaxation of the gateway to the closed configuration, G_1.

Thus, the reactions (17) and (18) "close" a reaction cycle which is formally "opened" with reactions (13) and (15).

Since under physiological conditions (i.e. without esterase inhibitor) no trace of AcCh is detectable outside the excitable membrane (axonal and synaptic parts), the sequence of events modelled in the above reaction scheme, is suggested to occur in a specifically organized structure of the cholinergic proteins; a structure that is intimately associated with the excitable membrane.

8.2 Basic excitation unit

Before proceeding towards a model for the organization of the cholinergic system, it is instructive to consider the following well-known electrophysiological observations.

In a great variety of excitable cells the threshold potential change to trigger the action potential is about 20 mV. This voltage change corresponds to an energy input per charge or charged group within the membrane field of only about one kT unit of thermal energy (k, Boltzmann constant; T, absolute temperature) at body temperature. If only one charge or charged group would be involved, thermal motion should be able to initiate the impulse. Since random "firing" is very seldom, we have to conclude that several ions and ionic groups have to "cooperate" in a concerted way in order to cause a suprathreshold permeability change.

Furthermore, there are various electrophysiological data which suggest at least two types of gateways for ion permeation in excitable membranes (for summary see Cole, 1968): a rapidly operating ion passage normally gating passive flow of Na^+ ions (into the cell interior) and permeation sites that normally limit passive K^+ ion flow.

There are various indications such as the direction of potential change and of current flow, suggesting that the rising phase of the action potential has predominantly contributions from the "rapid gateway"; the falling phase of the overall permeability change involves larger contributions of the K-ion gateways (see also Neher and Lux, 1973). There is certainly coupling between the two gateway types: electrically through field changes and possibly also through Ca^{2+}-ions transiently liberated from the "rapid gateways". As explicitly indicated in Eqs. (16) and (18), Ca^{2+} ion movement precedes and follows the gateway transitions.

Recent electrophysiological studies on neuroblastoma cells confirm the essential role of Ca^{2+} ions in subthreshold potential changes and in the gating phase of the action potential (Spector et al., 1973).

At the present stage of our model development we associate the direct cholinergic control of permeability changes only to the rapidly operating gateway, G. As seen in Fig. 3 the rising phase of the conductance change caused by the permeability increase is rather steep. This observation, too, supports a cooperative model for the mechanism of the action potential.

The experimentally indicated functional cooperativity together with the (experimentally suggested) locally limited excitation sites suggest a structural anchorage in a cooperatively stabilized membrane domain.

In order to account for the various boundary conditions discussed above we have introduced the notion of a basic excitation unit (BEU). Such a unit is suggested to consist of a gateway G that is surrounded by the cholinergic control system. The control elements are interlocked complexes of storage (S), receptor (R), and esterase (E), and are called SRE-assemblies. These assemblies may be organized in different ways and, for various membrane types, the BEUs may comprise different numbers of SRE assemblies. As an example, the BEU schematically represented in Fig. 6 contains 6 SRE units controlling the permeation site, G.

The core of the BEU is a region of dynamically coupled membrane components with fixed charges and counterions such as Ca^{2+} ions. Fig. 6 shows that the receptors of the SRE assemblies form a ring-like array. We assume that this structure is cooperatively stabilized and, through Ca^{2+}-ions, intimately associated with the gateway components. In this way the Ca^{2+} dependent conformational dynamics of the receptors is coupled to the transition behaviour of the gateway.

The receptor ring of a BEU is surrounded by the "ring" of the storage sites and (spatially separated) by the "ring" of the AcCh

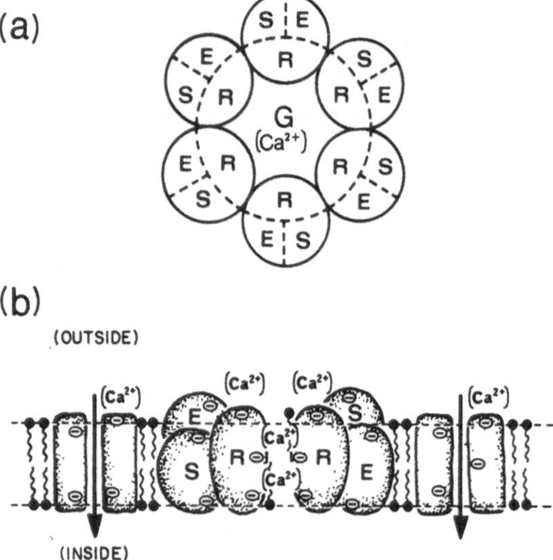

Fig. 6. Scheme of the AcCh-controlled gateway, G. (a) Basic excitation unit (BEU) containing in this example 6 SRE-assemblies, viewed perpendicular to the membrane surface. S, AcCh-storage site; R, AcCh-receptor protein; E, AcCh-esterase. (b) Cross section through a BEU flanked by two units which model ion passages for K^+-ions; the arrows represent the local electrical field vectors due to partial permselectivity to K^+-ions in the resting stationary state. The minus signs, ⊖ , symbolize negatively charged groups of membrane components.

esterases. The interfaces between the different rings define local reaction spaces through which AcCh is exchanged and translocated.

The BEUs are assumed to be distributed over the entire excitable membrane; the BEU density may vary for different membrane parts. The high density of cholinergic proteins found in some examples may be due to clustering of BEUs.

Different membrane types may not only vary in the number of SRE assemblies per BEU but also in the type and organization of the gateway components, thus assuring permselectivities for various ion types, particularly in synaptic parts of excitable membranes. It is only the cholinergic control system, the SRE-element which is assumed to be the same for all types of rapidly controlled gateways for passive ion flows.

(a) Action potential

In the framework of the integral model, the induction of an action potential is based on cooperativity between several SRE assemblies per BEU. In order to initiate an action potential a certain critical number of receptors, $\bar{m}^c$ per BEU, has on average to be activated within a certain critical time interval, Δt^c, (impulse condition). During this time interval at least, say 4 out of 6 SRE assemblies have to process AcCh in a concerted manner. Under physiological conditions only a small fraction of BEUs is required to gnerate and propagate the nerve impulse.

(b) Subthreshold responses

Subthreshold changes of the membrane are seen to involve only a few single SRE assemblies of a BEU. On average not more than one or two SRE elements per BEU are assumed to contribute to the measured responses (within time intervals of the duration of Δt^c).

The (small) permeability change caused by Ca^{2+} release from the receptor thus results from only a small part of the interface between receptor and gateway components of a BEU: the ion exchanges $AcCh^+/Ca^{2+}$ and Na^+ are locally limited.

In the framework of this model, spatially and temporally attenuating electrical activity such as subthreshold axonal-, postsynaptic-, and dendritic potentials are the sum of spatially and temporally additive contributions resulting from the local subthreshold activity of many BEUs.

Although the permeability changes accompanying local activity are very small (as compared to those causing the action potentials) the summation over many contributions may result in large overall conductivity changes. Such changes may even occur (to a perhaps smaller extent) when the core of the gateway is blocked. It is suggested that compounds like tetrodoxin and saxitoxin interact with the gateway core only, thus essentially not impeding subthreshold changes at the interface between receptor ring and gateway.

Influx of Ca^{2+} ions particularly through pre- and postsynaptic membranes may affect various intracellular processes leading, e.g., to release of hormones, catecholamines, etc.

8.3 Translocation flux of AcCh

As discussed before, the excitable membrane as a part of a living cell is a nonequilibrium system characterized by complex chemo-diffusional flow coupling. Although modern theoretical biology tends to regard living organisms only as quasi-stationary, with oscillations around a steady average, our integral model for the subthreshold behaviour of excitable membranes is restricted to stationarity. We assume that the "living" excitable membrane (even under resting conditions) is in a state of continuous subthreshold activity (maintained either aerobically or anaerobically). However, the nonequilibrium formalism developed later on in this section can also be extended to cover non-linear behaviour such as oscillations in membrane parameters.

In the frame of the integral model, continuous subthreshold activity is also reflected in a continuous sequential translocation of AcCh through the cholinergic system. The SRE elements comprise reaction spaces with continuous input by synthesis (Ch-T) and output by the virtually irreversible hydrolysis of AcCh. Input and output of the control system are thus controlled by enzyme catalysis.

(a) Reaction scheme

Since AcCh is a cation, translocation may most readily occur along negatively fixed charges, may involve concomitant anion transport or cation exchange. The reaction scheme formulated in Sect. 8.1 gives therefore only a rough picture. Storage, receptor and esterase represent macromolecular subunit complexes with probably several binding sites and the exact stoichiometry of the AcCh reactions is not known. As far as local electroneutrality is concerned, it appears in any case more realistic to assume that for one Ca^{2+}-ion released (or bound) there are two AcCh-ions bound (or released).

The conformationally mediated translocation of the AcCh-ion, A^+, may then be reformulated by the following sequence:

(1) Storage reaction

$$S_1(2A^+) + 2C^+ = S_2(2C^+) + 2\,A^+ \qquad (19)$$

(C^+ symbolizes a cation, 2 C^+ may be replaced by Ca^{2+}.)

(2) Receptor reaction

$$A^+ + R_1(Ca^{2+}) = R_2\,(A^+) + Ca^{2+} \tag{20}$$

(3) Hydrolysis reaction

$$R_2\,(A^+) + E_1 = R_2 + E_2\,(A^+) \rightarrow (Ch^+, Ac^-, H^+) \tag{21}$$

As already mentioned, the nucleation of the gateway transition (causing the action potential) requires the association of a critical number of A^+, $\bar{n}^c$, with the cooperative number of receptors, $\bar{m}^c$, in the Ca^{2+} binding form $R_1(Ca^{2+})$, within a critical time interval Δt^c. This time interval is determined by the life time of a single receptor-acetylcholine association.

Using formally $\bar{n}^c$ and $\bar{m}^c$ as stoichiometric coefficients the concerted reaction inducing gateway transition may be written:

$$\bar{n}^c A^+ + \bar{m}^c R_1(Ca^{2+}) = \bar{m}^c R_2(\bar{n}^c A^+) + \bar{m}^c Ca^{2+} \tag{22}$$

Storage- and receptor reactions, Eqs. (19) and (20), represent gating processes preceding gateway opening ("Na-activation") and causing the latency phase of the action potential. The hydrolysis process causes closure of the cholinergically controlled gateway ("Na-inactivation"). In the course of these processes the electric field across the membrane changes, affecting all charged-, dipolar and polarizable components within the field. These field changes particularly influence the storage site and the membrane components controlling the K permeation regions (see Adam, 1970). Fig. 7 shows a scheme modelling the "resting" stationary state and a transient phase of the excited membrane.

The complexity of the non-linear flow coupling underlying suprathreshold potential changes may be tractable in terms of the recently developed network thermodynamics covering inhomogeneity of the reaction space and nonlinearity (Oster et al., 1973). An attempt at such an approach, which formally includes conformational metastability and hysteretic flow characteristics (Katchalsky and Spangler, 1968; see also Blumenthal et al., 1970) is in preparation (Rawlings and Neumann, 1974).

Resting stationary state

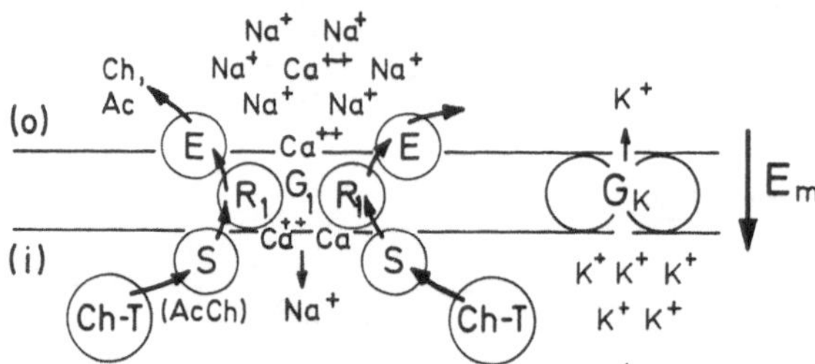

Excited state

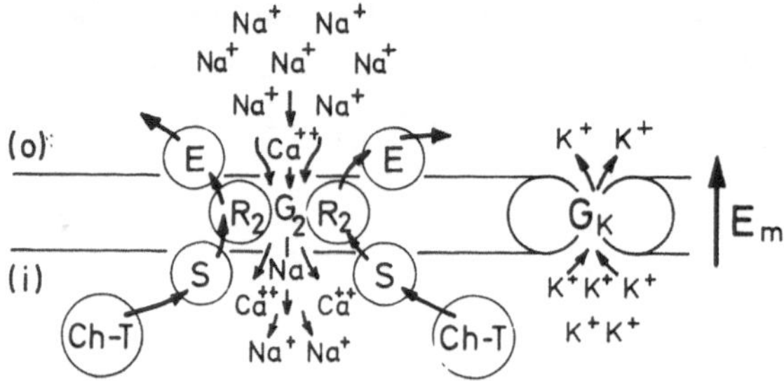

Fig. 7. Schematic representation of a membrane section (a) in the "resting" stationary state and (b) in a transient phase of excitation. In (a), the majority of the acetylcholine-receptors (R) is in the Ca^{2+} ion-binding conformation R_1; the cholinergically controlled rapidly operating gateway (G) is in the closed state G_1 and the permeability for Na^+ (and Ca^{2+}) ions is very small as compared to the permeability for K^+-ions through the slow gateway G_K. The electric field vector, E_m, pointing from the outside boundary (o) to the inside boundary (i) of the membrane is largely due to the K^+-ion gradient.
In (b), most of the receptors are in the acetylcholine-binding conformation R_2 and the rapid gateway is in its open configuration G_2 (Na-activation phase). The change in the electric field (directed outward during the peak phase of the action potential) accompanying the transient Na^+ (and Ca^{2+}) influx causes a transient (slower) increase in the permeability of G_K thus inducing a (delayed) transient efflux of K^+-ions. Hydrolysis of acetylcholine (AcCh) leads to relaxation of R_2 and G_2 to R_1 and G_1 restoring the resting stationary state.
Translocation of AcCh, occasionally in the resting stationary state and in a cooperatively increased manner after suprathreshold stimulation, through a storage site (S) of relatively large capacity, receptor and AcCh-esterase (E) is indicated by the curved arrows. The hydrolysis products choline (Ch) and acetate (Ac) are transported through the membrane where intracellular choline-O-acetyltransferase (Ch-T) may resynthetise AcCh (with increased rate in the refractory phase).

(b) Reaction fluxes

For the nonequilibrium description of the translocation dynamics we may associate reaction fluxes with the translocation sequence, Eqs. (19) to (21).

(1) The release flux is defined by

$$J(S) = \frac{d[\bar{n}_r]}{dt} \tag{23}$$

where $\bar{n}_r$ is the average number of A^+ released into the reaction space between storage- and receptor ring.

(2) The receptor flux including association of A^+ and conformation change of R is given by

$$J(R) = \frac{d[\bar{n}]}{dt} \tag{24}$$

where $\bar{n}$ is the average number of A^+ associated with R.

(3) The esterase (or decomposition flux) is defined by

$$J(E) = \frac{d[\bar{n}_e]}{dt} \tag{25}$$

where $\bar{n}_e$ is the average number of A^+ processed through AcCh-esterase.

Stationary states of the cholinergic activity are characterized by constant overall flow of AcCh; neither accumulations nor depletions of locally processed AcCh occur outside the limit of fluctuations. Thus, for stationary states,

$$J(S) = J(R) = J(E) = \text{constant}. \tag{26}$$

Statistically occurring small changes in membrane properties such as the so-called miniature-end-plate-potentials (caused by artificially reduced external Ca^{2+} ion concentration at synaptic parts) are interpreted to reflect amplified fluctuations in the subthreshold activity of the cholinergic system.

Oscillatory excitation behaviour observed under certain conditions (see, e.g., Cole, 1968) may be modelled by periodic accumulation and depletion of AcCh in the reaction spaces of the BEUs.

8.4 Field dependence of AcCh storage

In the simplest case, a change of the membrane potential affects the chain of translocation events already at the beginning, i.e. at the storage site. Indeed, the observation of AcCh release by electrical stimulation or in response to K^+-ion induced depolarization support the assumption of a field-dependent storage site for AcCh.

Denoting by $\bar{n}_b$ the amount of AcCh bound on average to S_1 we may define a distribution constant for the stationary state of the storage translocation by $K = [\bar{n}_b]/[\bar{n}_r]$. This constant (similar to an equilibrium constant) is a function of temperature T, pressure P, ionic strength I, and of the electric field E. A field dependence of K requires that the storage translocation reaction involves ionic-, dipolar-, or polarizable groups.

The isothermal-isobaric field dependence of K at constant ionic strength may be expressed by the familiar relation:

$$\left(\frac{\partial \ln K}{\partial E} \right)_{p,T,I} = \frac{\Delta M}{RT} \tag{27}$$

where ΔM is the reaction moment; ΔM is proportional to the difference in the (permanent or induced) dipole moments of reaction products and reactants. If a polarization process is associated with a finite value of ΔM, K should be proportional to E^2 (for relatively small field intensities up to 100 kV/cm). Furthermore, a small perturbation of the field causes major changes in K only on the level of higher fields (see, e.g., Eigen, 1967). It is therefore of interest to recall that, under physiological conditions, excitable membranes generate action potentials only above a certain (negative) potential difference.

The suggestion of a field-induced conformational change in a storage protein to release AcCh derives from recent studies on field effects in macromolecular complexes and biomembranes. It has been found that electric impulses in the intensity similar to the depolarization voltage changes for the induction of action potentials are able to cause structural changes in biopolyelectrolytes (Neumann and Katchalsky, 1972; Revzin and Neumann, 1974) and permeability changes in vesicular membranes (Neumann and Rosenheck, 1972). In order to explain the results, a polarization mechanism has been proposed that is based on the displacement of the counterion atmosphere of polyelectrolytes or of oligo-electrolytic domains in membrane organizations.

If the conformational dynamics of the storage site involves indeed a polarization mechanism, we may represent the dependence of bound AcCh, $\bar{n}_b$, on the electric field of the membrane as shown in Fig. 8. Increasing membrane potential increases the amount of bound AcCh and thus also the number of AcCh ions that, after fast reduction of the membrane potential, is translocatable to the receptor.

8.5 Relaxations of AcCh translocation fluxes

It is recalled that the receptor reaction (cf. Eqs. (20) and (22)) plays a key role in coupling the control function of AcCh with the permeability change of the gateway. Uptake of AcCh from the storage ring, conformational transition, and Ca^{2+} release comprise a sequence of three single events. It is therefore assumed that the processing of AcCh through the receptor is slower than the preceding step of AcCh release from the storage. The receptor reaction is thus

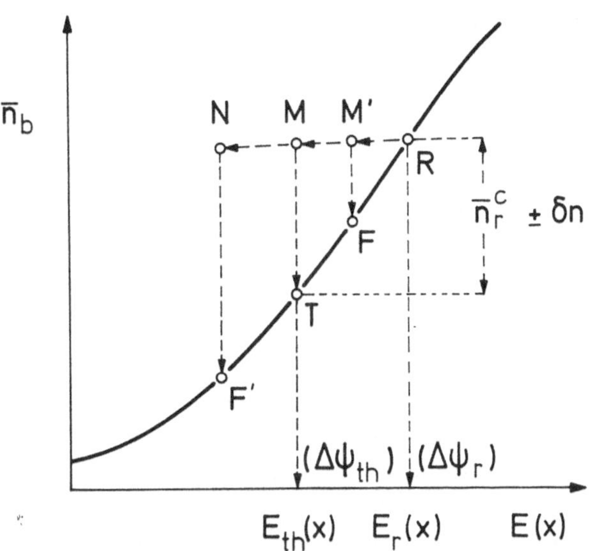

Fig. 8. Model representation of the field-dependent stationary states for AcCh storage. The mean number, $\bar{n}_b$, of AcCh ions bound to the storage site at the membrane site, x, of the release reaction, as a function of the electric field, E(x), (at constant pressure, temperature and ionic strength). The intervals M'F, MT, and NF' correspond to the maximum number of AcCh ions released, n_r, for 3 different depolarization steps: (a) a subthreshold change from the resting state R to F, (b) a threshold step R to T, releasing the threshold or critical number n_r^c ($\pm\delta n$, fluctuation), and (c) a suprathreshold step R to F' with $n_r > n_r^c$.

considered to be rate-limiting. Therefore, any (fast) change in the membrane field will lead to either a transient accumulation or a depletion of AcCh in the reaction space between S-ring and R-ring of a BEU.

In the case of a fast depolarization there is first a transient increase in the storage flux J(S) causing transient accumulation of AcCh in the S-R reaction space.

Since flux intensities increase with increasing driving forces (see, e.g., Katchalsky, 1967), J(S) will increase with increasing perturbation intensity, thus causing an increase in the rate of all following processes.

It is recalled that in the framework of our integral model the time course of changes in electrical membrane parameters such as the membrane potential is controlled by the cholinergic system and the gateway dynamics.

(a) Subthreshold relaxations

Subthreshold perturbations do not induce the gateway transitions and are considered to cause membrane changes of small extent only. The time constant, τ_m, for subthreshold relaxations of membrane potential changes (see Eq. (11)) is thus equal to the time constant, τ_R, of the rate-limiting receptor flux. For squid giant axons $\tau_R = \tau_m \simeq 1$ msec, at 20° C.

The relaxation of J(R) to a lasting subthreshold perturbation (e.g. current stimulation) is given by

$$\frac{d\,J(R)}{dt} = -\frac{1}{\tau_R}\left(J(R) - J'(R) \right) \tag{28}$$

where J'(R) is the stationary value of the new flux. Equivalent to Eq. (28), we have for $\bar{n}$,

$$\frac{d[\bar{n}]}{dt} = -\frac{1}{\tau_R}\left([\bar{n}] - [\bar{n}]' \right) \tag{29}$$

describing an exponential "annealing" to a new level of AcCh, $[\bar{n}]'$ processed through the receptor-ring.

It is evident from the reaction scheme, Eqs. (23) to (25), that the time constant τ_R is the relaxation time of a coupled reaction system.

Applying a few simplifying assumptions and using normal mode analysis (see Eigen and DeMaeyer, 1963), we may readily calculate τ_R as a function of the various reaction parameters (Neumann, 1974).

(b) Parameters of suprathreshold changes

It is recalled that the induction of an action potential is associated with three critical parameters:

$$\bar{n}^c,\ \bar{m}^c,\ \Delta t^c = \tau_R\ .$$

For a perturbation the intensity of which increases gradually with time, the condition $\bar{n} \geq \bar{n}^c$ corresponding $\bar{n}_r \geq \bar{n}_r^c$ (within a BEU) can only be realized if the minimum slope condition leading to

$$\frac{d\,J(S)}{dt} \geq \frac{d\,J(S)^{min}}{dt}$$

and to (the equivalent expression for J(R))

$$\frac{d\,J(R)}{dt} \geq \frac{d\,J(R)^{min}}{dt} = \frac{d}{dt}\,J(R)_{th} \tag{30}$$

is fulfilled.

Eqs. (30) represent the "chemical minimum slope condition" and the threshold receptor flux $J(R)_{th}$ may be considered as the equivalent to the rheobase; see Sect. 4.2.

For the fraction of BEUs necessary to evoke an action potential, the "chemical rheobase" may be specified by

$$J(R)_{th} = \frac{[\bar{n}^{c}]}{\tau_R} \tag{31}$$

For rectangular (step) perturbations the threshold condition is

$$J(R) \geq \frac{[\bar{n}^{c}]}{\tau_R} \quad . \tag{32}$$

Since J(R) increases with the intensity of the (step) perturbation, the time intervals Δt ($< \tau_R$) in which $\bar{n}^{c}$ AcCh ions start to associate with the receptor become smaller with larger stimulus intensities.

We may write this "strength-duration" relationship for suprathreshold perturbations in the form

$$J(R) \cdot \Delta t = [\bar{n}^{c}] \quad . \tag{33}$$

Compare Eq. (6). The time intervals Δt of the receptor activation correspond to the observed latency phases.

Since $\bar{n}_r^c$ and $\bar{n}^{c}$ are numbers describing functional cooperativity, the strength-duration product Eq. (33) does not depend on temperature. The fluctuations $\pm\delta n$ for $\bar{n}_r$, however, increase with increasing temperature (and may finally lead to thermal triggering of action potentials). The chemical rheobase, Eq. (31), is a reaction rate which in general is temperature dependent (normally with a Q_{10} coefficient of about 2).

There are further aspects of electrophysiological observations which the integral model at the present stage of development may (at least qualitatively) reproduce.

If the membrane potential is slowly reduced, subthreshold flux relaxation of the ratio $\bar{n}_b/\bar{n}_r$ may keep $\bar{n}_r$ always smaller than $\bar{n}^{c}$ Thus corresponding to experience slow depolarization does not (or only occasionally) evoke action potentials.

In order to match the condition $\bar{n}_r > \bar{n}^{c}$ starting from the resting potential, the depolarization has in any case to go beyond the threshold potential, where

$$\bar{n}_b(M) - \bar{n}_b(T) = \bar{n}^{c}. \qquad \text{(See Fig. 8)}$$

For stationary membrane potentials $\Delta\psi < \Delta\psi_{th}$, the maximum number of AcCh-ions that can be released by fast depolarization within Δt^c is smaller than $\bar{n}_r^c$. Thus, corresponding to experience, below a certain membrane potential, near below $\Delta\psi_{th}$, no nerve impulse can be generated.

(c) Refractory phenomena

After the gateway transition to the open state the receptors of the BEUs have to return to the Ca^{2+} binding conformation $R_1(Ca^{2+})$ before a second impulse can be evoked (cf. Eq. (22)). Even if $\bar{n}^c$ AcCh ions would already be available, the time interval for the transition of $\bar{m}^c$ receptors is finite and causes the observed absolutely refractory phase.

Hyperpolarizing prepulses shift the stationary concentration of $\bar{n}_b$ to higher values (see Fig. 8). Due to increased "filling degree" the storage site appears to be more sensitive to potential changes (leading, among others, to the so-called "off-responses"). On the other hand, depolarizing prepulses and preceding action potentials temporarily decrease the actual value of $\bar{n}_b$ thus requiring increased stimulus intensities for the induction of action potentials.

The assumptions for the kinetic properties of the storage site mentioned in the discussion of Eq. (15) are motivated by the accommodation behaviour of excitable membranes. The observation of a relatively refractory phase suggests that the uptake of AcCh into the storage form $S_1(A^+)$ is slow compared to the release reaction. Therefore, after several impulses there is partial "exhaustion" of the storage site. If during the (slow) refilling phase there is a new stimulation, $\bar{n}_b$ may still be lower than the stationary level. Therefore, the membrane has to be depolarized to a larger extent in order to fulfill the action potential condition $\bar{n}_r \geq \bar{n}_r^c$.

8.6 The AcCh control cycle

The cyclic nature of a cholinergic permeability control in excitable membranes by processing AcCh through storage, receptor, esterase, and synthetase is already indicated in a reaction scheme developed 20 years ago (see Fig. 11 of Nachmansohn, 1955). The complexity of mutual coupling between the various cycles directly or indirectly involved in the permeability control of the cholinergic gateway is schematically represented in Fig. 9. In this representation it may be readily seen that manipulations such as external application of AcCh and its inhibitory or activating congeners may interfere at several sites of the AcCh cycle. In particular the analysis of pharmacological and chemical experiments faces the difficulty of this complexity.

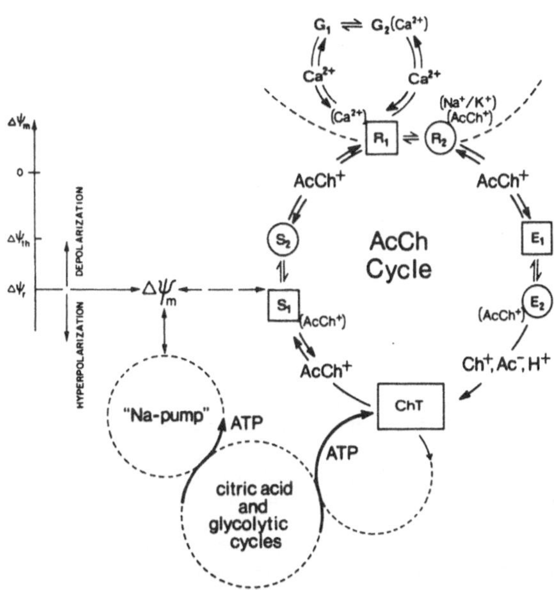

Fig. 9. AcCh-cycle, for the cyclic chemical control of stationary membrane potentials $\Delta\psi$ and transient potential changes. The binding capacity of the storage site for AcCh is assumed to be dependent on the membrane potential, $\Delta\psi_m$, and is thereby coupled to the "Na^+/K^+ exchange-pump" (and the citric acid and glycolytic cycles). The control cycle for the gateway G_1 (Ca^{2+} binding and closed) and G_2 (open) comprises the SRE assemblies (see Fig. 4) and the choline-O-acetyltransferase (Ch-T); Ch-T couples the AcCh synthesis cycle to the translocation pathway of AcCh through the SRE-assemblies. The continuous subthreshold flux of AcCh through such a subunit is maintained by the virtually irreversible hydrolysis of AcCh to choline (Ch^+), acetate (Ac^-) and protons (H^+) and by steady supply flux of AcCh to the storage from the synthesis cycle. In the resting stationary state, the membrane potential ($\Delta\psi_r$) reflects dynamic balance between active transport (and AcCh synthesis) and passive fluxes of AcCh (through the control cycles surrounding the gateway) and of the various ions unsymmetrically distributed across the membrane. Fluctuations in membrane potential (and exchange currents) are presumably amplified by fluctuations in the local AcCh concentrations maintained at a stationary level during the continuous translocation of AcCh through the cycle (Neumann and Nachmansohn, 1974).

In the previous sections it has been shown that basic parameters of electrophysiological phenomenology may be modelled in the framework of a nonequilibrium treatment of the cholinergic reaction system. The various assumptions and their motivations by experimental observations are discussed and the cholinergic reaction cycle is formulated in a chemical reaction scheme.

In conclusion, the integral model at the present level of development appears to cover all essential pharmaco-electrophysiological and biochemical data on excitable membranes. The model is expressed in specific reactions subject to further experimental investigations involving the reaction behavior of isolated membrane components as well as of membrane fragments containing these components in structure and organization.

Summary: The mechanism of nerve excitability is still an unsolved problem. There are various mechanistic interpretations of nerve behaviour, but these approaches cover only a part of the known facts, are thus selective and unsatisfactory.

An attempt at an integral interpretation of basic data well-established by electrophysiological, biochemical and biophysical investigations was inspired by the late Aharon Katchalsky and a first essay had been given (Neumann, Nachmansohn and Katchalsky, 1973). The present account on nerve excitability is a further step towards a specific physico-chemical theory of bioelectricity. In order to account for the various pharmaco-electrophysiological and biochemical observations on excitable membranes, the notion of a basic excitation unit is introduced. This notion is of fundamental importance for modelling details of sub- and suprathreshold responses such as threshold behaviour and strength-duration curve, in terms of kinetic parameters for specific membrane processes.

Our integral model of excitability is based on the original chemical hypothesis for the control of bioelectricity (Nachmansohn, 1959, 1971). This specific approach includes some frequently ignored experimental facts on acetylcholine (AcCh)-processing proteins in excitable membranes. According to the integral model, acetylcholine-ions are continuously processed through the basic excitation units within excitable membranes: axonal, pre- and postsynaptic parts. Excitability, i.e. the generation and propagation of nerve impulses is due to a cooperative increase in the rate of AcCh translocation through the cholinergic control system.

At the present stage of the model, the cholinergic control is restricted to the rapidly operating, normally Na^+-ion carrying permeation sites. The variations in the electric field of the membrane, caused by the cholinergically controlled rapid gateway, in turn, affects the permeability of the slower, normally K^+-ion carrying permeation sites in the excitable membrane.

The basic biochemical data suggesting a cyclic cholinergic control (AcCh-cycle) of the ion movements are presented in a concise form and some of the controversial interpretations of biochemical and electrophysiological data on excitability are discussed.

Acknowledgements

This study is based on numerous discussions with Prof. David Nachmansohn whom I thank for the many efforts to reduce my ignorance in the biochemistry of excitable membranes. Thanks are also due to Prof. Manfred Eigen for his critical interest and generous support of this work. Finally, I would like to thank the Stiftung Volkswagenwerk for a grant.

References

Abbot, B.C., A.V. Hill and J.V. Howarth: The positive and negative heat production associated with a single impulse, Proc. Roy. Soc. B 148, 149-187 (1958).

Adam, G.: Theory of nerve excitation as a cooperative cation exchange in a two-dimensional lattice. In: Physical Principles of Biological Membranes, ed. by F. Snell et al., pp. 35-64. Gordon and Breach,

New York, 1970.

Agin, D.: Electroneutrality and electrodiffusion in the squid axon, Proc. Nat. Acad. Sci. USA. 57, 1232-1238 (1967).

Agin, D.: Excitability phenomena in membranes. In: Foundations of Mathematical Biology, ed. by R. Rosen, pp. 253-277. Academic Press, New York, 1972.

Armett, C.J. and J.M. Ritchie: The action of acetylcholine on conduction in mammalian non-myelinated fibers and its prevention by anticholinesterase, J. Physiol. 152, 141-158 (1960).

Barrnett, R.J.: The fine structural localization of acetylcholinesterase of the myoneural junction, J. Cell Biol. 12, 247-262 (1962).

Bartels, E.: Relationship between acetylcholine and local anesthetics, Biochim. Biophys. Acta 109, 194-203 (1965).

Blumenthal, R., J.-P. Changeux, and R. Lefever: Membrane excitability and dissipative instabilities, J. Membrane Biol. 2, 351-374 (1970).

Brzin, M., W.-D. Dettbarn, Ph. Rosenberg, and D. Nachmansohn: Cholinesterase activity per unit surface area of conducting membranes, J. Cell Biol. 26, 353-364 (1965).

Calabro, W.: Sulla regolazione neuro-umorale cardiaca, Riv. biol. 15, 299-320 (1933).

Carnay, L.D. and I. Tasaki: Ion exchange properties and excitability of the squid giant axon. In: Biophysics and Physiology of Excitable Membranes, ed. by W.J. Adelman, Jr., pp. 379-422. Van Norstrand Reinhold Co., New York, 1971.

Cole, K.S.: Membranes, ions and impulses, ed. by C.A. Tobias, University of California Press, Berkeley, Calif., 1968.

Cole, K.S.: Dielectric properties of living membranes. In: Physical Principles of Biological Membranes, ed. by F. Snell et al., pp. 1-15. Gordon and Breach, New York, 1970.

Denburg, J.L., M.E.Eldefrawi, and R.D. O'Brien: Macromolecules from lobster axon membranes that bind cholinergic ligands and local anesthetics, Proc. Nat. Acad. Sci. USA 69, 177-181 (1972).

Dettbarn, W.-D.: The effect of curare on conduction in myelinated, isolated nerve fibers of the frog, Nature 186, 891-892 (1960).

Dettbarn, W.-D.: New evidence for the role of acetylcholine in conduction, Biochim. Biophys. Acta 41, 377-386 (1960).

Dettbarn, W.-D. and Ph. Rosenberg: Effects of ions on the efflux of acetylcholine from peripheral nerve, J. Gen. Physiol. 50, 447-460 (1966).

Dettbarn, W.-D.: The acetylcholine system in peripheral nerve, Ann. N.Y. Acad. Sci. 144, 483-503 (1967).

Eigen, M. and L. DeMaeyer: Relaxation methods. In: Technique of Organic Chemistry, ed. by S.L. Friess, E.S. Lewis, and A. Weissberger, Vol. 8, p. 895. Interscience Publishers, Inc., New York, 1963.

Eigen, M.: Dynamic aspects of information transfer and reaction control in biomolecular systems. In: The Neurosciences, ed. by G.C. Quarton, T. Melnechuk and F.O. Schmitt, pp. 130-142. The Rockefeller University Press, New York, 1967.

Evans, M.H.: Tetrodotoxin and Saxitoxin in neurobiology, Intern. Rev. Neurobiology 15, 83-166 (1972).

Fox, J.M.: Veränderungen der spezifischen Ionenleitfähigkeiten der Nervenmembran durch ultraviolette Strahlung. Dissertation, Homburg-Saarbrücken, 1972.

Gruber, H. and W. Zenker: Acetylcholinesterase: histochemical differentiation between motor and sensory nerve fibres, Brain Research 51, 207-214 (1973).

Hamprecht, B.: Cell cultures as model systems for studying the biochemistry of differentiated functions of nerve cells, Hoppe-Seyler's Z. Physiol. Chem. 355, 109-110 (1974).

Harris, A.J. and M.J. Dennis: Acetylcholine sensitivity and distribution on mouse neuroblastoma cells, Science 167, 1253-1255 (1970).

Hodgkin, A.L.: The Conduction of the Nervous Impulse, C.C. Thomas, Springfield. Ill., 1964.

Hodgkin, A.L. and R.D. Keynes: The potassium permeability of a giant nerve fibre, J. Physiol. 128, 61-88 (1955).

Hoskin, F.C.G., Ph. Rosenberg, and M. Brzin: Re-examination of the effect of DFP on electrical and cholinesterase activity of squid giant axon, Proc. Nat. Acad. Sci. USA. 55, 1231-1235 (1966).

Hoskin, F.C.G., L.T. Kremzner and Ph. Rosenberg: Effects of some cholinesterase inhibitors on the squid giant axon, Biochem. Pharmcol. 18, 1727-1737 (1969).

Howarth, J.V., R.D. Keynes and J.M. Ritchie: The origin of the initial heat associated with a single impulse in mammalian non-myelinated nerve fibres, J. Physiol. 194, 745-793 (1968).

Huneeus-Cox, F., H.L. Fernandez, and B.H. Smith: Effects of redox and sulfhydryl reagents on the bioelectric properties of the giant axon of the squid, Biophys. J. 6, 675-689 (1966).

Julian, F.J., and D.E. Goldman: The effects of mechanical stimulation on some electrical properties of axons, J. Gen. Physiol. 46, 197-313 (1962).

Katchalsky, A.: Membrane thermodynamics. In: The Neurosciences, ed. by G.C. Quarton, T. Melnechuk and F.O. Schmitt, pp. 326-343. The Rockefeller University Press, New York, 1967.

Katchalsky, A. and R. Spangler: Dynamics of membrane processes, Quart. Rev. Biophys. 1, 127-175 (1968).

Katz, B.: Nerve, Muscle, and Synapse, McGraw-Hill, New York, 1966.

Koelle, G.B.: Current concepts of synaptic structure and function, Ann. N.Y. Acad. Sci. 183, 5-20 (1971).

Landowne, D.: Movement of sodium ions associated with the nerve impulse, Nature 242, 457-459 (1973).

Lewis, P.R. and C.C.D. Shute: The distribution of cholinesterase in cholinergic neurons demonstrated with the electron microscope, J. Cell Sci. 1, 381-390 (1966).

Lissák, K.: Liberation of acetylcholine and adrenaline by stimulating isolated nerves, Amer. J. Physiol. 127, 263-271 (1939).

Masland, R.L. and R.S. Wigton: Nerve activity accompanying fasciculation produced by Prostigmine, J. Neurophysiol. 3, 269-275 (1940).

Muralt, A.v. and R. Stämpfli: Die photochemische Wirkung von Ultraviolettlicht auf den erregten Ranvierschen Knoten der einzelnen Nervenfaser, Helv. Physiol. Acta 11, 182-193 (1953).

Nachmansohn, D.: Metabolism and function of the nerve cell. In: Harvey Lectures 1953/1954, pp. 57-99. Academic Press, New York, 1955.

Nachmansohn, D.: Chemical and Molecular Basis of Nerve Activity, Academic Press, New York, 1959.

Nachmansohn, D.: Actions on axons and the evidence for the role of acetylcholine in axonal conduction. In: Cholinesterases and Anticholinesterase Agents, ed. by G.B. Koelle, pp. 701-740. Handb. d. exp. Pharmakol., Ergw. XV, Springer-Verlag, Berlin-Heid., 1963.

Nachmansohn, D.: Proteins of excitable membranes, J. Gen. Physiol. 54, 187 s - 224 s (1969).

Nachmansohn, D.: Proteins in bioelectricity. Acetylcholine-esterase and -receptor. In: Handbook of Sensory Physiology, ed. by Loewenstein, Vol. 1, pp. 18-102. Springer-Verlag, Berlin, 1971.

Narahashi, T., J.W. Moore, and R.N. Poston: Anesthetic blocking of nerve membrane conductances by internal and external applications, J. Neurobiol. 1, 3-22 (1969).

Neher, E. and H.D. Lux: Rapid changes of potassium concentration at the outer surface of exposed single neurons during membrane current flow, J. Gen. Physiol. 61, 385-399 (1973).

Nelson, P.G., J.H. Peacock and T. Amano: Responses of neuroblastoma cells to iontophoretically applied acetylcholine, J. Cell Physiol. 77, 353-362 (1971).

Neumann, E. and A. Katchalsky: Long-lived conformation changes induced by electric impulses in biopolymers, Proc. Nat. Acad. Sci. USA 69, 993-997 (1972).

Neumann, E. and K. Rosenheck: Permeability changes induced by electric impulses in vesicular membranes, J. Membrane Biol. 10, 279-290 (1972).

Neumann, E., D. Nachmansohn and A. Katchalsky: An attempt at an integral interpretation of nerve excitability, Proc. Nat. Acad. Sci. USA. 70, pp. 727-731 (1973).

Neumann, E.: Molecular hysteresis and its cybernetic significance, Angew. Chem., intern. Edit. 12, 356-369 (1973).

Neumann, E.: Thermodynamic analysis of the "thermal spike", Proc. Roy. Soc. B, in press (1974).

Neumann, E.: Towards a molecular model of nerve excitability, Hoppe-Seyler's Z. Physiol. Chem., in press (1974).

Neumann, E. and D. Nachmansohn: Nerve excitability - towards an integrating concept. In: Biomembranes, ed. by L. Manson, in press, 1974.

Oster, G.F., A.S. Perelson and A. Katchalsky: Network thermodynamics: dynamic modelling of biophysical systems, Quart. Rev. Biophys. 6, 1-134 (1973).

Plonsey, R.: Bioelectric Phenomena, McGraw-Hill, New York, 1969.

Porter, C.W., T.H. Chiu, J. Wieckowski and E.A. Barnard: Types and locations of cholinergic receptor-like molecules in muscle fibres, Nature New Biol. 241, 3-7 (1973).

Prigogine, I.: Thermodynamics of Irreversible Processes, Thomas Publ., Springfield, Ill., 3rd ed., 1968.

Rawlings, P.K. and E. Neumann: in preparation.

Revzin, A. and E. Neumann: Conformational changes in rRNA induced by electric impulses, Biophys. Chem., in press (1974).

Riker, W.F. Jr., G. Werner, J. Roberts and A. Kuperman: The presynaptic element in neuromuscular transmission, Ann. N.Y. Acad. Sci. 81, 328-344 (1959).

Ritchie, J.M.: The action of acetylcholine and related drugs on mammalian non-myelinated nerve fibres, Biochem. Pharmacol. 12(S)3 (1963).

Robertson, J.D.: The ultrastructure of synapses. In: The Neurosciences, ed. by F.O. Schmitt, Vol. 2, pp. 715-728. The Rockefeller University Press, New York, 1970.

Rosenberg, P. and F.C.G. Hoskin: Demonstration of increased permeability as a factor in the effect of acetylcholine on the electrical activity of venom-treated axons, J. Gen. Physiol. 46, 1065-1073 (1963).

Seeman, P.: The membrane actions of anesthetics and tranquilizers, Pharmacol. Rev. 24, 583-655 (1972).

Segal, J.R.: Surface charge of giant axons of squid and lobster, Biophys. J. 8, 470-489 (1968).

Spector, I., Y. Kimhi and P.G. Nelson: Tetrodotoxin and cobalt blockage of neuroblastoma action potentials, Nature New Biology, 246, 124-126 (1973).

Takeuchi, A. and N. Takeuchi: Actions of transmitter substances on the neuromuscular junctions of vertebrates and invertebrates. In: Advan. in Biophys. 3, ed. by M. Kotani, pp. 45-95. University Park Press, Baltimore, 1972.

Tasaki, I.: Nerve excitation, C.C. Thomas, Springfield, Ill., 1968.

Tasaki, I., I. Singer and T. Takenaka: Effects of internal and external ionic environment on excitability of squid giant axon, J. Gen. Physiol. 48, 1095-1123 (1965).

Tasaki, I. and I. Singer: Membrane macromolecules and nerve excitability: a physico-chemical interpretation of excitation in squid giant axons, Ann. N.Y. Acad. Sci., 137, 793-806 (1966).

Tasaki, I. and T. Takenada: Ion fluxes and excitability in squid giant axon. In: The Cellular Functions of Membrane Transport, ed. by J.F. Hoffman. Prentice-Hall, Inc., Englewood Cliffs, N.J., 1964.

Traeuble, H. and H. Eibl: Electrostatic effects on lipid phase transitions: membrane structure and ionic environment, Proc. Nat. Acad. Sci. USA. 71, 214-219 (1974).

Whittaker, V.P.: The biochemistry of synaptic transmission, Naturwissenschaften 60, 281-289 (1973).

Note added in proof. Recently it has been found that acetylcholine induces a conformational change in the isolated acetylcholine-receptor protein (from Electrophorus electricus). This configurational change controls the binding of Calcium ions to the polyelectrolytic macromolecule. The kinetic analysis of this fundamentally important biochemical reaction (see Eq. 16) results in number values for apparent rate constants and equilibrium parameters of the participating elementary processes, but also reveals the molecularity of the interactions between receptor, acetylcholine, and Calcium ions. (H.W. Chang and E. Neumann, Proc. Natl. Acad. Sci. USA, in press.)

MOLECULAR INFORMATION PROCESSING IN THE CENTRAL NERVOUS SYSTEM

PART I: SELECTION CIRCUITS IN THE BRAIN

Michael Conrad
Institute for Information Sciences
University of Tübingen
Tübingen, Germany

Contents

1. Introduction

Information processing systems are systems which dissipate energy in certain interesting (or highly selective) ways. In the case of biological systems the particular pathways of dissipation are ultimately determined by individual molecular catalysts which speed certain processes and not others. There is indeed accumulating evidence that such molecular catalysts play a crucial role in the brain, the system which is without doubt the biological information processor par excellence.

In this paper I will describe the general principles which are suitable for understanding information processing in molecular systems and in particular for understanding how algorithms and learning processes can be embodied in such systems. The first part will deal with the nature of molecular specificity and will outline a quite definite model for the brain's ability to perform complex operations on the basis of relatively simple anatomical structures and also for its ability to learn on the basis of gradual change in the nature of these operations. In the second part I will introduce memory and memory-based learning into the model by showing how the brain can build and manipulate data structures on the basis of changes in molecular conformation.

2. The Conceptual Framework

The first thing we must do is to formalize the notions of algorithm and learning. There are many such formalizations, but the most transparent is without doubt the idea of a Turing scheme (see Figs. 1a and 1b). This is simply a finite automaton together with a tape which it can move and mark. The automaton itself is a system with finite sets of states, inputs, and outputs, along with transition functions which determine the next state and output given the present state and input. The inputs and outputs include the markings on the tape, and in the case of the outputs the tape moves as well. We will also suppose that our automaton is in some external environment, so that its inputs also include the states of this environment.

The transition functions are the rules (or program) generating the behavior of the Turing automaton. So far as is known, any effective procedure can be directly expressed in terms of such transition functions (or Turing quintuples, cf. Fig. 1c). In some cases the rule generating the behavior of the Turing automaton is sufficiently general that it can read and follow any particular rule encoded in the sequence of markings on the tape. In this case we will say that the automaton is universal.

What is interesting about the Turing scheme, from the standpoint of brain models, is the type of process of which it is a formalization.

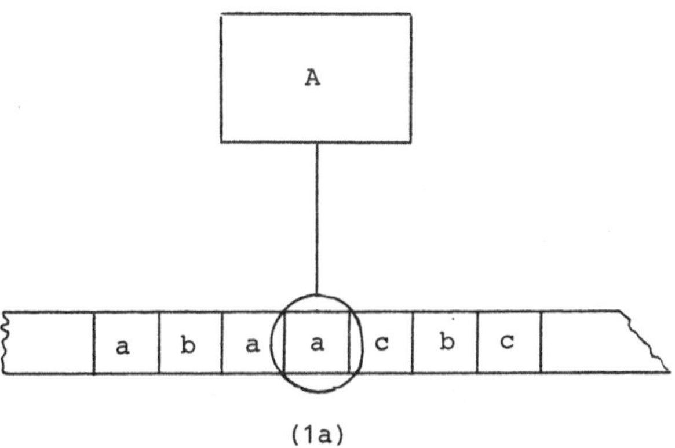

(1a)

$A = (X, Q, Y, \gamma, \lambda)$

X = finite set of inputs ($x_i \varepsilon X$)

Q = finite set of states ($q_i \varepsilon Q$)

Y = finite set of outputs ($y_i \varepsilon Y$)

γ: $X \times Q \rightarrow Q$ (next state function)

λ: $X \times Q \rightarrow Y$ (output function)

(1b)

q_1abRq_1
q_1baRq_1 } Turing quintuples
q_1ccNq_2

$X = \{a,b,c\}$ C = "clear" square

$Y = \{a,b,c,R,L,N\}$ R = move right

$Q = \{q_1,q_2\}$ L = move left

q_2 = stop state N = no move

(1c)

Fig. 1. (a) Turing scheme. A is the finite automaton. (b) Finite automaton (definition) (c) Turing quintuples. The (trivial) set of quintuples shown here describes a system which moves to the right, interchanging a's and b's on the tape, and which goes to the halt state (q_2) as soon as it reaches a clear square.

Imagine (as Turing originally did) a schoolchild working an arithmetic problem on a notepad marked with rows and columns (cf. Minsky, 1967). The child reads a particular symbol on the pad, rewrites this symbol and moves to a neighboring square, according to his state of mind, and then switches to a new state of mind. Any arithmetic or automatic symbolic procedure can be expressed in these terms, from which it follows that the Turing scheme is at one and the same time a formalization of an algorithmic process and a formalization of a particular kind of psychological process.

3. Learning in the Turing Scheme

The Turing scheme should also provide a means for describing learning processes. This is because the scheme is an abstract formalization of behavior, from which it follows that it should encompass any psychological process, whether algorithmic or not.

Roughly speaking, we can regard learning as the development of behavior either adapted to novel environments or better adapted to given environments. The Turing scheme allows four possible mechanisms for "second order" behavior processes of this type:

1. Inherent potentiality. In this case the transition functions are such that the environment induces appropriate responses in the automaton, e.g. pushes it into one of a number of suitable, but preexisting modes of behavior.
2. Memory based learning. In this case the automaton develops suitable behavior by performing computations, using the tape as a work (or memory) space. The power of this "problem solving" type learning is of course increased if the automaton is capable of recording environmental events on the tape.
3. Programmability from input. In this case the automaton is universal and the rule encoded on its input tape is such that it develops more suitable behavior.
4. Modification. In this case the automaton is replaced by one which follows a more suitable rule, i.e. the state set, output set, or transition functions are changed.

Notice that mechanisms 1), 2), and 3) are of the algorithmic type for which the Turing machine was originally devised. This is not so for mechanism 4). Learning mediated by this mechanism is not an effective process.

4. Classical Realizations

The Turing scheme is a formalism, a paper device, and not a real machine as it is often called. However, it is always possible to realize such schemes by concrete systems, and according to a definite (i.e. algorithmic) procedure. These (man-constructed) realizations all have one feature in common: they are synthesized out of certain base components by appropriately linking and setting the states of these components. In short, they use the <u>building block principle</u>.

The base components are of course canonical automata and the synthesis algorithm is a procedure for encoding the transition functions of the original automaton into the state settings and interconnections of the canonical automata (for which we already have concrete realizations). I will call any system with this property <u>structurally programmable</u>.

There are many examples of structurally programmable systems--indeed all digital computers are structurally programmable. The simplest of these, and the one which has probably had the most influence on nervous system biology, is the McCulloch-Pitts neural net (McCulloch & Pitts, 1943). This is simply a net of formal neurons, i.e. canonical automata which fire when the sum of their inhibitory and excitatory inputs exceeds some threshold (see Fig. 2). Any finite automaton can be realized in a definite way by a network of such formal neurons (see, e.g., Minsky, 1967).

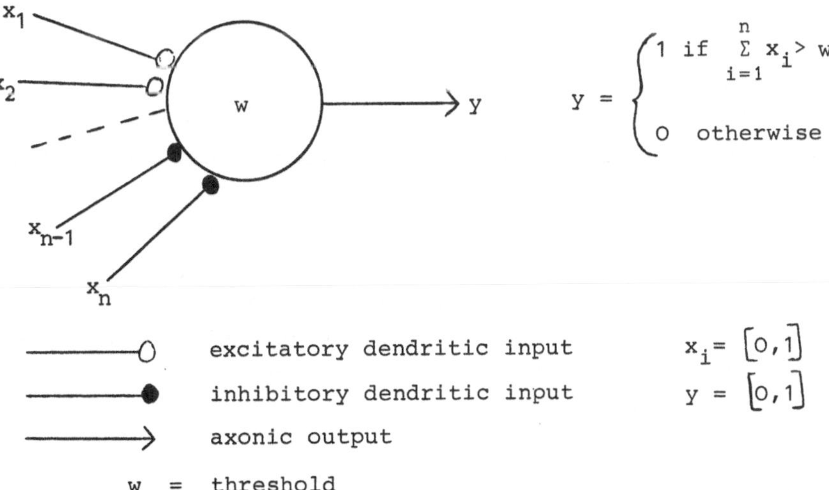

Fig. 2. McCulloch-Pitts formal neuron. The neuron fires if the sum of the inputs exceeds the threshold.

Structurally programmable systems have two basic disadvantages: inefficiency and inamenability to gradual change in behavior with gradual change in structure. The inefficiency arises from the canonical nature of the components, i.e. from the fact that the components are not in general specially suited for the particular task which the system is constructed to perform. The inamenability to gradual change in behavior with gradual change in structure comes from the fact that the rule which generates this behavior (the generating rule) is itself embodied in the structure according to a rule (the construction rule). Under certain conditions (to be described later) slight changes in the generating rule are associated with slight changes in behavior. In general, however, there is no predicting in advance how different the generating rule will be if the system is modified, for if this were possible it would be possible to answer the question: will the construction rule applied to some generation rule ever produce a particular structure. If we could answer this question we could answer the question as to whether the machine which embodies the construction rule will ever halt. But this is theoretically impossible since the halting problem is in general unsolvable.*

The sensitivity of computers and computer programs to slight changes in structure is, of course, a fact of experience. It is sometimes possible to reduce this sensitivity by encoding the rule in a highly redundant structure (cf. Winograd & Cowan, 1963). This allows for reliable execution of a particular function despite slight changes in structure. But it does not allow for adaptation to slightly changed environments through such changes. This is because a large number of structural changes may be required to produce a slight change in behavior in a structurally programmable system. But, of course, it is highly improbable that all of these changes will occur simultaneously (Conrad, 1972).

These difficulties are particularly clear in the case of the McCulloch-Pitts networks. Here the inefficiency arises from the fact that the formal neuron is an extreme averaging device, losing all the detailed information about its patterns of inputs. The sensitivity to structural change arises from the fact that changing the threshold of the neuron or reconnecting it to other neurons will in general have a radical effect on the pattern of information flow through the system. In fact networks of such neurons cannot perform functions comparable to natural

---

*In real systems we might of course put a finite restriction on the set of generating rules. However, what is unsolvable with no finite restriction is in general extremely difficult even when such a restriction is imposed (cf. von Neumann, 1966).

systems (on the same time scale) without using a much larger number and connectivity of neurons (cf. Minsky, 1967). They are thus too complicated to be models of the real brain, and too inamenable to gradual modification of behavior with step by step changes in structure to support learning through modification. Furthermore, they provide for no "tape" (i.e. mechanism for manipulating memory) and therefore are incapable of supporting memory-based learning as well.

5. Structural Nonprogrammability: the House Analogy

Most neuroscientists claim not to take the McCulloch-Pitts model too seriously because of its high simplification of the neuron. However, the model made a big impression because it provided a link between Turing schemes and anatomy. In particular, I think it is fair to say that what has been accepted, at least subliminally, is the idea of structural programmability--that the arbitrarily complex behavior of the brain is somehow embodied in a set of more or less complicated canonical components. However, any system with this property will have the fundamental defects of the McCulloch-Pitts model; indeed, the real value of the model is that it shows this so clearly.

Are there any alternatives? More precisely, is it possible to construct a structurally nonprogrammable system to realize any Turing automaton? In fact it is, but before describing my construction I would like to introduce the idea with an analogy.

Imagine a contractor commissioned to build houses suitable for a continuous range of environmental conditions. The contractor builds whatever house is described in a blueprint, which therefore serves as a rule for building the house. One way in which he could construct houses suitable for slightly different conditions would be to use a different rule. In general this would require a large number of changes in the blueprint. Alternatively, he could build slightly different houses by using the same rule but different components, i.e. use the same blueprint, but with the provision that it specifies the use of different components. This allows the contractor to produce alternative physical realizations of the same plan, and furthermore, realizations which vary gradually with gradual variation in the specification of the components.

Our second mode of adapting the construction to different environments is totally outside the framework of the rule for building the house, and therefore outside of the framework of programmability. This mode of adapting, at once so simple and powerful, has not been overlooked by biological systems (see Fig. 3). The blueprint of course corresponds to the sequence of bases in DNA. These code for the sequence of amino acids

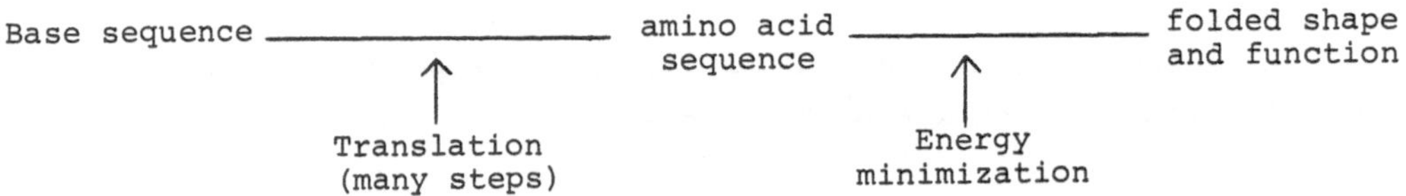

Fig. 3. Protein shape and function. The three dimensional shape of proteins determines their structural and catalytic functions (including translation and transcription of the DNA). The fact of folding means that the amino acid sequence (or the corresponding base sequence) is not itself the transition function (or program) of the enzyme, i.e. the function of the enzyme is not expressible in terms of its amino acid sequence in advance, without experiment or calculation (in contrast to the Turing quintuples of Fig. 1c).

in protein. These sequences in turn fold into definite three dimensional shapes on the basis of the weak (e.g. van der Waal's) interactions among the amino acids. The function of the protein (e.g. the particular reactions it catalyzes or the particular structures which it forms with other molecules) is determined by this folded shape (i.e. by its specific fit to the substrate or to the molecule with which it aggregates).

In principle, the folded shape might be determined by calculation, by minimizing the energy of the system and also considering the initial influence of the ribosome. Also, the function might be determined (in principle) by calculation, considering the interaction with the folded shapes of other molecules. This is important since it means that the sequence of amino acids is not a program; for if this were the case it would be possible to effectively express any enzyme function in terms of its amino acid sequence, without calculation, just as it is possible to directly express a rule in terms of the sequence of symbols on the Turing tape. In short, each new amino acid sequence is a de novo component, from which it follows that changing the sequence of bases in DNA is tantamount to changing the specification of the components.

Now, it may happen that changing the specification of the components (e.g. changing the enzymes) changes the rule according to which the cell is constructed; but it may also happen that this rule stays the same, or essentially the same, and what changes is only its physical realization. Furthermore, we may end up with only slightly different physical realizations if slight changes in the amino acid sequence produce only slight changes in protein shape and therefore in the function implicit in this shape (Conrad, 1973).

How gradual will the change in function (or behavior) be with slight changes in sequence (or primary structure)? The answer is that the degree of gradualism is itself an adjustable and evolved property. This is because enzymes may use redundant weak bonding (arising from the addition of otherwise unnecessary amino acids) to increase the liklihood that slight alterations in their primary structure produce only slight changes in some feature of their three dimensional shape. This is a controllable property, crucial for the rate of evolution, and therefore automatically adjusted to an optimal level in the course of evolution. Furthermore, this optimal level may vary with the type of function performed by the enzyme.

We can summarize the above considerations in what might be called the principle of _molecular adaptability_: _gradual changes in the primary structure of enzymes are often associated with gradual changes in the functions they perform, where the degree of gradualism is itself adjusted in the course of evolution_.

6. Molecular Realizations

The principle of molecular adaptability (along with changes in the number of already evolved proteins) is the basis for gradual change in function at all levels of biological organization. However, we still have not shown how this principle can be built into arbitrary Turing automata, and also in a way that makes a connection to the electrical activity of neurons.

This joint requirement (for dependence on both molecular adaptability and the nerve impulse) is satisfied by a single, simple postulate: that enzymes control the nerve impulse. To this end we introduce what we call the enzymatic neuron, i.e. an abstraction of the neuron in which enzymes (which we call excitases) catalyze events leading to impulse formation if the weighted sum of the inputs to the neuron assumes a suitable value at the location of this enzyme on the neuron surface.

Our picture of a highly simplified enzymatic neuron is illustrated in Fig. 4. The behavior of such neurons is described by the following general equation:

$$y_k(t+1) = \Theta\left[\sum_{j=1}^{n} \delta(\xi_{jk} - \sum_{i=1}^{m} w_{ijk}x_{ik}(t)) \pm \Delta\xi_{jk}\right]$$

where

t = time

x_{ik} = dendritic input i to neuron k (0 or 1)

y_k = axonic output of neuron k (0 or 1)

ξ_{jk} = electric field which activates excitase species located at position j in neuron k

$\Delta\xi_{jk}$ = tolerance factor of excitase species located at position j

w_{ijk} = weighted value of dendritic input i at position j of neuron k

n = number of excitase species

m = number of dendritic inputs

$$\delta(u \pm \xi_{jk}) = \begin{cases} 1 & \text{if } u < \pm\Delta\xi_{jk} \\ 0 & \text{otherwise} \end{cases}$$

$$\Theta(v) = \begin{cases} 1 & \text{if } v > 0 \\ 0 & \text{otherwise} \end{cases}$$

and we set $\xi_{jk} = \infty$ if no excitase is located at position j. The tolerance factor $\Delta\xi_{jk}$ (always finite) specifies the range of values of the electric field to which the excitase at position j responds.

The fundamental decision making elements in the enzymatic neuron are individual molecules. This, and also the independence of these molecules, seems to be necessary in order to take full advantage of the principle of molecular adaptability. Notice that in any real case the morphology and conductivity properties of the enzymatic neuron might be quite complex. This is important, indeed an essential feature of the model, since these properties serve as symmetry breaking mechanisms, i.e. mechanisms which exaggerate the weighting coefficients (the w_{ij}) and therefore increase the differences between the effects of different input patterns at given points in the cell. In general, each excitase recognizes (because of the tolerance factor $\Delta\xi_{jk}$) a family of closely related input patterns. This family may actually be quite large, since in fact the number of dendritic inputs may be quite large. As the tolerance factor decreases, the size of the family also decreases.

The activation parameter (ξ_{jk}), the tolerance ($\Delta\xi_{jk}$), and the capability of binding at region j are all properties of the excitase molecule. How do we know that by appropriate choice of amino acids we can run over the requisite values of these parameters? This is inherent in our principle of molecular adaptability. Indeed, for the excitase this may be rephrased as: gradual change in (certain segments) of the primary structure of the excitase are often associated with gradual changes in the

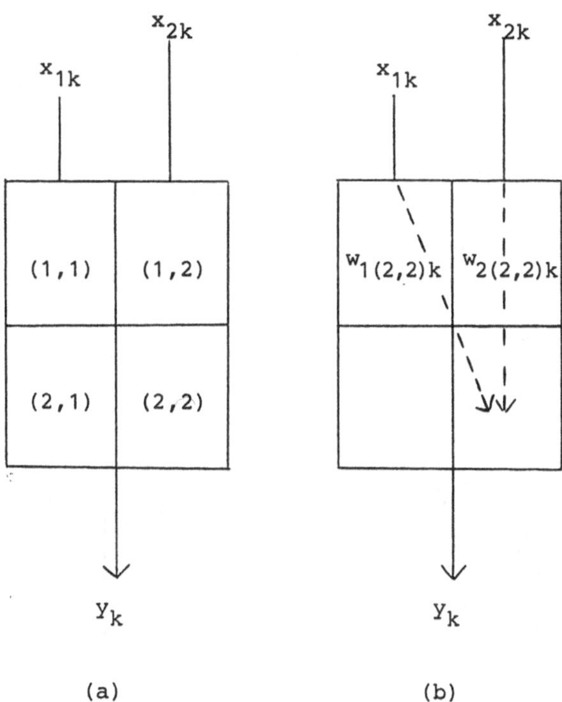

Fig. 4. Model enzymatic neuron. (a) The possible locations of excitases are represented by squares on a grid, with $j=(u,v)$. In the illustration we use four grid squares and suppose there are only two dendritic inputs. (b) The different lengths and positions of the dendritic inputs indicate that the neuron is really highly asymmetrical, so that the weighting coefficients $w_{i(u,v)k}$ may be very different for different i, u, v. If the symmetry breaking associated with these weighting coefficients is sufficient, it is possible to choose excitases so that the neuron computes any logical function of the two inputs. Notice that in general many possible choices of excitase parameters make it possible to compute any particular logical function. Also, if the weighting coefficients are themselves functions of the input patterns, the symmetry may be broken by the dynamical properties of the membrane, and therefore even more strongly. Moreover, in this case it is only necessary to vary the j and allow each excitase species to assume a number of j values (if we want to include the tolerances).

The general equation of the above neuron is:

$$y_k(t+1) - \Theta\left[\sum_{u=1}^{2}\sum_{v=1}^{2} \delta(\xi_{(u,v)k} - w_{1(u,v)k}x_{1k}(t) - w_{2(u,v)k}x_{2k}(t) \pm \Delta\xi_{(u,v)k}\right]$$

Suppose that $w_{1(1,1)k}= 1$, $w_{1(1,2)k}= .75$, $w_{1(2,1)k}= .50$, $w_{1(2,2)k}= .25$, $w_{2(1,1)k}= .80$, $w_{2(1,2)k}= .60$, $w_{2(2,1)k}= .40$, $w_{2(2,2)k}= .20$. By appropriate choice of the $\xi_{(u,v)k}$ and $\Delta\xi_{(u,v)k}$ the neuron can be made to compute any logical function of two inputs. For example, it computes an "exclusive or" when $\xi_{(1,2)k}= .75$, $\Delta\xi_{(1,2)k}< 1.5$ and both $\xi_{(u,v)k}$ and $\Delta\xi_{(u,v)k}= 0$ for all other values of u and v. Real neurons may of course receive a very much larger number of dendritic inputs.

location at which they bind on the neuron surface, in the value of the electric fields to which they respond, or in the range of values over which they respond. Furthermore, the degree of gradualism is adjusted in the course of evolution so that the system is sure to inherit or generate (ontogenetically) a sufficient variety of excitases.*

7. Enzymatic Networks

Enzymatic neurons may of course be interlinked to form networks. One of the simplest and most instructive configurations is illustrated in Fig. 5. This consists of r+1 neurons, with r of them receiving dendritic inputs from the outside. The (r+1)th neuron receives its inputs

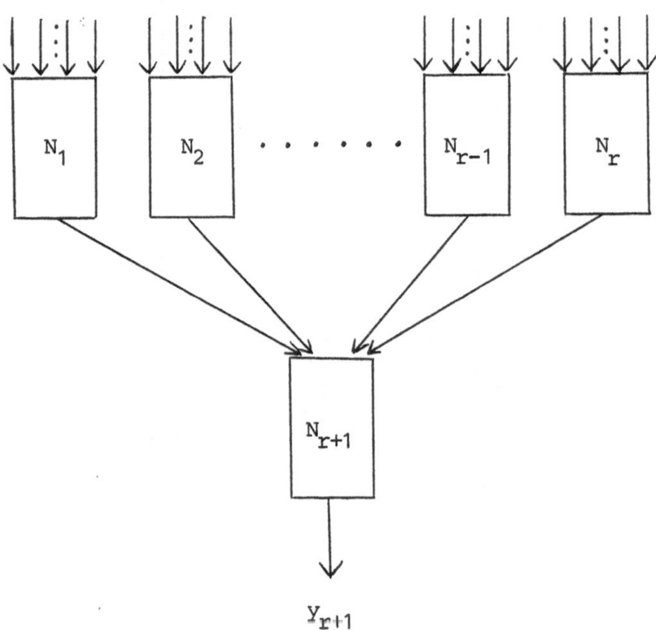

Fig. 5. Simple enzymatic network

*If the weighting coefficients are themselves functions of the input pattern, it is sufficient to vary only the locations and dispersions around these locations (see section 10). In this case the differences in the ξ_{jk} and the $\Delta\xi_{jk}$ are absorbed into these coefficients, which, however, must now be written $w_{ijk} = w_{ijk}(x_{1k},\ldots,x_{mk})$. This is simpler from the standpoint of physics, but of course more complicated from the standpoint of formal description.

from these and fires whenever all of its inputs fire. The equation of this neuron is

$$y_{(r+1)}(t+1) = \delta(\xi_{1(r+1)} - \sum_{i=1}^{r-1} w_{i1(r+1)} y_i(t))$$

where

$$y_i(t) = \Theta\left[\sum_{j=1}^{n} \delta(\xi_{ji} - \sum_{u=1}^{m} w_{uji} x_{ui}(t-1)\right]$$

and

$$\xi_{1(r+1)} = \sum_{i=1}^{r-1} w_{i1(r+1)} y_i(t)$$

and we assume, for simplicity, that all the $\Delta\xi_{uv} = 0$. Notice that this network is capable of recognizing distinct families of input patterns, that it is possible to traverse the space of possible input patterns step by step by adding or deleting particular excitases in the first r neurons, and that the size of the families can be increased or decreased by varying the tolerances of these excitases. In short, the network acts as if it were a single enzymatic neuron, except that it is capable of recognizing arbitrarily complex patterns without extreme symmetry breaking on the part of any neuron or extreme sensitivity on the part of any particular excitase.

Naturally we can add in more layers of neurons and also arbitrary interactions among the neurons (by specifying various equalities among the inputs and outputs). The resulting networks share the same general capabilities indicated above. To describe these capabilities more precisely, and also more generally, we introduce the following definitions:
Definition 1 (simulation). According to Fig. 2 the transition functions of the Turing automaton A are given by

$$\gamma: I \times Q \to Q$$
$$\lambda: I \times Q \to Y$$

where I ($i_j \varepsilon I$) is the set of possible inputs, Q ($q_j \varepsilon Q$) is the set of possible states, and Y ($y_j \varepsilon Y$) is the set of outputs. We will say that A' simulates A if there are maps

$$h_1: I' \to I$$

$h_2: Q' \to Q$

$h_3: Y' \to Y$

such that

$$\text{(i)}\quad h_2\left[\gamma'(q_j', i_k')\right] = \gamma\left[h_2(q_j'), h_1(i_k')\right]$$

$$\text{(ii)}\quad h_3\left[\lambda'(q_j', i_k')\right] = \lambda\left[h_3(q_j'), h_1(i_k')\right]$$

for all j, k. (Note: what we call simulation is sometimes called machine homomorphism.)

Definition 2 (inputs, states, and outputs).

(i) The input set of an enzymatic neural net is given by

$$I = \left\{x_1(t) \ldots . x_r(t) \mid x_u(t) = x_{ij}(t) \text{ for some i, j and } x_1 = 0 \text{ or } 1\right\}$$

This will sometimes be called the input pattern to the net (since it is the pattern of 0's and 1's on the input lines at any given time).

(ii) The state set of an enzymatic neural net is given by

$$Q_E = \left\{\delta(\xi_{1,1} - \sum_{i=1}^{m} w_{i11}x_{i1}(t) - \Delta\xi_{11}), \ldots, \delta(\xi_{mr} - \sum_{i=1}^{m} w_{imr}x_{ir}(t) - \Delta\xi_{mr} \mid \xi_{ik} \neq \xi_{jk} \neq \infty \text{ for all j, k}\right\}.$$

This is just the set of states (active or inactive) of all the excitase species in the network; in other words, states of distinguishable electrical excitation of all these neurons. However, the set of distinct states is given by

$$Q = \left\{y_1, \ldots y_r\right\}$$

i.e. the firing pattern of all neurons in the net.

(iii) The output set of an enzymatic neural net is

$$Y = \left\{y_k\right\},$$

i.e. the output of some selected neuron k. The complete output set is the same as the set of distinct states, i.e.

$$Y_i = \left\{y_1, \ldots, y_r\right\}$$

Notice that the number of states is 2^{nr} (where n is the average number of excitase species per neuron), but that the number of distinct states is given by 2^r. Thus the excitase statistics are: all patterns of excitase activation in a single neuron are indistinguishable from the standpoint of the pattern of network firing. Nevertheless, we will see that the extra states play an important role.

Theorem 1 (simulation of conventional networks). Any McCulloch-Pitts network can be simulated by some enzymatic network.

Proof. Consider the following enzymatic neuron:

$$y_r(t+1) = \delta(\xi_{1r} - w_{11r}x_{1r}(t) \pm \Delta\xi_{1r}) \qquad (*)$$

Now recall (from Fig. 2) the definition of the McCulloch-Pitts neuron:

$$y_r(t+1) = \delta'\left[w_r - \sum_i x_{ir}(t)\right] \qquad (**)$$

where

$$\delta'(u) = \begin{cases} 1 \text{ if } u > 0 \\ 0 \text{ otherwise} \end{cases}$$

Now choose $w_{11r} = 1$. Then (*) is precisely the same as (**) when $\xi_{ir} = \Delta\xi_{1r} = \frac{1}{2}w_r$ (i.e. h_1 and h_2 are identity maps). Thus the McCulloch-Pitts neuron is just a special case of the enzymatic neuron, from which it follows that we can simulate any McCulloch-Pitts net with an enzymatic net.

Theorem 1 implies that enzymatic neural networks have the same computational capabilities as finite automata (since any finite automaton can be simulated by some McCulloch-Pitts network). However, it is worthwhile to prove some direct simulation theorems, which, although very inefficient, give some insight into the design features of enzymatic neural nets.

Theorem 2 (simulation of automata). Any finite automaton with r (or fewer inputs, states, and outputs can always be simulated by an enzymatic neural network of at most r neurons, with at most $2^{(r+\log_2 r)}$ excitase species per neuron, and in such a way that h_1, h_2, and h_3 are either one-one or many-one.

Proof. According to Definition 2(i) the r inputs code for the firing pattern of $\log_2 r$ input lines. (This is h_1.) According to Definition 2(ii) a network of r enzymatic neurons with $2^{(r+\log_2 r)}$ excitases per neuron has 2^r distinct states (or $2^{r2^{(r+\log_2 r)}}$ if patterns of excitase firings are considered). These can be divided into r nonintersecting classes in an arbitrary way, but subject to the condition that no two classes include states associated with the same pattern of neuron firing. The h_2 and h_3 maps assign each of these groups to a unique automaton state and output, respectively. Now, suppose that each neuron receives all $\log_2 r$ input lines from the outside and also inputs from all r neurons (including itself). These $r+\log_2 r$ input lines encode all information about the external input pattern and firing pattern of the net in $2^{(r+\log_2 r)}$ input patterns to each neuron. Since there are up to $2^{(r+\log_2 r)}$ excitase species per neuron, these are be chosen such that the network undergoes a transition to any other class of states given the present class and (external) input pattern, from which it follows that we can choose these transitions so that they fulfill the $r2^r$ simulation conditions required by equation (i) in Definition 1. Furthermore, this choice automatically fulfills the conditions required by equation (ii) because h_3 is just a special case of h_2.

Theorem 3 (simulation of automata). Any finite automaton with r or fewer inputs, states, and outputs can always be simulated by an enzymatic neural network of at most r^2 neurons, each with one excitase species per neuron, and in such a way that h_1, h_2, and h_3 are either one-one or many-one.

Proof. The r inputs are coded by the firing pattern of $\log_2 r$ input lines (the h_1 map); the r states are each coded into patterns of network firing in which any one of r neurons fire, but subject to the condition that these sets of neurons are noninteracting, and likewise for the outputs (h_2 and h_3). Again we assume that each neuron receives all input lines from the outside and also inputs from all other neurons, including itself. These $r^2+\log_2 r$ input lines carry all information about the state of the net and the external input pattern to each neuron, but encoded in r^2 possible input patterns (because of the restriction that no two neurons fire at once). Thus the simulation may be implemented by choosing the excitases so that each neuron fires in response to only one of its input patterns and no other neuron fires in response to this pattern. Since the neuron which fires can be chosen arbitrarily, the network can undergo a transition to any other set of states given the present set of states and external input pattern, from which it follows that it is always possible to fulfill the simulation conditions (Definition 1, equations (i) and (ii)).

The above simulations are at opposite extremes of the design spectrum, therefore neither efficient nor biological. The following theorem gives a clearer indication of the real power of enzymatic neural nets.

Theorem 4 (pattern recognition). Any subset of 2^{nr} possible input patterns (containing at most nr of these patterns) can always be recognized by an enzymatic neural net containing r+1 neurons, with at most n excitase species for the first r neurons and at most 1 excitase species for the (r+1)th neuron.

Proof. The net illustrated in Fig. 5 does the job, assuming that each neuron receives all nr inputs (i.e. each of the r neurons can handle at most n of the 2^{nr} patterns). The final neuron (N_{r+1}) fires whenever one of these neurons fires, i.e. it is a threshold neuron of the type described in theorem 1.

The final neuron can of course be eliminated if we agree that recognition is equivalent to the firing of any single one of the first r neurons. Also, notice that if some of the neurons recognize the same input this can be transformed into a highly patterned output. Naturally any single input pattern or family of input patterns closely related by Hamming distance can be recognized by a single neuron with appropriate weighting coefficients and whose excitases have suitable tolerance. The network can recognize an arbitrary subset of n patterns (out of 2^{nr}) if each of the neurons receives only n of the inputs and the (r+1)th fires only if all its inputs fire.

The pattern recognition and generation network described above is still not as efficient as it could be since it does not use interactions among the neurons to take advantage of special features of the input, nor does it use the $\Delta\xi_{ij}$ to take advantage of similarities among the input patterns. However, it is quite efficient in comparison to, e.g. networks of the McCulloch-Pitts type, in which the neurons average away the detailed patterning of their inputs. This has the consequence that each state of the enzymatic network generally corresponds to many states of the McCulloch-Pitts network which simulates it, and even more so if we require the recognition of an input pattern to stimulate the firing of a single neuron. This is why such networks require so many neurons.

Finally we turn to the question of gradualism.

Theorem 5 (gradualism). The set of input patterns recognized by enzymatic neural nets can be changed by making single alterations in the net, and in such a way that only one pattern is added to or deleted from the set. The state or output patterns generated by the net in response to a given input can also be changed by making single changes in the net, and in such a way that the new pattern differs from the original by a Hamming distance of only one.

Proof. The net illustrated in Fig. 5 again does the job. The deletion of one excitase from any one of the neurons deletes one pattern from the subset of recognizable patterns; whereas the addition of an excitase with suitable ξ_{jk} adds an arbitrary pattern to the subset. The state of the net changes by Hamming distance one (i.e. the firing pattern differs from the original by the firing of one neuron) if many neurons respond to the same input pattern (including inputs from other neurons) and an excitase which recognizes one of these patterns is added to or deleted from one of the neurons. This is also trivially true for the output pattern if the state is the output; on the other hand, if there are distinct output neurons the same considerations can be applied to these.

Naturally the pattern recognition and generation behavior could also be changed by changing the binding properties of the excitases rather than the activation parameter; they could also be changed by changing the tolerances, especially if the weighting functions are such that slightly different inputs give rise to a slightly different electric field at a given location on the cell surface.

We note that a single McCulloch-Pitts neuron with excitatory and inhibitory inputs can recognize linearly independent input patterns, but that by single changes in the net (i.e. either changes in the threshold or in the excitatory or inhibitory nature of the inputs) it is impossible to span the entire space of input patterns (Sheng, 1969).

The gradualism property may also be extended to single changes in the next state or output functions, i.e. to single changes in the sets $\{\gamma(q_i, i_j) \mid \text{all } i, j\}$ and $\{\lambda(q_i, i_j) \mid \text{all } i, j\}$. For example, this is always possible by the addition or deletion of single excitase species in networks of the type constructed in theorem 2. But, of course, though the original and final transition functions may look similar, the behavior may be quite different.

The gradualism properties are important because they make enzymatic neural networks especially amenable to phylogenetic evolution and modification-based learning. The significance of this for the development of motor control and perception is discussed in Part II of this paper. Here we only point out that structurally programmable systems which simulate enzymatic neural nets in the sense of Definition 1 cannot in general simulate them in the broader sense that changes in structure comparable from the standpoint of probability produce comparable changes in function (Conrad, 1974).

8. The Artificial Selection System

The remarkable structural simplicity, pattern recognition capability, and modifiability of enzymatic neural nets are not free, for the structural nonprogrammability of such nets means that they can only be constructed by trial and error selection of the excitases.

Darwinian evolution is clearly one mechanism for this type of selection. Indeed, according to the principle of molecular adaptability the gradualness with which the function of the network changes with changes in molecular sequence is itself adjusted (in the course of evolution) for optimal amenability to further evolution. Furthermore, this evolution requires no change in the interconnectivity or weighting coefficients of the neurons, but only the appearance and selection of excitases which utilize the given (symmetry breaking) constraints in a suitable way. The excitase based nervous system is thus especially amenable to the phylogenetic development of various innate information processing capabilities.

We still have not explained how the organism can use the gradual modifiability property for learning, or the ontogenetic development of novel information processing capabilities. According to our argument this must also require trial and error selection of excitases. This is possible only if the brain contains

1. Interchangeable replicas of any given enzymatic net;
2. A selection circuit system which regulates the production of excitases on the basis of their functional value.

Our essential picture is that modification-based learning is mediated by processes analogous to those which mediate phylogenetic evolution, except that selection mechanisms are internalized in the circuitry of the brain. This is possible if the excitase enzymes are coded by nucleic acids which transmit adaptive changes in the nature or distribution of these excitases from the networks in which they arise to structurally identical networks by transforming them, for example, in the fashion of an RNA virus. The interchangeable replicas are necessary because the function of the excitases must be the same in the regions which they transform as in the region in which they were originally produced (see Fig. 6).

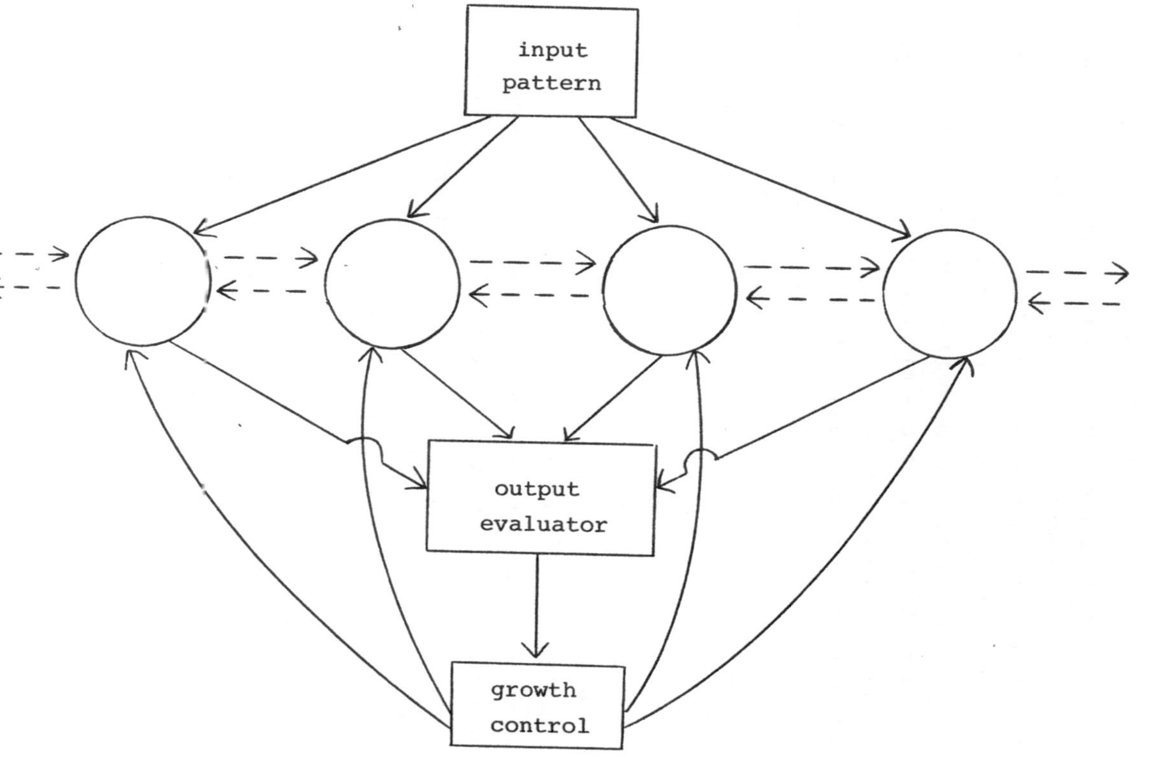

Fig. 6. Selection circuit scheme. Each local (interchangeable) network receives the same input pattern, the output evaluator assigns a fitness to each such network on the basis of its functional value to the organism and sends the information to the growth control. The growth control controls the culturing of excitase genes which transmit adaptive changes from one local network to another (the transformation process). Circles represent local networks, solid arrows inputs or outputs, and dotted arrows the transformation process.

9. The Implied Biology

The evolutionary selection scheme described in the previous section has certain important biological implications. The most important of these are:

1. Excitase "genes". These are the nucleic acids which code for the excitases which are cultured by the selection circuits, and which diffuse from local network to local network. They may either preexist in small numbers, in which case the culturing only controls their distribution, or they might arise through somatic genetic processes. Some excitase genes may be innate, i.e. associated with innate information processing capabilities. These must be distinguished from culturable excitase genes. The latter are not transmitted to offspring; or, if they preexist, their distribution is not heritable. The excitase genes could be either DNA or RNA. They must diffuse in association with some protein which is capable of recognizing the correct neuron type.
2. Excitases. These must have a control site which enables them to respond to fairly specific electric fields and a binding site which enables them to position themselves on a particular region of the cell surface. Neurons whose dynamics are controlled by excitases are in general multithreshold elements.
3. Selection circuits. The selection circuits must have definite anatomical correlates, including fibers which carry signals from the growth control to the various interchangeable networks. The evaluation system itself would comprise all of the organism's mechanisms for accessing the advantages or disadvantages of different forms of behavior.
4. Control enzymes. The operation of the selection must be mediated by control enzymes. These must either induce or activate enzymes which in turn control the culturing of the excitase genes. Actually, it is possible that the excitase genes are produced in glial cells, since there seems to be some evidence for transfer of RNA between glia (Hyden, 1967a).

The above implications are either testable or lead to further, testable predictions. They also provide a coherent account of a number of otherwise contradictory facts. In particular, the excitase genes are consistent with the evidence that transfer of learning can be mediated by RNA, that the strength of the transfer depends on the amount of training (McConnell et al., 1970), and that such transfers are only possible when they involve comparable regions of the brain (Albert, 1966) The excitase enzymes are consistent with the fact of many brain specific proteins and polypeptides, although the molecules so far investigated may not correspond to the ones we have postulated (see, e.g. Ungar, 1973). Also, the requirement that stimulation of particular regions by the selection circuit be mediated by control enzymes is consistent with the evidence for the appearance of specific, but primarily nuclear protein in learning situations (Hyden, 1967b). The high degree of specificity which has been observed in individual neurons is also consistent with the model, though to my knowledge there is no direct evidence that individual neurons respond to different bands of electrical stimulation (the multithreshold property). I would like to emphasize that successful descriptions of the phenomenological dynamics of the nerve impulse in the peripheral nervous system are not inconsistent with enzymatic triggering, nor do they even close the question as to the chemical nature of the impulse.

The selection circuits themselves are certainly consistent with the apparent anatomical homogeneity of some of the functionally most complex tissue in the CNS and also the physiological evidence for the high redundancy of local tissues, each with the same relation to function (Chow, 1967). This homogeneity, which is equivalent to our requirement of interchangeability of local regions, allows for both highly specific neural connectivity and behavioral plasticity and also explains why repair or growth processes are infeasible after a certain stage of development (since this would destroy the usefulness of the excitases unless the same structure could be regenerated). The selection circuit scheme also predicts the occurrence of "setting" phenomena in learning. This follows from the fact that excitase genes of increasingly functional value are more likely to appear as the genes from which they can be derived transform an increasing number of local networks. The scheme gives an extremely good explanation for Lashley's "mass action" laws, i.e. for the retardation of learning with degree of cortical ablation (Lashley, 1929). This follows directly from the importance of the number of local networks for the rate of learning and from their independence and equivalence as regards their relation to other types of networks in the nervous system.

The possibility for increase in the number of excitase genes and enzymes, the requirement for control enzymes, and also the connection between rate of learning and rate of culturing all depend on the development of increased metabolic support with learning and increased rate of learning. This is consistent with the evidence for gross changes in the nervous system with learning, in particular as regards glial cells (cf. Bennett et al., 1970) as long as these changes do not interfere with the weighting functions. These should not be affected by plastic changes in glial cells (one reason for introducing them into the model), or even by change in size or number of synapses, as long as these subserve the necessary function of maintaining the effects of given input patterns on individual excitases in the face of increasing activity. (We return to this point in Part II, where we discuss memory and its relation to the opening and closing of particular synapses.)

The model is also compatible with the resistance of learned behavior to radiation damage, since the high redundancy of local networks ensures that radiation would destroy vital functions before it could damage the excitases in each of these networks. However, this radiation resistance should decrease with the degree of ablation involving networks of a given type.

10. Conclusions

The theory put forward in this paper addresses itself to one simple problem: what properties must the brain have to support modification-based learning and at the same time support any information processing operations of which it is actually capable, subject to the additional constraint that it be amenable to phylogenetic evolution.

Actually, we have not quite shown that our model is capable of performing any possible information processing operation (or computing any computable function), but rather that it is potentially capable of simulating any finite automaton (what we have called the Turing automaton). We still have not accounted for the "tape" in the brain, i.e. for rapidly acquired memories and memory-based learning. Thus what we have constructed is a molecular automaton, but not yet a molecular computer. This is possible, however, and in a way that fits very neatly into the excitase model. According to the Turing thesis the resulting model (described in Part II) is capable of performing any information processing operation of which the brain is capable.

The gradual modifiability of our realizations of Turing automata is crucially dependent on the size of molecules (cf. Conrad, 1972). It is the special adaptability properties of individual proteins which makes it possible for step by step modifications in structure to produce realizations of Turing automata whose behavior is only slightly different. The question naturally arises as to whether the model necessarily requires individual enzymes to act as the fundamental decision making elements. In fact, the scheme is workable if the excitases affect only particular synapses (i.e. change the weighting coefficients of a threshold neuron). However, models of this type lose the efficiency and gradualism properties of the present model. The scheme is also workable if the excitases have indirect effects on the dynamics of the neuron; but in this case the system cannot complete nearly as many learning trials, since the development or elimination of the effects of such genes would require considerable delay. More important, the effects of the genes would not in this case be independent, so that the addition or dilution of excitase species from the enzymatic neuron would change the weighting coefficients and therefore the functional value of all other excitases. These factors are of crucial importance for the rate of evolutionary learning, and in the case of independence for the rate of phylogenetic evolution as well.

It is worth pointing out that all of the constructions described in section 7 work just as well if the weighting coefficients are functions of the input patterns rather than constants, i.e. if the dynamical properties of the membrane help to break the symmetry of the inputs and in a way which depends on the particular input. In this interpretation the neuron would only fire in response to a particular input pattern if it carries an excitase species which binds to a locus on the membrane which responds strongly to this pattern. This is perhaps important since it means that only the binding properties of the excitases have to change in order for them to take advantage of localized centers of excitation associated with different input patterns.

Of course it is not necessary for the excitase to be an individual molecule--it might be a molecular assembly. Also, the model works just as well if the probability of neuron firing depends on the number of excitases which are activated. What is crucial is only that each of the enzymatic units acts independently.

Does the selection circuit model actually make transformation processes a feasible basis for learning on the time scale of human life? Naturally it is not possible to answer this question without having some reliable numbers for a starting point. For the sake of argument, however, suppose that there are about 10^{10} glial cells involved in the transformation process and that on the average each one transcribes one strand of exportable excitase genes every hour (naturally some will be transcribing at an enormous rate and others not at all). This allows the brain to try out at least 10^{15} excitase genomes in a lifetime of about twenty years. This is roughly the same as the number of human genomes which have been tried out in the last ten million years of human evolution, assuming a (rather high) birth rate of 10^8 infants per year. Naturally the two situations are not directly comparable because of the smaller size of the excitase genomes and also because of the wide number of different types which must develop (i.e. their adaptive radiation).

However, I think the large number of possible trials is striking, especially in view of the fact that evolutionary learning processes in the brain will be much faster because of the special circuitry which detects and amplifies slight functional differences, because of the relatively small number and independence of the genes on the excitase genome, and also because many of the different types of excitases may preexist in small numbers anyway.

References

Albert, D.J.: Memory in mammals: evidence for a system involving nuclear ribonucleic acid, Neuropsychol. 4, 79-92 (1966).

Bennett, E.L., Rosenzweig, M.R. and Diamond, M.C.: Time courses of effects of differential experience on brain measures and behavior in rats. In: Molecular Approaches to Learning and Memory, ed. by W.L. Byrne, pp. 55-90. Academic Press, New York, 1970.

Chow, K.L.: Effects of ablation. In: The Neurosciences, ed. by G.C. Quarton, T. Melnechuk and F.O. Schmitt, pp. 705-713. Rockefeller University Press, New York, 1967.

Conrad, M.: Information processing in molecular systems, Currents in Modern Biology 5, 1-14 (1972).

Conrad, M.: Is the brain an effective computer?, Intern. J. Neuroscience 5, 167-170 (1973).

Conrad, M.: The limits of biological simulation, J. theoret. Biol. 45, 585-590 (1974).

Hydén, H.: RNA in brain cells. In: The Neurosciences, ed. by G.C. Quarton, T. Melnechuk and F.O. Schmitt, pp. 248-266. Rockefeller University Press, New York, 1967a.

Hydén, H.: Biochemical changes accompanying learning. In: The Neurosciences, ed. by G.C. Quarton, T. Melnechuk and F.O. Schmitt, pp. 765-771. Rockefeller University Press, New York, 1967b.

Lashley, K.S.: Brain Mechanisms and Intelligence, University of Chicago Press, Chicago, 1929. (Reprinted by Dover Publications, New York, 1963)

McConnell, J.V., Shigehisn, T. and Salive, H.: Attempts to transfer approach and avoidance responses by RNA injections in rats. In: Molecular Approaches to Learning and Memory, ed. by W.L. Byrne, pp. 245-274. Academic Press, New York, 1970.

McCulloch, W.S. and Pitts, W.: A logical calculus of the ideas immanent in nervous activity, Bull. Math. Biophys. 5, 115-133 (1943).

Minsky, M.: Computation: Finite and Infinite Machines, Prentice-Hall, Englewood Cliffs, N.J., 1967.

Sheng, C.L.: Threshold Logic, Academic Press, New York, 1969.

Ungar, G.: Molecular approaches to neural coding. In: The Physical Principles of Neuronal and Organismic Behavior, ed. by M. Conrad and M. Magar, pp. 169-176. Gordon and Breach, New York and London, 1973.

von Neumann, J.: The Theory of Self-reproducing Automata, ed. by A.W. Burke, University of Illinois Press, Urbana, 1966.

Winograd, S. and Cowan, J.D.: Reliable Computation in the Presence of Noise, M.I.T. Press, Cambridge, Mass., 1963.

MOLECULAR INFORMATION PROCESSING IN THE CENTRAL NERVOUS SYSTEM
PART II: MOLECULAR DATA STRUCTURES

Michael Conrad
Institute for Information Sciences
University of Tübingen
Tübingen, Germany

1. Introduction

The brain is a complicated system--too complicated to describe in detail and too detailed to describe statistically. What we can do, however, is describe its capabilities, e.g. its capabilities for computing a certain class of functions or for learning. Any model of the brain must of course fulfill these capabilities and at the same time operate in accordance with basic biological principles. Another way of saying this is: a brain model is no good unless it can do what the brain does and in a way which is amenable to phylogenetic evolution.

This is the kind of constructive approach we took in Part I, except that there we only considered realizations of Turing automata amenable to modification-based learning. Now we have to complete our construction by accounting for the Turing "tape", i.e. for the brain's general powers of memory and memory-based learning.

2. Fundamentals of Memory

a. Tapes

The Turing tape has the following fundamental properties:

1. Markability. The tape consists of components (squares) whose states can be changed by the automaton and in a single step. These states are the tape markings.
2. No superposition. Marking one square does not affect the marking on any other square.
3. Accessibility. The automaton must be able to exert some control over which square it marks or reads, e.g. by moving the tape.
4. Forgetting. In all real systems the tape is finite, so there must be some erasure mechanism enabling the system to re-use tape squares. (In the Turing scheme this is automatic since a rewritten square retains no

trace of its earlier markings.*)

The tape, of course, is just a metaphor for some much more sophisticated memory space in the brain. However, this memory space, and indeed any memory space, must satisfy properties 1-4, except that in the case of the brain the components are neurons and the accessing structures are very much richer--indeed so much so that entire scenes can be reconstructed and manipulated on the basis of time order, content, or associations with other stimuli.

b. Memory cycle of the brain

Now we describe a memory cycle which satisfies the above boundary conditions, deferring for a moment the considerations which lead to this particular cycle.

The brain contains neurons which receive inputs from receptors and other preprocessing neurons. We will suppose that high level neurons of this type (to be called primaries) also receive inputs from reference neurons. The property of the primaries is that their firing sensitizes certain of their dendrites (e.g. by an antidromic wave of excitation) and in such a way that these dendrites are opened (or loaded) by the firing of any reference neuron which contacts them. The property of the reference neurons is that they are activated by other reference neurons in sequence, are never activated for loading primary dendrites more than once (except under conditions to be specified), and are themselves sensitized when they are activated, so that their dendrites are also loaded by the firing of any primaries which they contact.

Clearly the second (or further) firing of each reference neuron calls all the primaries previously loaded by it, thereby reconstructing the original pattern of primary activity (see Fig. 1). For example, suppose that the organism is exposed to a particular scene and that this causes certain primaries to fire. Some of these primaries are loaded by the reference neuron active at the time and some also load this reference neuron. When the latter is activated at some future time it fires all the primaries which loaded it, thereby recalling the original scene as a whole (or at least with a fidelity which depends on the number of primaries which loaded it). The reference neuron may itself be called by the reference neuron active at the previous instant of time, in which case the organism recalls the original sequence of scenes in proper temporal order; or it may be called by one of the primaries which

---

*In principle the class of functions computable by a Turing scheme with a finite tape is no wider than that computable by finite automata. In practice, however, a system with a manipulable memory space, even if finite, is much more powerful than one without this feature.

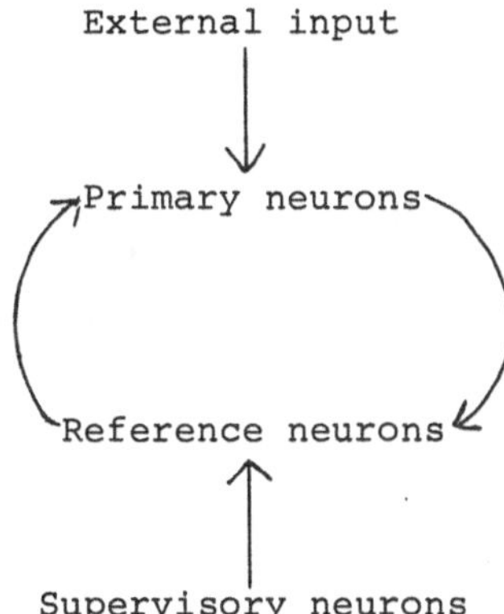

Fig. 1. Memory cycle. External inputs (ordinarily coming from preprocessing neurons) fire the primaries. The fired primaries are loaded and later called by some reference neuron. Also, some of the primaries may load and later call this reference neuron. The supervisory neurons manage the activation of reference neurons for loading and calling.

it loaded, and therefore by content (e.g. by re-presentation of part of the original scene). Furthermore, memories can be recombined into new structures, or parts of memories recombined into new memories, if two or more reference neurons are activated for calling at the same time, but with the provision that these calls are inhibited from reaching certain of their primaries. The resulting pattern of activity is memorized in just the same way as the original patterns if some reference neuron is active for loading at the time. Indeed this type of partial suppression and rememorization allows for general powers of memory manipulation, including the construction of time re-ordered, content re-ordered, and associative access structures. (The associative structures are just content re-ordered structures, i.e. structures in which primary neurons whose firing is associated with one input load reference neurons whose firing is associated with another.)

The recollection of an entire scene with some degree of fidelity clearly requires the reference neuron to load a large number of the primaries whose firing is associated with that scene. This in turn requires many contacts and therefore a lot of neural connectivity. The amount of connectivity can be reduced, however, if reference neurons use party lines, i.e. if a number of reference neurons can contact the same primaries using common intermediary (or <u>party line</u>) neurons (see Fig. 2). However, this is only possible if the reference neurons load the primaries differently, so that they do not call primaries loaded by other reference neurons which use the same party line. In other words, each reference neuron using the same party line must use a different code (or temporal sequence of pulses) to load and call primaries. Also, a number of reference neurons may be active for loading at any given time, so that no single party line has to contact an unreasonable number of primaries. Primary neurons which always fire in response to the same input may use party lines to contact reference neurons.

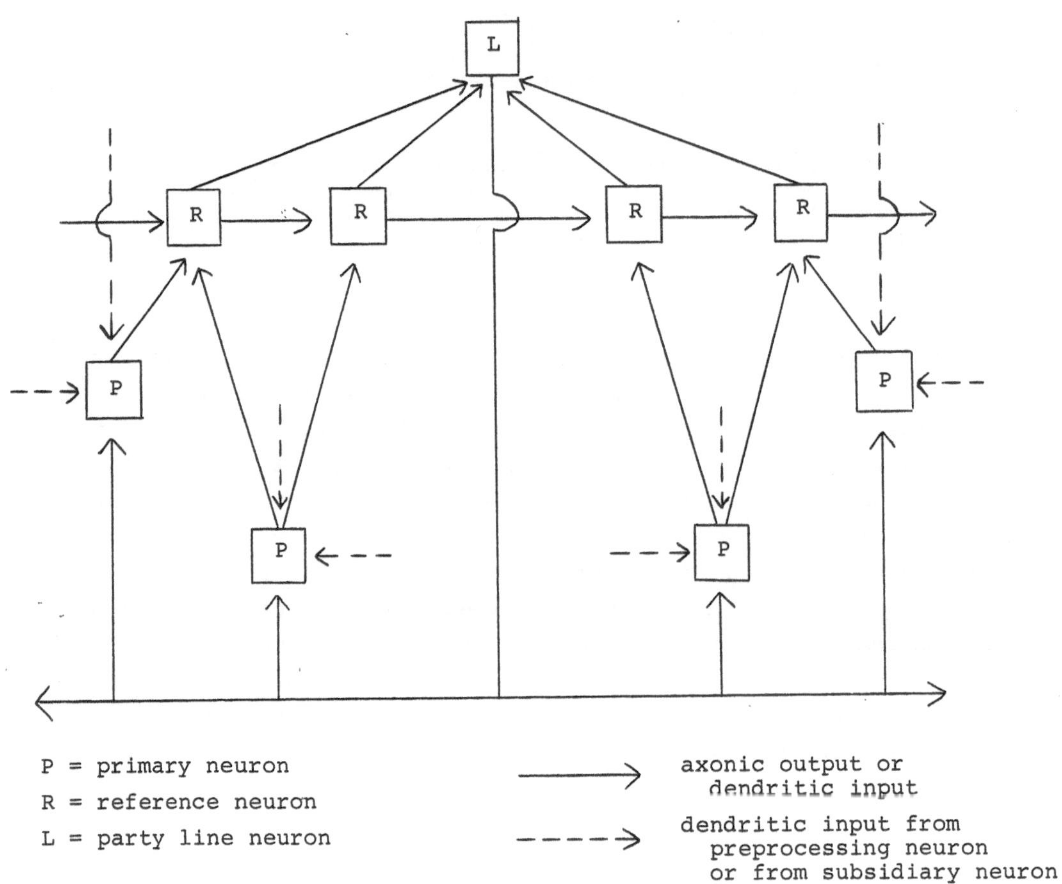

Fig. 2. Party line principle. The reference neurons send loading and calling sequences over party lines to contact the maximum number of primaries. The ordering of the reference neurons allows for time ordered memory structures. The loading of reference neurons by primaries allows for content ordered and associative structures. Not shown: preprocessing neurons, supervisory neurons, subsidiary neurons modulating the behavior of the primaries, direct interactions among the primaries, and long axonic outputs of the primaries. The layered structure of the diagram provides an interpretation for the (much more complex) layered structure of the cerebral cortex (see section 6).

So far we have described our memory cycle in terms of primary, reference neurons, and party line neurons. This is not sufficient, however, since there must be some mechanism for determining whether reference neurons are to be activated for calling or loading, whether they are to be activated by reference neurons earlier in the sequence, whether they are to be activated by the primary neurons, and also which output lines are to be inhibited. This requires certain <u>supervisory</u> neurons. Roughly speaking, these supervisory neurons correspond to the device which moves the Turing tape relative to the read-write head, except that in this case the activation of different reference neurons for loading corresponds to the choice of different writing heads.

The question also arises as to what controls the supervisory system. The answer is: the same thing which controls the moving and marking of the Turing tape, i.e. the transition functions of the Turing automaton. Thus, we do not have to chase this question any further up the line, for we know from Part I that any such transition functions can be embodied in networks of neurons, even purely conventional neurons. Indeed the network which embodies the rule may and generally would include the primaries themselves.

c. Why this cycle?

The storage units in the above model are the sensitizable dendrites. These dendrites are marked by the loading process, after which they become accessible by repetition of the signals which loaded them (except that the organism may use some extra information to distinguish calling inputs from loading inputs). The accessibility is controllable because of the triple function of the reference neurons, i.e. their ability to call primaries, to be called by primaries, and also to be called by other reference neurons.

The triple function of the reference neurons makes them particularly suitable as a memory device in the brain. However, there is a much more fundamental reason for introducing neurons of this type, viz. that they solve the problem of superposition. This is automatic because each reference neuron is associated with only one memory, from which it follows that the assimilation of new memories does not destroy old ones (though it may interfere with access).

This would not be true if the primary neurons sensitized and loaded each other directly. For example, suppose that we want to build an association between two stimuli. Certain primaries fire in response to S_1. These are sensitized and loaded by primaries which fire in response to S_2. The reoccurrence of S_2 thus reconstructs the pattern of firing resulting from S_1. Now, suppose that we repeat the process, except that this time we build an association between S_0 and S_2. In this case the reoccurrence of S_2 reconstructs a pattern of firing which is a superposition of the patterns associated with S_1 and S_0.

Of course there may be other ways of escaping the superposition problem. However, the reference neuron solution has the advantage that it not only solves the superposition problem but at the same time accounts for time ordered, content ordered, and associative access, and also for the reconstruction of "scenic" memories.

What we haven't shown so far is the mechanism of forgetting. This will become clear, however, as soon as we describe the molecular mechanisms underlying (indeed imposed by) the flow of information in the model.

3. Molecular Mechanism

We will assume that the flow of information in the memory cycle is mediated by conformational changes at the molecular level, deferring for a moment our reasons for making this assumption. These conformational changes must support sensitization of dendrites, loading of dendrites by different spike sequences (to distinguish different reference neurons using the same party line), and firing of the loaded neuron in response to the recurrence of the sequence which loaded it. This type of switching behavior is possible if the dendrites contain molecules (to be called "call molecules") each of which has:

1. a regulation site which controls its competence to be loaded;
2. storage sites whose conformation records the presence or absence of an impulse;
3. an active site which catalyzes events leading to impulse formation when the dendrite receives a sequence of inputs corresponding to the sequence which set the storage sites.

When the neuron is sensitized its dendrites release a sensitizing substance, putting some of the call molecules in the dendrite into the loadable state. After the loading process is completed the dendrite releases a fixative substance which prevents the call molecules just loaded from being cleared by future releases of sensitizer. Once the call molecule is loaded it assumes the callable state, i.e. the state in which it catalyzes events leading to impulse formation in response to the input sequence which loaded it. This occurs automatically if each of the storage sites activates the next, but only if the dendrite receives an input corresponding to its setting during a definite time interval after it was activated. If the last site in the chain receives such an input, it activates the active site.

Our picture of the call molecule is illustrated in Fig. 3. Notice that the scheme works with single site call molecules (i.e. call molecules which only open or close the dendrite). However, multisite call molecules are necessary if the system uses the party line principle. Also, notice that the fixative substance and also the intrinsic lifetime of the conformation provides the mechanism of forgetting. For example, if the call molecule has only a certain chance of binding or remaining bound to the fixative it will have a certain probability of being overwritten by future loading calls.

Conformational models have sometimes been criticized on the grounds that changes in molecular shape are not stable enough to mediate long term storage. However, the organism can maintain call molecule conformations indefinitely by duplicating them through the process of rememorization. This happens automatically if reference neurons reload any primary dendrites which recognize their call; indeed, the reloading process may take place in parallel as long as no two reference neurons sharing the same party line reload at the same time. The reference neurons may reload each other and are automatically reloaded by primaries

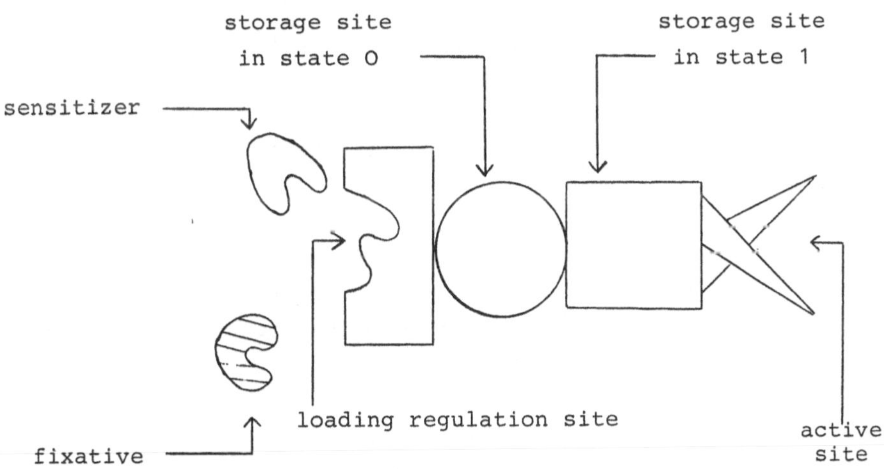

Fig. 3. Schematic diagram of two site call molecule. When sensitizer binds to the loading regulation site, the first storage site may be loaded (put into state 1) for a certain interval of time, after which its conformation is frozen and the second storage site becomes loadable. Later the recurrence of this sequence (with a suitable start signal) will trigger the active site and hence impulse formation. If the molecule binds with fixative, it is no longer sensitizable, and therefore will not be reset by new loading sequences. The loading and calling sequences might of course be encoded in redundant form. Also, the molecule may have more storage sites, or only one storage site.

whenever they call these. In principle, call molecules could also duplicate their conformations intradendritically, assuming they can go into a state in which they generate the sequence with which they are loaded, and at the same time some unloaded molecules are sensitized.

4. Why This Mechanism?

The call molecules are essentially allosteric molecules, except that they respond to electrochemical rather than purely chemical influences. These molecules certainly do the job required by the memory cycle. The job might of course also be done by some quite different mechanisms. There is no way of excluding this possibility a priori. Nevertheless, there are some good arguments against the major alternatives.

a. Reverse transcription. The first alternative, and one that has frequently been put forth, is that information is stored by encoding it into molecular sequence. In the present case this would mean that the loading inputs were somehow encoded into the sequence of monomers in nucleic acid or protein, and in such a way that these molecules could later recognize the sequence which loaded them. This violates the so-called Central Dogma, that reverse transcription never occurs, and also presumes an improbable relationship between the encoded sequence and the function which must be implicit in this sequence. Furthermore, the reverse transcription process would have to be controlled enzymatically and therefore ultimately by conformational changes, from which it follows that this mechanism is just a more complicated form of the conformational model. Moreover, nucleic acid and protein synthesis do not seem to take place in the dendrites.

b. Enzyme induction. A second proposal, also one that has been discussed by a number of authors (cf. Smith, 1962; Hydén, 1967), is induction of enzymes by incoming pulses. In the present case this would mean that the sequence of incoming pulses would induce the production of enzymes which are able to recognize the further occurrence of these sequences. The difficulty with this mechanism is again that protein synthesis does not seem to take place in the dendrites. Also, induced enzyme synthesis ultimately depends on conformational changes, except that these are more complicated than those required by the call molecule model, and also not as satisfactory from the standpoint of explaining the stability of the memory trace.*

---

*This does not mean that inducible enzymes do not play an important role. In particular, they mediate the operations of the selection circuit scheme described in Part I, i.e. determine whether or not a local network produces excitase nucleic acid for export.

c. Morphological changes. The problem with morphological change (e.g. increase in size or number of dendrites, or growth of spines) is that it is too slow to account for memories acquired in a single exposure and also does not allow for the party line principle.

The functional properties of the call molecules might also inhere in the membrane, in particular if the organism does not use the party line principle. However, if the organism uses this principle the dendrites must contain a number of independent switches--hence our assumption of individual molecules or molecular aggregates. However, these would presumably be bound to the membrane and indeed those which are differently loaded might very well be spatially isolated, e.g. in different spines.

5. The Importance of Rememorization

The call molecule scheme is actually capable of simulating the Turing scheme, disregarding, of course, the finitistic restriction on any real system. We will not bother to show this because the simulation is quite clumsy. This is to be expected of a brain model since humans are clearly not very good at simulating Turing schemes (without an external notepad). What they are good at is building time ordered, content ordered, and associative memory structures, including structures which store complex (e.g. scenic) memories. This, of course, is just what the call molecule scheme is suited for.

The most important feature of the call molecule scheme, from the standpoint of memory manipulation, is the process of rememorization. This process, the simplicity of which is of course due to the reference neurons, provides the organism with general powers of memory manipulation and data structure construction. The rememorization process is also important because it means that the call molecule theory is capable of accounting for both short and long term memory, with the only difference between the two being that "long term" reference neurons reload primaries which recognize their call while "short term" reference neurons do not. Short term memory may be transferred to long term storage by rememorization; or some very important experiences may be directly assimilated under the control of long term storage neurons. The theory is also capable of accounting for immediate (or very short term) memory. Indeed in this case we might expect the very short reference neurons to load lower level primaries (to retain a more accurate picture of the scene) and also to use single site call molecules (so that the storage and retrieval times are very fast).

The simplicity with which rememorization takes place is of course possible because of the reference neurons. These neurons also make it possible to give an easy account of classical conditioning and classical trial and error learning in terms of the call molecule scheme. In the case of classical conditioning all that is necessary is for the primaries firing in response to a conditioned stimulus (e.g. ringing of a bell) to load the reference neuron which itself loads primaries firing in response to an unconditioned stimulus (e.g. where food is the unconditioned stimulus and the unconditioned response associated with this is salivation). In the case of classical trial and error (or instrumental) learning the organism must guess a response. Whether or not this is bound to the stimulus depends on reward and punishment. However, this is no more than associative structure formation at the simplest level, from which it follows that it is mediated by partially suppressed rememorization.

Finally we turn to the question of forgetting. This is clearly associated with the degree of rememorization (since this accounts for the stability of the memory trace). However, it is also controlled by the degree of fixation. This is because call molecules in neurons in which the degree of fixation is incomplete will have some chance of being cleared by loading calls. For example, suppose that the unconditioned stimulus is repeated but not the conditioned stimulus. In this case the primaries firing in response to the unconditioned stimulus will not load the reference neuron which loads primaries firing in response to the conditioned stimulus, with the result that the association will extinguish if the degree of fixation in this reference neuron is incomplete. Furthermore, this means that it is possible for variation in the degree of fixation to result in variation in the strategy of forgetting. Indeed, this property can be adjusted by the species in the course of evolution.

6. Interfacing the Models

Now we must consider how the call molecule scheme relates to the excitase model described in Part I. The two models are clearly complementary in a number of senses:

1. In the excitase model the neuron responds to the integrated effect of its inputs, with the response being determined by the excitases and with the integration process being biased by the structural and dynamic properties of the neuron. In the call molecule model the neuron responds to the presence or absence of an input or to the temporal sequence of inputs, with the response being determined by the loading of call molecules.

2. The excitase model mediates modification-based learning, with adaptive changes in local networks being transmitted to other (typically adjacent) regions of the brain by transforming excitase "genes". The call molecule model mediates memory-based learning, with any transmission of learned behavior from one part of the brain to another being mediated by rememorization.*

3. The excitase model allows complicated pattern recognition and output control capabilities to be embodied in a relatively small number of neurons. The call molecule model accounts for the ability of the brain to store, manipulate, and retrieve memories, including memories of complex patterns.

These two modes of operation must of course communicate with one another. However, instead of discussing this in general it is much more interesting and valuable to look at the design of a particular system (viz. the cerebral cortex) in which it is reasonable to assume that both schemes play an important and interconnected role.

The cortex is typically pictured as a layered structure with an outer horizontal layer and inner layers of complexly interconnected neurons, but with many pyramidical cells and other large neurons sending apical dendrites into the horizontal layer (cf. Elliott, 1969). It is clear that the neurons corresponding to the primaries must be fairly high level neurons, with outputs running to the outside, and wide dendritic inputs, catching inputs not only from preprocessing and modulating neurons, but in particular from the reference neurons. This suggests that the primaries include various pyramidicals, particularly those of the outer layers, and that the inputs from reference neurons reach these through the apical dendrites. The latter dip into the outermost layer of the cortex, where they make contact with the party lines. These run horizontally along this layer, but originate from the white matter, or inner layer of axons. The reference neurons themselves are presumably in the cortex, perhaps in the outermost layer, where they can both receive input from the primaries, and send outputs into the white matter. There are of course many types of cells between the inner and outer layers, with diverse interactions, and also numerous inputs which emanate from the outside and other parts of the brain. Any of these cells may be governed by excitase dynamics. However, it is striking that the most numerous cells, the microneurons, could not possibly be call molecule neurons since their inputs and outputs are too local. These are often

---

*In the case of transmission from hemisphere to hemisphere this would require some direct communication between comparable primaries.

regarded as being of high importance for modulating and patterning the interactions among the more globally interconnected neurons. Thus it is plausible that it is the microneurons which are the most significant from the standpoint of modification-based learning, particularly in view of their large number (of critical importance for the rate of evolutionary learning) and also their small size and relatively small number of inputs (which offers the maximum advantage in terms of symmetry breaking and recognition of details of the input pattern). Moreover, the fact that the number of such neurons reaches a particularly high level in man (cf. Altman, 1967) hardly detracts from the possibility that they play a major role in learning.

7. Two Systems of Learning

The call molecule and excitase systems together comprise an electromolecular computer with all the learning and computing capabilities which we originally set out to account for (cf. Part I, section 2). Actually, these capabilities are more general than those of any present day, man-made computer. This is because the structural nonprogrammability of the excitase based nervous system allows for the economical embodiment of powerful rules; but, more important, because it allows for modification-based learning.

In general, the two systems, or really modes of operation, occur together. However, each is suited for particular functions, and therefore may dominate in certain species or in certain parts of the brain. In particular, the excitase based system is especially suited to feature detection, recognition of patterns built out of these features, and for fine control of motor output. This is because of the simplicity with which pattern recognition and generation capabilities can be embodied in this system (cf. Part I, theorem 4), and also because the gradual modifiability allows for the development of altered and even novel feature detection capabilities, the addition or deletion of patterns from the set recognized by the organism, and for quantitative variation in the motor output. The call molecule system, on the other hand, is especially suited to building up a "model" of the world by accumulating and ordering records of environmental events and also records of the consequences of the organism's responses to these events. This is also useful for pattern recognition, in particular for learning to recognize patterns on the basis of only one or a few of their features (because the content ordered memory makes it possible for these features to fire the reference neuron which reconstructs other features of the pattern).

It is worth pursuing this last process a bit further. Suppose that the organism re-exposes itself to some class of environmental inputs, each time loading the reference neuron associated with some particular memory by primaries firing in response to these inputs. The particular memory may be associated with some member of the class. However, it may also be associated with a name for the class, i.e. with a pattern of firing whose association with the class is more or less arbitrary. This is of special importance since it enables the organism to name classes of inputs, to call up the appropriate name in response to particular inputs or to certain features of these inputs, and also to add new members to these classes.

The interfaced scheme also has certain unique properties. In particular, the ability of the call molecule data structures to store what in some cases amount to representations of the environment would be of no use if it were not for the extreme efficiency with which the excitase system embodies rules for recognizing and generating complex patterns, in the first instance for responding appropriately to these representations and in the second instance for using this response for supervising the reference system.

We have also seen, in Part I, that the excitase system is potentially universal, i.e. programmable from input. Humans of course can (with some struggle) follow rules written on some external notepad. In principle, the call molecule scheme makes it possible to dispense with the notepad by memorizing the rules. In practice, organisms, including man, are more likely to use the classical trial and error mechanism to learn rules for ordering calls to reference neurons which themselves call various routines embodied in the primary system.

In concluding this section, I want to emphasize that the special virtues of the excitase and call molecule systems should not obscure the fact that certain of the same functions can be performed by both systems individually. For example, responses may be bound to stimuli either by classical trial and error learning (based on the call molecule mechanism) or by evolutionary trial and error learning (based on the selection circuit mechanism). This is without doubt important for interpreting learning experiments involving either different species or different parts of the brain. This is especially so in the simpler organisms since it is unlikely that these could afford to utilize more than one type of learning system. The particular choice depends on phylogenetic history and also on whether it is more important for the system to develop rapid, qualitatively different responses to experience or slow, quantitative variation in motor control and perception. But for some simple tasks the two systems can work just as well.

Higher organisms, functioning in more complex environments, are more likely to utilize both schemes, and also the advantages of the interfaced scheme. But even here specialization of labor in different parts of the brain would be expected, e.g. with the cerebral cortex utilizing the interfaced scheme and the cerebellum relying more on the selection circuit mechanism.

8. The Problem of Language

The molecular basis of language is too big a question to pursue here. However, the information processing properties described in the previous section, in particular the individual and interfaced schemes, are particularly clear in the case of language. It is therefore worthwhile to outline some of the major points.

The human brain must of course have the capability of learning any particular language (cf. Chomsky, 1968). This means that:
(i) It must be capable of learning to recognize and generate the special sounds (or other physical properties) of any particular language.
(ii) It must be capable of adding new words and interfacing these with the previously acquired data structure.
Following Chomsky (very loosely) we also suppose that:
(iii) The brain embodies innate rules which form the basis for handling all language, and also embodies innate capabilities for learning special rules which enable it to handle particular languages.

The discussion of the previous section suggests that the special sounds of the language are learned through the selection circuit mechanism, i.e. through modification-based learning. This is because their recognition and generation involves perception of basic features and motor control. In children this learning process is remarkably rapid, from which it might be inferred that a base set of excitases sufficient for all the difference possibilities are inherited, but that some are eliminated by the selection process. This would also explain why the evolution of these various functions in later life is slow, and apparently impossible to perfect after a certain stage.

Words, on the other hand, are learned through the memory mechanism of the call molecule scheme. According to the previous section words are represented in the brain by patterns of primary activity whose relation to some class of memories is more or less arbitrary. The word acquires its "meaning" by being built into the data structures of the system, e.g. through its reference neuron being loaded by primaries firing in response to various memories or through primaries (whose firing represents the word) loading reference neurons associated with appropriate memories. Also, it is possible that certain environmental inputs fall into innate classifications, that these classifications are themselves more or less arbitrary patterns of primary firing, and that in this case the learning process associates words with the classifications rather than with the memories directly.

The innate rules which form the basis for handling any language are presumably economically embodied in some innate enzymatic neural net. Roughly speaking, these must enable the supervisory system to manipulate or search through the call molecule data structure in a way which can be transformed into any particular language by any particular set of special rules. For example, imagine that the organism is exposed to some input, that the primaries firing in response to this input call the reference neuron associated with an appropriate word, that this in turn calls up some memory, and that the primaries associated with this memory call some other reference neuron. In this way the manipulation of the call molecule nervous system by the rest of the brain is constrained so that it both generates sequences of words and is itself constrained by this word generation process.

The special rules themselves must of course be learned. In principle this could involve modification of the innate rules (based on the selection circuit mechanism) or the assimilation (through the call molecule mechanism) of either encoding and decoding operations which determine how certain innate routines are used. The advantage of such assimilation processes is that they allow the system to utilize innate learning algorithms, e.g. to memorize particular linguistic experiences and perform computations on these to extract the special rules. Moreover, special rules associated with different languages can be isolated in this case since they are accessed through reference neurons.

The innate rules, the special rules, their relation to one another, and also the mechanism by which the special rules are acquired is a complex and controversial problem in modern linguistic theory. Thus it would seem to be particularly valuable to look at this problem from the standpoint of our already constructed molecular models, in particular by paying close attention to their implications for the rate at which different types of language skills are acquired. But, of course, we have here only outlined the approach.

9. Predictions of the Theory

The predictions of the excitase theory were described in Part I, section 8. Now we add to these some testable predictions of the call molecule theory:

1. <u>Nontransferability of memory-based learning</u>. The call molecules are the basic memory units in the brain. The conformations of these molecules can be set by the sequence of pulses impinging on the dendrite and in such a way that they later recognize these sequences. Thus, memory and memory-based learning cannot be transferred by macromolecules, either from one organism to another, or from one part of the brain to another. This is not so for the excitase theory, which allows transfer under certain restricted circumstances. This distinction is important since it means that attempts to transfer may succeed if conditioning is modification-based, but not if it is memory-based.

2. <u>Differential effects of chemical agents on long and short term memory</u>. The stability of the memory trace is based on duplication of call molecule conformations, e.g. through rememorization. Thus the call molecule theory accounts for both short and long term memory, the difference being that long term storage neurons reload the primaries which recognize their call. This means that call molecules in dendrites loaded by long term reference neurons are in short supply in the loadable form (since they are periodically reloaded and presumably also strongly fixed). This is important since it explains why agents which inhibit protein synthesis interfere with assimilation into long term memory but not into short term memory (Barondes, 1970; Agranoff, 1973). Inhibition of nucleic acid and protein synthesis should also interfere with modification-based learning (which is, of course, another form of long term change). However, the relative slowness of this type of learning might result in the interference being swamped by other effects.

3. <u>Effects of cooling</u>. Extreme reduction in temperature should not destroy either short or long term memory, since both depend on macromolecular conformations which are intrinsically stable at low temperatures (assuming that the shock associated with cooling does not cause the dendrite to clear weakly fixed call molecules).

4. <u>Locus of memory</u>. Memories may be accessed by stimulating reference neurons or primaries which call reference neurons. However, the memory trace is in general distributed over many primaries, and ordinarily with redundancy (since this is a requirement of the excitase theory and also necessary for subconscious rememorization). Furthermore, the redundancy may be increased by assimilation under the control of more than one reference neuron. Thus the call molecule theory accounts for both the distributed character of memory (cf. Lashley, 1963) and the accessibility of memories by stimulation of specific brain loci (cf. Penfield and Perot, 1963).

5. <u>Lesions of the brain</u>. The distributed character of the memory trace,

particularly the redundancy of primaries and the use of multiple reference neurons, makes the data structures of the brain resistant to ablation, radiation, and other forms of damage. This is of course not true for lesions affecting the supervisory system. Also, in the cerebral cortex vertical cuts through the outermost layer (severing the party lines) should interfere with recall and, if sufficiently complete, with future acquisition of memory. The effects of ablation on modification-based learning are quite different, since this does not depend on the transfer of information over long distances (but it is sensitive to lesions affecting the flow of information from the evaluation and growth control systems).

6. Mass action. As with modification-based learning ablation will often retard the rate of learning relative to a specific function (in agreement with the mass action effects of Lashley). In this case, however, the reduction arises because the probabilities for particular reference-to-primary and primary-to-reference loadings are reduced.

7. Plasticity and specificity. The excitase theory required interchangeability of adjacent networks (in a given type of tissue), and therefore at least statistical homogeneity. Thus the plasticity resides in the excitases in such networks and not in their morphological properties. This means that the particular structure of the networks makes no difference, as long as they are well connected and satisfy the requirements of homogeneity. The call molecule theory requires wide connectivity, but other than this imposes no requirements besides those associated with the party lines. This means that there is no requirement for interchangeability in this case, and therefore the model allows for plastic changes in the morphological properties of the synapse and even the connectivity. These differing requirements suggest that parts of the brain which are more likely to be excitase based (such as the cerebellum) should be more homogeneous and less plastic, whereas those which are based on call molecules should be quite plastic and perhaps less homogeneous. Regions based on both call molecules and excitases would appear plastic or not depending on which neurons and which dendrites are being examined. For example, long axon macroneurons could utilize excitases as well as call molecules provided that the plasticity is restricted to the call molecule containing dendrites. The situation should be reversed as regards glial cells, however, since the number of these would be more variable in the excitase based regions (cf. Part I, section 8). In fact, these varying requirements are quite consistent and indeed given an interpretation for the contradictory reports about plasticity and specificity.

The model also has the general information processing capabilities discussed earlier in the paper, including: general powers of memory manipulation, manipulation of scenic memories, capabilities for developing time ordered, content ordered and associative memory structures, capabilities for classical conditioning, instrumental learning, rule assimilation, and algorithmic learning. Many of these properties would not be feasible on the basis of the call molecule theory alone since they require the power and economy of the excitase-based nervous system.

Finally, the modifiability of the excitase based nervous system means that we have a model of the brain with general powers of computation, but one which also admits of alternative physical realizations. What happens if the organism loads certain patterns of primary firing and then undergoes a modification in its excitase nervous system? The reference neuron which did the loading will still be able to fire these primaries. Thus the modification will not interfere with memory reconstruction as long as the pattern to be reconstructed does not depend on the interaction with neurons whose excitase populations are modified. However, suppose that the organism learns to respond to a particular stimulus by the classical trial and error mechanism. It may then adjust either the stimulus which it recognizes or the response which it has associated with this stimulus by the evolutionary trial and error mechanisms, i.e. by changing excitase populations so that the inputs which fire certain primaries are changed or the neurons firing in response to certain patterns of primary firing are changed.

In short, the interfaced model implies that the brain is capable of learning rapidly on the basis of memory, and then slowly modifying what is learned on the basis of evolution, without interfering with the already assimilated data structure (cf. Conrad, 1973).

10. Conclusions

There is increasing, though still controversial, evidence that nucleic acids and proteins play a crucial information processing role in the brain (e.g. Ungar, 1973; McConnell and Shelby, 1970; Hydén and Lange, 1971; Cohen, 1970). But it is of course also virtually axiomatic that the electrical behavior of neurons plays a role as well. Thus, at the present time, it would seem to be particularly useful to take the hypothesis of crucial molecular function seriously and see what kind of brain models can be built using known molecular processes in the most natural way, but at the same time keeping a role for the neuron and satisfying the conceptual boundary conditions (appropriately formalized) which ought to be satisfied by any brain model.

In both our models macromolecules control the nerve impulse. In the excitase theory this control is based on the inherent specificity of molecules (which, in particular, determines where they bind on the nerve membrane). This theory allows for networks of high economy, and moreover networks which are capable of learning on the basis of mechanisms which are analogous to those which mediate natural evolution. In the call molecule theory, on the other hand, the control is based on conformation changes which are themselves regulated by the occurrence or nonoccurrence of nerve impulses. This theory accounts for memory acquisition and memory-based learning, and also interfaces with the excitase theory to give the brain information processing powers which exceed those of any present day artificial device.

The theory also utilizes the control of enzyme synthesis (e.g. enzyme induction), but not to control the nerve impulse. Instead, this (and perhaps also other forms of regulation) controls the operations of the selection circuits required by the excitase theory.

The question may arise as to whether we have done justice to the classical electrophysiology of the neuron. The author is certainly aware that the molecular functions postulated in his theory have not been observed (as yet) in electrophysiological investigations of the neuron, nor do they arise in any natural way out of current electrophysiological theories based on studies of the peripheral nervous system. However, it does not seem to him that this is sufficient reason to deform either the fundamental principles of molecular biology or our fundamental formal descriptions of the brain's functional capabilities in order to preserve the present neuron concept, especially at the level of the brain.

References

Agranoff, B.W.: Biochemical approaches to learning and memory. In: Macromolecules and Behavior, ed. by G.B. Ansell and P.B. Bradley pp. 143-149. MacMillan, London, 1973.

Altman, J.: Postnatal growth and differentiation of the mammalian brain with implications for a morphological theory of memory. In: The Neurosciences, ed. by G.C. Quarton, T. Melnechuk, and F.O. Schmitt, pp. 723-743. Rockefeller University Press, New York, 1967.

Barondes, S.H.: Some critical variables in studies of the effect of inhibitors of protein synthesis on memory. In: Molecular Approaches to Learning and Memory, ed. by W.L. Byrne, pp. 27-34. Academic Press, New York, 1970.

Chomsky, N.: Language and Mind, Harcourt, Brace, and World, New York, 1968.

Cohen, H.D.: Learning, memory and metabolic inhibitors. In: Molecular Mechanisms in Memory and Learning, ed. by Georges Ungar, pp. 59-70. Plenum Press, New York, 1970.

Conrad, M.: Is the brain an effective computer? Intern. J. Neuroscience 5, 167-170 (1973).

Elliot, H.C.: Textbook of Neuroanatomy, J.B. Lippincott, Philadelphia, 1969.

Hydén, H.: Biochemical changes accompanying learning. In: The Neurosciences, ed. by G.C. Quarton, T. Melnechuk and F.O. Schmitt, pp. 965-971. Rockefeller University Press, New York, 1967.

Hydén, H. and Lange, P.W.: Time sequence analysis of proteins in brain stem, limbic system and cortex during training, Biochimica e Biologia Sperimentale 9, 275-285 (1971).

Lashley, K.S.: Brain Mechanisms and Intelligence, Dover, New York, 1963.

McConnell, J.V. and Shelby, J.M.: Memory transfer experiments in invertebrates. In: Molecular Mechanisms in Memory and Learning, ed. by Georges Ungar, pp. 71-101. Plenum Press, New York, 1970.

Penfield, W. and Perot, P.: The brain's record of auditory and visual experience: a final summary and discussion, Brain, 86, 595 (1963).

Smith, C.E.: Is memory a matter of enzyme induction? Science 138, 889-890 (1962).

Ungar, G.: Molecular approaches to neural coding. In: The Physical Principles of Neuronal and Organismic Behavior, ed. by M. Conrad and M. Magar, pp. 167-176. Gordon and Breach, New York and London, 1972.

DISCRETE AND CONTINUOUS PROCESSES IN COMPUTERS AND BRAINS

H.H. Pattee
Center for Theoretical Biology
and
Department of Biophysical Sciences
State University of New York at Buffalo

1. Nerve Cells and Switches

Theories of computation and theories of the brain have close historical interrelations, the best-known examples being Turing's introspective use of the brain's operation as a model for his idealized computing machine (Turing, 1936), McCulloch's and Pitts' use of ideal switching elements to model the brain (McCulloch and Pitts, 1943), and von Neumann's comparison of the logic and physics of both brains and computers (von Neumann, 1958).

The basis for this historical relationship, as well as for the vast literature on computers and the brain produced since then, is the assumption, for better or for worse, that the nerve cell functions something like a discrete switch. On the experimental side, this assumption is supported by anatomical and electrophysiological measurements showing that nerve cells form complex networks and communicate only by "all or nothing" discrete input and output pulses that can be related by a relatively simple, but modifiable, switching or transfer function. On the theoretical side, this assumption is supported by the universality of simple switching elements for representing any finite symbol manipulating process which can be translated into an effective algorithm or program.

Adding significantly to the credibility of the switch-neuron analogy are the specific observations of neural coding networks such as feature detector zones in the optical system, (e.g., Hubel and Wiesel, 1968), whose general type of functional behavior can be simulated by artificial sensing arrays coupled to discrete computers, (e.g., Guzman, 1968; Duda and Hart, 1970). All this recent progress in neurobiology has led many to conclude that "the immense task of understanding the neural basis of perception is only immense; it is no longer incomprehensible" (Handler, 1970). Similarly we find many computer scientists who believe that, given enough switches, creating a truly intelligent, "thinking" computer is only a matter of programming these switches to behave in an intelligent way.

2. Are Discrete Switches Sufficient?

The question I want to raise is whether this concept of the discrete switch will prove sufficient as a basis for understanding intelligence in brains and creating intelligence in computers. There is no question that the analogy of the neuron and the discrete switch is fruitful and very likely sufficient for all permanently established functions or models in the brain, where no creative intelligence would serve any purpose; (e.g., in sensory and neuromuscular systems which interact directly with the real physical world). One has no need of learning to move or see in another physical world. I am only talking about intelligence as generation of new internal models. In other words, I am talking about how internal models of the world are first created and interpreted, not how they are finally represented or embodied. One we have a well-defined model, it certainly can exist as a discrete structure (i.e. as a symbol structure), but I agree with Emil Post (1965) that symbols are <u>created</u> in continuous dynamical time, and are only <u>preserved</u> in discrete, aribtrary structures.

Therefore, I have no doubt that the discrete automaton can be used to represent or embody all well-defined sensory, neuromuscular or computational functions. In other words, when we can say exactly what we mean, then we are in the discrete symbolic mode, characteristic of the inputs and outputs of switching nets. This mode is programmable by discrete symbol systems, such as punched cards or tapes, and can approximate the behavior of many continuous dynamical systems. I expect, then, that this discrete symbolic mode is essential for clear descriptions and constructions and will continue to be one essential form of description for nervous systems. It is also clear that automata theory and real computers will continue to depend on formal and concrete switches respectively. My question is whether there are other important modes of behavior that can not be adequately simulated by discrete events in automata. In other words, does all interpretation and intelligence involve only discrete events? Can we expect to account for, or simulate, the more intelligent problem-finding, pattern-interpreting or linguistic performance of the brain simply by more immense networks of discrete switches and discrete memories?

Now one might ask immediately, "What other type of description is there?" All concrete statements of our mathematical languages including continuous functions and variables are specified by a finite, discrete set of symbols and rules. Consequently, even a continuous dynamical system, such as the motion of several mass points in a potential field is "solved" in practice by approximating the values of the continuous variables over a discrete mesh, and representing the mesh behavior by an automaton. Furthermore, even our natural languages are made up of discrete, finite elements so that one could argue that all descriptions of continuous processes must be representable in some form by a finite, discrete sequence of finite elements.

One could also look at the phenomenological level of either the nerve cell or the switch and argue that the microscopic continuous dynamics is totally suppressed by the discrete functions of these units. Even cellular functions at the biochemical level appear to be adequately described by discrete sequences of specific reactions catalyzed by individual molecular subunits.

Thus both the automata theorist and molecular biologist provide very convincing evidence that discrete switching behavior has universality as an elementary concept for building abstract descriptions as well as a basic unit for building and controlling complex organisms.

To most physicists and mathematicians, on the other hand, the switch has never appeared as a fundamental unit, although the question of whether continuity or discreteness is more fundamental has always created problems and paradoxes. Von Neumann (1958) was one mathematician who took switches and neurons seriously. He said in the introduction to the Computer and the Brain

> "I suspect that a deeper mathematical study of the nervous system - 'mathematical' in the sense outlined above [as a real computer] - will affect our understanding of the aspects of mathematics itself that are involved. In fact it may alter the way in which we look on mathematics and logic proper."

Furthermore, he did not regard computer theory as abstract automata theory, but as a theory of real, physical devices. He says in his Theory of Self-Reproducing Automata,

> "By axiomatizing automata in this manner one has thrown half of the problem out the window and it may be the more important half. One has resigned oneself not to explain how these parts are made up of real things, specifically, how these parts are made up of actual elementary particles, or even of higher chemical molecules. One does not ask the most intriguing, exciting, and important question of why molecules or aggregates which in nature really occur in these parts are the sort of things they are..."
>
> (von Neumann, 1966)

Von Neumann was well-aware of the basic principles that prevent physicists from regarding the machine, including the switch, as just a simple objective physical device (e.g., Pattee, 1972b). Many mathematicians and biologists still do not understand the reasons for this point of view. Even Turing (1956) in his article, Can a Machine Think?, and Gödel (1964) in equating "finite procedure" with "mechanical procedure" imply that relating thought or formal logic to a "machine" would constitute some fundamental reduction or explanation of higher mental processes.

3. A Physical View of a Switch

To a physicist, however, machines of all kinds, including switches, cannot be a fundamental explanation of anything. The physical parts of machines do, of course, obey laws of physics, but the concept "machine" is defined by its function, i.e., a machine performs useful work, or is a kind of prosthetic device for making man's work easier or simpler. The essential conditions are "useful,""easy," and "simple," and these are concepts that in no way can be explained by physical laws. In other words, a machine can only be defined by its constraints, not by equations of motion (cf. Polanyi, 1968).

Insofar as a physical system can be recognized as a switch it is not an objective element, but is itself a model - a subjective interpretation made through some external system with which it interacts. There is no question that all switching behavior must possess an underlying physical dynamics, but the switching function itself cannot be an inherent property of this dynamics (Rosen, 1969). In fact, it is the selective disregard of the detailed dynamics (selective dissipation), which must originate through an outside "selector," that encodes the dynamics into the switching function. It is this outside selective encoding that creates the switching behavior and suppresses the dynamical details. This outside agent has, in effect, created its own internal model of the dynamics, and it is this model which we recognize as the switch. The discrete switch and its continuous dynamics therefore comprise _complementary_ descriptions of a two-level relation which cannot be understood by either description alone. This relation is essentially an "epistemological" relation of object-to-subject, matter-to-symbol, event-to-measurement or system-to-model, however you wish to express it.

4. A Philosophical View of a Switch

One could argue, on the other hand, that a switching brain is more fundamental than continuous dynamics because the concepts of continuous motion, and continuous space and time are really the abstract models created by the patterns of discrete switching activity of the brain. According to this point of view, the discrete switch is a fundamental objective element out of which is built the subjective interpretations of experience which includes the abstraction of continuity and hence physical equations of motion.

The logic of this somewhat solipsistic view appears sound, but I find this latter interpretation contrary to the more or less commonly accepted empirical facts of evolution. In particular, it presupposes the existence of brains as switching networks, whereas I find it more consistent with the historical view of the universe to presuppose the existence of space, time and matter. In other words, I find evidence which associates switching functions with living systems in an essential way, and I also find evidence that life did not exist on earth five billion years ago. Therefore I regard the origin and evolution of discrete switches as an evolutionary step that may be explained beginning from our dynamical view of non-living matter. I also find it natural to think of switching descriptions as a simplification or abstraction of the underlying dynamics, whereas I cannot conceive of the relation the other way around.

Before going into more detail, let me restate my general argument why discrete automata may be insufficient to generate intelligent behavior in brains or computers. By intelligent behavior, I mean the ability to create internal models of the world, on which predictive decisions are based. I see the discrete switch function as the primary <u>result</u> of this creative process, not the cause of it. The essential act of intelligence is the abstracted, simplified encoding of continuous, real-time dynamical trajectories into a discrete, switching <u>function</u>. Discrete programs instructing these discrete switches can, therefore, function as models of continuous dynamical systems. However, their function as models is not an inherent property of these automata, but require an interpretation to relate the automata to the dynamical system. For this reason, I would not expect a formal switching net, nor a real switching net with "hard" switches (i.e., isolated from their dynamical matrix) to be capable of recognizing or interpreting dynamical systems. This relating of discrete symbols to continuous dynamics is a crucial form of intelligence.

What I feel must be distinguished is the end <u>result</u> of intelligent activity, which is usually a formal symbolic structure capable of being preserved by a discrete switching network or "mechanical device," and the intelligent activity itself - the <u>creation</u> and <u>interpretation</u> of these symbolic structures.

5. An Operational View of a Switch

To make these general ideas clearer I shall take the simplest example of a switch that I can imagine. I shall define it operationally then represent it as a formal, discrete function, and then describe its simplest underlying dynamics as far as that is possible. Next, I consider the necessary conditions for its physical embodiment and the stages of its evolution. Finally, I suggest biological examples at several levels of organization where I believe complementary discrete and continuous descriptions are necessary to understand the creative aspects of both evolution and intelligence.

What are the operational characteristics of a switch? First, it must have at least one time-independent stable position or state. Of course, this "time independence" is only with respect to the switching variable, which in macroscopic switches may be better described as a stable equilibrium. Second, there must be some selective degree of freedom, which when perturbed by an external force, called the trigger, results in a relatively sudden change of state called the action. The <u>threshold</u>, which depends on the sensitivity and selectivity of the switch to the trigger perturbation, is a crucial aspect of the interpretation or <u>function</u> of the switch. Typically, a good switch will not respond to perturbations below a certain threshold which is well above noise, while on the other hand, the strength of the trigger is less than the strength of the action produced. Also, perturbations of the non-selective degrees of freedom must not trigger the action. Finally, the switch should be useful more than once, which means it must either reset itself or be resettable to the original time-independent state by an external force. It is implied, of course, that there is a definite "rule" of operation, that is, that all previous or subsequent triggers would produce the same action. A switch function is summarized operationally, then, by the following:

1) At least one time-independent stable configuration.
2) A selective, sensitive, but reliable trigger which causes a sudden change of state or action.
3) A return to the initial state - reset, repeatability.

6. A Discrete Logical View of a Switch

The formal behavior of a switch might be abstracted from this description by calling the resting state, S_o, the action, S_1, the absence of a trigger, 0, and the presence of a trigger, 1. Then one particular switch is defined abstractly by the following Table 1:

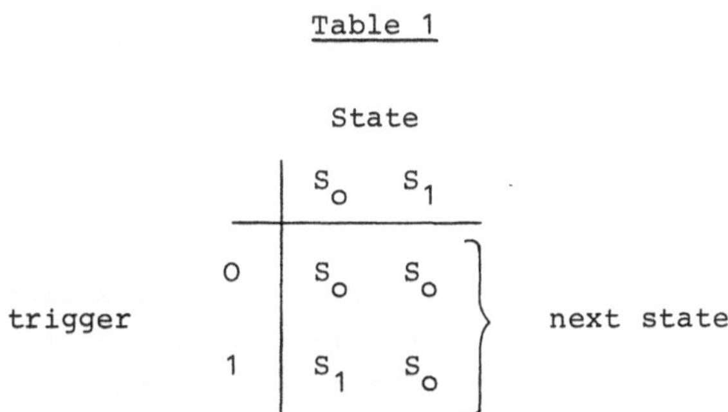

Table 1

		State	
		S_o	S_1
trigger	0	S_o	S_o
	1	S_1	S_o

(rows for trigger 0 and 1: next state)

This switch resets itself since if it is in the action state S_1 it returns to S_o with or without a trigger. Therefore, during the time it takes the switch to reset, it is not sensitive to the trigger.

Next we might ask what the physical embodiment of such a discrete, formal switching function could be. Clearly this is not itself a formal question. There are innumerable physical systems which can be _interpreted_ as functioning in this way, including mechanical, electrical, hydraulic, or molecular devices of all sizes and shapes. So this is not a very useful question as it stands. But what is it we would like to know beyond the formal switching table? Is the formal discrete abstraction the only universal description of this switch. Are all other more detailed descriptions arbitrary and hence extraneous to the function? This is the important question. If we are using the language of formal logic or computers we find that all details lose their significance below the functional level. How a truth table is actually executed, or how a switch is actually constructed simply has no bearing on the outcome. But the situation is more drastic than this. If in fact a formal switching function were not completely independent of its detailed embodiment, that is, if we had to know some additional details before we could define the switching behavior, then we would either call it a bad switch, or we would say that we have misrepresented it (i.e., it really has a different switching function than what we thought).

What we have, in fact, done in writing this formal description of a switch is to define what we are going to mean by _input_ and _output_, _trigger_ and _state_. We have created this switch by this definition alone, without reference to any physical device. If any real dynamical system behaved like this switch it is because of our _interpretation_ relating selected physical observables to the symbols in the formal definition. But the formal switch by itself has no underlying dynamics, hence the formal switching function is independent of dynamics by definition.

On the other hand, it is a meaningful question to ask what kind of continuous dynamical system could be reasonably interpreted as such a switch. What are the general conditions which allow a continuous real-time dynamical system to be interpreted as a formal switch with a particular switching function, such as given by Table 1?

7. A Continuous Dynamical View of a Switch

It is significant that the oldest dynamical descriptions of switching behavior (Blair, 1932; Rashevsky, 1933; Hill, 1936) are in fact models of nerve cell excitation. These are phenomenological models with no direct attempt to describe the underlying chemical variables as in the Hodgkin-Huxley equations. More recently E.C. Zeeman (1972a), at the suggestion of Francis Crick, applied the theory of the cusp catastrophe of René Thom (1970) to derive the simplest dynamical description of switching behavior, again, using the nerve cell as one embodiment of the switch function.

It is also significant that these continuous dynamical models of discrete behavior of nerve cells were motivated by the assumption that the discrete switching function is the basic element of all nervous activity and that once this function is successfully modeled by a dynamical system, then one can dispense with the dynamics in all higher level models of nervous function. In other words, the problem in these treatments is limited to the explanation of how continuous dynamical processes that occur in the soft chemical cell can result in the discrete functions of a hard switch. To the extent that the brain functions like a discrete switching network, these dynamical models of switches are a useful reductionist description, and this is what they were intended to be. Our aim, on the other hand, is to suggest why the discrete switching description, while simpler and often more practical than the continuous dynamical description, cannot stand alone as the basis for its own creation or its own interpretation.

Zeeman (1972a) has presented the simplest possible continuous dynamical system that can be interpreted as a switch, satisfying both the operational conditions described in Sec. 5 and the logical switching function given in Sec. 6. To appreciate in what sense this is the simplest dynamical system with these switching properties and its mathematical relation to Thom's canonical cusp catastrophe, one must read Zeeman's paper.

The system consists of a "fast foliation" equation, $\varepsilon\dot{x} = -(x^3+ax+b)$, where $\dot{x} = dx/dt$, and the two parameters a and b are interpreted as coordinates; x,a,b forming a 3-dimensional space. A "slow manifold" is formed by giving a and b dynamics, $\dot{a} = -2a-2x$ and $\dot{b} = -a-1$. This slow manifold forms the folded surface with a cusp (Fig. 1) which is a single-sheeted attractor except for the three-sheeted region inside the cusp with the middle sheet a repellor. The interpretation is as follows:

The equilibrium or time-independent stable state is given by $\dot{x} = \dot{a} = \dot{b} = 0$. A trigger is an external force moving b over the edge of the top attractor sheet. This edge is the threshold beyond which the fast foliation equation produces a sudden action, since the middle sheet inside the cusp is a repellor. When the trajectory reaches the bottom attractor, it returns slowly to the initial state by winding up around the cusp, resetting the switch.

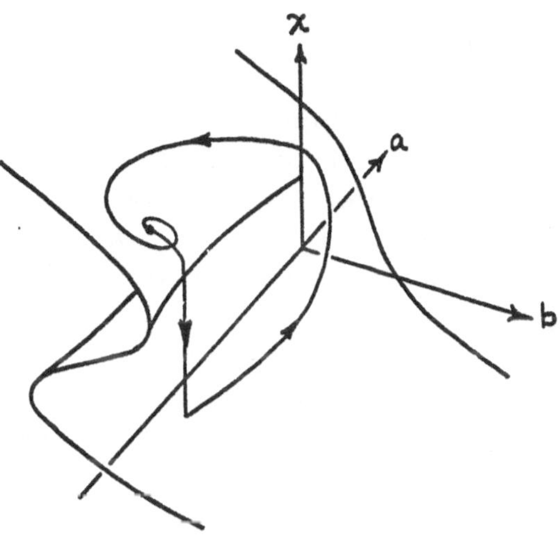

Fig. 1

The formal interpretation requires a definition of the states S_o and S_1. These states are the "resting" state and the "action" state, so we may simply divide the slow manifold into two regions such that one, S_1, contains the upper edge of the cusp where the action takes place, its boundary being such that all trajectories leaving S_1 enter S_o; and the other region, S_o, containing the equilibrium point, its boundary being such that no unperturbed trajectory leaves that region. A trigger is any displacement leading from the S_o region across the upper edge of the cusp. This leaves a large arbitrariness in the shape of the two regions corresponding to S_o and S_1. Perhaps more important is the arbitrariness in the details of the dynamical equations that can be interpreted as such a switch. The basic requirements are the region of equilibrium, the cusp topology, and the reasonably simple, fast foliation and slow manifold trajectories. This arbitrariness is important for any practical, reliable switch which must be relatively insensitive to design or construction variations, as well as external perturbations. This approximates the ideal switch which is sensitive only to the trigger.

8. A Dissipative View of a Switch

The mention of an "ideal switch" raises fundamental physical and logical problems. We have explained how a continuous dynamical system can be interpreted as a discrete switch with a logical behavior as described in Sec. 6. Now we must go one step further and ask, "Can any real physical system be described by the dynamical equations of Zeeman that we have been interpreting?"

Zeeman himself describes a very simple catastrophe machine as an appendix to his mathematical description (Zeeman, 1972b). It consists of a disk with a fixed pivot, with two elastic bands attached at one point of the disk's circumference. The other end of one elastic band is fixed two disk-diameters away from the pivot. As the other band is pulled over a two dimensional surface, parallel to the disk, the switching behavior of the disk will generate the cusp.

If one constructs this machine with great care, using fine bearings and elastic springs, so that there is very little friction, it will not return to equilibrium very quickly after being triggered but will undergo damped oscillations. Under these conditions, we must restrict the interpretation of S_o to a region very close to the equilibrium point, otherwise the trigger threshold will be phase or time dependent. In other words, the switch will be sensitive to a great variety of input triggers that are not independent of the detailed dynamics. Such a device that does not effectively suppress the continuous dynamics would not be interpreted as a reliable switch.

This illustrates the fundamental physical condition imposed on all switches, or more generally, on all logical operations, that for every discrete switching event (or equivalently, for any record, measurement, classification or decision process) there must be a corresponding dissipation of energy. The absolute minimum cost is kT of dissipation for each binary switching event or bit: but all real switches, including neurons, dissipate orders of magnitude more energy. This is to assure greater reliability. It also turns out that dissipation increases with the speed of the switch (e.g. Landauer, 1961; Brillouin, 1962; Keyes, 1970).

Therefore, the concept of the ideal "mechanical device" or "formal procedure" is no more realizable than the perpetual motion machine, and it is a serious physical and logical error to base any evolutionary theory of biological function only on such ideal elements, since natural selection does not operate on ideal switching functions, but on the real speed and reliability of the decision making dynamics. The same type of selection would also appear necessary for intelligent learning.

The only condition where the concept of the ideal switch is sufficient is in an ideal universe where space, time, matter and energy play no direct role. The most remarkable, and indeed, the most frightening aspect of modern computer technology is how closely it has approximated such an unnatural universe. However, one should constantly bear in mind that the only link between a computer and the real world (other than through the power mains and the air conditioner) is with the brain of a human programmer; and this link is accomplished only with alternative modes of description that presuppose a high level of creative intelligence on the side of the programmer. Yet we know that the practical and universal behavior of computers depends on this isolation of its hardware from everything but the programmer. The discrete mode of symbolic description which is characteristic of formal automata theory therefore appears to be necessary for precise description.

In my opinion, the trouble with discrete automata models of intelligence is that theorists have mistaken this necessity of isolation for its sufficiency. The interpretive or dynamical mode, since it is in the brain of the programmer, is left out.

9. An Evolutionary View of a Switch

What has happened with the evolution of computers is that the discrete sequential and switching mode has been artificially selected as the only significant behavior. It is then up to the programmer to translate every other mode of behavior into the particular discrete mode of the computer. The discrete symbolic mode in artificial computers has, in effect, been lumped into one box called the hardware where no continuous dynamical interactions or interpretation is allowed. Only as programs are written or as the output is read can the activity be interpreted or interact with the physical world.

Living systems also keep their discrete switching modes separate, but do not lump them all together. Instead, we find the discrete modes interspersed with dynamical modes at many levels. In fact, an essential requirement of evolution is the clear separation of the discrete genetic instructions from the phenotypic constructions. But this discrete genetic mode is also interpreted dynamically even at the subcellular level. For example, the cell's enzymes are constructed by a discrete code that translates the sequence of DNA bases to a sequence of amino acids; but the linear discrete string of amino acids does not function as a selective catalyst until it folds up into a three dimensional active form. This folding is not programmed, and can only be described by continuous dynamical interactions of the entire string. Going one stage further, the recognition of the substrate by the enzyme also requires a dynamical description with many degrees of freedom, but once the substrate is bound there follows the sudden action of the catalytic event, which has all the characteristics of the discrete switch we have been discussing.

This complementary alternation and interaction of discrete symbolic modes with the continuous dynamical modes goes on at all levels of biological organization from the enzyme structure-function levels up through the sensory and neuromuscular structure-function levels. One is inclined then to ask if this complementary discrete and continuous interaction is not also essential for the higher forms of intelligent behavior.

In particular, since the creative aspects of evolution are our best analogy to creative intelligence, we should ask if the process of evolution illustrates some general principles relating discrete and continuous modes. Indeed, a fundamental fact of Darwinian evolution is that the mutation or search process begins in the discrete symbolic mode, but that the interpretation of these symbols (i.e., selection) is a dynamical interaction in real space and time. For example, consider the enzyme again. Its evolution requires a discrete change in the linear sequence of its genetic <u>description</u>, not in the enzyme itself. There is no real-time or continuous process, such as the folding of its DNA description or <u>rate of reading</u> the sequence that has any bearing on this discrete mode. The genetic hardware is well isolated from real-time dynamics. On the other hand, whether or not such a change in description is selected depends first on how the new enzyme folds up. The folding process, as we said, is not a part of the enzyme's description but is a real-time continuous dynamical interaction involving many degrees of freedom. Of course, this folding is only one step in the selection process, but we may also call it one step in the interpretation of its description, the translation by the genetic code being the first interpretive step. At the next functional level this same enzyme may be interpreted further as a discrete switch in a network of catalysed reactions and hence may also be selected on the speed, specificity and reliability of its switching function in this network (e.g., Kauffman, 1969).

Thus the evolutionary view of switching function is complementary to the physical view. We see that from the physical view the switch function is a discrete interpretation of an underlying continuous dynamical system. From the evolutionary view, on the other hand, we see that the selection process is a dynamical interpretation of the underlying discrete switch-like genetical descriptions. Zeeman's description of the nerve pulse is one example of how continuous dynamics creates discrete activity, and the synthesis of the enzyme is one example of how discrete activity constrains the continuous dynamics of folding. This folding in turn creates the dynamical interactions that result in the switch-like selective catalysis characteristic of enzyme function at the next level.

10. Dynamics, Language and Intelligence

These few simple examples illustrate very general properties of discrete systems and their relation to continuous systems. As we pointed out in Sec. 2, all types of descriptions from the gene to our highly evolved languages are represented by sequential elements and sequential operations. Formal automata theory and real computers represent extreme efforts toward isolation from real dynamical behavior. Since formal automata processes are performed in the brain as well as in machines there must be switching behavior in the brain which also appears isolated from real-time dynamics.

On the other hand, we have argued that switching behavior, while suppressing continuous dynamical interactions cannot be entirely isolated from dissipative fluctuations, and even more important, cannot originate or have any evolutionary significance without interfacing with lower and higher levels of dynamical interactions. Specifically, we associate the origin of switching behavior with singularities in a continuous dynamical system and the significance or interpretation of switching behavior as a dynamical "folding" process that is essentially time-dependent and continuous. This interpretive dynamical mode may itself exhibit singularities that generate a new level of discrete description, and so this process may be repeated, producing a hierarchy of discrete symbolic levels interfacing on the lower generative side as singularities of a continuous matrix and on the upper side as constraints within a larger dynamical system (Pattee, 1972a).

We have used only the most primitive levels of biological organization, the gene and enzyme levels as examples. How would this discrete-continuous interplay apply to the highest levels, that is, to intelligence? How do discrete and continuous interactions apply to the higher levels of linguistic performance, for example? At this level we have very much less knowledge than we have of the molecular levels, and almost all of this knowledge is derived from the discrete mode of symbol manipulation and the transformation or coding to other discrete symbol structures. In other words, most of our knowledge of language is based on observations at the syntactical level. When we look on either side of this level, that is, at the origin and generation of language on the one side, or the interpretation of language on the other, our knowledge becomes vague or non-existent.

What little we do know of the creative level, however, does not support any theory based on the manipulation of well-defined discrete symbols. The creative act in all fields, according to the introspection of those who create, appears as a sudden flash, coherence, or harmony of indistinct psychical entities that, at first, cannot consciously be recognized or manipulated by the creator at the discrete verbal level. Only after much effort can these creations be transformed into formal expressions. We recall the well-known letter of Einstein to Hadamard (1945) where he says, "The words or the language as they are written or spoken do not seem to play any role in my mechanism of thought," and Einstein describes the pre-linguistic entities with which he plays as "of visual or muscular type." Poincaré (1913) was also clear about the essential role of the non-logical level that generates the "sudden illumination" only after an extended period of "non-mechanical" activity. The recognition of "harmoniously" disposed elements Poincaré attributes to the "delicate sieve" of "esthetic sensibility."

Although these examples can be multiplied over and over, and are characteristic of creation in music, literature and art as well as physics and mathematics (e.g., Ghiselin, 1952), one cannot prove from introspection alone that the underlying unconscious level is not some kind of discrete network. All we can say is that taken at their literal value, these descriptions of the creative process suggest a picture of singularities arising in a dynamical sea of ideas rather than the output of a discrete automaton.

On the other side of the discrete symbol level of language - the reading and interpreting of statements - we know that translation of such discrete symbols most often results in the performance of some dynamical action. For example, when we see the muscles actually execute a task according to instructions, we know that some discrete pulse patterns have been converted through an intricate set of constraints to a continuous dynamical action. We have evidence that the message in this case lies in the statistical distribution of discrete pulses and not in the detailed pulse timing, and this leads to continuous dynamical models of nerve pulse interpretation depending on our choices of how pulses are averaged (e.g. Griffith, 1963; Harth et al., 1970; Anninos et al., 1970; Cowan, 1972).

However, as we have said, motor control is relatively simple and fixed. What we know of the interpretation of higher levels of language does not fit either a discrete automaton model or the statistical dynamical models with only simple averaging. Our present automata and statistical models are actually extreme cases of interpretation themselves. At the automata extreme we assume that each elementary symbol is completely interpreted by its input, output or state specification as we indicated in Sec. 6. One, therefore, cannot discover new observables within such automata (Rosen, 1969).

On the other hand, to arrive at any statistical dynamics of a discrete system also requires a definite procedure for generating observables from the discrete elements. The other extreme case is when all elements are lumped together by a fixed and uniform averaging process, that is entirely independent of the sequence of elements.

We know that the interpretation of natural language is not this simple and cannot be explained or simulated by either of these extreme models. The elements of our language, the words, cannot be represented as fixed elements of a discrete automaton nor as the observables of a fixed, uniform averaging process. The way we interpret strings of words appears to be much closer to the way the cell interprets strings of DNA, where the meaning of the discrete elements is first created by a highly context-sensitive folding of these elements into functional or meaningful units.

Furthermore, we know that interpretation does not stop at one level either for a single sentence or a single enzyme. For example, even if an enzyme is synthesized and folds up locally in the proper way to function, it is likely that this is only one enzyme in a string of enzymes where the end-product is the global function for which this string has been selected. Furthermore, the control of the organism as a whole may require that this end-product act like a switch, inhibiting the reading of the genes producing this string.

In a similar sense, a sentence such as "Turn left at the second stop sign," may be interpreted quite clearly at the local level, but may be only one of a string of sentences which has a more global function of taking you, say, to a gasoline station. Finally, to complete the analogy, the meaning of the entire string of sentences may be overridden by the sentence, "But I do not need gasoline."

We do not know, of course, that this is a good analogy, since we have almost no evidence about how natural language is interpreted in the brain. However, we know enough about the folding and control of enzymes to know that both discrete and continuous descriptions are necessary for modeling their behavior.

One point needs to be clarified here. The claim that it would be possible to describe the folding process by a discrete program in a computer does not mean that continuous dynamical processes are not essential for the origin or interpretation of symbols. The reason for this, as we have noted, is that the process of evolution occurs in real dynamical time (i.e., natural selection is not a symbolic process), and therefore, the speed and reliability of interpretation is crucial. It follows that a discrete symbolic representation of a dynamical process is selectively competitive only if it can *predict* results faster than the result itself occurs. In the case of the enzyme, the evidence is very strong that the folding is not a programmed or discrete process, and no one has even imagined a predictive simulation which would take less time.

11. Brains, Computers and Intelligence

We have proposed that the hierarchical levels of organization in living systems are based on an interaction of both continuous dynamical modes and discrete switching modes at each level. We associate the discrete switching modes with the *informational*, *programmable*, or *linguistic* functions, such as the linear sequences of nucleotides in a single DNA molecule or the sequences of pulses generated by a single neuron. We associate the singularities of continuous dynamical modes with the *creation* of these discrete sequences and the parallel dynamical interactions or "folding" of these sequences with their *interpretation*. This continuous model is not effectively programmable, but results from the non-integrable physical constraints that couple the rate dependent sequential events of the discrete mode with the continuous real-time dynamics of the system.

We have suggested that this same type of complementary interplay of continuous and discrete activity is also the key to intelligent behavior, since intelligence implies not only a symbolic representation or model of the world, but even more, the *creation* and *interpretation* of this model.

What evidence do we have from our study of the brain that supports this suggestion? As we said at the beginning, there is not the slightest doubt that the discrete, all-or-none, pulse mode of activity exists and can account for many signaling and data-processing functions of the brain. The only doubtful assumption is that this mode alone is adequate for all forms of memory, learning and intelligent behavior.

The important part of the question, then, is what types of continuous dynamical behavior can we find in the brain that can be associated with the creation and interpretation of patterns of these discrete pulses. I shall mention only two possibilities, one at the molecular genetic level, and one at the whole brain level of organization.

At the molecular genetic level, there is growing evidence supporting Sperry's (1970) hypothesis that the nerve net circuits over which the pulses travel are constructed by utilizing the same type of detailed chemical specificity that we associate with enzyme-substrate recognition. In the case of the nerve cell growth, the process is undoubtedly even more complex, involving the interactions of highly specific membrane structures, and perhaps microfibrils or microtubules. In spite of our ignorance of the many intricate details of this growth and specificity, it is reasonable only from our present knowledge that several levels of discrete and continuous interaction processes are involved. What does <u>not</u> seem reasonable is that the growth of the brain is only a discrete switching process, and it is equally unreasonable to expect that such a highly evolved level of continuous chemical dynamics, growth and morphogenesis should at a later stage be entirely suppressed in favor of the discrete switching mode. Indeed the chemical evidence now makes the idea that all signaling in the brain is by all-or-none nerve impulses untenable, and as Eccles (1973) says, "We can now postulate that the whole nervous system has communication not only by impulses...but also chemically by transport of specific proteins or other macromolecules." Since the only known mechanism of specific recognition involves the "folding" of strong-bonded sequences by highly parallel dynamical interactions of weak bonds, we may reasonably postulate that this type of continuous dynamics plays an essential and complementary role in the interpretation and generation of the discrete nerve pulses at many levels (Conrad, 1974).

As a second possible example of the complementary interfacing of discrete and continuous modes, we may look at the highest level of brain organization, and consider the functions of the so-called dominant and minor cerebral hemispheres. Of course, at this level we are observing only the broadest general functions. We have no idea how many intervening levels of generative, symbolic or interpretive modes exist. We have no idea how many times the discrete pulse patterns at one level have been coded into a statistical dynamical continuum whose singularities have in turn generated a new discrete switching mode at a higher level. In any case, at the functional level the dominant hemisphere has been characterized by its verbal, analytic, arithmetical and sequential activities, while the minor hemisphere has been characterized by its musical, synthetic, geometrical and holistic activities (Sperry, 1970; Eccles, 1973). I would suggest simply that at the level we observe them the dominant hemisphere is operating primarily in the discrete switching mode while the minor hemisphere is operating primarily in a continuous dynamical mode. I would also infer, therefore, that the minor hemisphere is primarily generative and interpretive with respect to the discrete symbolic and linguistic activities of the dominant hemisphere.

One very significant fact should be borne in mind, that in spite of some degree of gross anatomical difference between the two hemispheres, there is no evidence of any basic difference in the nerve cells or the way in which they grow or are connected. This may appear puzzling if one assumes that discrete or continuous behavior should be observable as a structural property of the brain. But one point that I am trying to emphasize is that discrete and continuous modes are not intrinsic physical or logical properties of structures, but are themselves an interpretation of how these structures interact with each other. Again I return to the enzyme as the simplest paradigm example. Whether the enzyme is interpreted as a discrete or continuous device depends on which interactions are significant at any given time. The discrete, linear sequence of amino acids is only distinguishable from the continuous, three-dimensional folding because of the distinction between strong and weak bond interactions, since the strong bonds effectively define the discrete linear sequence while the weak bonds constrain the continuous dynamics of substrate recognition and catalysis.

In a more general way, Thom's catastrophe models, as exemplified here by Zeeman's dynamical view of a nerve pulse (Sec. 7), show how the discrete and continuous modes may be related. Here we interpret the discrete switching function as the "fast foliation" at the cusp singularity (the nerve pulse), which can exist only because of the continuous dynamics of the "slow manifold" in higher dimensions (the biochemical matrix).

Bear in mind that these are only the simplest conceivable examples illustrating the complementary relation of discrete and continuous modes. The enzyme molecule is, after all, only a functional unit within the context of an immensely more complex cell, and Zeeman's dynamical model of a switch was explicitly derived as the simplest possible representation of a cusp catastrophe with given switching characteristics. There is, therefore, little justification for assuming that any activity of the nervous system could be realistically simulated with model neurons of greater simplicity.

Finally, this brings me to the questions of why computers are not more like brains, and how we can design computers to be more intelligent. If what I have said about the generative, symbolic and interpretive roles of continuous and discrete modes has some degree of truth, then the answer to the first question is quite clear: Even the largest imaginable computer, if restricted only to the discrete switching mode of present-day computers, can at best approximate only half a brain, and this will be the sequential, analytic half, not the generative, interpretive half.

Even in comparison with enzyme function, present computers appear impoverished; for they have no natural equivalent of strong and weak interactions, hence no sequence folding operations, no simple procedure of pattern or substrate recognition and no corresponding mechanisms for selectively catalyzing the <u>rates of growth</u> of their own sequences. While these inherently continuous dynamical processes can be programmed to some degree, it is fair to say that what small successes have been obtained are the result of very large programs generated by the brains of very intelligent programmers.

A common remark intended to account for the functional discrepancy of brains and computers is that brains perform "parallel" computations while computers are, so far, only "sequential" machines. This may be part of the problem, but as long as we mean by "parallel" only more simultaneous <u>discrete</u> operations, I do not think it is the basic problem. As I pointed out in Sec. 9 on the evolution of computers, the discrete switching mode has been artificially selected as significant while all continuous dynamical interactions have been artificially suppressed. Consequently, the generation of new observables or discrete elements as singularities in an underlying continuum, and the self-interpretation of linear sequences by folding processes are precluded.

The answer to my second question of how to design more intelligent computers is very simple to state, but very difficult to implement. As we have argued, the problem is not in the programs or the organization of the switching elements, but in the limited functions of the elements themselves. By abstracting the switch too far we have, as von Neumann said, "...thrown half the problem out of the window and it may be the more important half." The trick will be to learn how to reintroduce a continuous dynamical mode into an artificial computer element at a simple enough level to be practical. To do this, I believe we must first have simpler, clearer theoretical models and interpretations of "intelligent behavior."

Acknowledgement

This work was supported in part by grant from National Aeronautics and Space Administration, No. NGR 33-015-002.

References

Anninos, P.A., Beek, B., Csermely, T.J., Harth, E.M. and Pertile, G.: Dynamics of neural structures, J. theoret. Biol. 26, 121-148 (1970).

Blair, E.: On the intensity-time relations for stimulation by electric currents, Gen. J. Physiol. 15, 709 (1932).

Brillouin, L.: Science and Information Theory, Academic Press, New York, 1962.

Conrad, M.: Molecular information processing in the central nervous system. Part I, Selection circuits in the brain, 1974, this volume.

Cowan, J.D.: Stochastic models of neuroelectric activity. In: Towards a Theoretical Biology, vol. 4, ed. by C.H. Waddington, pp. 169-188. Univ. of Edinburgh Press, Edinburgh, 1973.

Duda, R.O. and Hart, P.E.: Experiment in scene analysis, Proc. First Natl. Symp. on Industrial Robots, Chicago, April, 1970.

Eccles, J.C.: The Understanding of the Brain, McGraw-Hill, New York, 1973.

Ghiselin, B. (Ed.): The Creative Process, University of California Press, Berkeley, 1952 . (Reprinted 1955 by Mentor Books, New York)

Gödel, K.: In The Undecidable, ed. by M. Davis. Rowen Press, Hewlett, New York, 1964.

Griffith, J.S.: A field theory of neural nets: I. Derivation of field equations, Bull. Math. Bioph. 25, 111; II. Properties of field equations, Bull. Math. Bioph. 27, 187 (1963).

Guzman, A.: Decomposition of a visual scene into three dimensional bodies, Proc. Fall Joint Comp. Conf. 291-304 (1968).

Hadamard, J.: The Psychology of Invention in the Mathematical Field, Princeton University Press, Princeton, N.J., 1945.

Handler, P.(Ed.): Biology and the Future of Man, p. 360, Oxford University Press, New York, 1970.

Harth, E.M. Csermely, T.J., Beek, B. and Lindsay, R.D.: Brain function and neural dynamics, J. theoret. Biol. 26, 93-120 (1970).

Hill, A.V.: Excitation and accommodation in nerve, Proc. Royal Soc. B119, 305 (1936).

Hubel, D.H. and Wiesel, T.N.: Receptive fields and functional architecture of monkey striate cortex, J. Physiol. 195, 215-243 (1968).

Kauffman, S.A.: Metabolic stability and epigenesis in randomly constructed nets, J. theoret. Biol. 22, 437-467 (1969).

Keyes, R.W.: Power dissipation in information processing, Science 168 796-801 (1970).

Landauer, R.: Irreversibility and heat generation in the computing process, IBM Jour. July, 183-191 (1961).

McCulloch, W.S. and Pitts, W.: A logical calculus of the ideas immanent in nervous activity, Bull. Math. Bioph. 5, 115-133 (1943).

Pattee, H.H.: Laws and constraints, symbols and languages. In: Towards a Theoretical Biology, vol. 4, ed. by C.H. Waddington, pp. 248-258. University of Edinburgh Press, Edinburgh, 1973.

Pattee, H.H.: Physical problems of decision-making constraints, Int. J. Neuroscience 3, 99-106 (1972b).

Poincare, H.: Mathematical creation, from The Foundations of Science 1913, 1946, Science Press (1913).

Polanyi, M.: Life's irreducible structure, Science 160, 1308-1312 (1968).

Post, E.: Selections from diary of E. Post. In: The Undecidable, ed. by M. Davis, p. 420. Rowen Press, Hewlett, New York, 1965.

Rashevsky, N.: Outline of a physiomathematical theory of excitation and inhibition, Protoplasma 20, 42-56 (1933).

Rosen, R.: Hierarchical organization in automata theoretical models of the nervous system. In: Information Processing in the Nervous System, ed. by K.N. Leibovic, pp. 21-35. Springer-Verlag, New York, 1969.

Sperry, R.W.: Cerebral dominance in perception. In: Early Experience in Information Processing in Perceptual and Reading Disorders, National Academy of Sciences, Washington, D.C., 1970.

Thom, R.: Topological models in biology. In: Towards a Theoretical Biology vol. 3, ed. by C.H. Waddington, pp. 89-116. Edinburgh University Press, Edinburgh, 1970.

Turing, A.M.: On computable numbers with an application to the Entscheidungs problem, Proc. London Math. Soc. Ser. 2, 42, 230-265 (1936).

von Neumann, J.: The Computer and the Brain, Yale University Press, New Haven, 1958.

von Neumann, J.: Theory of Self-Reproducing Automata, ed. by A.W. Burks, University of Illinois Press, Urbana, 1966.

Zeeman, E.C.: Differential equations for the heartbeat and nerve impulse. In: Towards a Theoretical Biology, vol. 4, ed. by C.H. Waddington, pp. 8-67. Edinburgh University Press, Edinburgh, 1972a.

Zeeman, E.C.: Appendix: A catastrophe machine. In: Towards a Theoretical Biology, vol. 4, ed. by C.H. Waddington, pp. 276-282. Edinburgh University Press, Edinburgh, 1972b.

PART III

Cellular and Sensory Biophysics

POSTSYNAPTIC CELL CHARACTERISTICS DETERMINING MEMBRANE POTENTIAL CHANGES

P. Fatt
University College London
London WC1E 6BT, England

The aim of this paper is to review and re-appraise some of the available experimental information concerning the mechanism by which excitatory and inhibitory synaptic transmitters, which are released from presynaptic terminals, evoke membrane potential changes in the postsynaptic cell. These transmitter-induced changes will in turn determine the possible further transmission of a signal, usually in the form of an action potential. Attention will be focussed on the motoneurone in the spinal cord of the cat as representative of a nerve cell in the vertebrate central nervous system. Information from other kinds of experimental material, especially the nerve-muscle preparation will also be considered, as it serves to elucidate processes occurring in the central neurone.

Detailed study of the postsynaptic cell response has relied heavily on the use of intracellular microelectrodes, for observing membrane potential, for passing current, and for injecting ions into the cell. The description of processes operating in the postsynaptic cell membrane arrived at by these methods has involved concepts from both electrical circuit theory and from electrochemistry. As will appear in what follows, in the application of these concepts it is necessary to give consideration to the geometric characteristics of the cell, in particular to the location of areas of synaptic contact and to any spatial variation in electrical properties of the cell membrane.

1. Excitatory Synaptic Potentials

A. Neuromuscular transmission

Fig. 1 shows some records of membrane change evoked in a fibre of the frog sartorius muscle in response to a stimulus applied to the nerve supplying the muscle. These changes occur as displacements to less negative values (depolarization) of membrane potential from a resting level of about -90mV (inside of cell with respect to outside). In order to obtain records of postsynaptic potential changes without the complication of a superimposed action potential, the nerve-muscle preparation was exposed to a solution containing tubocurarine which served to reduce the effectiveness of the transmitter substance, acetylcholine, released from the motor nerve terminals, on the post-synaptic receptive region of the muscle fibre. The maximum membrane potential change in the muscle fibre was thereby reduced below the level at which an action potential would be initiated. The several records of membrane potential changing with time were obtained by inserting the microelectrode at different positions along the muscle fibre separated by intervals of 1.0 mm. In the case of the uppermost record the microelectrode is presumed to have been situated within 0.1 mm of the centre of the region of synaptic contact between the motor nerve and the muscle fibre. This distance of 0.1 mm is comparable to the muscle fibre diameter as well as to the length of the region of synaptic contact, called the endplate in the case of skeletal muscle. From such records it appears that the effect of a presynaptic nerve impulse on the muscle fibre is to produce a wave of membrane potential change which spreads away from the synaptic region, undergoing a progressive diminution of amplitude and slowing of time course with increasing distance. It has been shown that the form of this potential wave, that is its distribution in both time and distance, can be accounted for by the passive electrical properties of the muscle fibre, apart from an initial period of 2.5-3.0 msec during which there occurs a net inward movement of positive charge across the membrane at the endplate (Fatt & Katz, 1951). The latter process signals the duration of action of synaptic transmitter on the endplate region of the muscle fibre.

The records in Fig. 2 show the response to nerve stimulation obtained from a muscle fibre in the absence of tubocurarine. The upper record gives the response recorded at the endplate. The initial phase of potential change corresponds to the rising phase of the endplate potential. At a displacement from the resting level of about 40 mV there occurs an abrupt increase in rate of change of potential indicative of the initiation of an action potential in the muscle fibre. The action potential recorded at the endplate displays a sharp peak following which the potential holds at a level about 10 mV below the peak for a time extending to about 2 msec from the beginning of the response. The lower record shows the response recorded in the same muscle fibre 2.5 mm away from the endplate. There is seen to be an initial, slowly developing potential change corresponding to the endplate potential recorded at a distance from the endplate. At about 2 msec from the

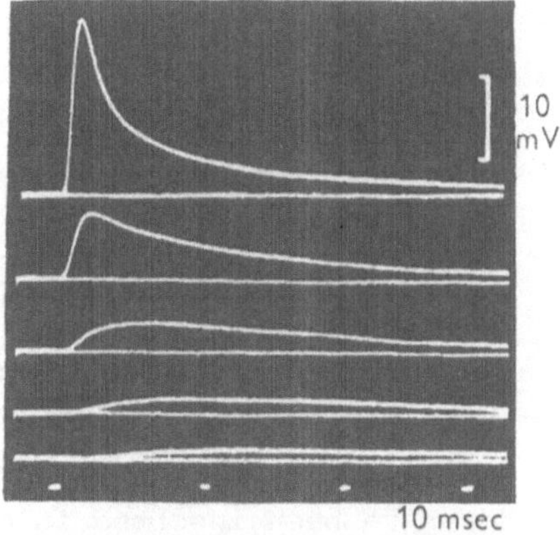

Fig.1. Records of endplate potential in a muscle fibre of the frog, obtained with a microelectrode placed successively at intervals of 1.0 mm along the fibre (from Fatt and Katz, 1951).

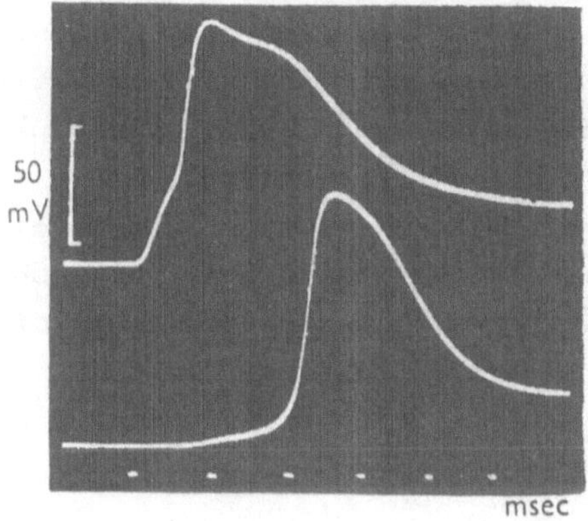

Fig. 2. Records of action potential in a muscle fibre evoked by nerve stimulation, obtained with the microelectrode at the endplate (above) and 2.5 mm away (below) (from Fatt and Katz, 1951).

beginning of electrical activity in the fibre, the action potential appears, delayed by propagation from its site of initiation at the endplate and preceded by a phase of exponentially rising potential (the so-called foot of the action potential).

One notes that the peak amplitude of the action potential away from the endplate is greater than at the endplate, both measured from the initial (resting) level of membrane potential. In addition, the action potential away from the endplate does not show the sharp peak followed by a hump as marks the action potential at the endplate. If the muscle fibre is stimulated elsewhere, so that the action potential propagates into the endplate region, the action potential at the endplate is then similar to that seen anywhere along the fibre. The conclusion drawn from these observations is that, apart from specific changes in electrical properties evoked in the synaptic region of the muscle fibre by the release of transmitter from the presynaptic nerve terminals, membrane potential changes in the synaptic region are influenced by the ordinary, non-junctional properties of the muscle fibre membrane which are effectively uniform along the length of the fibre.

Dependence of synaptic potential on membrane potential. In order to obtain further information on the change in electrical properties of the postsynaptic membrane which gives rise to the endplate potential and which produces the described modification in the peak of the action potential, two microelectrodes have been inserted into the muscle fibre close to the endplate. One of these electrodes is used to apply a steady current between the inside of the fibre and the outside so as to vary the membrane potential, while the other is used to observe the membrane potential including the response to nerve stimulation. Fig. 3 shows a plot of the relative amplitude of the endplate potential (i.e. its amplitude during the applied current divided by its amplitude in the absence of such current) against the relative level of membrane potential from which it arises (i.e. the steady, displaced level of membrane potential during the applied current divided by the membrane potential in the absence of current). As appears, the amplitude of the endplate potential varies linearly with the membrane potential. A straight line is drawn to pass through the point of unity endplate potential for unity

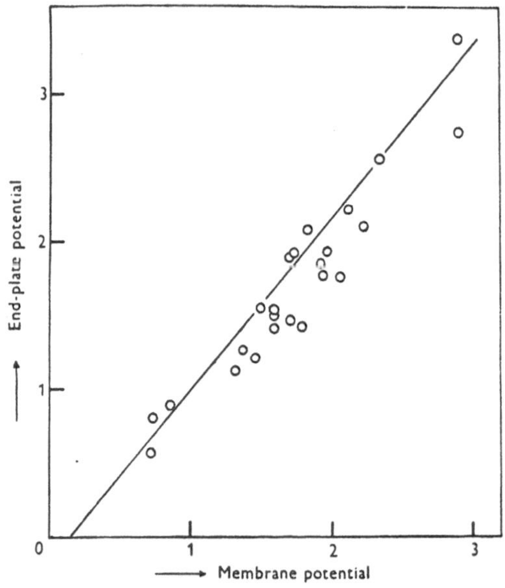

Fig. 3. Plot of relative amplitude of endplate potential, measured for a fixed interval on its rising phase, against relative level of membrane potential from which it arises. Displacement of membrane potential from unity was effected by current applied through a second microelectrode (from Fatt and Katz, 1951).

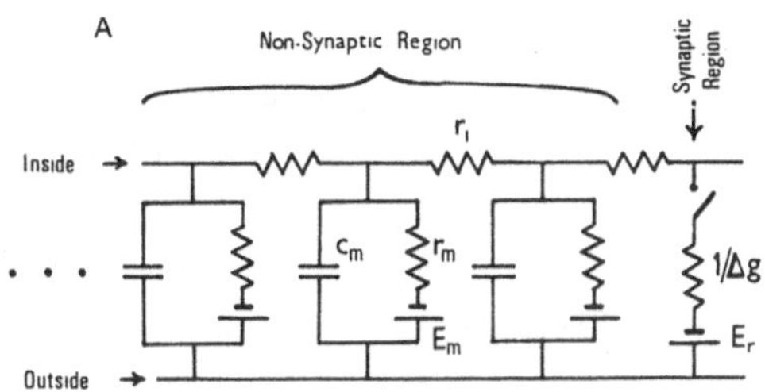

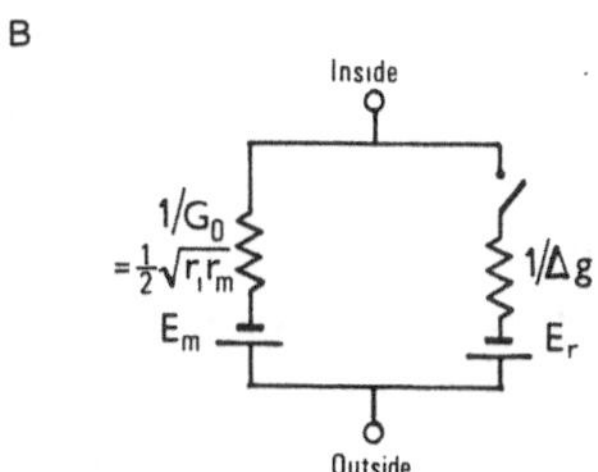

Fig. 4. A: Distributed (cable) model of the muscle fibre, with the effect of synaptic activity being represented by a switch-controlled conducting path at one position along the fibre. The model had been simplified by supposing there to be no potential gradient extracellularly and by ignoring complications in the capacitance associated with the transverse sarcoplasmic reticulum.

B: A lumped circuit model of the muscle fibre obtained by ignoring capacitance.

membrane potential and to give zero endplate potential at 1/6 unity membrane potential.

Equivalent circuit for synaptic action. The linear variation of the amplitude of the endplate potential with the level of membrane potential from which it arises suggests that the synaptic response follows from the addition to the electrical properties of the fibre of a conducting path between inside and outside at the endplate, which path can in electrical terms be represented by a resistor in series with a voltage source. The current which flows locally in the endplate region during the brief phase of transmitter activity is accounted for by the difference between the value of this voltage source and the pre-existing potential across the membrane in the neighbourhood of the synaptic region. A circuit model for the muscle fibre with distributed elements of surface membrane capacitance (c_m) and resistance (r_m) and of internal longitudinal resistance (r_i) and with a synaptically controlled resistance ($1/\Delta g$) at the end-plate is shown in Fig. 4A. The membrane potential of the non-junctional region is represented by E_m, while the voltage source in the synaptic path is represented by E_r. Synaptic activity can be envisaged as either a closing of the switch in series with $1/\Delta g$ or a variation in the value of Δg from zero upwards while the switch remains closed. If one ignores the membrane capacitance the model of Fig. 4A can be replaced by the model with lumped elements shown in Fig. 4B. The resistance between inside and outside of the fibre in the absence of synaptic activitiy is given by $1/G_0$ which is dependent as indicated on the core resistance of the fibre (r_i) and on the membrane resistance (r_m), both of these quantities being expressed for unit length of fibre.

The form of the action potential observed at the endplate when the action potential is evoked by nerve stimulation can now be explained. The reduction in peak amplitude of the action potential, the initial rapid decline giving the sharp peak to the action potential, and the delayed further decline producing the hump on the falling phase of the action potential (Fig. 2) can be accounted for by a tendency of the membrane potential to move toward and hold at E_r which in the case of the neuromuscular junction has a value of about -15 mV.

Non-linearity. A linear relation between membrane potential and synaptic potential would be expected from the scheme of Fig. 4 only if both the synaptic and the non-synaptic regions were to have linear electrical characteristics. However, as is well known, the conductance of the membrane of an electrically excitable cell is voltage-dependent. Indeed, it was possible to obtain the results illustrated in Fig. 3 only by operating within a limited range of membrane potentials, over which there is a linear relation between membrane potential and applied current. Fig. 5 illustrates what would be expected in a cell having a non-linear characteristic but in which the conductance increment introduced in the synaptic region is linear.

The continuous curve in Fig. 5 shows the relation of synaptic potential to membrane potential for a cell having the current-voltage

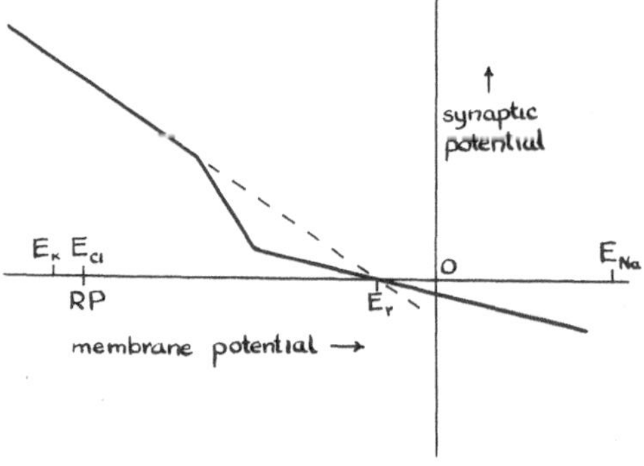

Fig. 5. Idealized plot of the relation between synaptic potential amplitude and level of membrane potential controlled by applied current. Solid line, with a jog between the resting potential and the reversal potential (E_r), gives the relation for a typical rectifying behaviour of the non-junctional membrane, its resistance decreasing abruptly on depolarization. Equilibrium potentials for the ions, Na^+, K^+ and Cl^-, are indicated along the abscissa.

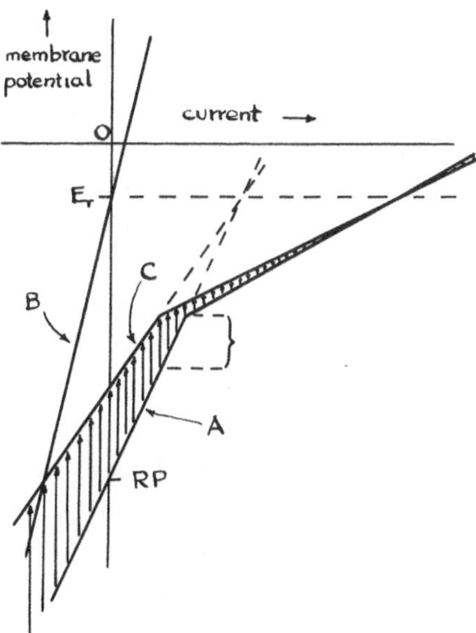

Fig. 6. Idealized plot of membrane potential against membrane current showing how the relation in Fig. 5 is arrived at. A is the voltage-current curve for the non-junctional membrane, B for the synaptically controlled path and C for the two in parallel.

relation represented by the line (A) in Fig. 6. For simplicity the non-linear behaviour of the cell membrane is represented by a line which has two regions of constant, but different slopes. The relation for the synaptically activated conductance increment is represented by a line of constant slope (B). Since these two contributions to the conductance across the membrane, given by A and B, are in parallel during transmitter activity, the total current in this state will be given by the sum of the currents through each path for any given value of membrane potential. The curve C which describes the membrane potential-current relation for the cell during synaptic activity is thus arrived at. (Time-dependent behaviour is ignored in this description in order to avoid a more complicated analysis involving the membrane capacitance.) The synaptic potential for a given value of applied current is obtained as the vertical displacement of C from A. The origin of the jog in the curve of Fig. 5 is evident. It occurs within the range of membrane potential (more precisely, the membrane potential from which the synaptic potential arises) indicated by the curved brackets in Fig. 6. In this range the synaptic potential is a displacement of potential between levels at which the potential-current relation for the cell has different slopes. It will be observed in Fig. 5 that there are two ranges of membrane potential in which the relation between synaptic potential and membrane potential is described by a straight line which passes through zero synaptic potential at a membrane potential, E_r.

Electrochemical considerations. The conducting paths across cell membranes described in this paper, including those controlled by chemical transmitters at synapses, are all dependent on the ability of ions present in the aqueous environment (both inside and outside the cell) to penetrate the membrane, with the rate of penetration being influenced by the membrane potential. The ion species present in high concentration either inside or outside the cell, which may thus be expected to contribute significantly to such paths, are sodium, potassium and chloride. From electrochemical considerations one may conceive an equilibrium potential for each species, corresponding to the level of membrane potential at which there is no net current carried by it, the unidirectional passive fluxes in the inward and outward directions across the membrane just balancing. This is given by the equation of Nernst (1889),

$$E_X = \frac{RT}{z_X F} \ln \frac{[X]_{outside}}{[X]_{inside}} \tag{1}$$

where X represents the particular ion species, z_X being its valency, $[X]_{inside}$ its concentration inside and $[X]_{outside}$ its concentration outside, while R, T and F have their usual connotations. (More exactly, the thermodynamic activities of the ions should be used in place of concentrations.) The estimated values of the equilibrium potentials E_{Na}, E_K and E_{Cl} relative to the resting potential (RP) are indicated by their placement along the abscissa in Fig. 5.

In the resting state the permeability of the membrane to Na^+ is very much less than its permeability to K^+ and, in some cells, to Cl^- as well. The low intracellular concentration of Na^+, which in turn is responsible for the large positive divergence of E_{Na} from the resting potential, is dependent on the extrusion of Na^+ from the cell interior by the operation of a metabolically driven ion-pump which moves Na^+ outward and K^+ inwards. The outward movement of Na^+ and inward movement of K^+ are coupled with approximately two-thirds as many K^+ moved as Na^+, under most conditions. The functioning of the pump is, within certain limits, independent of the relative values of the membrane potential, E_K and E_{Na}.

Now, the living cell in its resting condition is expected to be in a steady state, in which there is no net movement of any ion species across the cell membrane so that the concentrations of the various ions remain steady. In the case of K^+ this implies that the inward transfer of ions by pumping must be counterbalanced by the rate of passive penetration being greater in the outward than in the inward direction. According to a relationship derived on thermodynamic grounds by Ussing (1949), the ratio of the passive fluxes in the two directions will be dependent on the difference between the membrane potential and the equilibrium potential, this dependence being of the form

$$E_m - E_X = \frac{RT}{z_X F} \ln \frac{M^{(X)}_{\text{outward}}}{M^{(X)}_{\text{inward}}} \tag{2}$$

where $M^{(X)}_{\text{outward}}$ and $M^{(X)}_{\text{inward}}$ are the unidirectional passive fluxes in the indicated directions. In the case of K^+, taking into account the operation of the pump, this yields for the steady-state

$$E_K = E_m - \frac{RT}{F} \ln \left(1 + \frac{M_{K\text{-pump}}}{M^{(K)}_{\text{inward}}} \right) \tag{3}$$

where $M_{K\text{-pump}}$ is the rate of inward pumping of K^+. Thus on the knowledge that an inwardly directed pump exists one infers that E_K must have a value more negative than the resting potential. In the case of the frog muscle fibre it is probable that E_K is only a millivolt or so more negative than the resting potential. However, there is evidence that in the neurones of the central nervous system E_K is about 15 mV more negative than the resting potential, corresponding on the basis of the above equation to $M_{K\text{-pump}}/M^{(K)}_{\text{inward}} \approx 0.75$, the pumping of ions being appreciable in comparison with passive penetration. In Fig. 5, E_{Cl} is shown as being coincident with the resting potential, since there is no evidence for the existence of a Cl-pump in vertebrate muscle fibres. There is, however, reason to postulate its existence in central neurones, as will appear later, when inhibitory synaptic activity is considered.

By analogy with the equilibrium potential for a particular species of ion, the "equilibrium potential" for synaptic activity has been defined as that level of membrane potential at which there is no net current through the synaptically activated conductance paths. It is clear that the value of E_r of -15 mV for the excitatory synaptic potential at the neuromuscular junction does not correspond to an equilibrium potential for any of the above mentioned species. Studies carried out by Takeuchi and Takeuchi (1960) of the effect on E_r of changing the extracellular concentrations of ions has led to the conclusion that during synaptic activity at the frog neuromuscular junction permeability increases occur toward Na^+ and K^+, but not toward Cl^-. From the values of E_r, E_K and E_{Na} one obtains the relative conductance increments introduced by synaptic activity. The conductance increment for Na^+ amounts to 1.3 times that for K^+.

General inferences. A very general inference which may be drawn from these studies is that the synaptic transmitter exerts an excitatory postsynaptic action through its producing an increase in ionic permeability of the post-synaptic membrane, allowing ions to move passively across the membrane under the combined influence of the membrane potential and ionic concentrations on the two sides of the membrane. The permeability increase appears to apply in varying degree to a range of different cationic species, including sodium and potassium, but it does not apply to anions. Furthermore, those cations which are involved may interfere with one another's movement, whereas anions have no effect. A model may be envisaged according to which the permeability increase is due to the action of transmitter in opening pores across the lipid portion of the membrane. A discrimination between cations and anions, whereby only the former can penetrate, would follow from the presence of fixed anionic groups in the walls of the pore. The relative mobility of different species of cations through the pores is expected to depend on the size of the pore together with the distribution of charged groups, while the possibility of interference between different ions may be taken as an indication that they share the same pore.

B. Excitatory synaptic potentials of a central neurone

Dependence on membrane potential. Having obtained some understanding of the underlying mechanism of the postsynaptic response from observations on the nerve-muscle junction, one may turn one's attention to the somewhat more complicated case of the postsynaptic response of the motoneurone. Fig. 7 illustrates results obtained on the excitatory synaptic potential recorded from a motoneurone in the spinal cord of a cat in response to stimulation of a selected group of sensory nerve fibres which are known from studies on spinal reflexes to have an excitatory synaptic action on the motoneurone examined. In this study a double-barrelled microelectrode was used, one barrel for recording membrane potential and the other for applying current. The two orifices of the electrode (having a separation of no more than a few micrometers) are considered to be located within the cell body of the motoneurone. Each of the records shown on the left of Fig. 7 is formed by the superimposing of 10 or more sweeps, the aim of which was to reduce the influence on the measurements of uncontrolled fluctuations in level of membrane potential. The separate records, corresponding to different mean values of initial membrane potential as indicated in millivolts, were obtained on the application of different intensities and polarities of steady current. The record for an initial level of membrane potential of -66 mV represents the resting condition, that is the condition in the absence of applied current. This is somewhat lower than the value of -75 mV obtained from other motoneurones. In this study, outwardly directed currents were used of such intensity as to depolarise completely and even reverse the membrane by some tens of millivolts. It is noted that, on depolarization by applied current, the maximum rate of rise and peak amplitude of the synaptic potential decrease. The response vanishes when the level of membrane potential is displaced to zero, and then reverses in polarity when the initial level of potential is displaced further.

The level of membrane potential at which the synaptic potential reverses is seen to be displaced further from the resting potential than the level of about -15 mV where E_r is found to lie in the case of the postsynaptic response of muscle. There is no reason to suppose that this shift in the observed reversal potential is related to the lower resting potential of the motoneurone compared with the muscle fibre. A deviation of the level of potential recorded in the cell body at which the synaptic potential is observed to reverse from the expected value can be accounted for on the basis that the synaptically evoked conductance increment underlying the recorded response takes place mainly on the dendrites of the motoneurone. The membrane potential at the site of generation of the synaptic potential will in these circumstances undergo a smaller displacement by the applied current than that observed in the cell body owing to there being a voltage drop in the internal resistance of the neurone between the cell body and the synaptic sites on the dendrites. The idea that excitatory synapses occur to a large extent on the dendrites and that there is an appreciable core resistance along the dendrites is supported by the observation that a discrete reversal potential at which the synaptic potential would be completely abolished was usually not obtainable. Instead, the potential change in the early part of the phase of transmitter activity was abolished at a displacement of membrane potential from the resting level, several millivolts less than was the later part of the response. This type of behaviour is accounted for by the later part of the response being contributed by synaptic activity on more distant parts of the dendrites as compared with the early part of the response, the time difference being due to the membrane capacitance of the dendrites causing a delay in the recording of potential changes initiated at a distance, similar to what has been described for the muscle fibre (Fig. 1). The shorter time to the summit of the synaptic potential in the case of the reversed responses at initial membrane potentials of +9 and +34 mV compared with the responses at -32 and -66 mV is considered to have a similar origin. The synaptic potentials arising from the moderately depolarized levels of potential of -60 and -42 mV are seen to give rise to action potentials of which only the early part is recorded. On further displacement of the steady level of membrane potential in the direction of decreasing internal negativity, it was no longer possible for an action potential to be initiated by a transient depolarization owing to inactivation of the mechanism for the sodium permeability increase which generates the action potential. It appears that in the case of the motoneurone the mechanism for the increase in potassium permeability which is responsible for the falling phase and after-potential of the action potential is likewise inactivated during a maintained depolarization with the result that there was no appreciable decrease in resistance of the neuronal membrane on depolarization from the resting level.

On hyperpolarization by means of a steady applied inward current across the neuronal membrane, the changes effected in the excitatory synaptic potential were different from what was expected from observations on the nerve-muscle preparation. As is seen in the records of Fig. 7, the peak amplitude of the response fails to increase significantly although the maximum rate of rise of the response does increase. Thus the time course of the synaptic potential alters on hyperpolarization with the summit being reached progressively earlier and appearing less rounded. Plots of the maximum rate of rise and of the peak amplitude of the synaptic potential against the initial level of membrane potential are shown at A and B on the right of Fig. 7.

An electrical model of the motoneurone. The behaviour of the synaptic potential on hyperpolarization requires explanation. The most satisfactory explanation for the observed behaviour of the synaptic potential would seem to be one based on the electrical characteristics of the motoneurone membrane. The failure of the synaptic potential to behave at high levels of internal negativity in a manner predicted by extrapolation from the behaviour at lower levels would be due to some degree of rectification in the motoneurone membrane such that the dynamic (slope) resistance decreases on hyperpolarization. In fact, the variations in maximum rate of rise in peak amplitude and in overall time course of response with applied current are all precisely such as have been observed in a number of other postsynaptic cells for which the measured relation between steady membrane potential and applied current

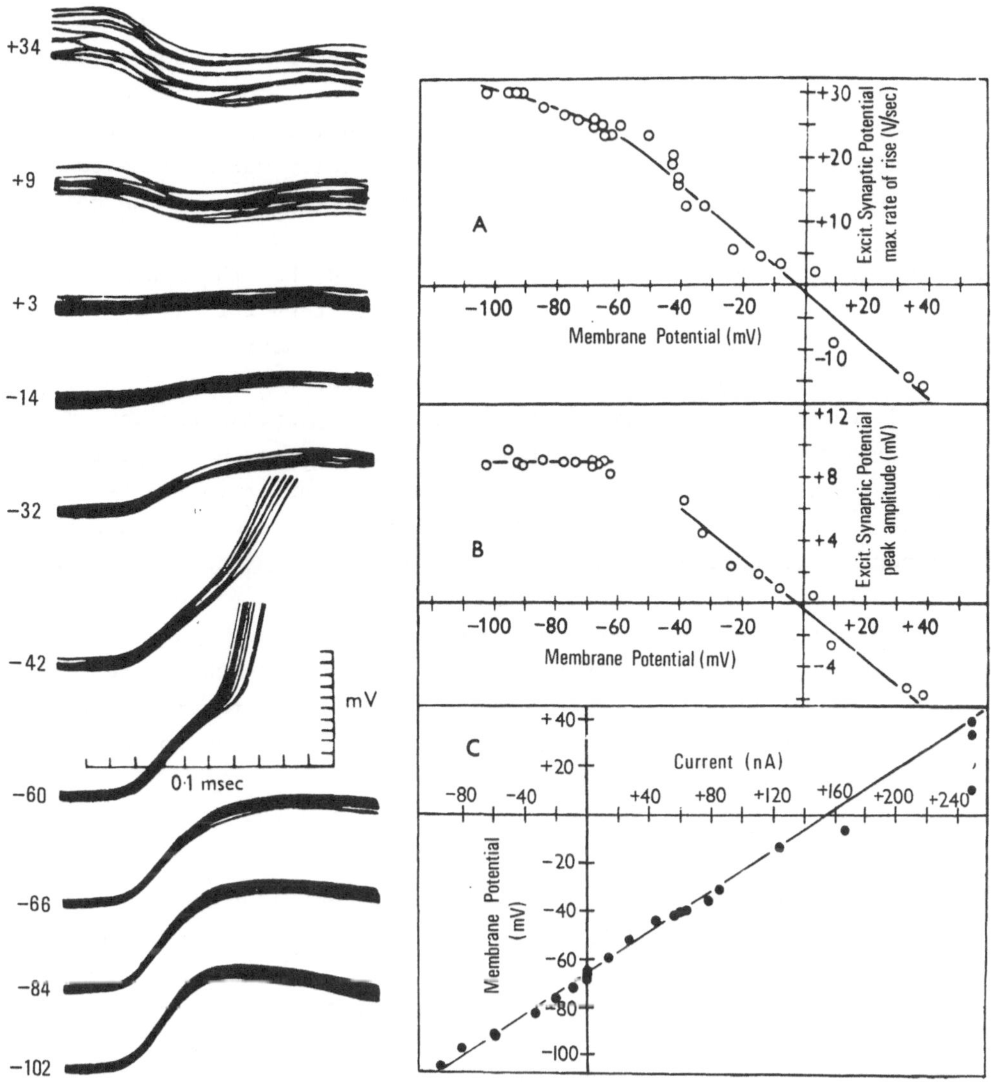

Fig. 7. On Left: Records of excitatory synaptic potential obtained from a motoneurone in the spinal cord of the cat. The steady level of membrane potential from which the response arises (given in millivolts) was controlled by applying current. A and B: Plots of maximum rate of rise and of peak amplitude of synaptic potential against level of membrane potential. C: Plot of steady level of membrane potential against applied current. The membrane potential represents the difference in potential recorded at the tip of the microelectorde, when it was inside the motoneurone compared with when it was immediately outside for the same current through the electrode (from Coombs, Eccles and Fatt, 1955 a).

shows a decreasing dynamic resistance on hyperpolarization (Boistel & Fatt, 1958; Kandel & Tauc, 1966; for a discussion see Ginsborg, 1967). The difficulty with the motoneurone is that, as is apparent from the measurements in C of Fig. 7, the membrane potential-applied current relation shows no change in slope. This difficulty can be removed if it is assumed first that the synaptic potential recorded at the cell body has its origin mainly in transmitter activity on dendrites, the membrane of which has rectifying properties such that the dynamic resistance decreases on hyperpolarization. It is then further assumed that the membrane of the cell body also has the behaviour of a rectifier, but of opposite sense to that in dendrites, the dynamic resistance of the cell body membrane increasing on hyperpolarization. The latter kind of behaviour can arise for any cell having a normal resting potential, even if the permeability toward ions is unaffected by membrane potential changes. It follows simply from the condition that the membrane conductance is determined by the movement of those species of ions which are close to equilibrium at the resting potential and such species will be so distributed across the membrane that an inward, hyperpolarizing current involves the net movement by each of them from that side of the membrane where they are in low concentration to where they are in high concentration, with the reverse applying for an outward, depolarizing current.

A sketch of the proposed electrical model for the motoneurone including rectification in the cell body membrane and in the dendrite membrane is shown in Fig. 8.

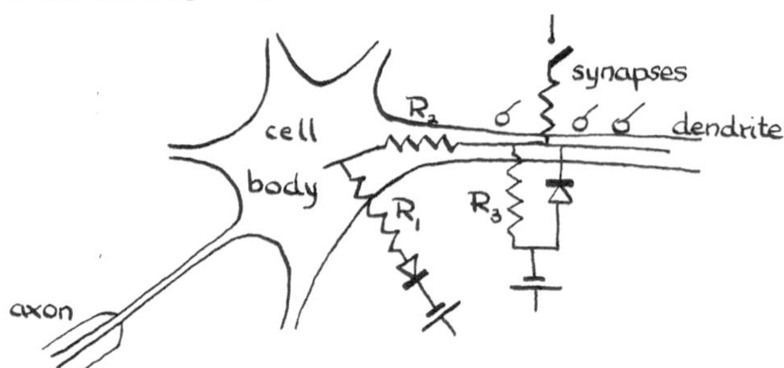

Fig.8. Scheme to explain the non-linear dependence of excitatory synaptic potential on membrane potential as shown in Fig. 7. It is based on there being different forms of rectifying behaviour in the membrane of the cell body and in the dendrites.

One may attempt to assign values to the three resistive elements representing the conducting path across the cell body membrane (R_1), the core resistance of all the dendrites in parallel as seen from the cell body (R_2) and the effective path across the dendritic membrane (R_3). An aim of this model is that the voltage-current relation measured between the interior of the cell body and the outside should be linear with a slope of 1 MΩ similar to what is observed experimentally. A reasonable set of values is R_1 = 2 MΩ, R_2 = 1 MΩ, R_3 = 1 MΩ. (The same value of resistance has been given to R_2 and R_3 on the grounds that for depolarizing currents the dendrites behave as long uniform cables, for which the input resistance, represented by $R_2 + R_3$, will be equally determined by the core resistance as by the membrane resistance; see Fig. 4.)

An additional matter of interest which arises from the present model is the amplitude of the synaptic potential in the dendrites at a distance from the cell body. On the basis of the assumption already made, that the effective core resistance of the dendrites is comparable to the resistance across the membrane of the cell body, one may infer that the synaptic potential in the dendrites is double or more the size of the response recorded in the cell body. Examination of the action potential recorded in the cell body indicates the existence of two thresholds for the initiation of an action-potential type of response in different regions of the motoneurone. One of the thresholds is reached at a depolarization of about 10 mV and is considered to correspond to a turning-on of a regenerative sodium permeability increase in the entire membrane of the cell body or possibly in some more restricted region where the axon emerges from the cell body. In the absence of excitatory synaptic activity in the dendrites, an action potential, which has been initiated by stimulation of the axon and which has propagated into the region of the cell body, is frequently found to attain a peak amplitude of about 85 mV, but there is then always a fairly abrupt change in rate of rise at a depolarization of about 40 mV indicative of a threshold at this level. This higher threshold is taken to correspond to that for the initiation of an action potential in the dendrites. It would thus appear that the excitatory synaptic potential may attain an amplitude up to four times greater in an individual dendrite than in the cell body without an action potential being initiated in the dendrite (rather than in the cell body). This condition for action potential initiation is essential in order that the depolarizing effect of synaptic activity on different dendrites should be capable of summing so as to determine the possibility of an action potential being generated in the motoneurone.

2. Inhibitory Synaptic Activity

Relation to membrane potential. One turns now to consider inhibitory synaptic activity. This can be elicited in the motoneurone by stimulation of a different group of sensory nerve fibres from those giving the excitatory synaptic potential, the particular group concerned arising from sensory organs in the opposing muscles. Fig. 9 illustrates the effect of inhibitory synaptic activity observed with a microelectrode in the cell body of a motoneurone. The method of investigation was similar to that used for the case of excitatory synaptic activity, the steady level of membrane potential being displaced by applied current and the effect on the response to nerve stimulation examined. At the resting potential, which in this case had a value of -74 mV, nerve stimulation is seen to give rise to an inhibitory synaptic potential in the form of a transient hyperpolarization. On depolarization of the membrane by an outward current, the synaptic potential is increased, the extent of this increase being approximately proportional to the initial displacement. On hyperpolarization by an inward current, the synaptic potential is diminished to the point of vanishing at a displacement of -7 mV from the resting level and then reversed in polarity on further displacement of membrane potential.

The inhibitory response is thus similar in its most important characteristic to the excitatory synaptic response, that is it appears as a transient displacement of membrane potential toward some fixed level of potential. In the case of the inhibitory response, the level of

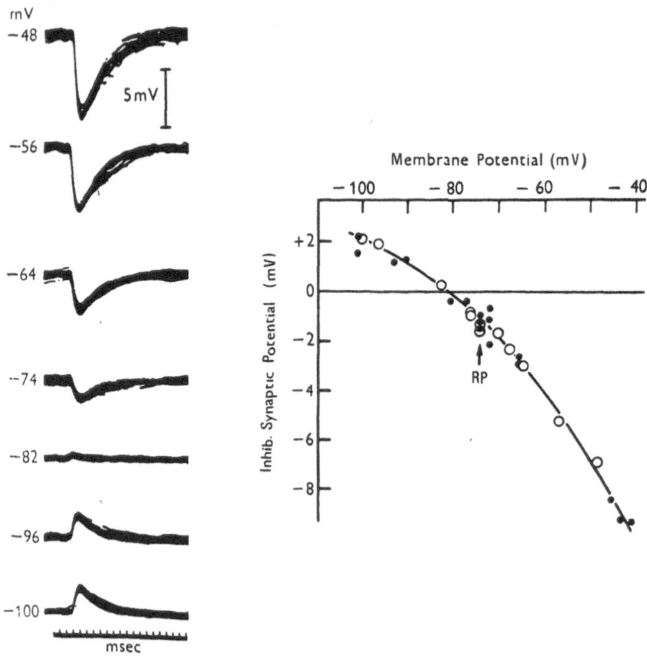

Fig. 9. Dependence of inhibitory synaptic potential of motoneurone on membrane potential (from Coombs, Eccles and Fatt, 1955b).

potential at which the response reverses is close to the resting potential, instead of close to zero membrane potential as for the excitatory response. A plot of the peak amplitude of the inhibitory synaptic potential against the initial level of membrane potential is shown on the right of Fig. 9. One respect in which this plotted relation for the inhibitory response differs from that plotted for the excitatory response is that the peak amplitude of the inhibitory potential continues to vary with membrane potential on increasing hyperpolarization, but with a gradual, progressive decline in the slope of relation. The observed nonlinearity of the inhibitory synaptic potential would be explicable on similar grounds to those invoked for the excitatory synaptic potential, i.e. a voltage-dependent variation of the conductance of the dendritic membrane. Several factors would operate to make the deviations from linearity less for the inhibitory synaptic potential than for the excitatory synaptic potential. One of these is a geometric one, since the inhibitory synapses may be located on dendritic sites which are closer to the cell body than are the excitatory synapses. A very important factor will be the position of the reversal potential in relation to the resting potential. For a given displacement of membrane potential to levels more negative than the resting potential, the close proximity of the reversal potential for the inhibitory response leads to the condition that there are large relative changes in the synaptically induced current across the membrane with changes in membrane potential so that the amplitude of the synaptic potential varies with membrane potential in a manner which is fairly insensitive to small variations in the dynamic resistance of the non-junctional membrane. In contrast, for the excitatory response the same displacement of membrane potential produces only small relative changes in the synaptic current with the consequence that the influence of variation in the dynamic resistance of the non-junctional membrane may outweigh the effect of membrane potential changes on synaptic current.

Electrochemical basis of the inhibitory response. It can be inferred from the value of the reversal potential that the inhibitory synaptic response involves an increase in membrane permeability toward either potassium or chloride ions, or both, since these ion species have equilibrium potentials close to the resting membrane potential. In the case of the motoneurone, the involvement of chloride ions in the inhibitory synaptic response is readily perceived. If one records from a motoneurone using a KCl-filled microelectrode, there is sufficient diffusion of KCl out of the electrode to alter significantly the Cl^- concentration inside the motoneurone and thereby to shift the equilibrium potential for Cl^-. The records on the right in Fig. 10 indicate

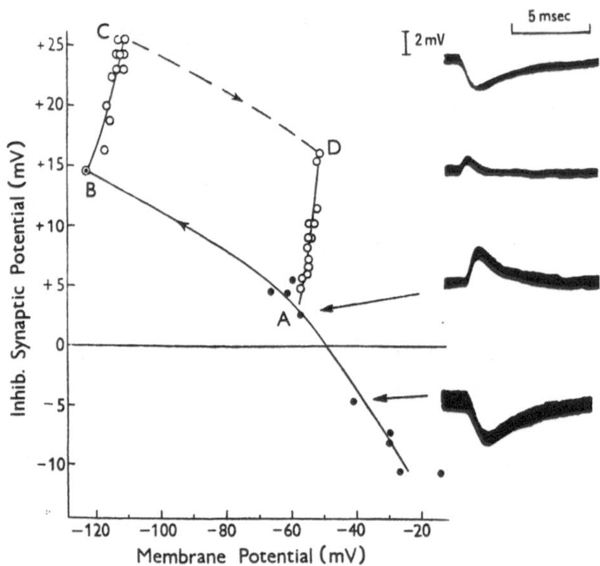

Fig. 10. Effect of injection of Cl^- on the inhibitory synaptic potential of motoneurone. Upper 3 records on right show the progressive change in the synaptic potential taking place over the course of a few minutes after inserting a KCl-filled micropipette (also serving as a microelectrode) into the neurone. Filled circles indicate the dependence of synaptic potential amplitude on membrane potential when the latter is rapidly displaced by applied current. Hollow circles extending from B to C indicate the progressive change taking place during a maintained inward current, causing Cl^- to enter the motoneurone from the microelectrode. Hollow circles from D to A indicate the change following discontinuation of the current (from Coombs, Eccles and Fatt, 1955b).

the changes in inhibitory synaptic potential which result. The uppermost record (a hyperpolarizing synaptic potential) was obtained very soon after the penetration of the motoneurone, the next lower record about a minute after penetration, and the third record (a depolarizing synaptic potential) a few minutes later by which time a steady condition had been obtained. The amplitude of the inhibitory synaptic potential at the resting potential, reached in this steady condition, is shown plotted at A in Fig. 10. It is possible to produce larger changes in Cl^- concentration inside the cell by passing current through the KCl-filled microelectrode. The measurement at B was obtained within 10 sec from the application of a steady inward current and shows the effect of a displacement of membrane potential before there was an appreciable further change in Cl^- concentration. On continuation of this current, the amplitude of the inhibitory synaptic potential increased while the level of membrane potential fell slightly as indicated by the hollow circles extending from B to C, to reach a steady state within about

3 min when the net outward movement of Cl^- across the membrane balanced the injection into the neurone from the microelectrode. On termination of the current, the inhibitory synaptic potential and the membrane potential were immediately displaced from C to D. During the next few minutes, the inhibitory potential and membrane potential shifted along the locus indicated by the hollow circles from D returning to A. The changes in response with chloride injection show that the reversal potential for the inhibitory synaptic potential is displaced more than is the resting potential by changes in the equilibrium potential for chloride. Hence the conductance increment produced by the transmitter must involve a larger proportional contribution from chloride ion movements than is the case for the conductance of the non-junctional membrane. The latter would be expected to include a major contribution from potassium ions together with a much smaller contribution from sodium.

Having ascertained that a movement of chloride ions is involved in the generation of the inhibitory synaptic potential, one may now inquire whether any other species of ions is also involved, in particular, potassium. This question is important in relation to the finding of an inhibitory synaptic potential in the form of transient hyperpolarization in the absence of applied current or injection of ions, that is, with the cell in a steady state with respect to the movement of ions across its membrane.

From observation of the after-hyperpolarization which follows the action potential of the motoneurone, it appears that E_K has a value about 15 mV more negative than the resting potential. Thus if the conducting path cross the membrane involved in the production of the inhibitory synaptic response included equal contributions of conductance from K^+ and Cl^-, the position of E_r for inhibitory synaptic action would be consistent with Cl^- being in equilibrium at the resting potential. However, despite the strenuous experimental attempts made to demonstrate an involvement of K^+ in inhibitory action, no convincing evidence has been produced (Coombs, Eccles & Fatt, 1955b; Eccles, Eccles and Ito, 1964a,b). Such effects of the displacement of cation concentrations inside the motoneurone as have been found can be explained by other familiar cellular mechanisms, including the acceleration of a coupled Na, K-pump by increased intracellular concentration of Na^+.

It seems reasonable to propose on the basis of available experimental information that the inhibitory synaptic activity of motoneurones consists of an increase in membrane permeability which is exclusive for anions and the occurrence of the reversal potential for the response at a level of membrane potential more negative than the resting potential may be taken as indicating the existence of a metabolically-driven Cl-pump, moving Cl^- outward across the membrane. Very clear evidence for the existence of such a Cl-pump has recently been obtained in neurones of an invertebrate, the seahare, Aplysia (Russell and Brown, 1972).

Acknowledgement

Thanks are due to Dr. Gertrude Falk for help in preparing this material for publication.

References

Boistel, J. and Fatt, P.: Membrane permeability changes during inhibitory transmitter action in crustacean muscle, J. Physiol. (Lond.) 144, 176-191 (1958).

Coombs, J.S., Eccles, J.C. and Fatt, P.: The specific ionic conductance and the ionic movements across the motoneuronal membrane that produce the inhibitory post-synaptic potential, J. Physiol. (Lond.) 130, 326-373 (1955a).

Coombs, J.S., Eccles, J.C. and Fatt, P.: Excitatory synaptic action in motoneurones, J. Physiol. (Lond.) 130, 374-395 (1955b).

Eccles, J.C., Eccles, R.M. and Ito, M.: Effects of intracellular potassium and sodium injection on the inhibitory postsynaptic potential, Proc. R. Soc. B. 160, 181-196 (1964a).

Eccles, J.C., Eccles, R.M. and Ito, M.: Effects produced on inhibitory postsynaptic potentials by the coupled injections of cations and anions into motoneurones, Proc. R. Soc. B. 160, 197-210 (1964b).

Fatt, P. and Katz, B.: An analysis of the end-plate potential recorded with an intracellular electrode, J. Physiol. (Lond.) 115, 320-370 (1951).

Ginsborg, B.L.: Ion movements in junctional transmission, Pharmacol. Rev. 19, 289-316 (1967).

Kandel, E.R. and Tauc, L.: Anomalous rectification in the metacerebral giant cells and its consequences for synaptic transmission, J. Physiol. (Lond.) 183, 287-304 (1966).

Nernst, W.: Die elektromotorische Wirksamkeit der Ionen, Z. physik. Chem. 4, 129-181 (1889).

Russell, J.M. and Brown, A.M.: Active transport of chloride by the giant neuron of the Aplysia abdominal ganglion, J. gen. Physiol. 60, 499-518 (1972).

Takeuchi, A. and Takeuchi, N.: On the permeability of the end-plate membrane during the action of the transmitter, J. Physiol. (Lond.) 154, 52-67 (1960).

Ussing, H.H.: Transport of ions across cellular membranes, Physiol. Rev. 29, 127-155 (1949).

LIMITATIONS TO SINGLE-PHOTON SENSITIVITY IN VISION

G. Falk and P. Fatt
University College London
London WC1E 6BT, England

1. Evidence for Single Photon Detection by Rods

Following the psychophysical studies of Hecht, Shlaer and Pirenne (1942) and van der Velden (1944), which involved the reporting by a dark-adapted human observer of his seeing a light stimulus, it has been generally accepted that the visual mechanism is capable under certain conditions of detecting a small number (less than 15) of photons absorbed in the retina. The only question that has arisen is over the precise number of photons within the range 2 to 15. Consideration of the experimental methods employed in these studies suggests that the threshold number of photons required to activate the visual mechanism is 5. In the experimental situation this corresponds to the mean number absorbed in the retina to cause the light flash to be seen in 60% of a large number of trials. The number of photons absorbed in individual trials necessarily fluctuates according to a Poisson distribution and it was shown in these studies that the major source of fluctuation in seeing the stimulus was due to this fluctuation in the light stimulus rather than in the visual detecting mechanism. The maximum duration of stimulus for constant threshold is about 100 msec and the maximum spatial extent amounts to a circular region in the visual field with a diameter of about 0.3 of a degree. The latter corresponds to an area of retina of about 0.004 mm^2 within which there are present about 500 rod receptors (Fig. 1). It is immediately evident that for 5 photons distributed at random within this area to be absorbed within this number of receptors, the probability of 2 or more photons being absorbed in the same receptor will be extremely small and cannot be involved in detection of the light stimulus. One arrives at the conclusion that the absorption of a single photon by the visual pigment contained within the outer segment of a single rod cell must, with a high degree of probability, lead to a signal being developed which can be transmitted further along the visual pathway.

Further studies (Barlow, 1956; 1957) have suggested that the threshold for seeing a stimulus, amounting to the absorption of 5 photons in

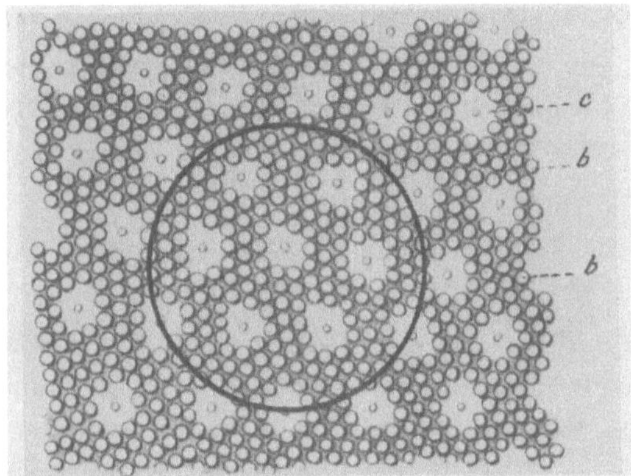

Fig. 1. Pattern formed by the outer segments of rods (b) and cones (c) in the receptor layer of the human retina in a region about midway from the fovea to the periphery of the retina. The large circle drawn on the figure is scaled to have a diameter of 48.5 μm and represents the limits of an optical image formed by a spot in the visual field which has a diameter of 0.17^{o} (angle subtended at the eye). This corresponds to the test spots used by Hecht, Shlaer and Pirenne (1942). (Drawing from Schultze, 1866 and Pirenne, 1961)

the dark-adapted retina, is determined by the presence of random fluctuations of excitation in the visual mechanism in the dark, the fluctuations being equivalent in their effect to 5 absorbed photons within the maximum of duration and spatial extent over which a stimulus may be applied and yield a constant threshold. If one assumes that the random fluctuations in excitability occurring in the dark have their origin in the photoreceptor cell and that the only signal transmitted onward to other cells in the visual pathway is one equivalent to that produced by the absorption of a photon, it can be calculated that for spontaneous signals to arise within the group of 500 receptors at the average rate of 5 per 100 msec, each rod must on the average generate a spontaneous signal once every 10 sec.

Now, it may be remarked that evidence for the detection of single photons has also been obtained among invertebrates. Studies of the locomotive behaviour of the fly, Musca, have shown an ability of the animal to respond to light flashes which result in the absorption of only a few photons in its visual pigment (Reichardt, Braitenberg & Weidel, 1968; Scholes & Reichardt, 1969). Furthermore, recording with microelectrodes from receptor cells of the compound eye of this animal has revealed discrete potential changes which are considered to represent the response to absorption of individual photons (Kirschfeld, 1966). Discrete potential changes attributable to the absorption of single photons have also been recorded from cells in the compound eye of the grasshopper (Scholes, 1964) and in the lateral eye of the horseshoe crab,

Limulus (Fuortes & Yeandle, 1964). In all these cases the observed potential change is in the form of a transient depolarization (decrease in internal negativity) of the receptor cell membrane close to the site of absorption of a photon by a pigment molecule situated in the membrane.

2. Structure of Vertebrate Retina

Receptor cell. Despite the similarity of the ultimate sensitivity of vision, its being limited by the absorption of a single photon in the receptor cell, the mechanism whereby this is accomplished may be very different in the vertebrate from what it is in the various species of arthropod where it has been examined. A structural feature of the receptor cell of fundamental significance in this respect is that in the rod cells of vertebrates practically all of the visual pigment is attached to membranous structures in the form of flattened sacs or discs which fill the outer segment and most of which are not continuous with the surface membrane of the cell (Fig. 2). The same situation applies in the cone cells of placental mammals. The relation of the visual pigment to the cell membrane is thus very different from that in the arthropod eye where the pigment is attached to the surface membrane of the receptor cell, which membrane is greatly expanded in the form of microvilli. No discrete potential changes that could be attributed to the absorption of single photons have been recorded from any cells in the vertebrate retina. A striking characteristic of the vertebrate retina is that the receptor cells, both rod and cone, as well as the nerve cells which are in direct synaptic contact with the receptor cells,

Fig. 2. Electronmicrograph showing a portion of a longitudinal section of a rod outer segment of frog retina. The surface membrane appears at the left of the figure, as a vertical line with local irregularities. The pair of horizontal lines correspond to the stacked flattened sacs (or discs) which occupy most of the interior of the outer segment and to which the visual pigment, rhodopsin, is attached.

are all found to be incapable of producing a regenerative electrical response in the form of a membrane depolarization: the mechanism for the generation of an action potential is absent in these cells.

Retinal organization. Since this paper is concerned with events taking place within and between the different cells of the vertebrate retina, it is appropriate to give a description of the structures concerned. The drawing in Fig. 3 represents a section of the retina of a dog, perpendicular to the plane of the retina. The structure shown is typical for the retinae of mammals, including that of man away from the fovea. The thickness of the retina varies but little between different species of mammals, being in the region of 0.20 to 0.25 mm. Light entering the eye reaches the retina from below (as the drawing is presented). It passes upward through the various layers of the retina from I through D (as designated at the right of the figure) within which there occurs very little absorption, but some scattering. Those photons that

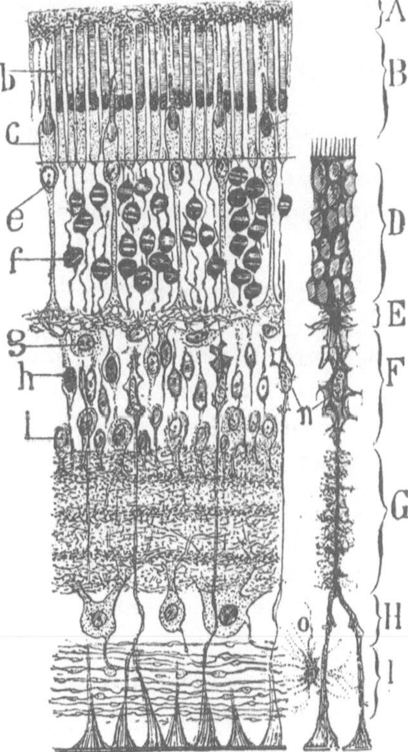

Fig. 3. Drawing of a section of the retina of a dog showing all the cellular elements composing it. The vertical direction in the figure corresponds to the direction normal to the retina, with the lower edge corresponding to the inner limit of the retina, bordering the vitreous humor. The drawing on the right (which is to the same scale and at the same vertical position) shows the structure of the large glial cell, the Müller cell. (From Cajal, 1911)

are effective in eliciting a sensation are absorbed in the outer segments of the receptor cells, shown as transversely striated structures in the upper half of B. Any light that has not been thus absorbed in the receptors will then be absorbed in the layer of pigment epithelium (labelled A) which is composed of closely fitting melanin-containing cells and forms an opaque screen on the outer surface of the retina. The major layers of the retina, distinguished by their neuronal content, are as follows: The receptor layer (B) includes the inner and outer segments of the two types of receptor cell, the outer segment of a rod (b) and the inner segment of a cone (c) being indicated. (The dense ovoid bodies shown in this region represent densely packed clusters of mitochondria which occupy the outermost part of the inner segments.) The outer nuclear layer (D) contains the nuclei of the rod (f) and cone (e) receptor cells and the vertically running processes of these cells. The outer plexiform layer (E) includes the synaptic contacts between receptor, bipolar and horizontal cells and the horizontally running processes of all these cells. The inner nuclear layer (F) includes the nuclei of horizontal cells (g), bipolar cells (h) and amacrine cells (i). It is spanned by the vertically running processes of the bipolar cells. The inner plexiform layer (G) contains the synaptic contacts between bipolar cells, ganglion cells and amacrine cells. The extensively ramifying processes of these cells form a number of secondary layers of increased density of neuronal elements within this region. The ganglion cell layer (H) includes the cell bodies of those neurones, the axons of which course horizontally in the optic fibre layer (I), from whence they pass out of the eye in the optic nerve.

In addition to these neural elements, the retina contains two types of glial cells: small cells of stellate form (o) which are present in the optic fibre layer and much larger cells, called Müller cells, which extend over most of the thickness of the retina with their nuclei (n) being located in the inner nuclear layer. A drawing of a single Müller cell, showing the relation of its various parts to the different layers of the retina, is included in Fig. 3 to the right of the complete section.

What may be envisaged as the direct pathway for signal transmission from rod receptor, through bipolar and ganglion cells to the optic nerve fibre, is indicated in Fig. 4.

3. Transmission of Signals

As mentioned earlier, investigations with microelectrodes have demonstrated that the receptor cells in the vertebrate retina, as well as

Fig.4. The direct pathway for visual signal transmission at low light levels (scotopic vision), including (from above downward) rod receptor cell, bipolar cell, ganglion cell with synaptic contacts between them. (From Cajal, 1911)

the horizontal and bipolar cells which make synaptic contact with them, are all incapable of generating action potentials. In contrast, the amacrine and ganglion cells do generate action potentials leading in the latter case to the propagation of impulses along the optic nerve fibers to the brain. If amplification does occur in the transmission of the visual signal from the photochemical conversion of a molecule of visual pigment in the outer segment of a receptor cell to the eventual generation of post-synaptic potential changes in the amacrine and ganglion cells, it appears not to involve regenerative electrical activity but it might involve some form of chemical triggering The term amplification is not restricted here to the idea of an increase in energy available for doing work on an outside body, but includes the idea of an increase in the number of similar chemical units, either molecules or ions, undergoing some change in chemical state or in position. Now in the provision of amplification, a grouping of chemical units into a packet which can then be triggered so that all the chemical units contained within it will act synchronously, would appear to be essential. One does in fact expect such a collective action to occur in the release of chemical transmitter from the pre-synaptic terminal of the receptor cell and that this will lead to the production of discrete postsynaptic potential changes in the bipolar cell which lies in the direct pathway for visual signal transmission.

A. Chemical Transmission at Synapses

(1) Miniature synaptic potentials: discrete units of synaptic activity

Evidence for the existence of discrete units of synaptic activity was first obtained by recording membrane potentials from skeletal muscle fibres of the frog (Fatt & Katz, 1952). The series of traces occupying the upper part of Fig. 5A was obtained at a high sensitivity of recording and with a slow sweep in the absence of any stimulus. The intracellular microelectrode used for recording of membrane potential was in this case situated close to the region of synaptic contact of the motor nerve fibre with the muscle fibre (the endplate), as indicated by the response to nerve stimulation recorded at lower amplification and faster sweep, and shown at the bottom of Fig. 5A. This record, obtained at the same site as the upper series of traces, shows the muscle fibre action potential to be initiated by an endplate potential which attains a height of about 40 mV before the abrupt increase in rate of rise, indicative of the transition to a regenerative action-potential type of response. The upper part of Fig. 5A shows the occurrence at the endplate of small, but discrete, fluctuations of membrane potential of strikingly uniform time course and of fairly uniform amplitude. These were not detectible when the microelectrode was situated more than about 1 mm from the endplate of the same muscle fibre (Fig. 5B). They have been designated miniature endplate potentials on the basis that their time course resembles that of the endplate potential evoked by nerve stimulation in a muscle fibre.

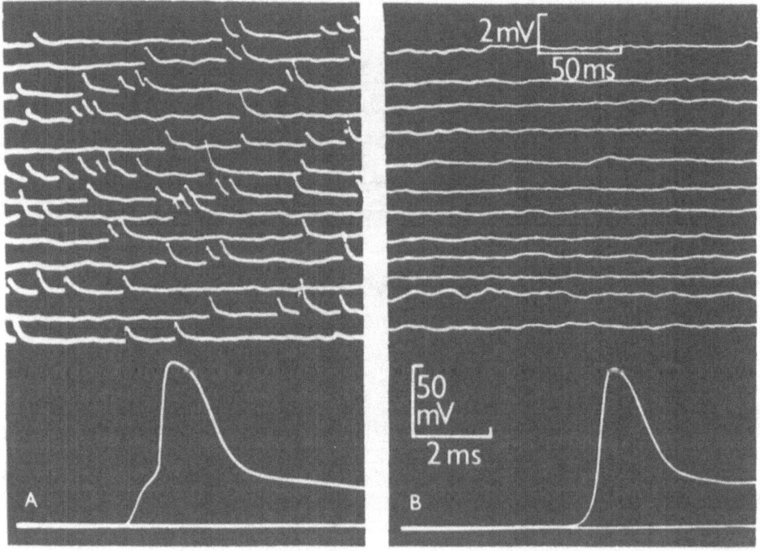

Fig. 5. Records of membrane potential of frog muscle fibre obtained with intracellular microelectrode. A: with microelectrode placed at the endplate as indicated by the response to nerve stimulation (lower part of figure), the several sweeps recorded at high gain in the absence of stimulation (upper part) show spontaneous, randomly occurring deflections known as miniature endplate potentials. B: with mircoelectrode placed 2 mm away from the endplate in the same muscle fibre, miniature endplate potentials are hardly detectible. Separate potential and time scales, as given in B, apply to the upper and lower parts of both A and B. (From Fatt and Katz, 1952)

The miniature endplate potentials occur in a random time sequence. There is some scatter in the amplitude of these potentials at any endplate, but under favourable conditions for observing them it can be seen that they do not grade down to negligible amplitude. Examined statistically, the standard deviation of their amplitude is usually about one-third of the mean value of amplitude. Treatment of the muscle with a curarizing drug causes a diminution of the miniature endplate potential, while treatment with a substance which interferes with the action of the enzyme, cholinesterase, which normally rapidly hydrolyses acetylcholine, causes a prolongation of the miniature endplate potential. Both these effects suggest that the miniature endplate potential, like the endplate potential, is generated by a quantity of acetylcholine being discharged from the nerve terminal into the extracellular space in the neighbourhood of the postsynaptic region of the muscle fibre and acting on the muscle fibre membrane to increase its permeability to sodium and potassium ions, and hence to produce a depolarization from the resting level of membrane potential.

Site of transmitter action on postsynaptic membrane. That the discrete form of the miniature potential is not due to a single pore for ionic passage across the membrane being activated by a single molecule of acetylcholine (or possibly by a few such molecules) is indicated by the graded depolarization produced by the application of acetylcholine to the junctional region of the muscle fibre, which shows much less fluctuation than would occur if it were the result of the superposition of rapidly recurring miniature endplate potentials. Only recently has evidence been obtained for the operation of individual pores, as a result of the examination of fluctuations in the depolarization produced by acetylcholine (Katz & Miledi, 1972). Analysis of these fluctuations have indicated that the activation of an individual pore, presumed to be due to a single molecule of acetylcholine, produces a conductance increment with a duration similar to that underlying the miniature endplate potential but with an amplitude only about 1/1000 times as great. While the individual miniature endplate potential is thus shown to depend on the near synchronous activation of about one thousand receptive units (or pores) in the postsynaptic membrane, the extent of the region involved is only a very small fraction of the entire region involved in the production of an endplate potential or in the production of a prolonged train of miniature potentials. The latter is demonstrated by recording with a microelectrode situated just outside the muscle fibre in the endplate region. Such an electrode can detect current flowing across the membrane in the immediate neighbourhood of its tip, this current producing a measurable voltage drop in the external fluid where it converges onto a small active region of membrane. Comparison of externally recorded miniature potentials with those recorded intracellularly from the same endplate shows that successive miniature potentials in a spontaneous train involve the activity of different small patches of the postsynaptic membrane, each such patch amounting to less than 1/100th (perhaps less than 1/1000th) of the total region capable of being activated by the train. This is important in that it reveals that there can be no interference in the postsynaptic action of separate packets of acetylcholine producing the miniature endplate potentials, except through the change effected by them in the membrane potential.

Evoked synaptic potentials. One field of investigation which has been opened up by the discovery of the spontaneously occurring miniature endplate potential is that concerned with showing that the response of the muscle fibre to nerve stimulation is the result of the superposition of units of synaptic activity, the individual unit having the characteristics of the spontaneous miniature potential (del Castillo & Katz, 1954; Boyd & Martin, 1956). The results of these and other studies have led to the conclusion that a presynaptic nerve impulse causes a large increase in the probability of release of packets of transmitter from the nerve terminal, each packet if released by itself being capable of producing a miniature endplate potential, and furthermore that this effect is probably dependent on the action potential inducing an entry of calcium ions into the nerve terminal.

Generality of miniature synaptic potentials and synaptic vesicles. Similar spontaneously occurring miniature synaptic potentials with the same relation to the evoked response have been detected in many other types of postsynaptic cells, including sympathetic ganglia (Blackman, Ginsborg & Ray, 1963) and motoneurons of the spinal cord (Katz & Miledi, 1963). Electron-microscopic examination of many kinds of synapses where chemical transmission is known to occur has revealed the presence of membrane-bound vesicles about 30 nm in diameter, concentrated in the presynaptic nerve terminal. It has been generally supposed that miniature synaptic potentials are a consequence of the release of transmitter from the nerve ending in discrete quantities corresponding to the contents of individual vesicles. Vesicles, isolated from the rest of the nerve cell following disruption of nerve tissue, have indeed been found to contain transmitter substance in high concentration, in agreement with the idea that they are the anatomical correlate of the miniature synaptic potential. Miniature synaptic potentials are not confined to excitatory synapses but can be observed at inhibitory synapses as well, involving chemically different transmitters, and indications have been found that the shape of the synaptic vesicles is dependent on the type of transmitter (Uchizono, 1965).

(2) Presynaptic-postsynaptic transfer function

Dependence of miniature potential frequency on presynaptic potential. Studies have been carried out of the effect of displacements of presynaptic membrane potential on the frequency of randomly occurring miniature synaptic potentials (del Castillo & Katz, 1954b; Liley, 1956). Liley applied current to the phrenic nerve of the rat close to its synapse with the diaphragm muscle. The results of applying currents of varying intensity, both anodic (in the direction to produce hyperpolarization of the nerve terminals) and cathodic (to produce depolarization of the nerve terminals), are illustrated in Fig. 6. The frequency of the miniature endplate potentials is seen to vary as an exponential function of the applied current over a 100-fold range of frequencies The scale for applied current given in Fig. 6 is in relative units. If it is assumed that the electrical behaviour of the nerve fibre is linear over the range of observation in which a straight-line relation is obtained in the plots of Fig. 6, then it would appear that throughout this range the rate of transmitter release (T) given by the mean frequency of

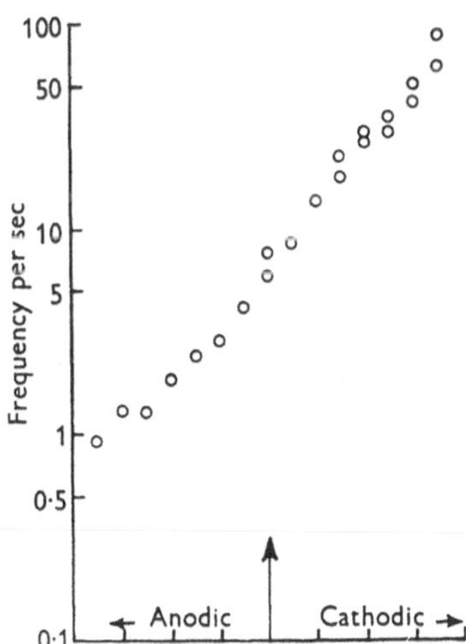

Fig. 6. Plot of the frequency of miniature endplate potentials (on a logarithmic scale) recorded from a fibre of the rat diaphragm muscle against current (on a linear scale) applied to the splanchnic nerve close to the point of entry of the nerve into the muscle. The vertical arrow indicates zero applied current; anodic current represents the situation with the electrode nearer the muscle made positive, so that current flows inward across the membrane of the nerve terminals, producing hyperpolarization; cathodic current represents the opposite. Relative values of applied current are given by the abscissal scale. (From Liley, 1956)

miniature endplate potentials is related to the change in membrane potential of the presynaptic nerve terminals (ΔE_{pre}) by an equation of the form

$$T = Ae^{b\Delta E_{pre}} \quad (1)$$

where A and b are positive constants. Confirmation of these ideas was obtained in another set of experiments in which the nerve terminals were depolarized by increasing the potassium ion concentration of the solution bathing the nerve-muscle preparation. These results indicated that a 10-fold change in transmitter release rate is elicited by a 15 mV change in presynaptic membrane potential, which thus gives a value for b in eqn. 1 of 2.3/15 mV = 0.15 mV^{-1}.

It must be emphasized that the relation given by eqn. (1) is an empirical one. It seems, however, to have a wide range of applicability. In the case of the nerve-muscle synapse, it is capable of describing the release of transmitter by a nerve impulse: the action potential having an amplitude of about 120 mV is predicted to produce a 10^6-fold increase in release rate for its duration of about 1 msec, and will thus elicit an endplate potential consisting of 200 to 300 miniature endplate potentials from a background spontaneous release rate of the order of one miniature endplate every second. Experiments carried out on the synapse between giant nerve fibres in the stellate ganglion in the mantle of the squid have employed intracellular microelectrodes situated in both the pre- and postsynaptic fibres. These have also indicated an exponential dependence of transmitter release rate, as measured by depolarization of the postsynaptic membrane, on changes of presynaptic membrane potential in accordance with eqn. (1), but with b = 0.20 mV^{-1} (Katz & Miledi, 1967).

Non-linear summation. It is to be noted that changes in the postsynaptic membrane potential will be proportional to the presynaptic rate of release of transmitter only for small displacements of postsynaptic potential. A non-linearity in the relation between membrane potential change and transmitter release rate is expected on the grounds that the immediate electrical effect of transmitter on the postsynaptic cell membrane is a conductance increase, and this effect will be proportional to transmitter release rate so long as increasing the rate results in a proportional increase in the population of transmitter-controlled pores being acted upon. The dependence of membrane potential on synaptic activity can be understood from consideration of the electrical network of Fig. 7 (see also the accompanying paper, Fatt, 1974). The effect of the transmitter on most postsynaptic cells is to introduce a conductance increment Δg which has in series with it a battery, E_r. This conductance path is in parallel with another conductance path between inside and outside of the cell, representing the properties of the cell membrane in the absence of transmitter activity. This latter path consists of the transmembrane conductance, G_O, in series with the voltage source, E_m.

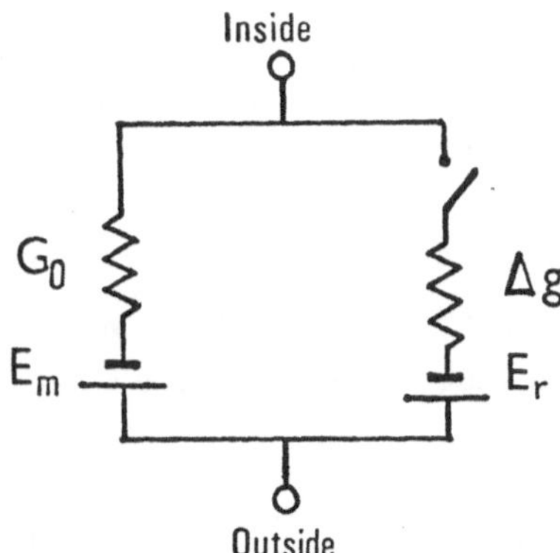

Fig. 7. Network model used to represent the action of synaptic transmitter on the postsynaptic cell. This model assumes linear electrical behaviour for both the membrane conductance in the absence of transmitter activity (G_O) and the conducting path introduced by transmitter activity.

Increasing the rate of transmitter release will result in a proportional change in the transmitter-induced conductance increment Δg so that

$$\Delta g = \alpha T \tag{2}$$

where α is a positive constant. The steady-state potential change produced by Δg will then be given by the hyperbolic relation

$$\Delta E_{post} = \frac{E_r - E_m}{1 + \frac{G_o}{\Delta g}} \tag{3}$$

Equation (3) assumes that G_O behaves electrically in a linear manner. It will be seen that the change in postsynaptic potential change is proportional to transmitter activity at low levels of activity, but is limited to $E_r - E_m$, approached at high levels of transmitter activity.

Dynamic voltage transfer function. Combining equations (1) to (3), one obtains the dependence of postsynaptic membrane potential change on presynaptic membrane potential change, that is

$$\Delta E_{post} = (E_r - E_m)(1 + ce^{-b\Delta E_{pre}})^{-1} \qquad (4)$$

where $c = G_o/\alpha A$. This relationship gives a sigmoidal curve, which is of the form of a hyperbolic tangent with its argument ranging from $-\infty$ to $+\infty$. One may differentiate the expression on the right of equation (4) with respect to ΔE_{pre} to obtain the dynamic voltage transfer functions (S) for the change in postsynaptic membrane potential with small change in the presynaptic potential. Since uncertainty exists regarding the quantity c, which may be viewed as a measure of the coupling between pre- and postsynaptic cell, it is more instructive to consider S as a function of ΔE_{post} rather than of ΔE_{pre}, in which case there is no longer an explicit dependence on c. One thus obtains the simple parabolic relation

$$S \equiv \frac{d\Delta E_{post}}{d\Delta E_{pre}} = b\Delta E_{post}\left(1 - \frac{\Delta E_{post}}{E_r - E_m}\right) \qquad (5)$$

From equation (5) it can be seen that the transfer function will be a maximum when the transmitter-induced change in postsynaptic potential, ΔE_{post} has a value equal to $\frac{1}{2}(E_r - E_m)$. This maximum will be given by $\frac{1}{4}\,b(E_r - E_m)$ which, for $b = 0.15\ mV^{-1}$ and $(E_r - E_m) = 75$ mV, will have a value of 2.8.

B. The Problem of Signal Transmission in the Retina[§]

(1) Membrane potential changes in receptor cells

It is intended to apply the above considerations to the synaptic contacts made in the outer plexiform layer of the retina, where the receptor cells act as presynaptic elements and the bipolar and horizontal cells act as postsynaptic elements. First, however, one wishes to consider the nature of the change in surface membrane properties of the receptor cell, which the absorption of light within the cell is able to effect. Investigations with intracellular microelectrodes have revealed that the sole effect on the membrane potential of any type of vertebrate receptor cell, produced by the absorption of light within that receptor, is a hyperpolarization (an increase in internal negativity) from the pre-existing level of membrane potential (Bortoff & Norton, 1967; Toyoda, Nosaki & Tomita, 1969; Werblin & Dowling, 1969; Baylor &

[§]Falk & Fatt, 1972, pp. 230-239.

Fuortes, 1970). The hyperpolarization produced by light, from a level of internal negativity in the dark of about -40 mV, is associated with a conductance decrease of the surface membrane. Other studies, with electrodes situated in the extracellular space, have shown that in the dark current leaves the inner segment and enters the outer segment of receptors and the absorption of light causes a diminution of this current (Hagins, Penn & Zoshikami, 1970). The observations suggest that in the dark the surface membrane of the outer segment is fairly permeable toward sodium ions, which pass inward across the membrane under the combined influence of a higher concentration of sodium outside than inside and of the membrane potential. The effect of light is to cause a reduction in sodium permeability of the surface membrane of the outer segments. Consistent with this idea is the observation that the light-induced change in receptor potential is dependent on the external sodium concentration. There is, furthermore, evidence for the existence of a metabolically driven pump which extrudes sodium ions from the receptor cell. This is required in order to compensate for the continual passive penetration of sodium ions into the cell which occurs in the dark, so that the cell will over long periods of time remain in a steady state with respect to its ionic content.

The hyperpolarization initiated by a reduction in sodium permeability of the outer segment will spread with decrement to other parts of the receptor cell by the cable properties of the cell, that is, by a light-induced component of current flowing longitudinally through the interior of the cell and the extracellular space and across the surface membrane. Because of the long duration of the potential change in the outer segment even with flashes of brief duration and low intensity (amounting in the mammalian retina to about 200 msec between points at one-half peak amplitude), the spread of membrane potential change will be affected only to a minor extent by the surface membrane capacitance. A severe attenuation of the light-induced hyperpolarization, is, however, expected to occur between the outer segment and the presynaptic terminals because of the surface membrane conductance of the intervening portion of the cell. Measurement of the variation of the longitudinal current with distance along the receptor cell, corresponding to the intensity of current crossing the membrane per unit length, are presented in Fig. 8. The measurements (Penn & Hagins, 1969) were made on the retina of the rat which appears to consist entirely or almost entirely of rods. (It should be noted that the intensity of the brief flashes employed in evoking the responses examined in this figure was in excess of that which would give a response with minimum duration and with an amplitude prop-

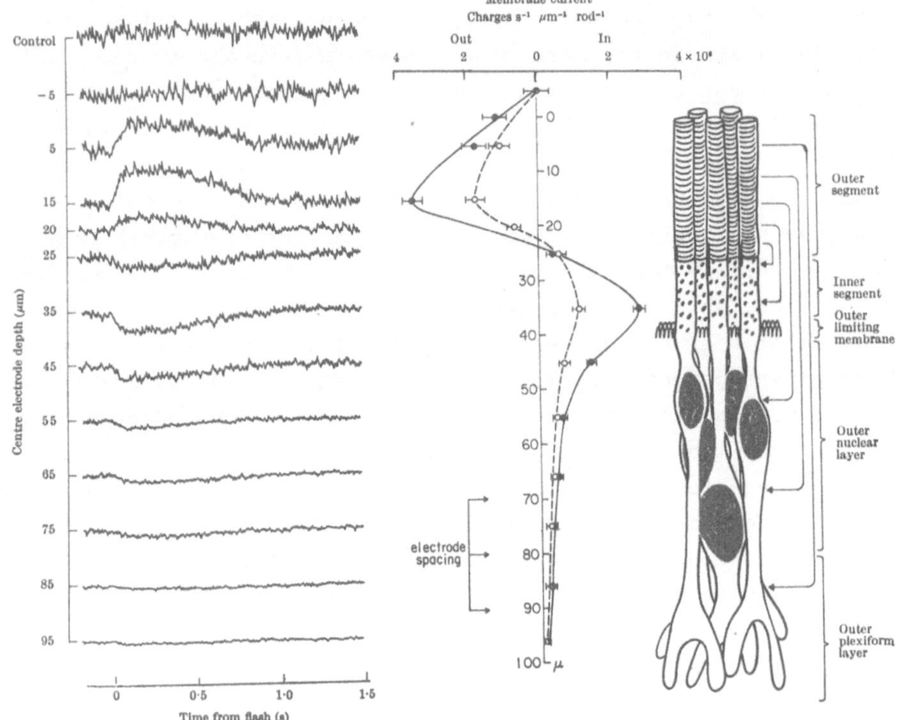

Fig. 8. Spatial variation (parallel to the rod axis) of the change in membrane current density following a brief flash of light delivered to rods in the rat retina. Middle: Plots of peak change in membrane current calculated for unit length of a single rod cell. Hollow circles give the peak response to a light flash leading to the absorption of 250 photons per rod (corresponding to the records on the left, obtained by means of 3 microelectrodes spaced 10 μm apart in the direction of the rod axis, used to measure the divergence of extracellular current at each retinal depth). Filled circles give the peak amplitude of the membrane current with a flash from which 10^3 photons per rod were absorbed. Right: Drawing of rod receptor cell of rat retina. Arrows indicate the flow of current transiently following a light flash.

ortional to flash intensity.) The indicated variation of membrane current along the length of the rod cell is superimposed on a current flowing in the dark which is of opposite sign. From such information it has been concluded that the hyperpolarization occurring in the presynaptic terminals will be about 1/10th of that in the outer segment.

Now, it is found in the rat retina that the absorption of 30 photons per rod is sufficient to produce a potential change of one-half the maximum amplitude obtainable with very bright flashes. Intracellular recording from rods and cones of other animals indicated a maximum hyperpolarization of about 30 mV. On the basis that the response up to half-maximum amplitude is proportional to the number of absorbed photons, one may compute that the absorbtion of a single photon would produce a hyperpolarization of 0.5 mV in the outer segment and thus of 50 μV in the presynaptic terminal. Other estimates that have been made have yielded even smaller values for the amplitude of hyperpolarization produced by a single absorbed photon, such values being smaller by a factor of 1/10.

(2) Signal to noise considerations

Random fluctuations due to thermal agitation. The question now arises as to whether the estimated potential change in the presynaptic terminal would be sufficient to account for single-photon detection. In the first place, it may be considered in relation to spontaneous random fluctuations of membrane potential, due to statistical effects in the movement of individual ions across the membrane. On the basis that the rate of change of membrane potential produced in this way is limited by the membrane time constant, with an assumed value of 20 msec, the root-mean-square amplitude of the fluctuations in membrane potential from the mean level is calculated to amount to about 3 μV. As described earlier, psychophysical studies have lead to the idea that the requirement for a minimum of 5 photons to be absorbed, in order for a visual sensation to be effected, is a consequence of each rod producing in the dark a single event equivalent to that caused by an absorbed photon on the average of one every 10 sec. The random fluctuations would be effective in yielding these signals in the dark if the threshold for evoking them corresponded to a displacement of membrane potential from its mean value of about 10 μV. (This is arrived at by considering that the deviations of potential from the mean level will follow a Gaussian probability distribution with a standard deviation equal to the root-mean-square deviation, and for this distribution there will be a probability of 0.002 for the potential to deviate in one direction from the mean by more than 3 times the standard deviation. The value of 0.002 is the probability of the occurrence of a signal within each integration period of 20 msec, if signals occur once in 10 sec.) It appears from the foregoing that with the generous estimate of 50 μV per absorbed photon, as the potential change effected in the rod cell terminal, single photon excitation can be accounted for, provided there is no important source of random fluctuation in the visual pathway other than that expected to occur in the membrane potential of the rod-cell terminals owing to thermal agitation.

Random fluctuations at chemical synapses. In fact, a much more important source of fluctuation in the transmission of a visual signal will be in the mechanism of release of chemical transmitter at the presynaptic terminals. Although miniature synaptic potentials have not been recorded from the postsynaptic cells (bipolar and horizontal cells) receiving synaptic contacts from receptor cells, let us assume that transmitter acts on them in the form of packets producing discrete increments of membrane conductance similar to those found at most other synapses. Apart from technical difficulties in recording from these postsynaptic cells with microelectrodes, the observation of individual miniature synaptic potentials would be made virtually impossible by the expected circumstance that the transmitter is released at a high rate. This expectation follows from the relatively low level of internal negativity of the receptor cell observed in the dark which would lead to a high rate of release of transmitter, if the mechanism of transmitter release is the same as has been found at all other kinds of chemical synapses. The effect of the transient hyperpolarization produced in the receptor cell terminal by the absorption of light by the receptor cell is expected to be a decrease in the rate of release of transmitter. Recordings have been made from horizontal cells, which are in a postsynaptic relation to the receptor cells but which also synapse back on to receptor cells (Svaetichin & MacNichol, 1958; Naka & Rushton, 1966; Steinberg,1969; Baylor, Fuortes & O'Bryan, 1971). Potential changes recorded from horizontal cells are consistent with the idea of a high rate of release of transmitter by the receptor cell in the dark and a decrease in release produced by light. Hyperpolarization of horizontal cells of up to 60 mV by light stimuli has been reported (from a membrane potential of -20 mV), consistent with the idea that the transmitter released from receptor cell terminals is an excitatory one. Bipolar cells, owing to their small size, are more difficult to record from than horizontal cells. One would expect that bipolar cells would likewise be in a depolarized state in the dark and undergo hyperpolarization on illumination of the retina, at least with a small spot of light centred on the bipolar cell. A class of bipolar cells obeying this expectation has been found (Werblin & Dowling, 1969; Kaneko, 1970). There is however another class of bipolar cells which depolarize in response to light. Both classes exhibit a centre-surround antagonism such that illumination of the surround gives rise to a potential change of opposite sign to that evoked by illumination of the centre of the receptive field of the bipolar cell. (The effect of the surround is likely to be mediated via the horizontal cells. A centre-surround organisation could arise if the horizontal cell acted as an interneurone between receptor and bipolar cell or if it fed back onto receptor cells, given that horizontal cells have a large receptive field, as has been observed. It is not known if horizontal cells make synaptic contact with bipolar cells. Studies of the effect of applying current to horizontal cells on the discharge of ganglion cells (Naka & Witkovsky, 1972) lead one to suppose that centre-surround organization at the bipolar cell level is the result of feedback from horizontal cells to receptors.)

Voltage fluctuations at bipolar cells. The following argument will apply to those bipolar cells which are hyperpolarized by light absorbed in the centre of their receptive field, as well as to horizontal cells. The high release rate of transmitter from the receptor cell in the dark will be effective in displacing the membrane potential of these postsynaptic cells a substantial fraction of the way from the potential which would exist in the absence of transmitter activity, toward the reversal potential for excitatory postsynaptic activity, E_r. It will be recalled that, when the displacement of membrane potential by transmitter activity is about half-way toward E_r, the maximum sensitivity for the transfer of a small signal from a presynaptic to a postsynaptic potential change occurs, as indicated by equation (4). Under these conditions the dynamic voltage transfer function will be a maximum with a value of about 2.8. However, a consequence of the presynaptic release of transmitter in packets will be the introduction of noise in the postsynaptic cell. The problem of fluctuations arising in this way and the limitation thus imposed on the transmission of small signals will now be considered. In the absence of quantitative experimental information on miniature synaptic potentials in the postsynaptic cells contacted synaptically by the receptor cells, one may reasonably assume that the time integral of the membrane conductance increment underlying a miniature synaptic potential is about the same as in other postsynaptic cells. Measurements at the neuromuscular junction of skeletal muscle indicate that a typical miniature endplate potential of peak amplitude 0.3 mV is generated by a local current across the postsynaptic membrane of peak amplitude 3 nA and with a time constant of exponential decay of 2 msec. For a driving force of $E_r - E_m = 75$ mV, this is seen to correspond to a conductance increment, the time integral of which is given by $\frac{3 \text{ nA} \times 2 \text{ msec}}{75 \text{ mV}} = 8 \times 10^{-11}$ mho sec.

Now, the membrane potential of the postsynaptic cell will be displaced from its resting level (that is, its level in the absence of transmitter activity) half-way to the reversal potential E_r (equation

(2)) when the mean conductance increment produced by transmitter activity is equal to the conductance between inside and outside of the resting cell, G_0. If the input conductance of the resting cell is 10^{-6} mho, then depolarization to the extent of $\frac{1}{2}(E_r - E_m)$ will occur when packets of transmitter impinge upon the cell at the rate of 10^{-6}mho/(8 x 10^{-11}mho sec)= 12,000/sec. (The input conductance of the bipolar and horizontal cells is probably much smaller than 10^{-6} mho. To the extent that this is so, a smaller number of packets of transmitter per unit time will be required to keep the cell depolarized in the dark, with the result that the difficulty in accounting for small signal transmission from pre- to postsynaptic cell, as described below, will be even more serious.)

There will be smoothing of postsynaptic potential changes due to the membrane capacitance, the effectiveness of which will be determined by the membrane time constant which may be reasonably assumed to have a maximum value of about 20 msec. There is a possibility, however, that smoothing might also be exerted by a delay in the diffusion of transmitter from the presynaptic release site to the postsynaptic membrane. The maximum value of integration time (a measure of the smoothing of potential changes) that one may assume is limited to about 200 msec by the requirement that the signal being transmitted across the synapse should not be attenuated to nearly the same degree as spontaneous fluctuations. Taking the integration time to be 200 msec, one calculates the level of membrane potential displacement by transmitter activity in the dark to be determined effectively by 2400 packets of transmitter. For a random time sequence, such as the release of packets of transmitter is known to follow, the number of events within many separate intervals of constant duration will have a probability distribution in accordance with a Poisson series. For this type of distribution the standard deviation of the number of events within the interval of given duration will be equal to the square root of the mean number of events. Thus the number of packets of transmitter determining the membrane potential can be described in terms of a mean and a standard deviation as 2400 $\pm$ 49. The relative fluctuations in the amount of transmitter acting on the postsynaptic cell amounts to 2.0%. It has been considered that for synapses in general, a 10-fold increase in transmitter release is produced by a 15 mV change in presynaptic membrane potential. The calculated fluctuations in postsynaptic transmitter action would thus be equivalent to fluctuations in presynaptic potentials with a standard deviation of 15 mV x $\log_{10}1.020$ = 0.13 mV. As was mentioned earlier, in order to account for the infrequent occurrence of spontaneous signals equivalent in their effect to the absorption of a photon, it is necessary that the presynaptic potential change produced by the absorption of a photon should correspond to a certain factor times the standard deviation of the random fluctuations in presynaptic potential, this factor being determined by the ratio of the integration time for the production of postsynaptic potential changes to the mean interval between spontaneous signals. In the present case this ratio will be 0.2 sec/10 sec = 0.02, which requires that the photon-evoked presynaptic potential change have an amplitude 2.1 times the standard deviation of the fluctuations (the latter is assumed to have a Gaussian probability distribution, with 0.02 corresponding to the area of a tail of the normalized curve extending beyond 2.1 standard deviations from the mean). This amounts to 2.1 × 0.13 mV = 0.27 mV. But as will be recalled, the magnitude of potential change produced in the presynaptic terminals by the absorption of a photon has been estimated to be only 0.05 mV, that is, too small by a factor of 1/5. Moreover, the values assumed for those quantities about which there is some uncertainty have been such as would tend to reduce the importance of spontaneous fluctuations due to the release of transmitter in packets. Yet one knows from the psychophysical studies that signal transmission from rods absorbing a single photon does occur.

Rod-bipolar convergence. One way out of the present difficulty might be to suppose that the presynaptic potential change induced by the absorption of a single photon in a rod receptor is much larger in the intact living animal than it is in the isolated tissue maintained under artificial conditions as used for electrical measurements of the receptor cell response. There is, however, a further condition of signal transmission from rod to bipolar cells which renders any such fudging of the amplitude of the presynaptic potential change of no avail in accounting for single-photon detection. Account must be taken of the fact that in the region of the human retina where single-photon detection by rod receptors has been demonstrated, each bipolar cell which is concerned in the transmission of signals between the rod and ganglion cells receives synaptic contacts from between 15 and 50 rod cells. Since there is no action potential mechanism in the bipolar cell, the postsynaptic actions of these numerous receptor cells will sum in the bipolar cell and thus produce the signal to be conveyed forward by passive spread of potential change within the bipolar cell toward the synaptic contacts of the inner plexiform layer. If one takes into account convergence of receptor cells onto bipolar cells, then the number of packets of transmitter (2400) which have been considered to determine the level of depolarization of the bipolar cell in the dark would be released, not from a single receptor cell, but from all those receptors making synaptic contact with the individual bipolar cell. Taking the average number of rod receptor cells contacting a bipolar cell to be 25, and assuming that each such receptor contributes an equal number of packets of transmitter, one arrives at the result that each receptor contributes 2400/15 = 96 packets to the total number determining the postsynaptic potential level in the dark. The standard deviation of the fluctuation in the number of packets acting on the bipolar cell will continue to be 49, with 2.1 times this number being the change required for signal transmission. The conclusion to be reached is that with the known convergence of receptor cells onto bipolar cells it would be just marginally possible for single-photon detection to take place, but with the requirement that the absorption of a single photon by a rod cell should produce a hyperpolarization sufficient to shut off completely the release of transmitter from its terminals!

Depolarizing-type bipolar cells. We have yet to consider those bipolar cells which depolarize when the centre of their receptive field is illuminated. It has been found recently by Toyoda (1973) that this depolarization is associated with a conductance increase of these cells, the reversal potential E_r being in the region of zero membrane potential. Two explanations can be offered: (1) that hyperpolarization of the receptor cell increases its rate of release of an excitatory transmitter, or (2) that the transmitter released in the dark is one which decreases the conductance of ionic channels. As to the first proposition, such a dependence of transmitter release on membrane potential has never been observed. It has recently been found, however, that at a number of synapses, transmitters act to decrease ionic conductance (Weight & Votava, 1970; Paupardin-Tritsch & Gerschenfeld, 1973). The membrane potential of a postsynaptic cell at which transmitter blocks conducting channels will be given by

$$E_{post} = E_m + \frac{E_r - E_m}{1 + \frac{G_o}{G_r + \Delta g}} , \quad -G_r < \Delta g < 0 \tag{6}$$

where G_r is the total conductance of the channels in the absence of transmitter. The postsynaptic potential change will be a hyperbolic function of transmitter release, but in contrast to equation (3) the change in postsynaptic membrane potential with transmitter release becomes

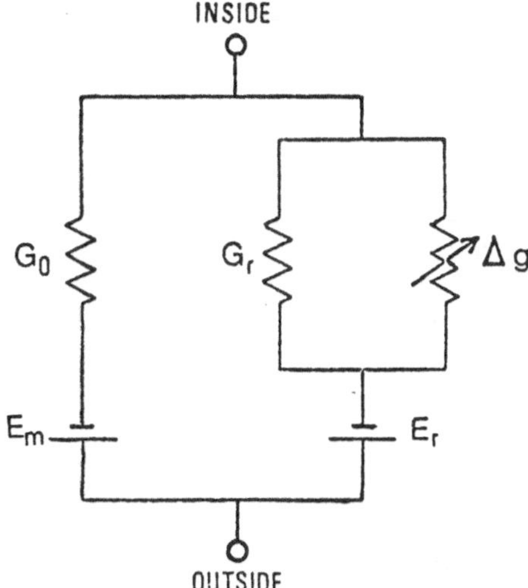

Fig. 9. Network model used to explain the behaviour of a class of bipolar cells (depolarizing on illumination of the centre of their receptive field), on the basis of a transmembrane conductance decrease being produced by a synaptic transmitter released from rod receptor cells in the dark. The synaptically controlled conductance element, Δg, is restricted to negative values from $-G_r$ to zero.

increasingly rapid until the release of transmitter approaches a rate at which all conductance channels are blocked and no further change is possible. For such a mechanism of synaptic transmission, it can be shown in the same way as used in deriving equation (5), that the dynamic voltage-transfer function will be given by

$$S = b(E_r - E_{post})\left[1 + \frac{E_{post} - E_r}{E_r - E_m}\ \frac{G_o + G_r}{G_o}\right] \qquad (7)$$

The absolute value of S can be considerably greater for a transmitter producing a conductance decrease rather than an increase if G_r is comparable to or greater than G_O. If the transmitter released from the terminals of retinal rods were of this type, then a value of about 14 (absolute value) is not unreasonable for the dynamic voltage-transfer function for transmission between rods and bipolar cells in the dark-adapted retina. This value is based on an estimate of $(E_r - E_{post})$ = 60 mV, $b = 0.15\ mV^{-1}$, $(E_r - E_m)$ = 75 mV, $G_r/G_O = 2$. A value for S of $|14|$ is about 5 times the maximum value obtaining for a transmitter operating to increase the ionic conductance of the postsynaptic cell.

(3) Other forms of signal transmission

Even with this large increase in sensitivity of the postsynaptic potential to change in presynaptic potential, a limitation to single photon detection will still be imposed by the release of transmitter in

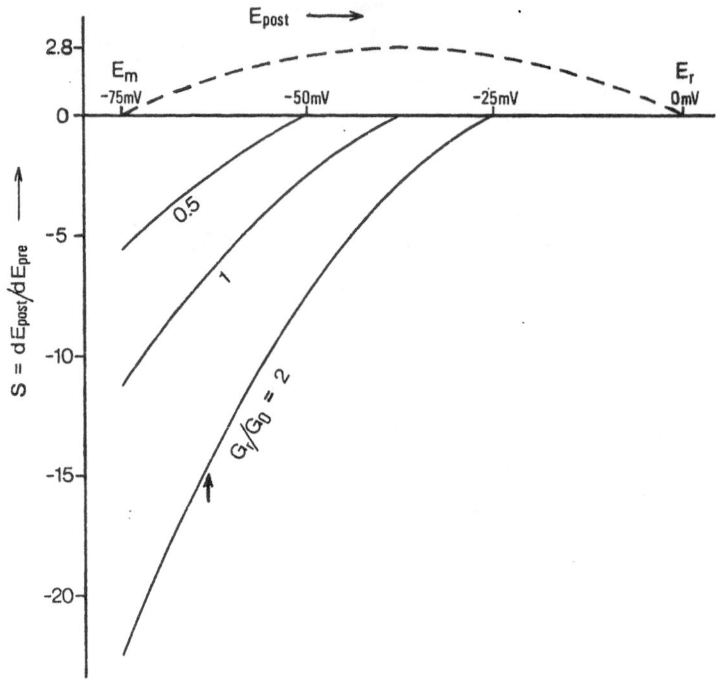

Fig. 10. Plots of the dynamic voltage-transfer function (the presynaptic potential controlling the postsynaptic potential) as derived to explain the response of bipolar cells which depolarize on illumination of the centre of their receptive field (the solid-line curves for different values of G_r/G_o) and of bipolar cells which hyperpolarize with the same stimulus (the dashed-line curve).

packets. A way out of this difficulty is to postulate that at very low levels of illumination signal transmission between receptor and bipolar cells involves some quite distinct mechanism of transmission between cells. What appears to be required is a mechanism having the following characteristics: a finer gradation of signal transmission than that determined by the release of transmitter in the usual packets, and/or a spread of membrane potential changes over a number of similar receptor cells which may then act in parallel at a subsequent stage of signal transmission between cells where the usual synaptic mechanism applies, so that limitations to signal transmission due to spontaneous fluctuations at that stage may be overcome.

Electroretinogram b-wave. At this point it is relevant to consider the b-wave of the electroretinogram, recorded between electrodes on the opposite sides of the retina. Following a brief flash or the onset of steady illumination, the b-wave occurs subsequent to the a-wave. The a-wave arises from the change evoked by light in the receptor cell and is a measure of the current flowing between different parts of the cell, which evokes a hyperpolarization of the receptor cell terminals, as discussed earlier. The b-wave which follows is of reverse polarity to the a-wave and arises by current flowing in the distal direction (upwards, for the orientation of the retina shown in Fig. 3). Using uniform illumination over a large area of retina, one finds that at low levels of illumination, the b-wave amplitude varies in proportion to the stimulus intensity, reaching an amplitude, as recorded from the rat retina, of about 60 μV for a flash from which on the average only one out of every 16 rods absorbs a photon (Cone, 1964). At low levels of illuminatio the amplitude of the b-wave may be 10-100 times that of the a-wave.

Although the b-wave has long been recognised as a distinct component of the electrical response of the retina to the absorption of light in the visual receptor, it is only recently that anything more than speculation has been possible regarding the type of cell within which the potential wave has its origin. It has been shown by the use of intracellular microelectrodes for recording potentials and for labelling the site of recording by the injection of dye (Miller & Dowling, 1970), that the b-wave arises, in part at least, as a result of a transient depolarization of Müller cells, the large non-neural, glial cells of the retina (Fig. 3). It is by no means clear yet how depolarization of the Müller cell arises. The Müller cell is not in the visual pathway and yet the b-wave has long been used as an index of visual function. The possibility exists that the Müller cells, each of which envelops groups of 20 or so receptor cells, might provide a non-synaptic interaction between receptor cells or between receptor cells and second order neurons.

Electrical coupling between cells. Evidence has also been obtained for a form of interaction, involving electrical coupling via low resistance pathways between cells of the same type in the retina. (This form of interaction is by no means unique to the retina.) Electrical coupling between horizontal cells has been shown in some species (Kaneko, 1971). In fish such coupling may extend for cells separated by as much as 1 cm. It has been shown that cones of the same spectral sensitivity may be electrically coupled within a radius of 50 μm in the retina of the turtle (Baylor & Hodgkin, 1973). Some form of summative interaction between neighbouring rod cells has also been observed in the retina of the turtle (Schwartz, 1973), but the nature of the interaction has not been established. If the coupling were electrical and each group of coupled rods converged onto the same bipolar cell, then the condition would be equivalent to that analysed on the basis of a single rod making contact with a bipolar cell. This would obviate the difficulty arising from the convergence of independent rods on a bipolar cell, with most of the rods contributing only noise to the visual pathway at low levels of illumination. However, this advantage would be offset, at very low levels of illumination, by those receptor cells which had not absorbed a photon acting as an electrical load on the single receptor (or the very few receptors) which had absorbed a photon. The problem of accounting for the high sensitivity of the visual process remains an intriguing one, although fraught with difficulties owing to the at present fragmentary nature of the experimental information required for its solution.

References

Barlow, H.B.: Retinal noise and absolute threshold, J. opt. Soc. Amer. 46, 634-639 (1956).

Barlow, H.B.: Increment thresholds at low intensities considered as signal-noise discriminations, J. Physiol. 136, 469-488 (1957).

Baylor, D.A. and Fuortes, M.G.F.: Electrical responses of single cones in the retina of the turtle, J. Physiol. 207, 77-92 (1970).

Baylor, D.A., Fuortes, M.G.F. and O'Bryan, P.M.: Receptive fields of single cones in the retina of the turtle, J. Physiol. 214, 265-294 (1971).

Baylor, D.A. and Hodgkin, A.L.: Detection and resolution of visual stimuli by turtle photoreceptors, J. Physiol. 234, 163-198 (1973).

Blackman, J.G. Ginsborg, B.L. and Ray, C.: Spontaneous synaptic activity in sympathetic ganglion cells of the frog, J. Physiol. 167, 389-401 (1963).

Bortoff, A. and Norton, A.L.: An electrical model of the vertebrate photoreceptor cell, Vis. Res. 7, 253-263 (1967).

Boyd, I.A. and Martin, A.R.: The end-plate potential in mammalian muscle, J. Physiol. 132, 74-91 (1956).

Cajal, S. Ramon: Histologie du Système Nerveux de l'Homme et des Vertébrés, Vol. II, A. Maloine, Paris, 1911.

Cone, R.A.: The rat electroretinogram I. Contrasting effects of adaptation on the amplitude and latency of the b-wave, J. gen. Physiol. 47, 1089-1105 (1964).

del Castillo, J. and Katz, B.: Quantal components of the end-plate potential, J. Physiol. 124, 560-573 (1954a).

del Castillo, J. and Katz, B.: Changes in end-plate activity produced by pre-synaptic polarization, J. Physiol. 124, 586-604 (1954b).

Falk, G. and Fatt, P.: Physical changes induced by light in the rod outer segments of vertebrates. In: Handbook of Sensory Physiology, Vol. VII/1, Photochemistry of Vision, ed. by H.J.A. Darnell, pp. 200-244. Springer Verlag, Berlin, 1972.

Fatt, P.: Postsynaptic cell characteristics determining membrane potential changes, this volume.

Fatt, P. and Katz, B.: Spontaneous subthreshold activity at motor nerve endings, J. Physiol. 117, 109-128 (1952).

Fuortes, M.G.F. and Yeandle, S.: Probability of occurrence of discrete potential waves in the eye of Limulus, J. gen. Physiol. 47, 443-463 (1964).

Hagins, W.A., Penn, R.D. and Yoshikami, S.: Dark current and photocurrent in retinal rods, Biophys. J. 10, 380-412 (1970).

Hecht, S., Shlaer, S. and Pirenne, M.: Energy, quanta, and vision, J. gen. Physiol. 25, 819-840 (1942).

Kaneko, A.: Physiological and morphological identification of horizontal, bipolar and amacrine cells in goldfish retina, J. Physiol. 207, 623-633 (1970).

Kaneko, A.: Electrical connections between horizontal cells in the dogfish retina, J. Physiol. 213, 95-105 (1971).

Katz, B. and Miledi, R.: A study of spontaneous miniature potentials in spinal motoneurones, J. Physiol. 168, 389-422 (1963).

Katz, B. and Miledi, R.: A study of synaptic transmission in the absence of nerve impulses, J. Physiol. 192, 407-436 (1967).

Katz, B. and Miledi, R.: The statistical nature of the acetylcholine potential and its molecular components, J. Physiol. 224, 665-699 (1972).

Kirschfeld, K.: Discrete and graded receptor potentials in the compound eye of the fly *Musca*. In: The Functional Organization of the Compound Eye, ed. by C.G. Bernhard. Permagon Press, Oxford, 1966.

Liley, A.W.: The effects of presynaptic polarization on the spontaneous activity of the mammalian neuromuscular junction, J. Physiol. 134, 427-443 (1956).

Miller, R.F. and Dowling, J.E.: Intracellular responses of the Müller (glial) cells of the mudpuppy retina: their relation to b-wave of the electroretinogram, J. Neurophysiol. 33, 323-341 (1970).

Naka, K.I. and Rushton, W.A.H.: S-potentials from luminosity units in the retina of fish (Cyprinidae), J. Physiol. 185, 587-599 (1966).

Naka, K.I. and Witkovsky, P.: Dogfish ganglion cell discharge resulting from extrinsic polarization of the horizontal cells, J. Physiol. 223, 449-460 (1972).

Paupardin-Tritsch, D. and Gerschenfeld, H.M.: Neuronal responses to 5-hydroxy-tryptamine resulting from membrane permeability decreases, Nature 244, 171-173 (1973).

Penn, R.D. and Hagins, W.A.: Signal transmission along retinal rods and the origin of the electroretinographic a-wave, Nature 223, 201-205 (1969).

Pirenne, M.H.: Light quanta and vision, Endeavour 20, 197-209 (1961).

Reichardt, W., Braitenberg, V. and Weidel, G.: Auslösung von elementarprozessen durch einzelne lichtquanten im fliegenauge, Kybernetik 5, 148-170 (1968).

Scholes, J.J.: Discrete subthreshold potentials from the dimly lit insect eye, Nature 202, 572-573 (1964).

Scholes, J. and Reichardt, W.: The quantal content of optomotor stimuli and the electrical responses of receptors in the compound eye of the fly Musca, Kybernetik 6, 74-80 (1969).

Schultze, M.: Zur Anatomie und Physiologie der Retina, Arch. mikr. Anat. 2, 175-286 (1866).

Schwartz, E.A.: Responses of single rods in the retina of the turtle, J. Physiol. 232, 503-514 (1973).

Steinberg, R.H.: Rod and cone contributions to S-potentials from the cat retina, Vis. Res. 9, 1319-1329 (1969).

Svaetichin, G. and MacNichol, E.F.: Retinal mechanisms for chromatic and achromatic vision, Ann. N.Y. Acad. Sci. 74, 385-404 (1958).

Toyoda, J.: Membrane resistance changes underlying the bipolar cell response in the carp retina, Vis. Res. 13, 283-294 (1973).

Toyoda, J., Nosaki, H. and Tomita, T.: Light-induced resistance changes in single photoreceptors of Necturus and Gekko, Vis. Res. 9, 453-463 (1969).

Uchizono, K.: Characteristics of excitatory and inhibitory synapses in the central nervous system of the cat, Nature 207, 642-643 (1965).

van der Velden, H.A.: Over het aantal lichtquanta dat nodig is voor een lichtprikkel bij jet menselijk oog, Physica 11, 179-189 (1944).

Weight, F.F. and Votava, J.: Slow synaptic excitation in sympathetic ganglion cells: evidence for synaptic inactivation of potassium conductance, Science 170, 755-758 (1970).

Werblin, F.S. and Dowling, J.E.: Organization of the retina of the mudpuppy, Necturus maculosus. II. Intracellular recording, J. Neurophysiol. 32, 339-355 (1969).

VISUAL SENSITIVITY

H.B. Barlow
Psychology Laboratory
Cambridge CB2 3EB

For a long time it has been recognized that the eye is very sensitive, and almost as soon as it was realized that light is absorbed in discrete quanta it was suggested that the resulting limits to performance might be important in vision. The measurements made in the 1940's by Hecht and his colleagues at Columbia, NYC, and by Bouman and Van der Velden in Holland, showed that this was indeed the case. The experiments are well known, but they are important and provide the basis for much of what has followed, so I shall describe them in some detail.

1. Absolute Threshold

The first step was to find the conditions under which a human subject is able to detect the smallest quantity of light. This involved optimising the stimulus and the subject's preparation. The stimulus parameters found to be optimal were: 507 nm wavelength; duration 1 msec; size 10' diameter; position 20° in nasal field; delivery was to one eye through an artificial pupil. The subject fixated a dim red light to ensure correct stimulus positioning, was in darkness for at least 40 minutes before thresholds were determined, and selected the moment of exposure himself by pushing a button when ready. On the other hand the order of presentation of stimuli was randomized, in the later experiments, in order to ensure that the subject had no prior knowledge of whether it would be a strong or weak stimulus. Trouble was taken over his comfort, convenience, and cooperation, and in Hecht's lab considerable emphasis seems to have been placed on reliability. The experimenters were themselves subjects, and there were few if any naive outsiders. Furthermore, reliability was checked by presenting blanks (zero stimuli); subjects were expected never to report these as seen, and some potential subjects were apparently excluded because of their unsatisfactory performance in this respect. Also in Hecht's lab very great pains were taken with the physical calibrations. We don't know as much on either of these points about Bouman and Van der Velden's experiments.

2. Frequency of Seeing Curves

As the intensity of the light flashes is reduced there comes a point where a subject sometimes fails to see it. Then a further reduction and he only sees half, still lower and he almost never sees it. The intensity change from 90% seen to 10% seen is quite large - a factor of about 3 X, or 0.5 log unit; what point in this range should one take as "threshold"? Hecht's laboratory took the intensity seen on 60% of trials, and to determine this point they did "frequency-of-seeing curves". Six fixed intensities spanning a 10 X change in intensity were presented in a random sequence, a total of 50 or 100 being presented at each intensity. Blanks were apparently not included but the lowest intensity was in most cases low enough to cause only a few "seen" responses or none at all.

When the appropriate calculations were made the thresholds (60% seen) came to values in the range 50 to 150 quanta entering the cornea. Hecht et al. then did some judicious estimating (subsequently pretty well justified and verified) that led them to conclude that about 10 out of the roughly 100 quanta entering the cornea were absorbed by rhodopsin in the receptors. 50% of the quanta never reach the retina, and only about 20% of those reaching it are absorbed by rhodopsin.

The detection of 10 quantal absorptions seemed quite a good performance in the 1940's (though photocells can do better now, and in energy terms the ear is nearly 10 X lower than the eye), but Hecht et al. pursued the analysis two steps further.

A. Quantum fluctuations. These account for most of the threshold variability, or range of uncertain seeing. Quantal absorptions are random independent events, and therefore follow Poisson statistics:

$$P(n/a) = a^n \frac{e^{-a}}{n!}$$

where $P(n/a)$ is the probability of exactly n absorptions occurring when the average number is a. If some physical event (such as a report that a stimulus flash was seen) requires that c or more quanta are absorbed, the probability that this will occur if the average number of events is a is given by

$$P(c^+/a) = \sum_{n=c}^{\infty} a^n \frac{e^{-a}}{n!} = 1 - \sum_{n=0}^{c-1} a^n \frac{e^{-a}}{n!}$$

Tables of this function are available and curves relating $P(c^+/a)$

and a can be plotted out for various values of the criterion c. Now the value of a is directly proportional to the average number of quanta delivered to the eye, though the constant of proportionality can only be estimated (vide supra). Now if these curves are plotted as a function of log a two things happen: (a) the curves become directly comparable with frequency of seeing curves, which are usually plotted as a function of log I, not I; (b) the unknown multiplicative constant which relates a to the average number of quanta at the cornea becomes an additive constant which is the same for all intensities; hence changing this constant will correspond to shifting the theoretical curve horizontally along the log I axis, without changing its shape. In fact eq. (1) plotted as a function of log a is an S-shaped curve whose slope increases with c. If one fits a theoretical curve to an experimental frequency-of-seeing curve, the slope required for a fit gives an estimate of 'c', the number of quantal absorptions required for 'seeing', and the lateral positioning gives an estimate of the proportion of quanta incident at the cornea that are absorbed.

The remarkable result Hecht et al. obtained was that the estimates of absorption and threshold number of quanta from this very indirect, statistical, argument were quite close to those obtained by their judicious direct estimates of the fraction absorbed. They therefore felt those estimates received some confirmation, and could also conclude that quantal fluctuations were an important determinant of threshold range and variability.

B. Receptor threshold. The second way they pursued their analysis was to ask, not "What is the threshold for seeing?", but "What is the threshold for receptor activation?" The argument here is also probabilistic, but in essence they showed that the 10 quantal absorptions almost certainly occurred in 10 different receptors. Hence one could not suppose that receptors required two hits to respond, but must accept that they were activated by every single absorption.

The statistical arguments described above must be used with caution, mainly for two reasons: (1) A biological system is typically erratic and unpredictable, and hence one must consider seriously what will happen if there is another cause of random behaviour in addition to the inevitable, irreducible quantum fluctuation. Such additional variability will add to that due to quanta, and cause experimental curves to be flatter than the theoretical ones for the actual number of quantal events that occurred. For this reason it is best to regard the statistical argument as giving a limit: "At least 6 (actually 5 - 8 in Hecht's experiments) quanta are absorbed from a threshold flash and participate in causing it to be seen". Bouman and Van der Velden did a very similar experiment, but found lower thresholds and flatter curves, so the statement they would make would be "At least 2 quanta...". Clearly Hecht's is a stronger statement, which is my reason for emphasizing their work. In addition, Bouman and Van der Velden's actual figures for threshold were extraordinarily low and it is difficult to square them with other workers. However, reliability is a possible reconciling variable which I shall return to. (2) The second reason for being cautious about statistical arguments is the eagerness and efficiency with which humans (and perhaps animals too) seek out and exploit flaws in the statistical design. The random order of presentation of stimuli is essential, and so is the avoidance of other non-visual cues to the nature of a stimulus being presented (experimenters comments or threats, noise of shutter or filters being changed, responses to subject's complaints, and so on). These can be avoided, but one must be convinced that the experimenter is aware of their importance.

3. Subsequent Developments

Since the 1940's these results have been taken up and developed in a number of ways. First, Hecht's "judicious estimates" of what happens to light after it enters the cornea have been confirmed by more direct reflexion densitometry (Rushton). I won't describe this here. The other main developments came from following up these questions: "Why are 6 quanta needed?" "Are quantum fluctuations important under non-threshold conditions?" And finally there are some interesting results obtained by following up the quantal question in neurophysiological preparations.

4. Intrinsic Noise and Reliability of Response

If receptors can respond to single quanta, why are 6 or more required for seeing? The lower limit of detection of a signal is usually set by random internally generated disturbances that obscure the signals one is trying to detect. This cannot be overcome by amplification, which will affect this intrinsic noise as well as the signal. A good deal of indirect evidence suggests that this is the case in the eye, and the maintained discharge of retinal ganglion cells, even in complete darkness, provides some direct support. The theory is worked out in my 1956 paper and I shan't go into it in detail here. Perhaps the two most interesting consequences are: (a) The extraordinarily high stability that must be achieved by rhodopsin and the photoreceptor activation system; there is room for an average false response rate only of about one per molecule per 300 years. (b) Reliability is shown to be a very important variable. If a subject is allowed to give "seen" responses to, say, 10% of blanks, then he can set his criterion much lower, and may detect just one or two quantal absorptions. But if false positive responses are thought of as deadly sins to be avoided at all costs, he must set his criterion much higher. This might reconcile Bouman and Van der Velden and Hecht et al.

The consequences have recently been worked out by Sakitt, who did a rather different type of experiment. Instead of seen-not seen, she used ratings (0 - 6). Instead of many intensities she used only 3. The results fit extraordinarily well the hypothesis that a subject says 1 for 1 quantal absorption, 2 for 2 absorptions, 3 for 3, etc. Another subject said 1 for 2, 2 for 3, etc., while a third said 1 for 3, 2 for 4, and so on. Thus it seems probable that the receptors signal centrally what happens to them, and the centres can correctly interpret these messages. The complication is that spurious events indistinguishable from absorptions are being signalled at a low rate all the time. A subject can detect and count 1 event, 2 events, etc., but it is only the number of them that enables one to tell if real light has entered the eye. We shall see more about this when we look at the neurophysiology.

5. Increment Thresholds

Does the threshold rise when a stimulus is superimposed on a background because that background increases the "noise level"? Such a hypothesis does not fit all the facts. The increment threshold usually rises more rapidly than quantum fluctuations would lead one to expect. These would predict that $\Delta I \propto I^{\frac{1}{2}}$ whereas it is usually closer to $\Delta I \propto I$. An exception occurs if the test stimulus is small and brief, where the square root law holds for a considerable range of backgrounds. However if this is because of quantum fluctuations, it appears that the background fluctuations over an area and time much larger than those of the test stimulus must be interfering with detection of the stimulus. To follow this up read about "Quantum Efficiency" and Increment Thresholds.

6. Threshold of Retinal Ganglion Cells

Obviously one would like to see how these weak lights are signalled to the brain. Are optic nerve fibres quiet or noisy? Is a threshold response a single impulse or many? Are threshold responses signalled by single ganglion cells or are the responses of many fibres averaged? Some of these questions can be answered by looking at cat retinal ganglion cells using techniques I shall describe in the next lecture. For a long time, published thresholds were extraordinarily high, but good preparations, with good optics, using <u>central</u> vision in the cat, now yield very low thresholds.

These measures gave 10 - 20 quanta at the cornea to obtain a reliably detectable response from a single ganglion cell. Our "judicious estimate" of absorption in cat is 25%, so the number absorbed from a threshold flash is about 2 to 5. We were therefore very surprised to count up the extra impulses these quanta had caused, and to find that it was often 2 or 3 times larger than the number of quanta absorbed. Apparently one quantum absorbed can cause several impulses.

This was confirmed by statistical tests of the Hecht, Shlaer, Pirenne type. One additional point is of interest. If one random independent event (quantal absorption) causes several output events (nerve impulses) one predicts that if one looks at the distribution of numbers of output events in some fixed interval of time, or at the distribution of intervals between output events, these will have some very unusual statistical properties. These were in fact what we based our statistical tests upon. But what was interesting was that the background maintained discharge also had those properties. One is therefore led to conclude that it, too, results from single random independent events (perhaps spontaneous, thermal, isomerizations of a rhodopsin molecule) each of which causes several output impulses.

Perhaps the most interesting conclusions of these studies of the sensitivity of the eye are, first that the eye does pretty well by physical standards, so that each receptor, and also the whole visual system, can respond to single quanta; and secondly that what is true of the whole eye is also true of the single ganglion cells: they are reliable and efficient components.

Selected References

Barlow, H.B.: Retinal noise and absolute threshold, J. opt. Soc. Amer. <u>46</u>, 634-639 (1956).

Barlow, H.B.: Increment thresholds at low intensities considered as signal/noise discriminations, J. Physiol. <u>136</u>, 469-488 (1957).

Barlow, H.B.: A method of determining the overall quantum efficiency of visual discriminations, J. Physiol. <u>160</u>, 155-168 (1962a).

Barlow, H.B.: Measurement of the overall quantum efficiency of visual discriminations, J. Physiol. 160, 169-188 (1962b).

Barlow, H.B., Levick, W.R. and Yoon, M.: Responses to single quanta of light in retinal ganglion cells of the cat, Vis. Res. 11, Supplement 3, 87-102 (1971).

Bouman, M.H.: History and present status of quantum theory in vision. In: Sensory Communication, ed. by W. Rosenblith, pp. 377-401. M.I.T. Press, New York, 1961.

Hecht, S., Shlaer, S. and Pirenne, M.H.: Energy, quanta, and vision, J. Gen. Physiol. 25, 819-840 (1942).

Pirenne, M.H.: Vision and the Eye, Pilot Press, London, 1948 and Science Paperbacks, London, 1967.

Rushton, W.A.H.: Rhodopsin density in the human rods, J. Physiol. 134, 30-46 (1956).

Sakitt, B.: Counting every quantum, J. Physiol. 222, 131-150 (1972).

DETECTION OF FORM AND PATTERN BY RETINAL NEURONES

H.B. Barlow
Psychological Laboratory
Cambridge CB2 3EB

From the previous lecture we saw that the eye's extraordinary sensitivity is probably not due to the averaging of responses from many much less sensitive channels, but is based on the high sensitivity and reliability of individual retinal ganglion cells. In this lecture I am going to talk about form vision and feature extraction. Here again we are led to a rather similar conclusion: single visual neurones are selectively sensitive to remarkably specific spatio-temporal patterns.

1. History

The work of Adrian and his colleagues showed that peripheral sensory nerves were selectively sensitive to types of sensory stimulation which corresponded, at least approximately, to the different subjectively separable sensory modalities. Further work, chiefly by Adrian, Bard, Marshall and Woolsey, using gross surface electrodes, showed that these modality-specific fibres relayed maps of the physical stimuli at the sensory surface to the cerebral cortex. Then microelectrode recording techniques were introduced and made it possible to record from individual cells at levels one or more synapses deeper than the initial sensory receptors. The earliest work was by Hartline and Granit on the retina, while Davies and Galambos in the auditory pathway got some of the earliest evidence for the inhibitory interactions which seem to be so important in obtaining pattern selectivity. I shall confine this lecture to vision, where I think progress has been most rapid and the most complete picture can be presented, and shall take Hartline's work on frog retina as the starting point.

2. Frog Retina

Hartline's three papers (1938a,b; 1940) showed that the retina did more than transmit a map of the retinal image in the form of impulse frequencies, and by following up the implications of some of his results it was found that it did even more than he said. He found (1) that optic nerve fibres were diverse in their response requirements. Some responded at "on", some at "off", and some at both "on" and "off". (2) Each fibre was excitable from a region of retina he called its receptive field, and this was surprisingly large so that the fields of different fibres overlapped extensively: a spot of light at one point could excite many fibres. (3) Excitatory influences from different places within a receptive field summated.

The combination of excellent optics with findings (2) and (3) is puzzling: why form a sharp image if it is immediately to be "blurred" neurally? The ganglion cells might be responding to edges, rather than total light, so I compared thresholds for different spot sizes with thresholds predicted from detailed maps of the sensitivity of the parts of the receptive field (Barlow, 1953). As the spot was enlarged to fill the whole receptive field the threshold fell, but instead of levelling out to a plateau when it spread outside the receptive field, the threshold rose again. This could have been explained as "edge sensitivity", but a simple more direct explanation was that the region surrounding the receptive field had an inhibitory influence, and direct experiments showed that this was the case. Nothing happens when this inhibitory surround is excited alone, but if it is illuminated at the same moment as the receptive field proper is stimulated, then the response is diminished or abolished. Furthermore, this occurs at both "on" and "off".

The paradox of Hartline's large receptive fields was not really explained by the inhibitory surround, but there were two other findings. First, not all receptive fields are as gigantic as Hartline originally described, and there are some rather small ones, especially in the more central retina. Second, some fibres are sensitive to small movements of black or white spots within their receptive fields. It was reasonable to suppose that good optical quality would aid the detection of a small moving object, even if the fact that the fibre detecting it had a large receptive field meant that the exact position of the object was lost.

Attention to this paradox thus revealed that quite complex processes occurred in the retina, and it was not solely concerned with sending a map of the retinal image to the brain. Indeed, the sensitivity of some ganglion cells to small moving objects suggested that they were "fly-detectors", that they were in fact the innate releasers for feeding that ethologists were postulating. The lateral inhibitory mechanism was also suggestive in relation to Mach Bands and contrast phenomena in human vision.

Lettvin and his colleagues took these notions a great deal further.

First, they found a greater variety of types of ganglion cells and optic fibres probably because their electrodes were finer. They actually recorded mainly from the fibre terminals in the superior colliculus, and described five types. They made the additional discovery that the depth of the electrode correlated with the type of fibre they were recording from, suggesting that there were five laminae of endings. The types are given slightly different names in different papers. These are the 1961 versions:

I. Boundary detectors. These give maintained responses at "on" if there is a boundary in the receptive field, but give no sustained response to uniform illumination.

II. Movement-gated dark convex boundary detectors. The darker area must, they say, be convex.

III. Moving or changing contrast detector: the same as Hartline's "On-Off".

IV. Dimming detectors: same as "Off".

V. A rare type that gives a sustained response inversely related to illumination.

There are some oddities in their accounts of the properties of these fibres, though much has been confirmed. But I feel the main contribution of their papers was to establish more firmly the idea of a retinal ganglion cell as a luminance-invariant feature detector which was specifically sensitive to some property of the visual space in front of the animal, and signalled it regardless of the luminance of the scene. Instead of the traditional view of the retina as a transducer of light to nerve impulses, they saw it as an object-detector which ignored the vagaries of the lighting as far as possible.

Historically, I should already have mentioned Kuffler (1953) on the cat retina, and Hubel and Wiesel started publishing on cat cortex at about the same time as Lettvin et al. But cat retina, LGN, and cortex are better told as one story, and I shall now go to other retinae.

3. Rabbit and Ground Squirrel Retina

The rabbit turns out to have a retina with an even greater number of ganglion cell types than the frog - about 18 types when last counted, and that does not include any for colour-vision, which the rabbit almost certainly has. As well as the relatively simple types which will be described in the cat, the rabbit has ganglion cells which are selectively sensitive for stimuli moving in a particular direction. Others are selective for speed of motion, and also for the orientation of a stimulus object. Sometimes one knows a cell has been isolated by the appearance of an action potential, but it may still take hours of patient exploration to characterize the stimulus that makes it discharge optimally - sometimes called its "trigger feature" (Barlow, Hill and Levick, 1964; Levick, 1967).

Directionally selective cells are found sufficiently often for it to be possible to design experiments to reveal the mechanisms. They show (1) that it is based on sequence-selectivity, but the parts of the receptive fields which must be excited in sequence are not fixed; it is as if any two adjacent points would excite if stimulated in the right sequence, but would not if excited in reverse sequence. (2) Sequence selectivity seems to result from inhibition preventing responses to the reverse sequence rather than from detection of the correct sequence.

Since the sequence selectivity was a property of a sub-unit of the receptive field, and since bipolar cells are the anatomical sub-units feeding retinal ganglion cells, it was at first thought that bipolar cells were the sequence selective elements. Intracellular recording techniques (to be considered elsewhere) and electron microscopy have shown that this is probably wrong (Dowling and Werblin). The mechanisms are probably in the inner plexiform layer and involve amacrine - ganglion cell "serial synapses", but testable models have not yet emerged.

Michael's work on ground squirrel confirms some of the results on rabbits. He also found colour selective cells (which will not be considered here) and has analysed the central projection targets of different ganglion cell types.

4. Cat

Kuffler (1953) used an improvement of Granit's technique in which he left intact the eye's optics and thus was able to explore the receptive fields of retinal ganglion cells. He found two types, on-centre with off surround, and off-centre with on surround. These surrounds are slightly different from those of frog; they evoke a discharge at the opposite phase of illumination from the centre, as well as counteracting or reducing excitation at the centre if illuminated in phase with it. In the frog on-off units the centre is excited at both on and off, whereas the surround inhibits at both on and off. The frog's could be called a "suppressive" rather than inhibitory surround, but it is also a good plan to avoid calling a region of a receptive field "inhibitory" simply because light in that region causes a discharge at off. The "on" surround of an off-centre cell inhibits the off discharge from the centre if it is dimmed synchronously with the centre. Terminology can be confusing.

Kuffler's simple story has been complicated by the discovery (Enroth-Cugell and Robson) of X and Y cells: this distinction has been followed up by Levick and colleagues, Fukada and Stone. Stone has lately suggested there is a third type - the W cells. One of the most interesting developments is that differences in receptive field properties have been correlated with (a) conduction velocity of optic nerve fibres; (b) destination: X go to LGN; Y go to LGN and colliculus; (c) fate after visual deprivation. I shall not take this up here (read Cleland, Dubin and Levick, 1972; Stone, 1973).

5. Conclusions

One should regard the retina as a device for abstracting information from the retinal image which is of special importance to the animal. In doing this it is the spatio-temporal patterns of luminance in the image that are important, whereas the overall luminance has to be discounted. After two (possibly three) synapses, a ganglion cell can respond to something as specific as, for example, a vertically oriented, slowly moving, edge. Detection of pattern features of this order of complexity can be done by a single neuron (with its connected network of bipolars, horizontals, and amacrines) and does not require cooperation of an ensemble of ganglion cells. A single ganglion cell is apparently as sensitive as the whole animal, and it can also selectively respond to specific pattern elements.

Selected References on Pattern Selectivity of Retina

Barlow, H.B.: Summation and inhibition in the frog's retina, J. Physiol. 119, 69-88 (1953).

Barlow, H.B., Hill, R.M. and Levick, W.R.: Selectivity to direction and speed...J. Physiol. 173, 377-407 (1964).

Barlow, H.B., Hill, R.M and Levick, W.R.: Mechanism of directional selectivity...J. Physiol. 178, 477-504 (1965).

Cleland, B.G., Dubin, M.W. and Levick, W.R.: Sustained and transient neurones...J. Physiol. 217, 473-496 (1971).

Enroth-Cugell, C. and Robson, J.G.: The contrast sensitivity of retinal ganglion cells of the cat, J. Physiol. 187, 517-552 (1966).

Fukada, Y.: Receptive fields and conduction velocity, Vis. Res. 11, 209-226 (1971).

Hartline, H.K.: The response of single optic nerve fibers of the vertebrate eye...Am. J. Physiol. 121, 400-415 (1938).

Hartline, H.K.: Receptive fields...Am. J. Physiol. 130, 690-699 (1940a).

Hartline, H.K.: Spatial summation...Am. J. Physiol. 130, 700-711 (1940b).

Kuffler, S.W.: Discharge patterns and functional organization of mammalian retina, J. Neurophysiol. 16, 37-68 (1953).

Lettvin, J.Y., Maturana, H.R., McCulloch, W.S. and Pitts, W.H.: What the frog's eye tell the frog's brain, Proc. Inst. Rad. Eng. 47, 1940-1951 (1959).

Lettvin, J.Y., Maturana, H.R., McCulloch, W.S. and Pitts, W.H.: Anatomy and physiology of vision in the frog, J. Gen. Physiol. 43, 129-171 (1960).

Lettvin, J.Y., Maturana, H.R., McCulloch, W.S. and Pitts, W.H.: Two remarks...In: Sensory Communication, ed. by W. Rosenblith, pp. 757-776. M.I.T. Press, New York, 1961.

Levick, W.R.: Receptive fields and trigger features of ganglion cells in the visual streak of the rabbit's retina, J. Physiol. 188, 285-307 (1967).

Levick, W.R.: Receptive fields of retinal neurons. In: Handbook of Sensory Physiology, ed. by M.G.F. Fuortes, pp. 531-566. Springer-Verlag, Berlin, 1972.

Michael, C.R.: Receptive fields. I. Contrast sensitive units, II. Directional selective units, III. Opponent color units, J. Neurophysiol. 31, 249-282 (1968a,b,c).

Michael, C.R.: Superior colliculus of ground squirrel, J. Neurophysiol. 35, 833-846 (1972).

Michael, C.R.: L.G.N. of ground squirrel, J. Neurophysiol. 36, 536-550 (1973).

Stone, J. and Dreher, B.: Projection of X- and Y cells of the cat's L.G.N...J. Neurophysiol. 36, 551-567 (1973).

AUDITORY PROCESSING IN THE NERVOUS SYSTEM

Juan G. Roederer
Department of Physics and Astronomy
University of Denver
Denver, Colorado 80210 USA

1. Introduction

In this review we shall discuss some relevant aspects of the auditory nervous system and the mechanisms which are believed to play a role in the processing of acoustical information. We will mainly focus on the effects of musical tones,[1] leaving out speech and noise forms. As compared to language, musical tones are simpler input signals, and their processing requires fewer cooperative actions by, and associations in, other regions of the brain. Hence, they are more appropriate stimuli for the exploration of the fundamental primary processes that operate in the auditory system from periphery to the cortex.

Let us point out some characteristic features of the auditory system that make it particularly interesting. First of all, we must realize that the auditory tract represents the main input channel to the communications system whose development has elevated man above all other primates. Second, it is a sensory system that operates on one-dimensional patterns in time, with three characteristic time scales, each one of which has an equivalent operational time scale in the neural network: (i) A "microscopic" time scale represented by the incoming acoustical vibration patterns (~0.1-50 msec), which in its upper range is of the order of possible time intervals between neural spikes. (ii) An "intermediate" time scale represented by the transients in musical tones (buildup and decay) and by phonemes and syllables in speech (~10-200 msec[§]), which in the neural system corresponds to typical transfer times in the afferent pathway and propagation times through the cortical layer in the auditory receiving area. (iii) A "macroscopic" time scale ($\gtrsim$200 msec[§]), represented by the tone successions in melodic lines and by the words and sentences in speech, equivalent to memory storage and recall times, and to the duration of all the other complex operations that are required in the identification process of the actual acoustical message.

[§]These limits are only very approximate.

Third, although peripheral processing in the auditory system may be more primitive than in the retina, there are considerably more interconnections between the left and right ear pathways than in the visual system, leading to strong intraaural interaction of acoustical neural signals before they reach the pons. Fourth, in the auditory system there is a striking hemispheric partition of tasks at the cortical level: while language is primarily processed in the left cerebral hemisphere (in about 98% of all individuals), the right hemisphere executes control tasks (mainly checking on what the left is doing with language perception and speech control), and handles all non-language associated sound patterns (such as music).[2] Finally, fifth, there is an increasing amount of (indirect) evidence that the auditory system is ontogenetically the first "sophisticated" sense to be turned on - indeed, before an individual is born. The fetus is subject to a quite well-defined sound environment, with heartbeat, gastric sounds, and the (filtered) voice of the mother representing the first sound memory impregnations.[3]

2. Single Tone Processing

Auditory processing starts in the inner ear, the cochlea.[§] This is a complex transducer in which the incoming vibrations are converted into traveling waves in a strongly coupled system, the perilymph-endolymph fluid and the basilar membrane, and picked up by the actual neural sensors, the so-called "hair cells" of the organ of Corti on the basilar membrane. The response of a given hair cell is determined by the oscillation pattern of this membrane at the place of the hair cell. For a pure, sinusoidal tone of given frequency f, the basilar membrane behaves like a waving flag (Fig. 1),[4] with an oscillation amplitude envelope that reaches a maximum at a position $\underline{x}$ that is a nearly logarithmic

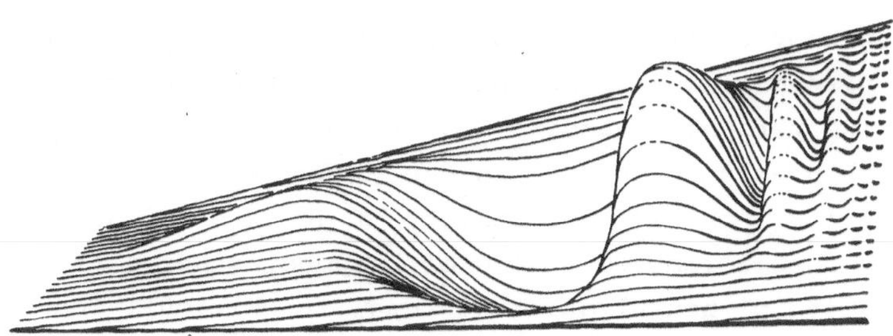

Fig. 1. Traveling wave pattern on the basilar membrane in a cochlear model (arbitrary scales).[4] By permission of the Acoustical Society of America.

[§]For high intensity levels, a non-linear signal distortion already occurs in the mechanical transmission line from the eardrum to the entrance to the cochlea (oval window).

function of f (Fig. 2).[5] The oscillation frequency of each point of the membrane is always equal to that of the stimulus, f, in the case of a pure tone. A shift in frequency leads to a shift in position of maximum response and hence to a change in the sensation of pitch. This is the basis of the so-called "place theory" of pitch perception.[6]

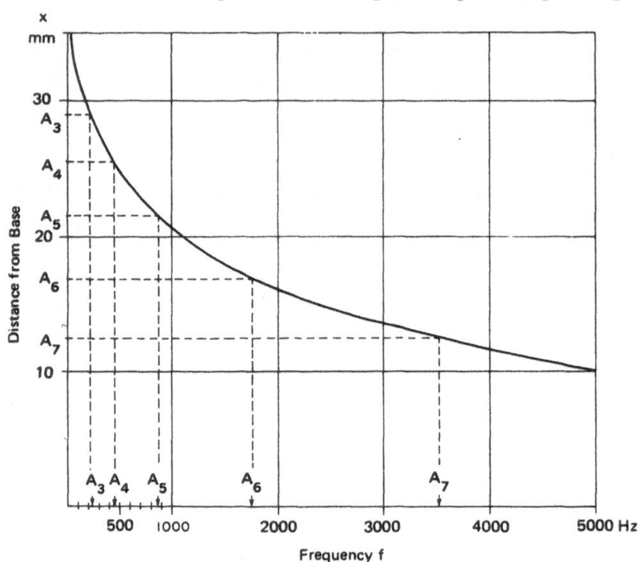

Fig. 2. Position of the resonance maximum on the basilar membrane for a pure tone of frequency f (linear scales). A_4: musical note f = 440 Hz (after ref.[5]).

For a monochromatic tone, the excited region of the basilar membrane is quite broad (Fig. 3),[7] and one must wonder why we perceive such a stimulus as truly single-pitched. Moreover, the width of the resonance region (over 1/3 of an octave) seems incompatible with the remarkable frequency resolution capability of the ear (less than 0.1% under certain circumstances). Obviously, a "sharpening" mechanism must be at work in the peripheral auditory neural network,[8] in analogy to the contrast enhancement mechanism in the visual system, whose main function is to funnel or "focus" the activity of an extended area of primary excitation into a limited bundle of responding nerve fibers surrounded by a region of neural quiet (lateral inhibition). The sensation of loudness is related to both, the average firing frequency of the fibers (at low sound pressure levels) and the total number of activated fibers.

Individual fibers in the auditory nerve are "tuned" to a given frequency. Indeed, their firing thresholds are a strong function of the frequency of a pure tone (Fig. 4).[9] Each tuning curve seems to be determined mainly by the neural wiring scheme, in particular, by the

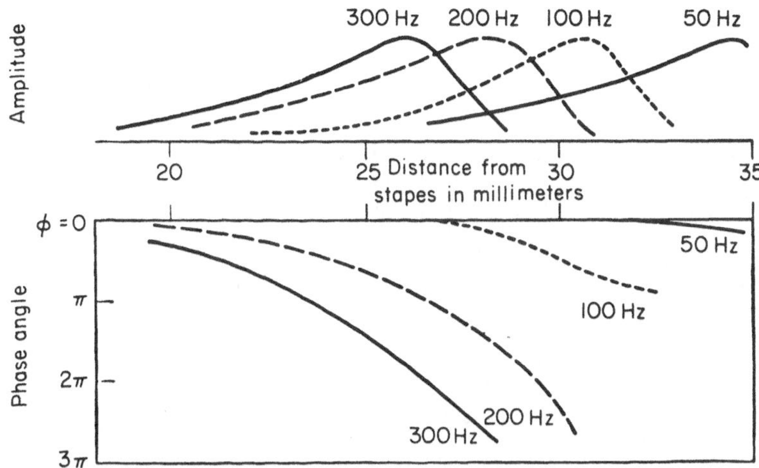

Fig. 3. Distribution of vibration envelopes along the basilar membrane (top) and cumulative phase lag of the membrane oscillations relative to the eardrum.[7] By permission of the Acoustical Society of America.

position x of the region of the basilar membrane to which the fiber in question is connected. There are indications that the tuning curves become sharper as one moves up the afferent pathway, indicating the action of a contrast enhancement mechanism (see, however, Section 4).

Neural coding in the auditory system does not only come in the form of tonotopic (spatial) distribution of neural activity. It has been found that for the musically most relevant frequency range ($\lesssim$ 2,000 Hz), information on frequency or repetition rate is also encoded in the form of interspike time interval distribution.[9] A given fiber is found to fire only when the region of the basilar membrane to which the fiber is wired is moving into <u>one</u> direction, remaining inhibited while it is moving into the other.[§] The result is a time-distribution of neural impulses that, for a pure tone, has sharp maxima of occurrence at integer multiples of the wave period (Fig. 5).[12] It is thus conceivable that information on the actual vibration pattern of a $\lesssim$ 2,000 Hz sound wave may be encoded in the form of a statistical "Morse Code" given by the particular interspike time-interval distribution. There is no direct evidence as yet that the auditory nervous system actually does utilize

[§]Until recently, evidence seemed to indicate that the interval of activation would correspond to the displacement of the basilar membrane toward the scala vestibuli (i.e., during pressure rarefaction in the acoustic wave, if phase shifts in the cochlea are neglected).[10] A recent study, however, points to the opposite, and furthermore indicates that both displacement and velocity of the membrane determine the intervals of activation and inhibition.[11]

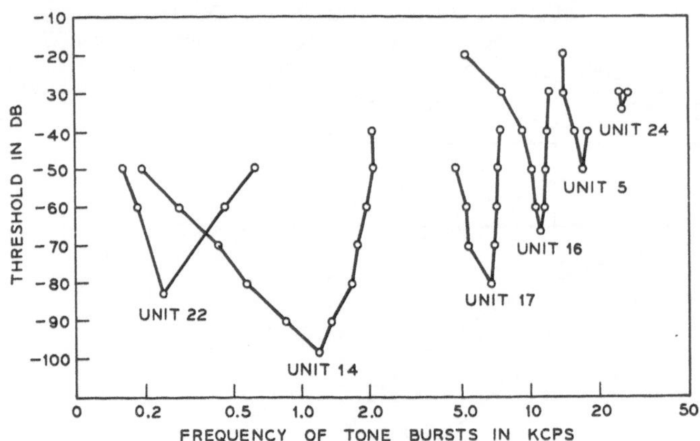

Fig. 4. Frequency sensitivity for different fibers in the auditory nerve of a cat (after ref. 9).

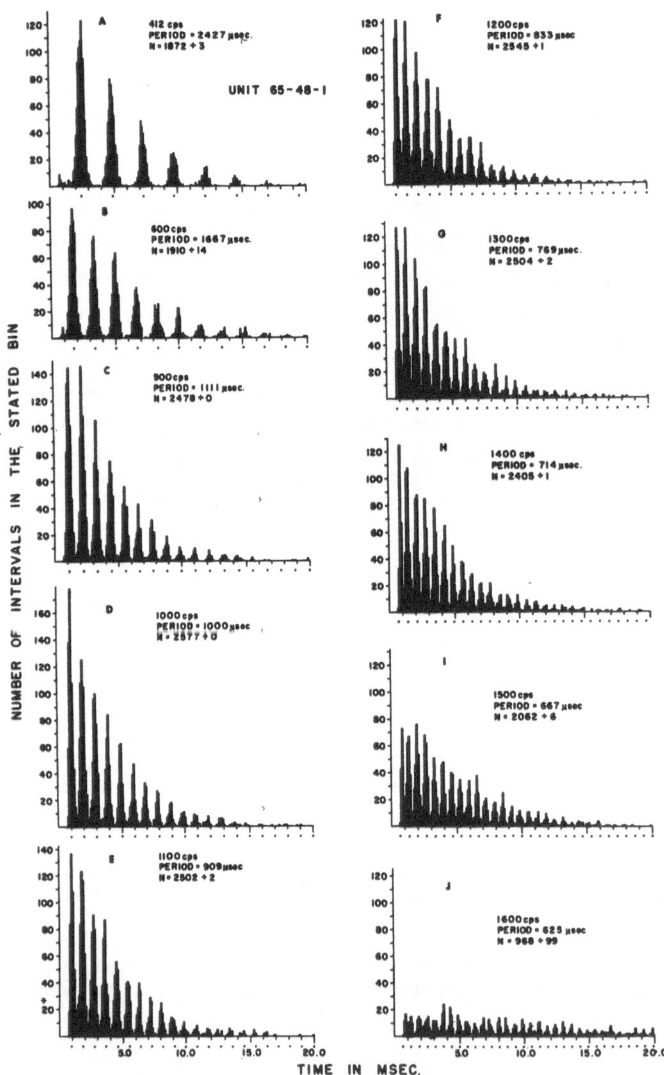

Fig. 5. Histograms of interspike time intervals when a pure tone of different frequency activated the fiber.[12] N: number of interspike samples for each case. By permission from the authors.

this information.[10] As a matter of fact, there is a (rather emotional) battle raging among psychoacousticians and neurologists regarding this point. Indirect evidence, however, seems pretty convincing that it does somehow use this time-interval distribution (next section).

3. Tone Superposition Effects

A familiar phenomenon is that of beating tones. When two slightly out-of-tune unisons of frequencies f_1 and $f_1+\varepsilon$ are superposed, we hear beats of frequency ε, caused by the amplitude modulation of the resulting wave (Fig. 6A). But when we listen to a slightly mistuned octave of pure (electronically generated) tones of frequencies f_1 and $2f_1+\varepsilon$, we <u>also</u> perceive beats at a rate ε (although it is much more difficult to describe <u>what</u> is beating).[13] In that case there is <u>no</u> modulation of amplitude (cycle-averaged energy flow); rather, it is the wave pattern that is being modulated (Fig. 6B). Other mistuned musical intervals of pure tones also lead to beat sensations. All this shows that the auditory nervous system is capable of detecting cyclic changes in vibration pattern forms. Note that Fourier analysis cannot account for this phenomenon: there is no amplitude beating at frequency ε in the original stimulus. Nonlinearities have also been ruled out.[13] Coding in the form of particular neural impulse time distributions, mentioned in the previous section, could account quite well for this effect.

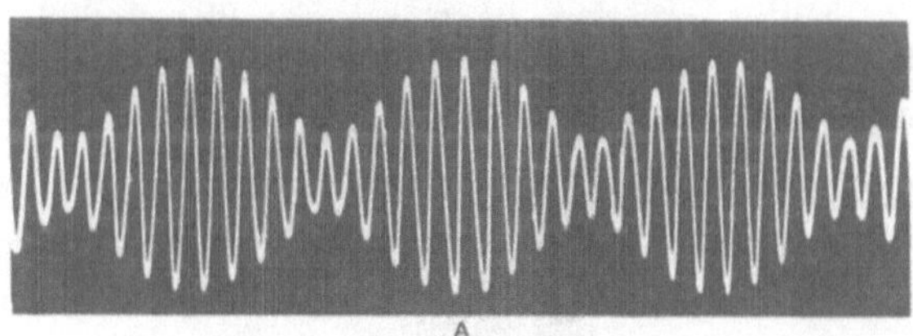

A

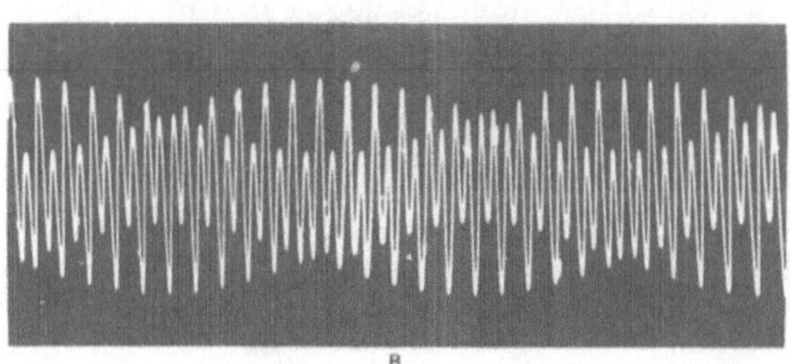

B

Fig. 6. A: Beats of a mistuned unison (amplitude modulation with no change in vibration pattern form). B: Beats of a mistuned octave (vibration pattern modulation with no change in cycle-averaged energy flow).[1]

More evidence for a non-Fourier-type wave pattern analysis comes with the study of the perception of complex tones, i.e., tones from real musical instruments. A complex tone, made up of a superposition of harmonics, elicits a quite complicated excitation pattern on the basilar membrane. Yet we perceive it as a single tone, of single pitch - that of the fundamental frequency. By changing the spectrum we change the sensation of timbre or quality of the tone[14] - but it still will come as a single tone sensation, a single pitch perception. It requires both a musical ear and a long lasting tone to be able to "hear out" individually some of the lower order harmonics. [15]

Even more strikingly, we may cut off completely the fundamental frequency component of a complex tone stimulus - and find that the pitch sensation remains unchanged![§] This phenomenon is called "fundamental tracking", the ensuing pitch sensation is called the "missing fundamental", "residue pitch" or "periodicity pitch". [16] Psychoacoustical experiments show that superposition of two neighboring harmonics of frequencies nf_o and $(n+1)\ f_o$ indeed lead to a pitch sensation corresponding to a tone of frequency f_o, although no energy is present at that frequency in the Fourier spectrum of the original stimulus. What happens is that the form of the stimulus vibration pattern has a repetition rate f_o, as can be easily shown (Fig. 7). Again, the auditory system seems capable of extracting information on the details of the vibration pattern - the repetition rate in this case. If we invoke neural impulse time distributions to represent the primary coding, a neural auto-correlation mechanism must be postulated, that would extract information on the pulse repetition rate (Fig. 8). [17]

A most astounding capability of the auditory neural system is that of discriminating the quality of two simultaneously sounding complex tones. No real music would be possible without such capability. This discrimination mechanism is not at all understood; a time element (non-simultaneity and transients) seems to play a key role. [1] Quite generally, the complex tone discrimination has its equivalent (and probably, its origin) in the mechanism that steers our auditory perception when we follow the speech of one given person among many different conversations conducted simultaneously at similar sound levels.[1] This ability has been pointedly called the "cocktail party effect", and very likely uses the same cues, primary and secondary, as the complex tone discrimination mechanism.

[§]Small transistor radios do this all the time with low frequency tones (basses)!

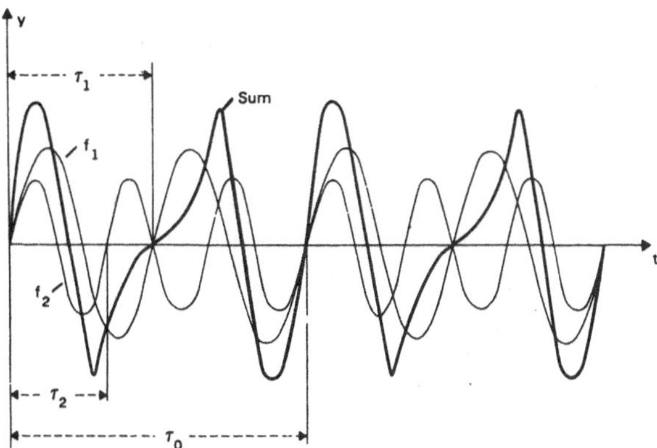

Fig. 7. Superposition of two pure tones that form a musical fifth ($f_2 = \frac{3}{2} f_1$). Note the repetition period $\tau_0 = 2\tau_1$.[1]

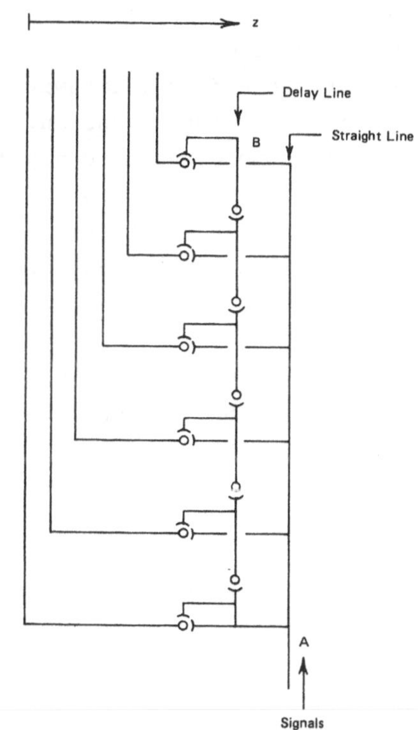

Fig. 8. Model for a neural autocorrelation mechanism (repetition rate detector). Impulses traveling up along axon A are intercompared with the same impulses that have been branched off into a neural "delay line" B (each synapse introduces an additional retardation and hence signals travel slower through the delay line than through the straight line). If the time distribution of the original signals has a certain periodicity, there will be coincident excitations at only certain endings of ascending fibers (those for which the delayed signals have fallen an integer number of periods behind the original signals). Assuming that two such excitations are simultaneously required to fire an ascending fiber, the location z of all activated neurons will bear information on vibration pattern characteristics, such as phase and repetition rate. (After ref.[17]).

Visual discriminatory and "lock-in" tasks are performed at a lower level in the superior colliculus of the midbrain,[18] a nucleus of neurons of the sensory pathway, where incoming signals from the retina and outgoing signals from the cortex can interact with each other, giving feedback instructions from the latter a chance to influence or control the processing of the former. Very likely, a similar mechanism handles the cortex-feedback-controlled lock-in operations in the auditory pathway.

4. Processing at the Cortical Level

The wiring scheme of the sensory areas of the cortex with its local feedback loops (e.g. the Martinotti cells) seems ideally fitted to accomplish autocorrelation and other time-pattern analyses. It is highly questionable, however, that this does apply to the processing of the "microscopic" time scale features of the original sound wave (Section 1). Indeed, as one moves up the auditory pathway (Fig. 9), the neural firing rate decreases considerably for a constant auditory input[19] and individual responses seem to be more and more tuned to complex features of the input stimulus. Wave pattern analysis, if any, should thus be accomplished well before the signals reach the cortex. On the other hand, a most puzzling fact is that both, beats of mistuned intervals and the missing fundamental (Section 3) are perceived even if the individual tone components are fed in dichotically, i.e., into separate ears.[13,20] In that case, the superposed sound wave pattern never arises - the only superposition is that of neural impulses from both cochleas, after the decussations in the auditory tract (Fig. 9). This poses a serious problem to the interpretation of these psychoacoustical effects in terms of neural impulse time-interval coding.

Finally, another complicating factor is the fact that many psychoacoustical magnitudes - even the "simple" ones, such as pitch and loudness - may under certain circumstances depend strongly on the <u>context</u> in which they appear. For instance, the missing fundamental may not be perceived at all, if the component tones of higher harmonics are presented as isolated events. Yet when they are part of a tone succession, a meaningful melody, the fundamental pitch sensation appears in uncontested fashion! Of course, this only may be an indication that for short tone durations our auditory system pays more attention to the "periodicity pitch" mechanism, while for lasting, isolated tones it tends to focus on the (multiple) output of the "place pitch" mechanism (Fourier components in the cochlea). Other scientists, however, believe that the central nervous system actually must <u>learn</u> at an early age to

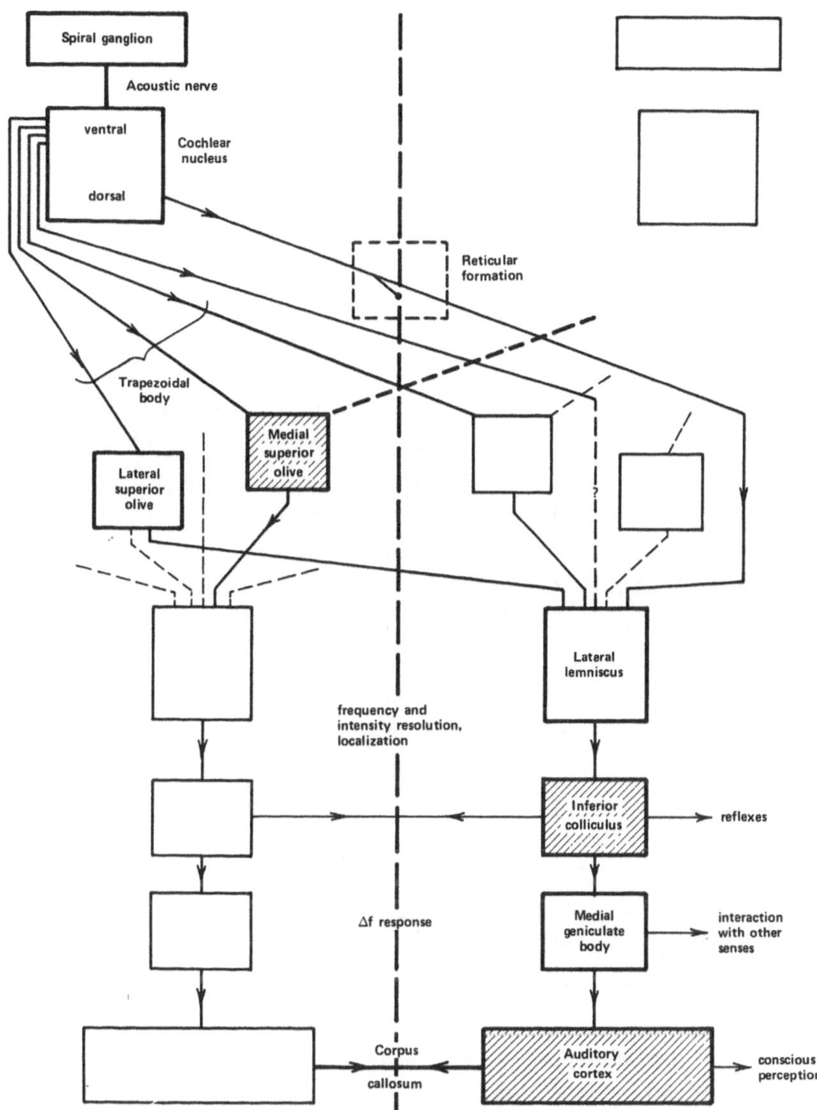

Fig. 9. Flow chart of neural signals from one ear through the brain stem to the auditory cortex. Note the stages at which binaural interaction occurs. The wiring scheme in the medial superior olive seems to be especially fitted to perform intra-aural crosscorrelation (spatial sound source localization). Lateral inhibition (and perhaps some autocorrelation--Fig. 8) could possibly be performed in the highly elaborate structure of the cochlear nucleus. The reticular formation--a "switchboard" controlling information flow to and from the cortex--receives fibers from one of the acoustic routes, but also sends out fibers to some of the nuclei. Together with the superior colliculus (not shown) it may play a key role in gating operations. Contralateral pathways are dominant: in case of conflicting signals from both ears, the information carried in the ipsilateral route tends to be ignored. Spatial mapping (tonotopical organization) of the instantaneous acoustical "image" on the basilar membrane gets lost as the signals proceed through the higher stages (in the waking animal). Antiparallel to the channels shown in the figure is a system of efferent fibers (not shown) that may play a crucial role in lateral inhibition (contrast enhancement), particularly between the superior olives and the spiral ganglion (the olivocochlear bundle).

assign a single pitch sensation to the multiple resonance regions arising simultaneously and coherently on the basilar membrane when a complex tone is fed in. After this initial learning experience a single tone sensation would appear automatically whenever a series of harmonics is presented simultaneously in a meaningful context such as a melody.

Earlier we mentioned timbre as the psychophysical correlate of tone spectrum. As a matter of fact, timbre perception is just a first stage of the operation of tone source recognition - in music for instance, the identification of the instrument.[1] From this point of view, tone quality perception is the mechanism by means of which information is extracted from the auditory signal in such a way as to make it suitable for storage in the memory with an adequate label of identification, and comparison with previously stored and identified information. The first operation involves learning or conditioning. A child who learns to recognize a given musical instrument is presented repeatedly with a melody played on that instrument and told: "This is a clarinet". His brain extracts suitable information from the succession of auditory stimuli, attaches to this information the identity label with the qualification "clarinet", and stores it in the memory. The second operation represents the conditioned response to a learned pattern: when the child hears a clarinet play after the learning experience, his brain compares the information extracted from the incoming signal (i.e. the tone quality) with stored cues, and, if a successful match is found, conveys the response: "a clarinet". On the other hand, if we listen to a "new" sound, e.g., a series of tones concocted with an electronic synthesizer, our information-extracting system will feed the cues into the matching mechanism, which will then try desperately to compare the input with previously stored information. If this matching process is unsuccessful, a new storage "file" will eventually be opened up for this new, now identified, sound quality. If the process is only partly successful, we react with such judgments as: "almost like a clarinet" or "like a barking trombone".

Few details are known on how these identification processes work. First we note that neurons in the auditory cortex are feature detectors, responding to well-defined, but complex, physical features of the sensorial input. The exogenous, externally stimulated activity of a cortical neuron is the result of neural signal processing in the afferent pathway as well as in the neighboring cortical network. Repetitive stimulation with a given input pattern creates neural engrams that lead to a well defined cortical activity or cortical output - even if the input pattern fluctuates within certain error limits.

When a certain neural activity is "displayed" on the cortex, we report a given sensation. However, experiments have shown that the cortical activity evoked by a given stimulus is profoundly altered if the information carried by that stimulus has a certain meaning.[21] To the peripherally triggered firing pattern, an endogenous activity or readout signal is added (with a typical delay of 50-80 milliseconds), triggered internally in the brain by higher order operations of comparison and identification, somehow representing the conscient experience related to the act of perception. This readout signal is found as a common, coherent pattern of activity spread over many different areas of the brain. It is absent altogether if the original signal is meaningless (i.e., not related to previously stored or learned messages) or if no attention[§] is being paid to the stimulus. Experiments with animals have shown that a readout signal may sometimes be in error, i.e., may not correspond to the externally evoked activity - in those cases the behavioral response also comes out in error![21]

It is believed that the replay of a given endogenous readout pattern represents the elementary act of remembering. When this pattern is triggered externally while we listen to a tone, we "remember" that this tone comes, say, from a clarinet. When this activity is released internally (by some association or by a volitive command), we are remembering the sound of a clarinet in absence of a true external sound. This then represents the most simple form of activation of the "internal hearing mechanism". Experiments with vision in fact have shown that, for instance, the mere imagination of a geometrical form evokes activity in the visual cortex very similar to that externally evoked when the subject actually sees that form.[22] A similar effect is likely to occur with the imagination of musical tones and musical forms.

Nothing is known as yet on how the readout activity is actually produced. Obviously, it must involve a complex interaction between different centers of the brain. When we listen to a "new" musical instrument and at the same time are told what the instrument is, the primary exogenous auditory cortical activity evoked by the musical sound and the activity in other centers elicited by the awareness of the source identity, the name or the physical appearance of the instrument, interact somehow in the interconnecting neural tissue, giving rise to the formation of a readout pattern that is a signature of that particular experience.[1] During the initial phase of learning, this readout pattern

[§]"Attention" probably means to let a given readout pattern of neural activity spread over a large area of the brain by willfully inhibiting all other potential sources of readout patterns.

can be triggered only by the simultaneous concurrence of the different inputs that are part of the learning experience (e.g., the instrument's sound and the instrument's appearance). As a result of repetitive stimulation, however, some long lasting alterations occur in the intervening neural tissue, probably in the form of changes (increases) of synaptic connections between the participating neurons.[23] These alterations could be such that, in the future, the same readout patterns can be elicited even if only one of the originally concurrent input forms appears. Thus, if we hear tones from a "known" instrument, the activation of the corresponding readout pattern tells us what instrument it is. On the other hand, the same readout pattern can be triggered from the "other end" by "thinking" about the instrument - and the spread of this activity onto the auditory cortex will allow us to "hear" its tones internally. We may speculate[24] that the learning engrams are formed in those regions of the cortical association areas whose input and output fibers are wired to those source areas of exogenous activity that are stimulated simultaneously in the particular learning experience. All this is summarized in the model shown in Fig. 10.

Quite generally, the auditory system, like other sensory systems, operates on the principle of "minimum effort": in the identification process of musical (and other sensorial) messages, the system first discards all but a certain minimum of information cues. If the identification has been successful it proceeds ahead with the next message. If not, it goes back to the fast memory and searches for additional cues. This applies not only to a single one-tone input, but also to the musical message as a whole: the nervous system tries to use whatever information is available from previous experiences (e.g., memory stored messages) to anticipate the identification process of new incoming information. This "prediction" or "extrapolation" capability - quite generally, perhaps the most essential operation of the central nervous system helping to enhance the chance of survival for all higher living beings - has been confirmed through electro-physiological measurements: when a certain event in a previously learned succession of stimuli is expected but does not occur, cortical activity appears at a latency similar to that usually evoked by the expected event.[26] We may speculate that whenever this prediction mechanism fails during the perception of music (e.g., because of an "unexpected" passage or a goof in the performance), the extra work required for reidentification gives a particular sensation of "musical tension".[27]

In 98% of all humans, the left cerebral lobe preferentially executes temporal operations, that is, sequential time-pattern analyses

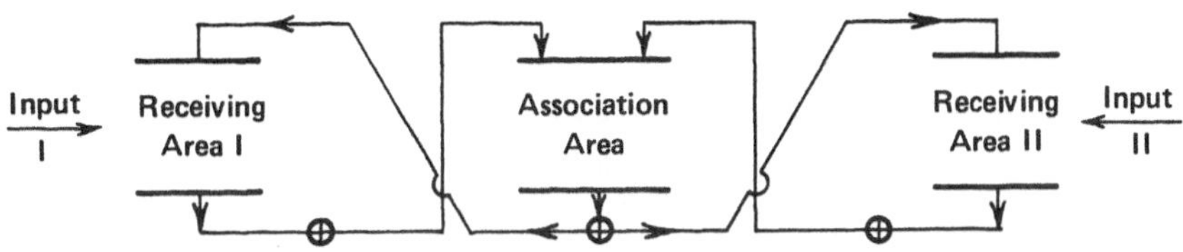

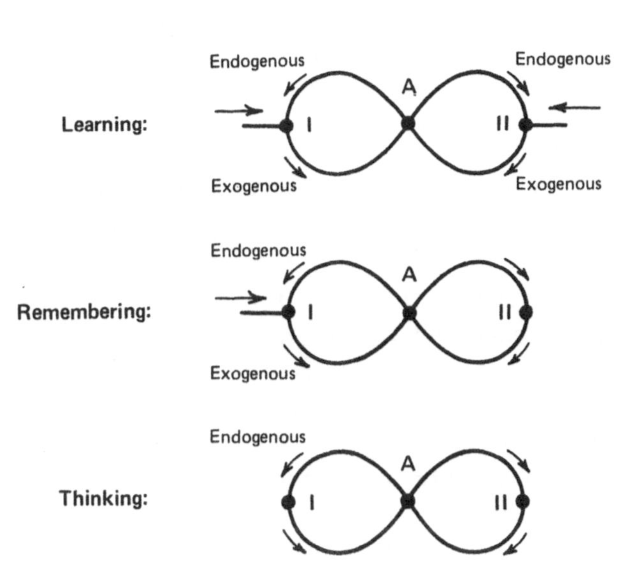

Fig. 10. Top diagram: A model of the learning-memory-thinking processes in the cortex.[24] The thalamic afferents corresponding to the two independent, simultaneous, external inputs that are part of a hypothetical learning experience, enter the respective sensory receiving areas halfway between the top and the bottom of the cortex in each area (IVth-layer).[25] The simultaneous outputs are assumed to give rise to a specific distribution of neural activity in the input layers of those association areas to which fibers from both receiving areas converge. It is further assumed that the output from these association areas is fed back to the first cortical layers of the original receiving areas. Repetition of the input events causes the formation of neural engrams[23] in the participating association areas. This happens in such a way so as to insure that the same endogenous firing pattern[21] is conveyed to the original receiving areas, whenever the particular input constellation of that learning experience arises (within certain error limits).
Lower diagrams: It is assumed that, after the engrams have been formed (in A), only one of the two independent inputs (I) is sufficient to trigger the endogenous firing pattern that is a signature of the original learning experience. The spread of this activity onto the other, initially silent, receiving area (II) represents the internally triggered replay of the original sensory stimulation: the act of remembering. Finally, it is assumed in this model that the thinking process involves an internally triggered stimulation of feedback loops that have been built up during previous learning experiences. See text for a "musical" example.

such as required in language[2] and, more generally, in thinking. This is why it is called the dominant hemisphere. The right lobe, instead, is mainly geared to perform spatial pattern analyses, geometric representations, and holistic syntheses.[28] Of course, both hemispheres must cooperate, particularly during the early stages of learning experience. They exchange information through the roughly 200 million nerve fibers of the corpus callosum.

A most fascinating fact is that music is processed in the minor hemisphere.[29] Complex tones, chords, polyphony, indeed require holistic, "geometric" pattern analyses--geometric in the sense of a spatial distribution of neural activity evoked on the basilar membrane! Of course, time evolution is an essential ingredient of music, yet the focal point seems to be the information contained in the momentaneous holistic pattern--be it just one complex tone from a musical instrument, or the simultaneously sounding orchestral tones of a symphony. It is important to note that in 65% of examined cases, a remarkable asymmetry has been found, with one of the main auditory processing areas (the planum temporale) being significantly and systematically larger on the left side (Wernicke's speech area!) than in the right lobe.[30]

Finally, another most intriguing question remains open: why are we rewarded by certain kinds of musical passages, why irritated by others? In other words, why is there an emotional content in the highly abstract, environmentally uncorrelated, sound stimuli of music? What kind of connection is there between music processing and the functions of certain components of the limbic system--our genetically built-in reward dispenser? Could this perhaps be somehow related to early memory storage of acoustical experiences during our fetal stage?

References

1. Roederer, J.G.: Introduction to the Physics and Psychophysics of Music, Springer-Verlag, New York, 1973.

2. Penfield, W. and Roberts, L.: Speech and Brain Mechanisms, Princeton University Press, Princeton, New Jersey, 1959.

   Gazzaniga, M.S.: The Bisected Brain, Appleton-Century-Crofts, New York, 1970.

   Geschwind, N.: Language and the brain, Sci. Am. 226, number 4, 76-83 (1972).

3. Benenzon, R.O.: Musicoterapia y Educación, Editorial Paidós, Buenos Aires, 1971.

4. E.g., Tonndorf, J.: Dimensional analysis of cochlear models, J. Acoust. Soc. Am. 32, 493-497 (1960).

5. von Békésy, G.: Experiments in Hearing, McGraw Hill Book Co., New York, 1960.

6. von Helmholtz, H.: On the Sensations of Tone as a Physiological Basis for the Theory of Music, English translation, Dover Publications, New York, 1954 (original 1863).

7. von Békésy, G.: The variation of phase along the basilar membrane with sinusoidal vibrations, J. Acoust. Soc. Am. 19, 452-460 (1947).

8. Houtgast, T.: Psychophysical evidence for lateral inhibition in hearing, J. Acoust. Soc. Am. 51, 1885-1894 (1972).

9. Kiang, N.Y.-S., Watanabe, T., Thomas, E. and Clark, L.: Discharge patterns of single fibers in the cat's auditory nerve, M.I.T. Press Research Monograph No. 35 (1965).

10. Whitfield, I.C.: Central nervous processing in relation to spatiotemporal discrimination in auditory patterns. In: Frequency Analysis and Periodicity Detection in Hearing, ed. by R. Plomp and G.F. Smoorenburg, pp. 136-152 A.W. Sijthoff, Leiden, 1970.

11. Zwislocki, J.J. and Sokolich, W.G.: Velocity and displacement responses in auditory-nerve fibers, Science 182, 64-66 (1973).

12. Rose, J.E., Brugge, J.F., Anderson, D.J. and Hind, J.E.: Some possible neural correlates of combination tones, J. Neurophys. 32 402-423 (1969).

13. Plomp, R.: Beats of mistuned consonances, J. Acoust. Soc. Am. 42, 462-474 (1967).

14. Plomp, R.: Timbre as a multidimensional attribute of complex tones. In: Frequency Analysis and Periodicity Detection in Hearing, ed. by R. Plomp and E.G. Smoorenburg, pp. 396-414. A.W. Sijthoff, Leiden, 1970.

15. Plomp, R.: The ear as a frequency analyzer, J. Acoust. Soc. Am. 36, 1628-1636 (1964).

16. Plomp, R.: Pitch of complex tones, J. Acoust. Soc. Am. 41, 1526-1533 (1967).

17. Licklider, J.C.R.: Three auditory theories. In: Psychology: A Study of a Science, Vol. I, ed. by I.S. Koch, pp. 41-144. McGraw Hill Book Co., New York, 1959.

18. Gordon, B.: The superior colliculus of the brain, Sci. Am. 227, No. 6, 72-82 (1972).

19. Katsuki, Y.: Neural mechanism of hearing in cats and insects. In: Electrical Activity of Single Cells, Igakushoin, Hougo, Tokyo, 1960.

20. Houtsma, A.J.M. and Goldstein, J.L.: Perception of musical intervals: evidence for the central origin of the pitch of complex tones, J. Acoust. Soc. Am. 51, 520-529 (1972).

21. John, E.R.: Switchboard versus statistical theories of learning and memory, Science 177, 850-864 (1972); and references therein.

22. Herrington, R.N. and Schneidau, P.: Experientia 24, 1136 (1968).

23. Eccles, J.C.: The Understanding of the Brain, McGraw-Hill, New York, 1973.

24. Roederer, J.G., unpublished.

25. E.g. see Braitenberg, V.: On the representation of objects and their relations in the brain, this volume.

26. Chistovich, L.A.: Temporal course of speech sound perception. In: Proc. Fourth Int. Congr. Acoust., Copenhagen, 1962.

27. Roederer, J.G.: La musique, l'oreille et le cerveau, Decouverte 41, (May, 1972).

28. Levy-Agresti, J. and Sperry, R.W.: Differential perceptual capacities in major and minor hemispheres, Proc. U.S. Natl. Acad. Sci. 61, 1151 (1968).

29. E.g., review by Eccles, J.C. and Scheid, P.: The human brain with respect to speech and musical abilities. In: Workshop on the Physical and Neuropsychological Foundations of Music, Ossiach, Austria, 1973.

30. Geschwind, N. and Levitsky, W.: Human brain: left-right asymmetries in the temporal speech region, Science 161, 186-187 (1968).

General Literature

Whitfield, I.C.: The Auditory Pathway, Monographs of the Physiological Society, No. 17. Arnold, London, 1967.

Tobias, J.V. (Ed.): Foundations of Modern Auditory Theory. Academic Press, New York, 1970.

Plomp, R. and Smoorenburg, G.F. (Eds.): Frequency Analysis and Periodicity Detection in Hearing. A.W. Sijthoff, Leiden, 1970.

Flanagan, J.L.: Speech Analysis, Synthesis and Perception, 2nd. ed. Springer-Verlag, New York, 1972.

Roederer, J.G.: Introduction to the Physics and Psychophysics of Music. Springer-Verlag, New York, 1973.

PART IV

Network Physiology

NEURAL NETWORKS AND THE BRAIN

J.G. Taylor
Department of Mathematics
King's College, London

The purpose of my talk, as far as I see it, is to give you the initial steps we have taken so far in our understanding of the brain and the way it controls behaviour. This is a very complicated story, so I will only try to sketch it briefly; other speakers will indicate in much more detail some of the beautiful results which have been and are being obtained. In all it is a very exciting story, with the main questions still unsolved and presenting an enormous challenge to science.

The brain itself can be regarded, with some legitimacy, as a very complicated machine. In particular it is the only part of the world we 'know' about in two ways - from the inside as well as from the outside. This gives us extra information, gained by means of introspection or similar techniques, which might help us. Sadly enough it has not done so to any degree, and we have had to rely to an ever greater extent on external probing of the brain by ever-more delicate instruments.

There are over a million animal species with different sorts of brains, and nature has been prodigous in provision of laboratory material for us to observe. This lavishness has not helped us sufficiently in our study of the brain; the brains of different species certainly have similarities of structure, but all present the same degee of mystery to our gaze.

One of the basic problems about the brain is that it is so complicated a system that it requires description, at least initially, by means of a few gross variables which allow for neglect of much of the fine detail. Yet these variables have not been discovered; the brain analogues of thermodynamic variables for unthinking matter - temperature, pressure, etc. - are absent. In this latter case it has proved possible to introduce an ever greater wealth of detail through the chain of molecules, atoms, electrons and nuclei, sub-nuclear particles, and now possibly even quarks and similar esoteric particles.

It has not proved possible to even commence this reductionistic or 'atomic exploration' program for the brain since we don't even have the gross variables to begin with. It might be thought possible to choose gross behavioural responses, but these are difficult to quantify, involving as they do a large number of possible actions. An alternative is to use the surface brain wave patterns, measured by the EEG, but this has also proved unsuccessful.

What we are faced with is a very complicated system which we have to understand in terms of its basic constituents from the very beginning - as if we had to understand the gross behaviour of matter immediately in terms of its atomic structure. It is as if we were only a millionth of a centimetre tall, so that we could not observe the thermodynamic variables of matter at all, yet wanted to understand its behaviour. I am going to plunge straight into the 'atoms' of the brain, the nerve cells, and build up from there.

Let me start by pointing out the basic features of the brain as shown in the figures - ones which you will see and hear about in much more detail in other talks and in the discussion groups. In Figure 1 is shown a cross section of the brain, with various parts of brain given their respective names. These regions have gross anatomical differences from each other, and so can be clearly distinguished.

Further differentiation is obtained by considering the various regions of 'grey matter', each composed of brain tissue with a higher density of cell bodies than the surrounding 'white matter'. Such grey regions are often called nuclei, and usually have different functions from each other in information processing. The second figure is a view of the brain from underneath, the spinal chord having been left off (it should have been a downward continuation of the medulla oblongata), as it was in the first figure.

We note the cerebellum, two hemispheres hanging over the back of the brain stem region and joined to each other by the pons. The cerebellum is thought to be crucial in the control of movement, since cerebellar lesions produce clear defects in such activity. The lower brain

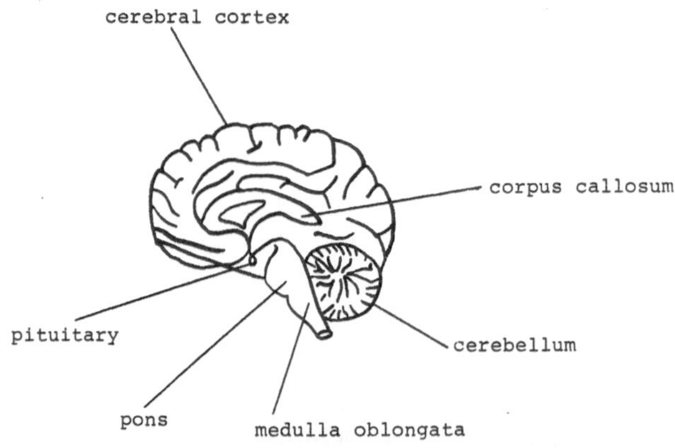

Fig. 1. Median surface of the brain.

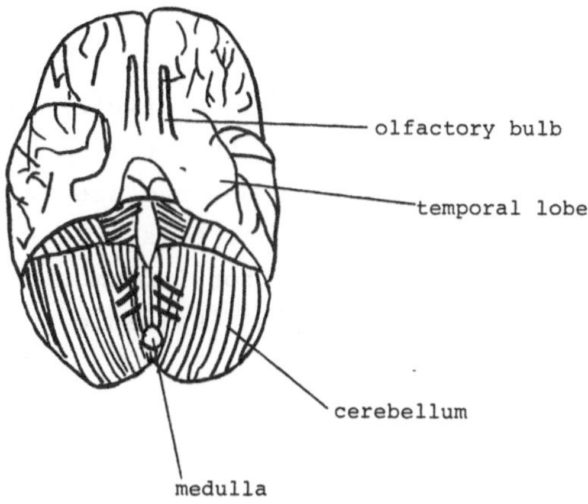

Fig. 2. The brain seen from the base.

stem itself is involved in very primitive reflex activity as well as acting as a way-station for information travelling up to the higher centres or away from them down to the body.

Further up the brain, in what is called the mid brain region, we come to the areas controlling the emotions as well as the primitive body functions. At the highest level of the brain are the cerebral hemispheres, large outgrowths which overhang the more primitive regions, and almost completely cover them. It is on the surface of the cortex, with its great folding in man, that we think that information processing is occurring which gives us our intelligence - and that is one of the puzzles we have to try and solve.

Let us now consider briefly the structures that are involved in the brain, looked at now in a more schematic fashion. Again going up the brain stem from the spinal cord through the pons, one has the cerebellum at the back, involved in the fine control of movement, and further up the thalamus, which is a main relay station for the cerebral cortex. Just in front of and below it, is the hypothalamus, which is the centre of the satisfaction drive centres; of sex, thirst and so on, and to the side of it the limbic system which is a control system for the emotions. Then there are the hemispheres of the cerebral cortex which are hanging over the upper brain stem in the fashion shown by the shaded outline in Figure 3. In the brain stem is a loose network of cells, called the reticular formation, which is known to be very important in behaviour. This reticulum is thought to be essentially the consciousness switch and is heavily involved in sleeping and in attention mechanisms. All these regions have very important functions, as one finds when one removes them, or when they are removed for you in an accident. We have therefore to try and understand the information processing by looking in detail at the connectivity and structure of each of these organs. And that is then the problem before us. However, as I said, this problem is not an easy one because it is so difficult to know the relevant gross variables. At this simple level, for example, one can think of drawing flow diagrams for information processing but one finds it very difficult to go beyond that unless one goes right down to the atomic level; to the nerve cell level.

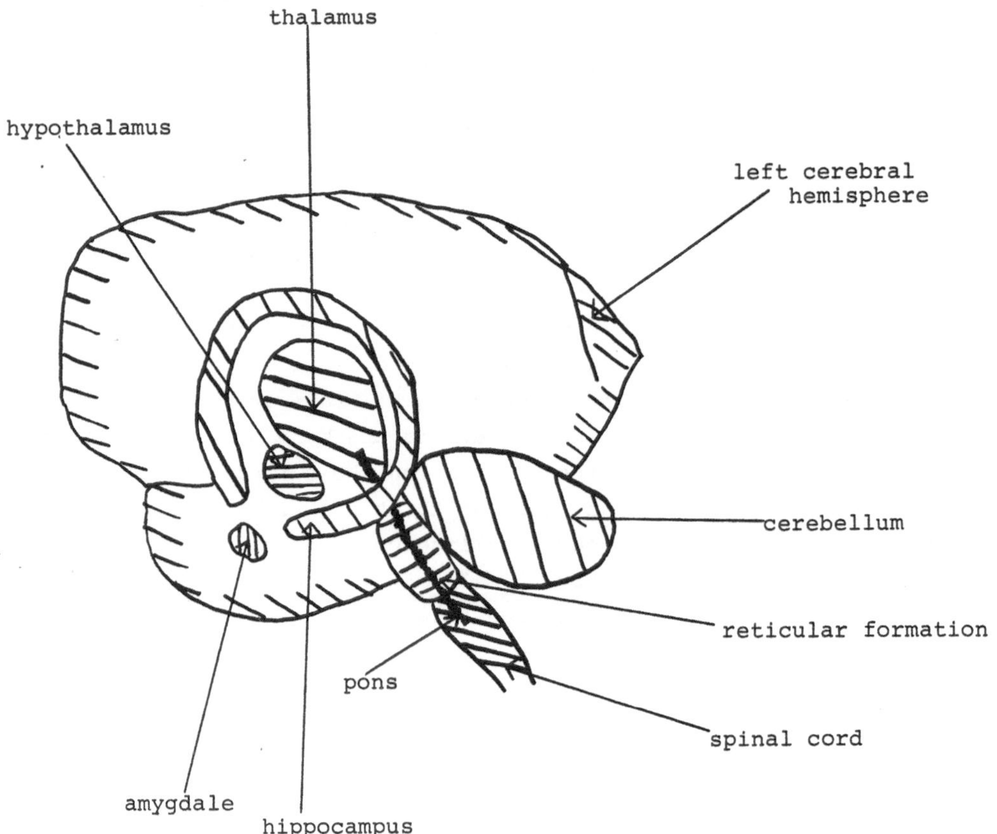

Fig. 3. Schematic representation of important organs in the brain.

I've shown in Figure 4 the nature of localisation of function on the surface of the cerebral cortex and that is another question as to the manner of information processing there. The frontal lobes, for example, are certainly involved with personality; there are some speech areas, the motor-sensory and the somatic-sensory regions are strips across the top, and then at the back is the visual processing region. The regions in between are the secondary or associative areas whose removal doesn't cause deficit of any appreciable or very noticeable form, but does reduce general powers of response and reasoning. Those regions shaded in Figure 4 are the primary regions; their presence is crucial for correct function of the appropriate sense perception. For example, if you remove the visual cortex a person will be blind. If you remove the secondary cortex around the primary visual region the patient won't be blind but may have difficulties in using what he sees and relating it to other sensory modes. So that is why these regions are called the associative regions, the areas in Figure 4 which are not shaded in. Perhaps they comprise 90% of the cortex and the old wives tale that we only use 10% of our brain essentially arises because one thinks of these associative regions as being unused. In fact one can conjecture that it is precisely these regions that makes Man, Man.

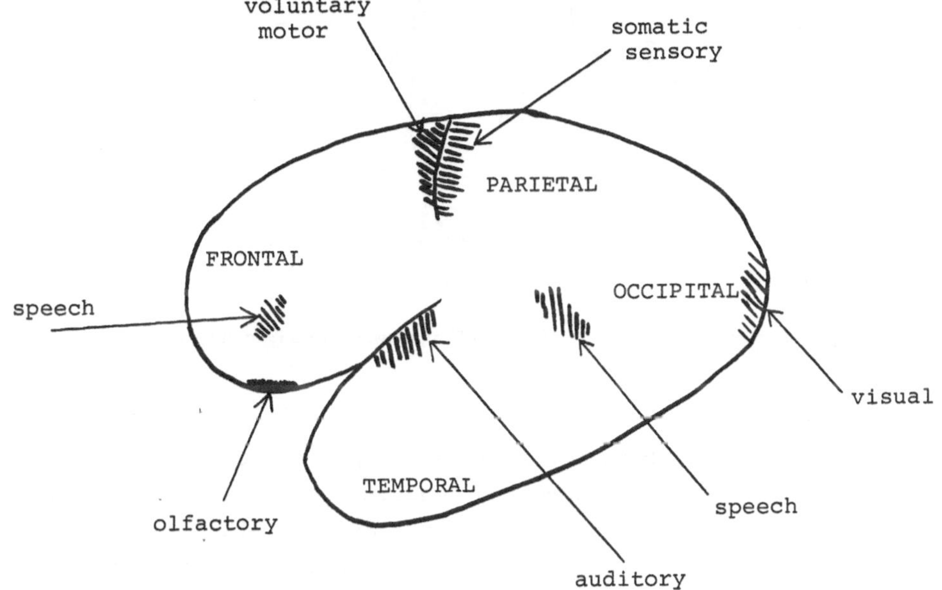

Fig. 4. Primary sensory areas on the cortex are shown shaded.

I want to leave that now, because as I say it is difficult at this stage of the game to understand the information processing that is going on in the various organs of the brain, and let me go now to the actual 'atom' that is used in making up the brain, the nerve cell. It was only recognized as the basic unit of the brain in 1890 - only just over 80 years ago - quite recent as far as our understanding of the material world goes. The typical nerve cell that I have shown in Figure 5 has a basic cell body with processes extending from it; the processes with spikes or little dots on them are called dendrites and the single process which is smooth with no apparent appendages to it is called an axon.

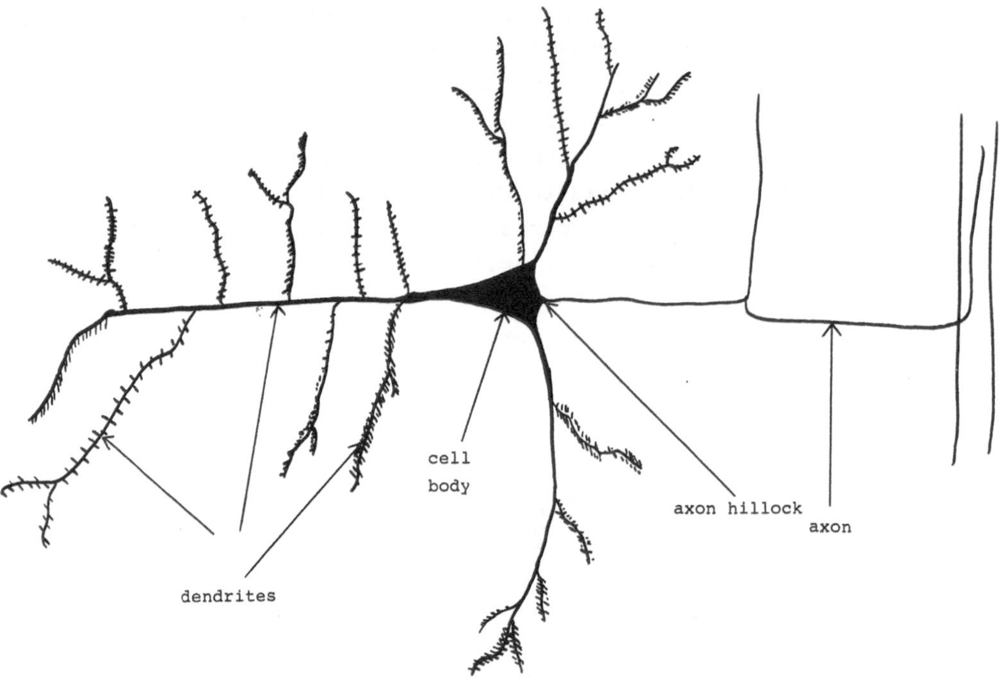

Fig. 5. Representation of a typical cell, showing axon, dendrites and the axon hillock.

The function of the nerve cell, as a specialised cell of the body, is to acquire information along its dendrites and all over its cell body and to send out information down its axon to other dendrites or cell bodies of other cells. The actual apposition of an axon onto a dendrite is at one of the small appendages or buttons seen in Figure 6. I haven't shown the axons from other cells that are attached to the dendrites of the cell; there are many of them, possibly as many as 100,000 on a particular cell, say in the cerebellum. It's clearly a very complicated situation inside the central nervous system.

The sizes that are indicated in Figure 6 are very rough. There is quite a large variation in sizes of cell bodies, in sizes of lengths of dendrites and in lengths of axons, particularly the latter. The axon is the process of the cell which carries the information away from the

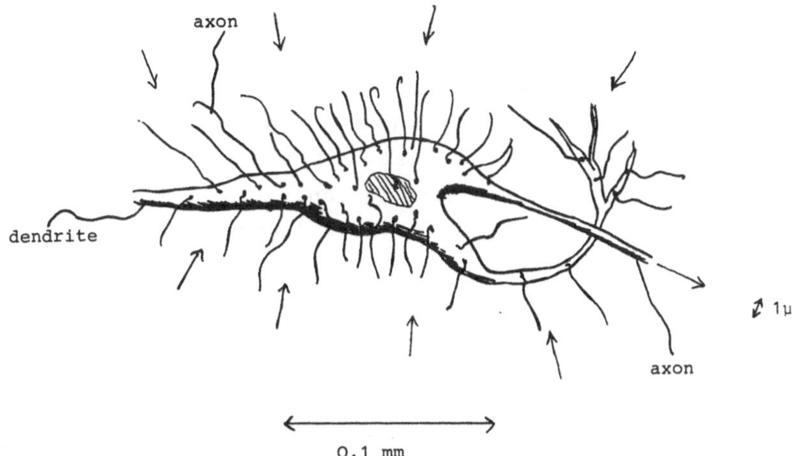

Fig. 6. Typical sizes for a motor nerve cell for the spinal chord of a cat (1μ = 10^{-3}mm).

cell, and it may be as long as the arm or down the leg from the spinal cord, in other words it may be 1 metre in length. It may, on the other hand, be as short as a hundredth or a thousandth of that. In the brain itself the axons are much shorter and the cell bodies are smaller. If one goes through the animal kingdom, of course, one gets very great variation in size of nerve cells and their processes; experiments can be performed on the larger ones much more easily. Such experiments become more difficult when one goes to the brain itself, since the nerve cells tend to be too small to put electrodes inside them to find out what is going on.

There are altogether about 10,000 million nerve cells in the human brain. In small animals there are far fewer, for example 1000 in typical worms. Luckily enough there do not seem to be 10,000 million different sorts of nerve cells. It's rather like in our explanation of matter in terms of elementary particles; there may be two or three hundred of them, that's too many. If quarks exist then there may only be a dozen different sorts of particles. In the case of nerve cells there do, in fact only seem to be about a dozen or so different shapes and sizes. I've shown, for example, a couple of very distinct types in Figure 7 that are very important and they obviously have very different functions in the brain. One is a stellate cell; it has dendrites which are uniformly distributed or tend to be, round the cell body. The pyramidal cell tends to have one large apical dendrite and a number around the base called basal dendrites as well. The cell body is again different in structure. These cells found, for example, in the cortex, tend to be final signalling cells, their axon carrying information to tell muscles what to do. Stellate neurons tend to be interneurons, that is, they are involved internally in the information processing. They may even be initial neurons in the information channelling, but don't tend to be final ones. They also are used in inhibition, which of course is just as important as excitation in nerve cell connections.

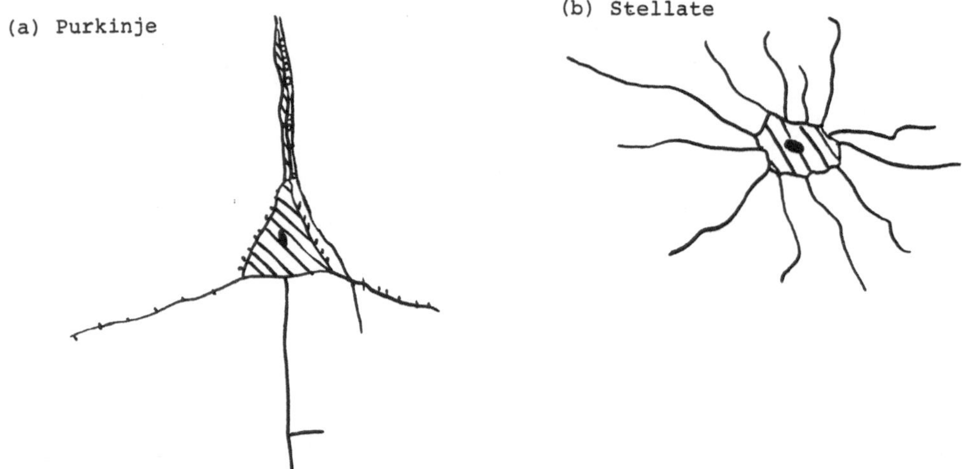

Fig. 7. Different types of cell, described in the text.

There are also specialised endings of cells. What happens at the ends of the dendrites and of the axons are going to be very important as to how the information gets from one cell to the next. Various sorts of specialised endings exist and the nature of the information transmission from the environment to the nerve cells will depend on the nature of the endings and is itself a very complete and a very full area of research. I gather we will hear a little bit about this from Professor Fatt later on. In Figure 8 is a typical set of endings of dendrites. I've shown next to them the sorts of function they perform though there is, I gather, some argument in detail as to whether they are unique. These ambiguous cases are the specialised endings for pressure, heat, touch or pain. One also has specialised endings for example, of a motor nerve, an axon going down to a muscle to cause motor activity. So we have to take account of the nerve cells with their specialised endings and the fact that there will be different sensory information coming in at the different senses.

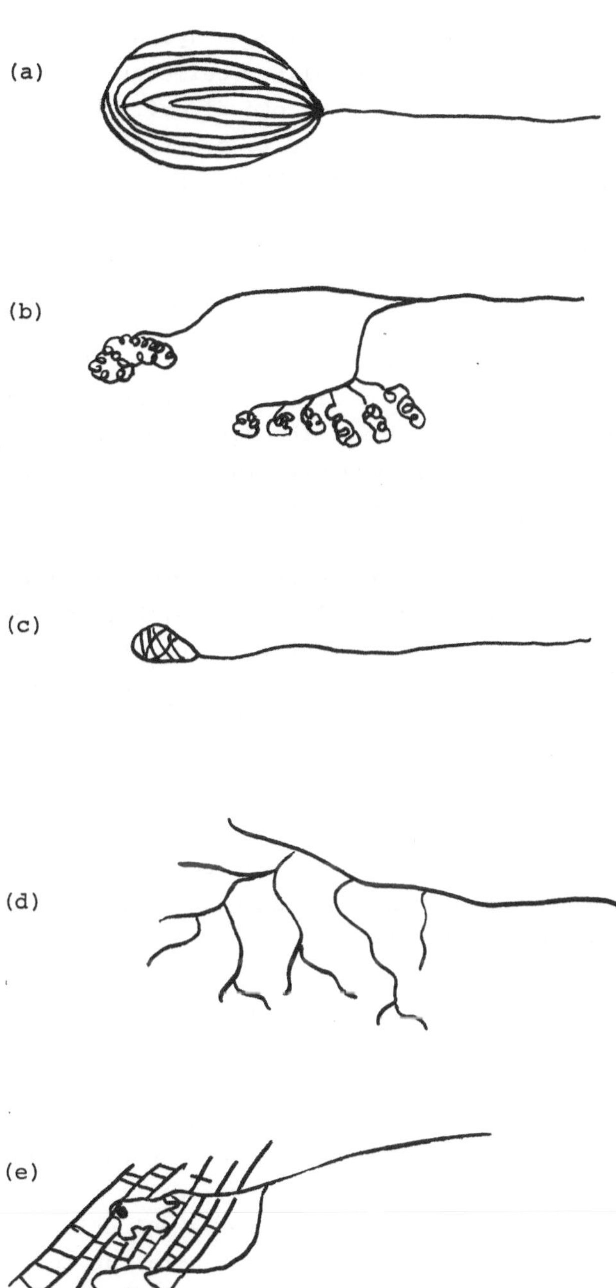

Fig. 8. Specialised nerve endings for peripheral cells. (a) pressure (b) heat (c) touch (d) pain (e) motor nerve, onto muscle fibre

The general structure of the nervous system of any animal tends to be along the line of input; specialised cells or affector cells, either going straight in to connect or synapse directly onto an affector cell, for example that will cause a muscle to contract, or to go throught inter-neurons and then to affector cells. The direct link affector-effector is a reflex path, and requires very little information processing. A little more information processing goes on in the spinal cord, in the presence of inter-neurons, but at the higher level of the brain there is a large number of inter-neurons and a correspondingly greater amount of information processing say at the level of cerebral cortex, where of course the most difficulties lie. So that is the general feature of the connectivity of nerve cells. The detailed connectivity is something we will hear about, I hope, much more in lectures; in the neural nets discussion group there will be questions on the cerebellum, cerebral cortex, reticular system and so on.

Let me turn now to the nature of the information that is carried by a nerve cell. Many of you will know this, but some may not, and I thought it would be useful just to talk about this. It is very introductory neurophysiology but at least we should get this said so that we all are at least on the same level at this stage. The nerve impulse is an electrical pulse that can be measured by putting micro-electrodes, for example, inside an axon if it is large enough. If one stimulates the dendrites of the nerve cell, one will measure an electrical change in the potential measured by the micro-electrode and will see the pulse going along one of the branches of the axon and onto the next nerve cell. On this nerve cell itself will be coming other pulses from other axons. The nature of the nerve impulse is obviously important and it arises from the resting potential inside a nerve cell. This is what is so special about the nerve cell: it has got a -70 milli-volt resting potential arising from the fact that sodium ions are pumped out of the cell, potassium ions pumped into it. Precisely what this pumping mechanism is, is not fully understood and is of great interest in research in this area. But certainly metabolic processes in the cell are allowing this pumping process to proceed. The nerve pulse itself is a rapid increase of this -70 milli-volt resting potential up to about 50 milli-volts in the order of a milli-second.

This is an extremely rapid process which is all or none and is the important feature of nerve cell information transmission. Suppose the nerve cell is at about -70 milli-volts, and that in some way, say by feeding an electrical current in, one depolarises the cell by at least 10 milli-volts or so. One will then find that the cell cannot stop further depolarisation and that there is a very rapid depolarisation up to about 50 milli-volts and then a re-polarisation back, to a little below the resting potential, in a total of about 2 milli-seconds. It is this process which is the nerve impulse - an all or none process, in the sense that if the cell gets above this 10 or so milli-volt threshold, the membrane will have such a rapid depolarisation then re-polarisation. The actual mechanics of this all-or-none activity are related to the sodium and potassium conductances; in the rising phase of the potential the sodium conductance suddenly increases and sodium ions are allowed to pour into the membrane and cause the potential to become even more positive. This conductance then decreases and potassium ions now enter the cell, due to an increase in the potassium conductance. The potential then reduces to its resting value, when the potassium conductance decreases. This rapid cycle of potential change will flow down the membrane. Because there is a certain resting time when the membrane is no longer excitable the nerve impulse will tend to flow from one end to the other of the axon. During the period of time when the membrane is active or just recovering there is a certain degree of refractoriness lasting in all for several milliseconds after the nerve impulse has passed. We have to include this refractory nature of the membrane in our discussion.

The neuron is thus a summation device; electrical pulses coming down the dendrites from other axons are fed to the axon hillock of the cell being considered (where the axon joins the cell body). A nerve pulse will go down the axon if there is enough excitation at that point. In other words the nerve cell is a threshold device which will allow the cell to fire provided it has had enough total excitation from all the axons of other cells that are actually apposing on it. It has spatial and temporal summation features, since it can sum in space over its cell body and over the dendrites that are in various regions, and it can sum even in time in the sense that provided the excitation comes soon enough, with respect to a previous excitation, the two excitations can add together in time and help to cause the cell to fire. This makes the nerve cell quite a complicated machine; we have to take account of both space and time effects. It is a four-dimensional object and not a one-dimensional one. It also possesses a refractory period so that we have to take account of the fact that if information comes in during a refractory period, that information may be lost.

The cell body is thus a decision-making device; it will fire if it has enough information, enough electrical pulses coming in from around it from other cell bodies down their axons. But the nature of the information that is being carried or being transmitted by the cell may not be just in one single pulse. If it were in one pulse alone it may be rather dangerous for the organism for if it decided to act on one pulse, this pulse may have been fired in error. Indeed we will see that there can be a great deal of error in the system. So one finds, in particular in peripheral nerve cells that the frequency of firing is proportional to the input. One single pulse is not important, but the frequency of pulses is important in that if the pressure increases, for example on a cat's paw, then one gets an increasing firing frequency of nearby pressure-sensitive cells. It is not known whether this frequency coding is also used in parts of the nervous system other than the periphery, though it is very likely in the initial stages of the visual cortex and in the motor sensory region of the motor cortex. Elsewhere it is not at all understood and in the cerebellum there may be evidence that straightforward frequency coding is not being used. That is one of the problems that is really facing setting up theories about the brain as to how the information as being coded. If you don't know that, it's very difficult to set out a model and one has to try different models which use different information coding to see which is going to be the best.

Naturally enough though, already the theorists have got enough to lay their hands on. They have the model of the neuron as a binary decision element and like any theorists anywhere, they rub their hands in glee and rush away into their rooms, lock their doors and think things possibly quite unrelated to the actual things that are going on inside their brain. One of the ideas which arose was the deterministic neuron. I now want to talk about this ideal neuron and where it has led theorists, because it is very relevant to the problem of the brain.

There is one critical question at the outset. If you are building up a neural network of 10^{11} neurons, you have to be pretty sure that you really know what the neuron is like you're using as the unit, as the 'atom'. It's rather like not knowing quantum mechanics properly, but still trying to build up the behaviour of microscopic matter from the property of the atom. The point about the decision element nature of the neuron is that if the input is larger than a certain threshold, which is a threshold unique for the cell, then the neuron will fire. In detail if the sum of the inputs i_1 to i_4 on the dendrites of the cell in Figure 9 is larger than a certain threshold then the output along the axon will be larger than zero, and if it is less than the threshold, nothing will happen; the nerve cell will just not notice its surroundings. This is the deterministic neuron, no probabilities being involved here. This is not what is seen by neurophysiologists; there is random firing all the time in the brain and the senses; in particular the retinal nerve cells, for example are apparently firing randomly even when they have no light falling on them. And throughout the animal world, spontaneous firing is a very inherent aspect of nervous activity.

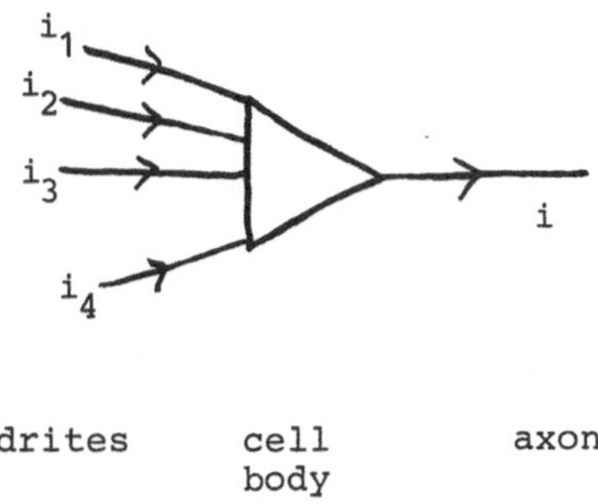

Fig. 9. The cell shown schematically, with four inputs $i_1, \ldots i_4$ and single output i.

Alright, so we've thrown away that very basic fact of life when we'd gone into our little cells, locked our doors and talked about deterministic neurons. But let us suppose we decided to close the doors and carry on. The first important property of networks composed of deterministic neurons was the theorem of McCulloch and Pitts in 1943, that a net of decision elements could perform any logical function of a fairly large class. This result led to the digital computer, and we shouldn't complain about that at all. However, for the brain it is difficult to see that the ideas of McCulloch and Pitts - the representation in terms of logical functions - is going to be very helpful. Let us consider, for example, the case of N inputs, each either 0 or 1. We have 2^{2N} different functions possible on those N inputs, because there are 2^N different states, and for each state we can have a 0 or 1 as a function. If I take N to be 10^6, as there are for example in the case of the optic nerve in a human, $2^{2^{10^6}}$ is rather a large number. It is clearly not helpful to try and sort out which logical functions are being performed by the processing just after the optic nerve in this manner; we have to go and look at the data more carefully. So I would say that at least for any worthwhile endeavour, one cannot use the logical approach. Very many people have tried to do that in the past, but the method has not proved of value and I don't think there are people working in the brain itself who are trying to use a logical approach. The method that has been used a little more successfully for deterministic neural nets, especially for small nets, uses analog procedures. We set up detailed neural net equations. The equations that have been set up are very straightforward, following immediately from the decision. Let us consider a particular neuron, the ith neuron, say, in terms of it firing or not firing. We assign to it a variable $u_i(t)$ which at a given time takes the value 1 if the ith neuron fires at that time and is 0 otherwise. To determine the state of the system, we need to know the value of $u_i(t)$ at a given time t for all values i describing the neurons of the net. The point about the net itself, is that excitation comes from the jth neuron down to the ith neuron and we want to assess whether that will affect the firing of the neuron or not. Let us suppose that excitation can be described by some co-efficient $a_{ij}u_j(t)$. This is important, because if $u_j(t)$ was 0, in other words the jth neuron doesn't fire, no excitation should come down the ith neuron. If the jth neuron does fire, we have a value 1 for $u_j(t)$ and so we get an amount of excitation on the ith neuron equal to this coupling co-efficient a_{ij}. Let us assume that there is some time delay; it takes time for the excitation to travel from the jth neuron to the ith neuron which obviously it must do; we'll assume the same time delay τ for all neurons for simplicity. One can make much more complicated models which don't have the same time delay, but at least to get off the ground let's make this simplification. Then the equation describing the neural net is a very simple equation, first properly discussed by Caianiello in 1961. It is

$$u_i(t+\tau) = 1 \quad \text{if} \quad \sum_j a_{ij}u_j(t) > \text{threshold } q_i \text{ for } i^{th} \text{ cell}$$

$$u_i(t+\tau) = 0 \quad \text{otherwise}$$

If we take the step function $\theta(x) = 1 \ (x>0)$
$= 0 \ (x<0)$

then we can write

$$u_i(t+\tau) = \theta(\sum_j a_{ij}u_j(t) - q_i) \quad .$$

This is a set of equations for each i and for each time t. We can solve it by iteration with respect to time. We start off with given values of the states of each of the neurons at time 0 and from the right hand side, which we now know, we can deduce the left hand side for $t = \tau$. Thus we can get the states of all the neurons at time τ, and at time 2τ, and so on and so we can build up at discrete time intervals, 0, τ, 2τ, 3τ,... the states of firing or non-firing of the neurons of the net.

The problem that is presented of course for a large net is that there are many choices of coupling co-efficients a_{ij} or of the corresponding coupling matrix A. In fact a very large number indeed, and to try and attack the problem of the brain by looking at all possible coupling co-efficients and seeing the different behaviour patterns and hoping that you'll get the right one sooner or later, will be the most inefficient way; it will be rather like trying to act out the old story of the monkeys playing the typewriter and hopefully getting Shakespeare's Othello. The point is of course that we have to use further information about the nature of the a_{ij}, such as whether they are positive or negative, and we have to put that in. This is a very crucial feature of the net because you have to be very careful of epilepsy; we don't want all the neurons to end up firing, if they do the net is in fact useless, it's killed itself. If of course, everything dies out, then again it's useless, it's dead. We know we are not in that state so we hope that our neural nets can be such that they are in some useful intermediate state. The programme that has been followed by Caianiello and his co-workers, and numbers of other people throughout the world, is to try and use these deterministic equations and either by hand waving or by detailed computation, obtain possible useful types of behaviour patterns that might be related to the way animals behave. We can't hope to programme the neural net equations with 10^{11} neurons. In fact the greatest number of neurons that have been looked at is 2,000.

If we think of a net of these binary decision neurons, which have closed circuits in them so that you can go from one neuron and come back to it in the end, going along axons onto cell bodies, onto axons and so on, we can think of activity going round and round and this which would form a reverberation or periodic motion. This latter was regarded as a thought or as a short-term memory, of the order of a second or so that we possess, and of course it could correspond to this. A possible reduction of the thresholds may also occur on experience, so as to make the firing easier, or there may possibly be an increase of the thresholds if you want to increase inhibition. This process can be regarded as long-term memory. We don't need to say in detail what the mechanism is that achieves this, though hopefully we'll find it. But at least we can regard this facilitation as the process of laying down long term memory. One can then disconnect the short term processes of on-going firing from the slower changes, the slower drifts of the thresholds. The slow changes of personality that we all experience over the years and which for example over a period of a year or so, can become recognisable, but certainly not over a period of a few minutes.

Facilitation then, corresponds to reduction of thresholds and facilitated reverberations will correspond to learned patterns. In this way it is possible to explain conditioning, self organisation, but one has to beware of epilepsy. Nerve net models of up to 2,000 or so neurons have been built which we might begin to recognize as behaving like worms, for example and not human beings. Well the difficulty about this of course, is to put it into real practice, especially to do with human beings because that's what we are most interested in. If we take, for example, 512 neurons, the numbers of ways of connecting them up and the numbers of choices of the a_{ij} is infinite; one tends to reduce the problem to taking very simple neural nets. The randomly connected neural net is the simplest. In this each neuron has a certain probability, say a tenth, of connecting with all the other neurons of the net. And in fact one can obtain analytic expressions for the firing pattern as it develops, so computer simulation is not required in a purely random net. You can have quasi-randomly connected nets in which you do need to do computer simulations; hopefully such nets are more related to experimental data, but I think that's still rather doubtful.

One finds the following feature of these randomly connected neural nets, which is a very embarrassing one and one which has essentially caused the activity of neural net modelling to grind to a halt, at least at this simple level. There are either of two possible activities of randomly connected neural nets: in a net with, say, 100 neurons, each neuron being connected on the average to, say, 5 others with the threshold for firing being three nerve pulses. If the initial number of active neurons is less than about 50 in a certain time all the neurons will have ceased firing; the net has died. However, if one starts with more than 50 neurons initially firing, the net excites itself ever more and more until it ends up in epileptic convulsions and is useless. Either extreme is all we have in purely excitatory randomly connected neural nets; this is shown graphically in Figure 10(a).

That is very sad but the lifetime for either extinction or epilepsy may be very long, such as 100 years or 200 years or even 1,000 years in, say, our own brains, we just don't know, but at least it's a bit dangerous to try and build real living organisms with this sort of behaviour. The behaviour patterns can actually be quite complicated in large nets. If we look at the patterns of activity in space at different times in a net, excitation started in the middle will spread out possibly dying away at the edges, or coming back again if it's reflected back. These

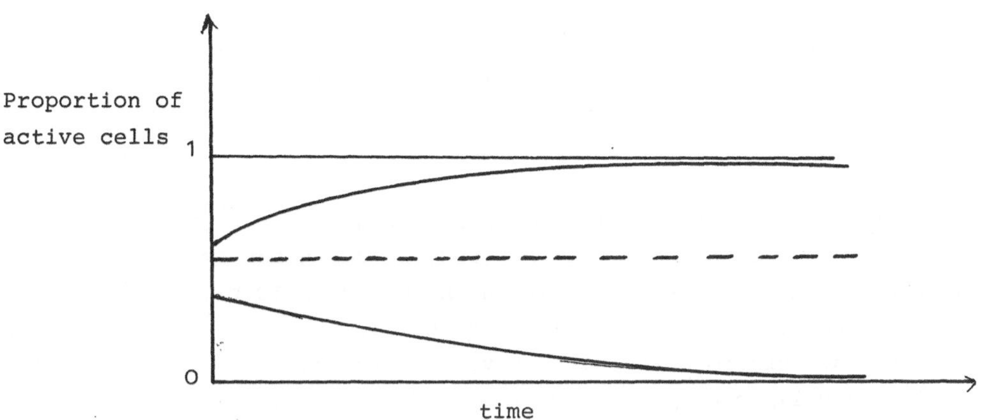

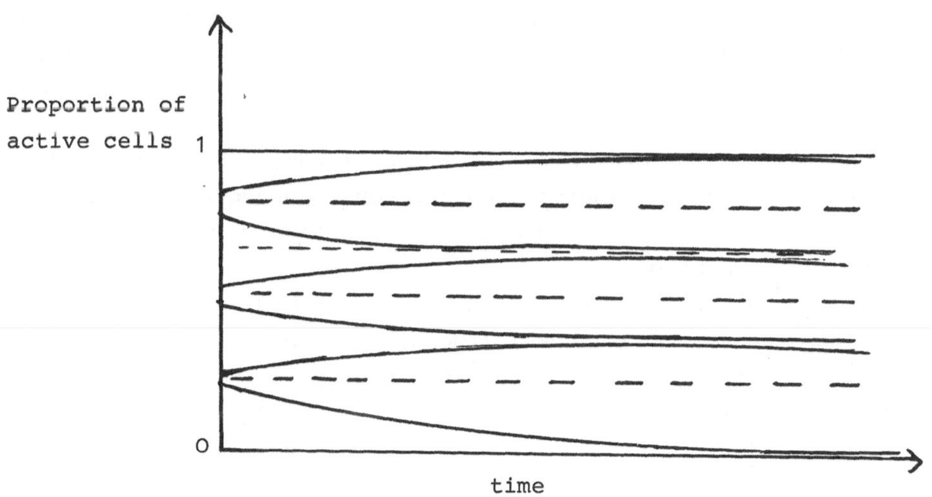

Fig. 10. (a) Randomly-connected neural net with only excitation. (b) Similar to (a) but now with some inhibition.

waves of excitation can interact in a very non-linear fashion. It was fashionable about 15 to 20 years ago to base the whole of the model of the nervous system on interacting waves of excitation flowing through. Many workers tried to see if one can get enough adaptability and all sorts of other behaviours out of these interacting waves of excitation. I think that one of the difficulties is presented by the fact that we have epilepsy or death ultimately in these neural nets.

We suppose that some of the nerve impulses coming on a neuron make the nerve cell upon which they are flowing a little less excitable and not more so. In other words they don't depolarise, they re-polarise, so that if we change from only excitatory connections to include inhibitory ones one doesn't only get epilepsy or death for such a net but also intermediate stages. This is shown in Figure 10(b). Computer experiments that have been done have not indicated however, that there are many alternatives; no more than 4 have been found. Clearly such a net is still very limited.

That is the situation with respect to determinisitc nerve nets, and it is not a very positive one. In any case we have to realize that we have based our model of nerve nets on models of neurons which are not realistic. There is a great deal of noise in the system; even when there is no signal from outside. Nor have we based our model on what goes on in information transfer from one neuron to the next. We have assumed that it does happen, but we haven't said how and that is really very important in fact, in the total net behaviour. So you should take account of that. It may be very important for understanding memory, since memory may be embedded at the synapses between neurons and not in the actual information channels themselves; the axons, the dendrites and the cell bodies. Of course the crucial changes that may be brought about by experience are really ones built by the cell body. But ultimately any new material to change what the connection is between one cell and the next ends up at the synapse and it is there that we must go and look and see what is happening.

Let us now go back to the real world; I've taken you on a slight excursion into theory. I've shown, I hope, that really that excursion into theory is a limited one and that this theory is very limited indeed. To get back to the real world, I want to talk a little about the synapse. In Figure 11 is a picture of synapses on a cell in the cerebellum. The synapses are the gaps from one cell body to the next; there is no continuity of nervous material. There has been a big battle as to whether in fact nerve cells were continuous, one with the other or whether there was a gap, and it was only 10 or 15 years ago that it was properly resolved. The gap has won.

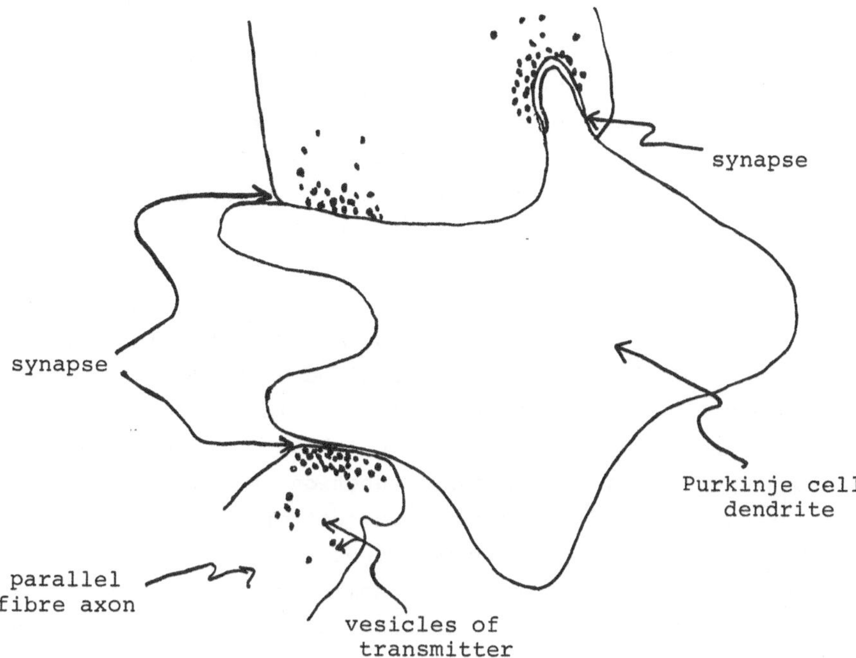

Fig. 11. Synapses of parallel fibres (axons) onto Purkinje cell dendrites in the cerebellum. The vesicles of chemical transmitter are clustered close to the synapses.

The synapse or gap has on one side of it a lot of vesicles, as is shown in Figure 12. It appears as if the vesicles are moving towards the synaptic cleft. Some of them may just be fusing with the synaptic cleft. These vesicles are pushed to the membrane surface, most likely by a pulse coming down the axon, and will release into the synaptic cleft chemical transmitter substance. The molecules of the transmitter diffuse across the synaptic cleft, hit the presynaptic membrane and cause it to be slightly depolarised. There may be 1,000 to 10,000 molecules in each vesicle and the actual nature of the chemical will vary from one region of the brain to another. Many peripheral nerve cells have acetylcholine as transmitter. In the central nervous system there are many other transmitter substances that have been found.

It is clear that there is a quantised transmission of signals. This quantum nature of information transmission has been very carefully studied, in particular by Professor Fatt and Professor Katz in the early

50's and has since been very seriously and thoroughly understood, at least at the level I'm talking about now. When a pulse comes down there may be 100 quanta which are released into the synaptic cleft to allow depolarisation on the post-synaptic membrane, causing the next cell to fire. In the central nervous system there may be only one quanta that is needed to cause the next cell to fire. These quanta are constantly banging against the pre-synaptic membrane and are constantly leaking out of it. This noise is always going on in the central nervous system and so we have to take account of it in building models of nerve nets. It may be that the noise is not true noise; it may be very crucial in information processing. In the cerebellum for example, a Purkinje cell has a firing rate of about 30 times a second when there are no signals; that will increase when signals are required up to 200 a second. There is spontaneous firing all the time, and this is a feature apparently of the whole nervous system. It could occur because activity must be going on all the time to keep the organism functioning and keep the central nervous system ticking over properly and effectively. At this time we just do not know. One doesn't even properly understand what is going on with respect to the vesicles of transmitter substances. Precisely how the release of transmitter substance is achieved is not fully understood, but heavily involves calcium ions. If these are depleted from the system, then there is a much reduced level of transmission of chemical transmitter substances.

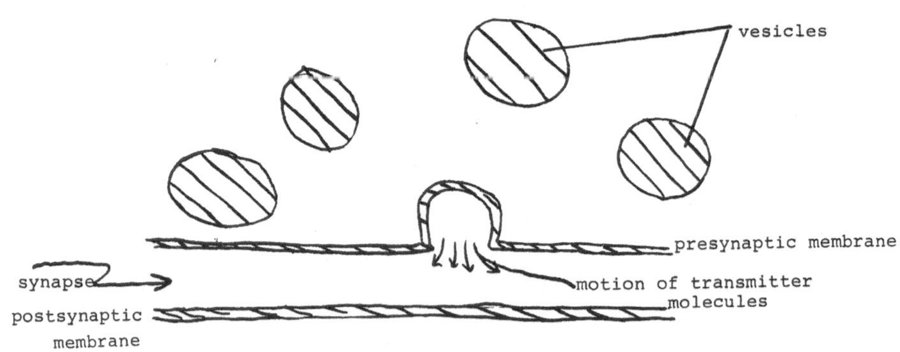

Fig. 12. A highly magnified schematic picture of a synapse, showing a vesicle fused with the presynaptic membrane and releasing its contents into the synapse.

The problem of memory also is involved in chemical transmission at synapses. They are going to change through experience, involving the rates of transmission, the number of quanta that is released and so on. We have to analyse and understand how synapses can be changed, at what rates and what determines their change.

It appears, then, that we must consider not just deterministic neural nets but noisy neural nets. Well we can do that in fact, and estimate the importance of noise in the nervous system, indeed we can even introduce parameters that may be important in memory in a very simple way. I have written down elsewhere equations that generalize Caianiello's equations to this noisy signal transmission. If we look at the problem of noise at the synapse, what are the crucial parameters that are involved? Well the first very important one, is the frequency of emission of quanta. I'll call it ν. It's the inverse of a time and that time t_1, is the mean time between one quanta leaking into the cleft and the next quanta leaking into the cleft. t_1 can be speeded up by the change in calcium ion concentration, by an axon pulse, for example, for then frequency goes up enormously.

The next parameter that is important is the decay time t_{dec} of each quanta. If this is long, then activity stays in the cleft and so can cause the next cell to go on firing. For example, fibrillation of muscles occurs by removing the chemicals that will destroy the chemical transmitters; in other words if one allows the chemical transmitter to just stay on and on and on, one can get in a very bad state.

We can estimate the importance of noise throughout the nervous system that arises from this type of leakage behaviour by looking at the mean amount of transmitter substance. It is not possible to consider the amount of transmitter at a given time since this varies. For example one quanta may be around and then another one appears and then suddenly three may be released. The important quantity is the mean amount of transmitter substance in a synaptic cleft. This will indicate whether there will be an important firing rate that comes from leaking transmitter substance. We can write down immediately the mean number of quanta; it is just a ratio t_{dec}/t_1 because if we increased t_{dec}, we have more quanta around. If we increased t_1 there are less quanta coming into the synapse because there is a longer time between each of them leaking into the cleft. When $t_{dec}= t_1$ the mean number should be 1, because each time that one quantum has decayed another one is formed. So spontaneity will be important if t_{dec}/t_1 is larger than the threshold number, m, the number of quanta that are required for firing the next cell. This is only for one synapse; there are many synapses on a cell. There may be 1,000 to 10,000 on cells in the cortex whilst there may be 100,000 in the cerebellar cortex. If there are N synapses on a given neuron and we assume simple addition of chemical transmitter substance, there will be spontaneity if m times this amount of quanta in each synapse is larger than the threshold value. We can introduce the spontaneity parameter $S = mt_{dec}/t_1$; if S is much larger than 1 we get spontaneity; if it is less than 1 we can forget about it. We would hope that in the peripheral nervous system there will be little spontaneity and signals come through with little noise added to them. Indeed it would be very bad if they did have a lot of extra noise. The sizes for S are of the order of 1/4,000 to 1/400 times m. I've put in certain numbers that have been taken from experimental data. So if m is small, say m of the order of 1 to 10, the spontaneity number is small and so much less than 1 that we can forget about leakage. But in the cortex and the cerebellum we get S of the order of 1 or higher and so may have to take account of noise. One of the problems is how such noise functions. We may find that it is not important, but since we don't really know what the information coding is in the central nervous system, there is still something that one can't get an obvious answer to.

These parameters that I've introduced, the decay time and the mean leakage time or the frequency are going to be important parameters that will vary in memory. One has to find mechanisms for this variation. This is a very local form of memory and one must ask how it is that one can have memories about very complicated processes, very complex events. They must involve more than one single neuron and certainly more than one single synapse. And it is here the connectivity of the nervous system must be crucial. There will be synaptic modifications over many neurons and over many synapses so that while we may look for memory at a local level, we also have to look for it at a connected level. There has been the suggestion that in fact certain synapses are more modifiable than others. The spine synapses on certain cells in the cerebral cortex and in the cerebellum may be more modifiable than others. They are very large synapses that are large spikey, spiny protruberances that may allow changes to occur which do not occur at other synapses which are much smaller. There is a conjecture that has been brought forward recently that the memory modifications are crucially going on at these points. There is experimental data on this modification in some cells, but it is not very strong.

So far, the moral of the story is that the central nervous system is a very complicated system; that we have no thermodynamic variables;

but that we have the atoms. We don't however, understand enough about the wiring diagram apparently. And it is at that that we need to turn. If we build models of nerve nets there are so many possible choices that we don't get very far and certainly the activity in nerve net theory of the last twenty to thirty years has shown that we have got nowhere. We have to turn to the data, and look very carefully at not only human beings but the other million species that we have and that we can put under the microscope; especially an electron microscope. Let me put up then a set of questions that I think we should try and realize may be the crucial questions that will help us in explaining the brain.

The first one of course, is the neuron itself; how simple is it? I've talked about it as a binary decision element initially, and then I showed that the information transmission was a little more complicated. But things may be much more complicated than that and earlier work has shown this, in particular the work of Rall. One may have the problem of whether the dendrites are more important than the cell body. How important are these long processes and the information coming along them? There are battles between dendrite excitation and inhibition; battles between the different sorts of information that are coming in; the apposition of different axons onto the dendrite and onto the cell body. The problems are very complicated indeed, but certainly we must take account of the fact that this may be where very important information processing is going on and not simply in the gross wiring diagram. There are also problems of axon interaction which may be arising; if axons are parallel to each other they can excite each other. Impulse going down one axon can excite another axon coming down and hitting the same synaptic button, and not excite the post synaptic cell membrane. This adds to the connectivity problems; the wiring diagram again is not as simple as we might have thought.

Then there is a question of fluctuations in membrane response. Being part of a living organism, membranes are constantly being repaired and do not necessarily always have the same response pattern. How important are fluctuations in this? Many neuro-physiologists will complain that a theorist never takes account of this. The trouble is that it is difficult to find out from experimentalists precisely what is involved here.

Then there is the problem of leakage from one neuron to the next. There may also be problems of coding in which if the frequency is important there may even be division of frequency; it may be that a certain frequency comes in from one neuron to the next and that's divided by a half; however this feature may not be necessarily part of the normal functioning of the neuron. By simulating neuron models it has been found that frequency division can occur.

There is also the problem of inhibition versus excitation. How do you know when one synapse on a neuron is excitatory or inhibitory? There are claims that spherical vesicles in excitatory synapses are excitatory and in inhibitory ones these are flattened. It is something that we must take account of. It will help build our wiring diagrams better because inhibition and excitation obviously have very different information transformation features.

The second problem is that of neural nets. If I take a number of neurons together what can they do? What are their capabilities? Various models have been suggested at which neurons can do remarkable things, they are much better than me at arithmetic, for example they can divide; they can multiply; they are magnificent computing machines; that is in a single neuron. But can a neural net or a couple of neurons do that? Is even a single real neuron like that; can it multiply? We need the answers to these questions. It is very crucial if we are trying to set up a theory of the brain to know what actual neurons and small nets capabilities they possess. In this endeavour we must keep to the data itself.

But what data: is it relevant data? Just going out to look for data, like collecting mushrooms, may give you indigestion. What is important to find out, that really is the crucial thing. Do we need to include such things as, for example, flicker fusion, where one has the inability to distinguish between two sharp light signals if they are spaced too short an interval in time apart? This is obviously telling us something about the information processing. A great deal has also been found out by the use of solar rays and sharp visual wave forms. There is an increasing amount of such results; it must be used carefully at the appropriate time.

In the visual cortex there appear to be some simple feature detectors. Some nerve cells function so that they can easily see illuminated rectangles for example. At the next stage in the cortex, a little further on, cells exist that respond to larger rectangles. You can find further on that there are moving edge detectors; for example if the moving illuminated edge moves up and down suitably the cell will respond to that. Now where do we stop in this process of complexity and feature detectors? Is there a face detector; is there a cat detector; is there a lunch detector; is there a watch detector? Since I actually have one, and have just used it, I'd better stop.

CEREBELLAR MORPHOLOGY AND PHYSIOLOGY: INTRODUCTORY REMARKS

W. Precht
Max-Planck-Institut für Hirnforschung
Neurobiologische Abteilung
Frankfurt/Main

In the following a summary will be given of the main topics which were treated in three lectures on the cerebellum. Since most of the material presented at the conference may be found in a detailed form in the neurobiological literature the present paper will restrict itself to a brief summary of the important data. For details the reader will mainly be referred to important books and reviews on the cerebellum. It is in these monographs that the reader will find numerous references to original papers. Since this summary intends to reach primarily theoreticians it is believed that this procedure of quotation will assure a fast access to the main topics of cerebellar neurobiology.

1. Morphology of the Cerebellum

A. Gross anatomy

The cerebellum is present throughout the vertebrate phylum; there are, however, great differences in relative size and structural differentiation between different species (for review see Larsell, 1967; Llinás, 1969). These varieties do in general not correlate with the position of a given species within the vertebrate kingdom. Thus, amphibia have a much smaller and less sophisticated cerebellum than fish. Among the latter, particularly the electric fish have a gigantic cerebellum making up some 70% of the total brain weight. It is possible that this enormous size is required to handle all the information obtained by the electroreceptors which monitor changes in the electrical field produced by the animals own electric organs. Less complex sensory information will require fewer cerebellar neurons for data handling. Thus, the auricular lobes of the cerebellum are relatively large in tadpoles as compared to adult frogs because in the former both labyrinthine and lateral line receptors reach this area of the cerebellum whereas in the latter only labyrinthine receptors go there.

The cerebellum may easily be recognized by its foliated surface, the folia being continuous across the midline. Anatomical and physiological studies of the cerebellum in higher vertebrates lead to the gross concept of four major regions (Jansen, Brodal, 1954; Chambers, Sprague, 1955 a,b): 1) a median zone consisting of the vermal cortex (lobulus I - X) and the fastigal nuclei. It regulates tone, posture, locomotion and equilibrium of the whole body; 2) the intermediate zone (paravermis, n. interpositus) is important for skilled movements and the muscle tone and posture of the ipsilateral limbs; 3) the lateral zone comprising the hemispheres and dentate nucleus controls skilled movements of the ipsilateral limbs without influencing tone and posture; 4) the vestibulo-cerebellum comprising nodulus, flocculus and parts of the deep cerebellar nuclei is the oldest part of the cerebellum and plays a role in the control of vestibular induced effects on eyes and body musculature. Each of these zones may vary in size in different animals depending on functional requirements in a given species.

B. Histology

Comparative studies of the cerebellar cortex (for ref. see Larsell, 1967; Llinás, 1969) show that a basic cerebellar circuit is present in all vertebrates. It consists of the Purkinje cell - the only output element of the cerebellar cortex - which receives two inputs, the climbing fibers which terminate directly on the Purkinje cell, and the mossy fiber system reaching the Purkinje cells through the granule cell-parallel fiber system. During phylogeny this basic cerebellar circuit is preserved but new elements are added, i.e., the number and types of interneurons is greatly increased. For example the frog cerebellum contains only a modest number of interneurons (stellate cells) and thus consists mainly of the basic cerebellar circuit. Alligators are the first species to have all cerebellar elements. The complete cerebellar circuit (Fig. 1) is characterized by the presence of the interneurons of the molecular layer (stellate and basket cells) and the interneurons of the granular layer (Golgi cells). The main structural features of the elements of the cerebellar cortex will be briefly described. For details and references the reader is referred to the extensive reviews of cerebellar morphology (Eccles et al., 1967; Llinás, 1969; Llinás, 1970; Eccles, 1973). Recent quantitative data of the cerebellum of higher vertebrates have been published in a series of important papers (Palkovits et al., 1971 a,b,c;1972).

Purkinje cells: These cells are characterized in all vertebrates by a complex dendritic ramification. Their dendritic trees are close to isoplanar and extend as thin leaflets (6μ) in the sagittal plane of

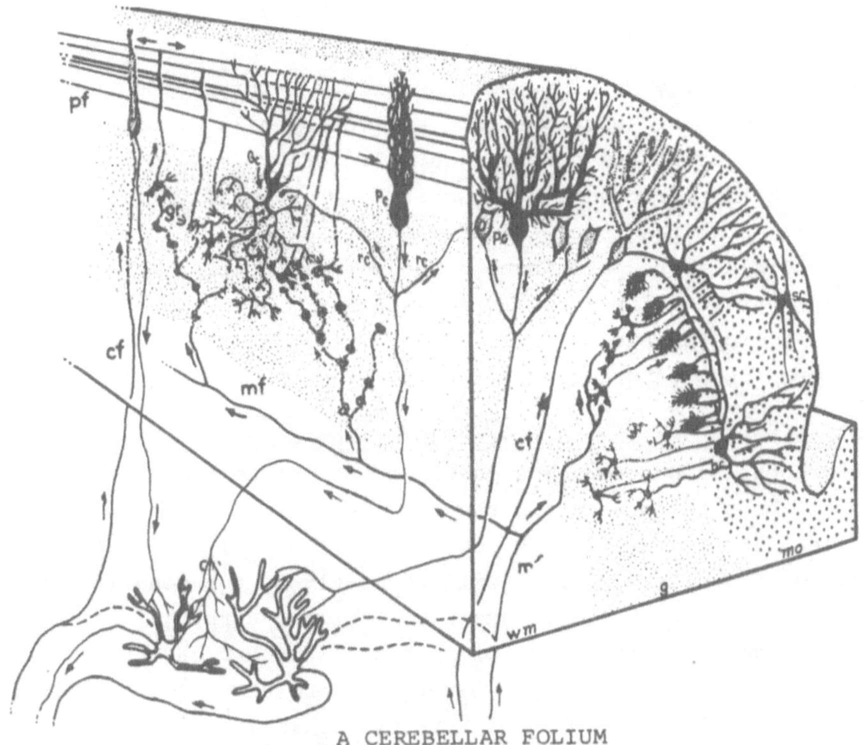

Fig. 1. Schematic drawing showing the neuronal connections of a cerebellar folium of higher vertebrates. Abbr.: Layers-wm, white matter; g, granular layer; mo, molecular layer. Cells-bc, basket cell; cn, cerebellar nucleus; Gc, Golgi cell; gr, granule cell; Pc, Purkinje cell; sc, stellate cell. Axons-cf, climbing fiber; mf, mossy fiber; pf, parallel fiber; rc, recurrent collateral (modified from Fox, 1962).

the body, i.e. at right angle to the direction of the parallel fibers. An extensive, illustrated review of the morphology of the Purkinje cell in various species may be found in Llinás (1969). Their absolute number in cats is in the order of 1.2×10^6. They are the only neuronal element in the cerebellar cortex whose axons leave the cerebellum to contact neurons in the cerebellar nuclei and to some extent in the brain stem. Axon collaterals of Purkinje cells contact basket and Golgi cells. Purkinje cell axons are inhibitory on target neurons (Eccles et al., 1967). It was estimated that the divergence and convergence numbers between Purkinje cells and cerebellar nuclear cells are approximately 20 - 50.

Granule cells: An enormous number (2.2×10^9) of small cells form the granular layer of the cerebellar cortex. In all vertebrates these cells produce ascending axons which divide in a T-fashion to form the so-called parallel fibers. After bifurcation these fibers run in the direction of the folium - 1mm in each direction - and establish contact with the spines of the Purkinje cell dendrites. This "crossing over" synapse is found in all vertebrates. A given parallel fiber passes ca. 200 Purkinje cells and contacts ca. 45 of them (divergence number). On the other hand the convergence number between parallel fibers and Purkinje cell dendrites is ca. 80,000 which indicates that more than 400,000 parallel fibers pass through the dendritic arbor of each Purkinje cell. Parallel fibers exert an excitatory action on Purkinje cells. Mossy fibers (excitatory) and Golgi cell axons (inhibitory) represent the input to granule cells.

Stellate and basket cells: interneurons of the molecular layer. These two cell groups may be treated as extremes of one homogeneous group. The axons of both of these cells run transversely at right angles to the parallel fibers (for details see Eccles et al., 1967; Llinás, 1969 a,b). Axons of stellate cells terminate mostly on the dendrites of Purkinje cells, and basket cell axons contact the Purkinje cell soma and initial segment. The action of both stellate and basket cell axons on Purkinje cells is inhibitory. Divergence and convergence numbers between basket cells (7×10^6) and Purkinje cells may be in the order of 8 and 50, respectively.

Both interneurons receive most of their input from parallel fibers (divergence number 20 and convergence number 20,000). This input is excitatory. In addition, both neurons receive terminals from other interneurons (inhibitory) and from collaterals of the climbing fibers (excitatory). The lower basket cells also receive inputs from axon collaterals of Purkinje cells.

Golgi cells: interneurons of the granular layer. They have a large dendritic tree extending in all three planes. The axons of these cells terminate on the dendritic digits of the granule cell, and their action is inhibitory. Golgi cells receive inputs from both parallel fibers and mossy fibers (excitatory). In addition, they receive recurrent collaterals of Purkinje cells (inhibitory) as well as collaterals of climbing fibers (excitatory).

Climbing and mossy fibers: Afferents to the cerebellum. Following this brief summary of the cellular elements of the cerebellar cortex the two afferent systems will be described.

a) climbing fibers: Ontogenetically the climbing fiber system is the first input that reaches the Purkinje cell. In the adult it establishes a monosynaptic contact with the soma and dendritic spines of Purkinje cells. Each Purkinje cell receives only one climbing fiber (convergence number one) which branches profusely and twines around its dendrite establishing many synaptic contacts (ca. 200 - 300). Its action is excitatory. The cells of origin of climbing fibers are located in the inferior olive although other sites of origin can not be excluded at the present time. Given that the total number of Purkinje cells is in the order of 1.2×10^6 and the number of inferior olive cells has been calculated to be 1.2×10^5 a divergence number of 10 may be assumed. In addition to Purkinje cells, climbing fibers get into contact with cerebellar interneurons and cerebellar nuclear cells.

b) mossy fibers: Like the climbing fibers, the mossy fibers are part of the basic cerebellar circuit, i.e. they are present throughout the vertebrate kingdom. They terminate on the dendrites of granule cells with the well-known glomerular arrangement. Their action on granule cells is excitatory. It has been calculated that ca. 4.8×10^6 mossy fibers reach the granular layer of the cerebellar cortex and that they supply 2.2×10^9 granule cells. Therefore, a divergence number of 460 has been calculated. On the other hand, each granule cell receives synaptic input from separate mossy fibers on each of its four dendritic claws (convergence number 4.2). Finally, it should be added that mossy fibers also reach Golgi cells in the granular layer (see above).

In concluding this brief summary of cerebellar morphology it should be emphasized that an enormous amount of knowledge about the detailed structure of the cerebellum has been accumulated in the past decade so that the cerebellum is most likely the best-studied region of the central nervous system. As described above, qualitative and quantitative data are available at this stage which may be sufficient to generate approximative computer models of the cerebellar cortex. In particular the relative simple cerebellum of the frog lends itself to such an approach (Llinás, 1971).

2. Physiology of the Cerebellum

In the past cerebellar physiology was mainly based on ablation and stimulation experiments which described the motor deficits and motor effects resulting from these procedures. Most of the important data derived from this approach may be found in the monograph by Dow and Moruzzi (1958) which summarizes experimental and clinical data. In the past decade, however, microphysiological techniques have been successfully applied in the study of the functional organization of the neuronal elements of the cerebellar cortex. Furthermore, the effects of various inputs on cerebellar neurons and the action of Purkinje cells (the only output of the cerebellar cortex) on their target neurons have been analyzed. Thus cerebellar physiology and morphology have experienced a joint development at the single neuron level. We know the sign of function (excitatory or inhibitory) of practically all synaptic connections in the cerebellum and of the Purkinje cells on target neurons (see above); we know for a variety of sensory inputs the mode of action on Purkinje cells and cerebellar nuclear cells and we know that Purkinje cell firing may precede or follow the initiation of movement. What we still don't know is the overall function of the cerebellum. In the following we will briefly summarize the important physiological facts, discuss briefly the different theories regarding specific functions of various subsystems (e.g. mossy and climbing fiber system, interneurons) and the overall function. For details of recent cerebellar physiology the reader is referred to Eccles et al., 1967; Llinás, 1969; Llinás, 1970; Eccles, 1973).

Climbing fiber system: Comparative physiological studies have shown that the climbing fiber system is an excitatory input to Purkinje cells. Its activation produces a burstlike response (several action potentials) in Purkinje cells. These typical responses are generated by a large, all-or-none, excitatory synaptic potential (EPSP) which reflects the all-or-none nature of the action potential arriving via the climbing fiber which belongs to a given Purkinje cell (convergence number 1) and the intense synaptic connection of a climbing fiber with a Purkinje cell (see above).

Several theories have been offered regarding the functional meaning of the climbing fiber system. The so-called "read out system" theory (Eccles et al., 1966) suggested that the climbing fiber due to its strong excitatory one-to-one relation to Purkinje cells may function as a testing probe by which it could measure the excitability level of Purkinje cells. The latter receive excitation and inhibition via the mossy fiber system and the intercalated inhibitory neurons so that the state of excitability of Purkinje cells may vary considerably with changing input situations. It has, indeed, been shown that the number of spikes per burst of a given Purkinje cell changes with its level of excitability. The "read out" theory, however, does not explain the function of such a "read out".

Another theory proposes that the climbing fibers function as a phasic system (derivative function operator) and that they use the Purkinje cell in a time sharing fashion with the tonic mossy fiber system (Llinás, 1969 b). Such a phasic control system would be well suited in cases in which a strong and localized inhibitory action is needed (step function like correction of motor performance).

Oscarson's (1969) hypothesis suggests that the climbing fiber system informs the cerebellum about ongoing activity in central motor systems (i.e. prior to motor execution) while the mossy fibers convey information regarding performed movements. Intended and performed movement are compared in the cerebellar cortex and error correcting commands are initiated thereafter. Whereas in the phasic control theory climbing and mossy fibers perform separately, Oscarson's concept implies a joint performance.

Finally, a hypothesis will be mentioned which has been widely discussed in the past years. It may be called the "learning" theory (Marr, 1969) and it suggests that the climbing fibers modify the synaptic efficacy between parallel fibers and Purkinje cells in such a way that the synaptic efficiency of activated parallel fiber synapses would be increased up to 100 msec after a climbing fiber activation of a Purkinje cell. In this way the Purkinje cell is supposed to learn to recognize patterns, which would be related to specific movements. Learning may occur provided a particular pattern of activation is repeated often enough. So far this theory has not been proven experimentally.

It is clear from the above that at present no final word can be said regarding climbing fiber function. This interesting problem, however, lends itself to a joint effort of theoreticians and experimentalists.

Mossy fiber system: As described in the morphological section the mossy fibers are excitatory to granule cells. The latter, by way of their axons, the parallel fibers, excite Purkinje cell dendrites and interneurons of the cerebellar cortex. The fact that the isoplanar dendritic tree of Purkinje cells is oriented orthogonally to the axis of the parallel fibers may assure a high degree of convergence, i.e. a given Purkinje cell receives as many inputs as possible. At the same time this spatial organization allows maximal divergence (Fox, Barnard, 1957). Similar morphological considerations have led Braitenberg (Braitenberg, Onesto, 1962; Braitenberg, 1967) to the assumption that the mossy fiber-granule cell - Purkinje cell system may function as a timing device in the millisecond range. Since a beam of parallel fibers enters into contact with many Purkinje cells along the line its activation would lead to a sequential activation of Purkinje cells. The minimum fraction of time would be given by the conduction time of parallel fibers between two adjacent dendritic trees (a fraction of a millisecond), and the maximum time range calculated on the basis of the length of the parallel fibers would be in the order of 5msec. Sequential activation of neighbouring Purkinje cells has been demonstrated physiologically by Freeman (1969). It should be pointed out, however, that the demonstration of such sequential activation of Purkinje cells does not necessarily mean that the cerebellar cortex and its mossy fiber system function as a timing device, e.g. in motor control.

In conjunction with the climbing fiber system it has already been mentioned that the mossy fiber system may also be viewed as a tonic activation system (Llinás, 1969) which would activate a large number of Purkinje cells owing to its widespread ramifications. In agreement with this concept is the high spontaneous activity of both granule cells and mossy fiber activated Purkinje cells (e.g. by rotation, Precht, Llinás, 1969). This tonic, widespread activation of Purkinje cells would impose a tonic inhibition on nuclear and brain stem neurons.

Finally, the "codon system" hypothesis deserves mention (Marr, 1969). It suggests that mossy fiber-granule cell-parallel fiber systems produce specific patterns of Purkinje cell activation (codons) which represent specific functional states in the motor system. The information is stored by the simultaneous activation of climbing fibers (see above).

Clearly, much more experimental work is needed to test the ideas expressed in various theories. Furthermore, not only the different input systems but also the properties of Purkinje cells deserve our interest. The complexity of the dendritic tree with its ability to generate spikes (Llinás, Nicholson, 1969) is functionally only poorly understood. It should be pointed out that the dendritic tree may be envisaged as having a vertical and/or transverse integrative function, and it probably depends on the particular situation which of the two is chosen by a given input.

Purkinje cells may have characteristics which determine the time course of the output. Such a possibility is strongly suggested by the serial plots of impulse frequency of many Purkinje cells of the frog which have been obtained in response to ipsi- and contralateral rotation (Fig. 2). Even though the amplitudes of the responses may be different, the general time course of the response to rotation in either direction is very similar for an individual cell in spite of a different input situation. The responses may be more phasic (left-hand plots) or more tonic (right-hand plots).

Interneuron systems: As pointed out above, cerebellar of higher vertebrates are characterized by two groups of interneurons, the stellate/

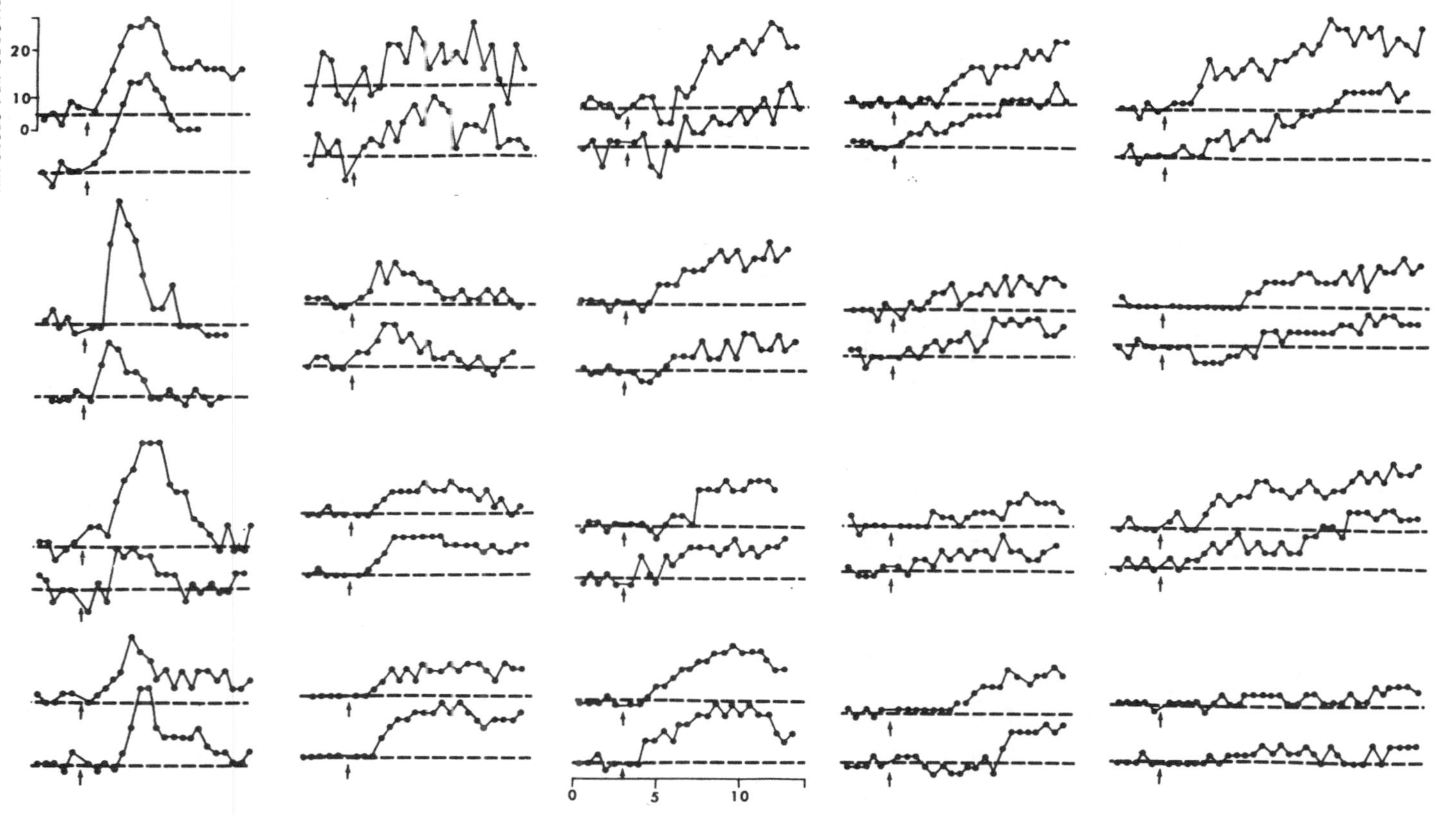

Fig. 2. Frequency diagrams of discharges of 20 type III Purkinje cells in response to constant angular accelerations in the ipsi- and contralateral directions. Each pair of frequency diagrams represents the responses of individual Purkinje cells to accelerations applied in the ipsi- and contralateral directions at the time indicated by arrows. Broken lines represent the resting level of spontaneous activity (from Llinás, Precht, Clarke, 1971).

basket cells and the Golgi cells. First, we shall consider the functional role of the interneurons of the molecular layer. Morphological and physiological evidence strongly suggests that basket and stellate cell inhibition serves as lateral inhibition (Eccles et al., 1967). Thereby activation of Purkinje cells is restricted to those which are located directly underneath a beam of excited parallel fibers. This view is valid only if one assumes that synchronous activation of a beam of parallel fibers is the physiological parallel-fiber function. It may also be assumed that in general stellate-basket cell inhibition functions to minimize the possibility of inactivation of Purkinje cells by depolarization (Llinás, 1970) and thus presents a sort of overflow-preventing mechanism.

Golgi cell inhibition has been assumed to present a recurrent inhibitory system (Eccles et al., 1967) which acts as a straightforward negative feedback system directed towards granule cells. Recent experiments (Precht, Llinás, 1969) suggest, however, that a Golgi cell inhibition may have also a gating function, which by a feed-forward control (direct mossy fiber input to Golgi cells), would prevent short-term sequential activation through one input, but would allow the activation of the same granule cell population by mossy fibers of different origins.

<u>The vestibulo-cerebellar system</u>: After this brief review of the concepts of cerebellar function or rather of the function of cerebellar subsystems we shall deal very briefly with the functional organization of the vestibulocerebellum, i.e. with the function of a complete cerebellar circuit. This part of the cerebellum comprises the nodulus, flocculus and some parts of the cerebellar nuclei and represents the oldest part of the cerebellum. As shown in Fig. 3 it is characterized by the presence of a <u>direct</u> sensory input (vestibular input) which arrives at the granular layer via <u>mossy</u> fibers (for details see Precht, Llinás, 1969; Llinás et al., 1971). In addition, Purkinje cells in the vestibulocerebellum receive a climbing fiber input which is activated by visual stimulation (Maekawa, Simpson, 1973). Thus, it was shown that single Purkinje cells in this region were activated in response to vestibular stimulation via the mossy fiber system and that the same cells received a climbing fiber activation when moving visual stimuli were presented or when the animal was rotated against a background with the eyes opened (Simpson, Precht, Llinás, unpublished observation). It appears that at least in this system mossy and climbing fibers carry different sensory information to the same Purkinje cells.

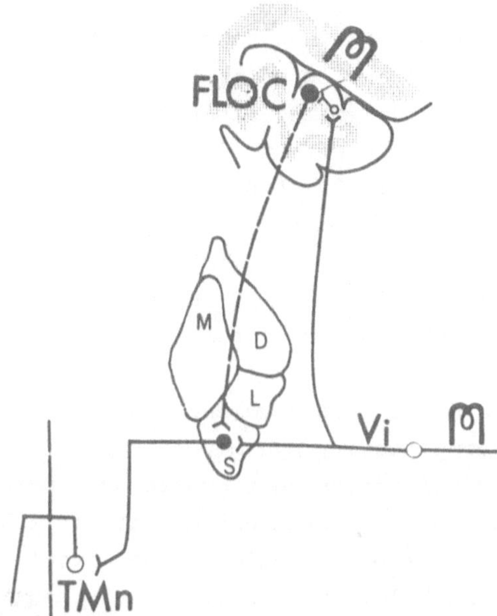

Fig. 3. Diagram of the inhibitory vestibuloocular and vestibulo-cerebello-vestibular connections. Inhibitory neurons are shown with filled circles. Abbr.: FLOC, flocculus; V_i, ipsilateral VIIIth nerve; TMn, trochlear motoneuron; D,L,M,S, descending, lateral, medial and superior vestibular nuclei. Note that vestibular axon (V_i) supplies both the vestibular neurons and the cerebellum via mossy fiber-granule cell - Purkinje cell pathway and that Purkinje cell axons (broken line) reach vestibular neurons monosynaptically. Same circuitry applies for excitatory vestibuloocular reflex (not shown).

Whether this is true for all vertebrates remains to be shown.

Purkinje cells of the vestibulocerebellum send their axons to secondary vestibular neurons where they meet with primary vestibular fibers which establish contact with the same neurons (Fig. 3). It has been demonstrated that by means of this projection floccular Purkinje cells inhibit vestibuloocular reflexes (Baker et al., 1972; Fukuda et al., 1972). These reflexes function to stabilize the retinal image when the head is moved by generating compensatory eye movements. Even if one were to look at the simplest situation, i.e. when an experimental subject is rotated in the dark, compensatory eye movements will be influenced by the pathway from the labyrinth to the cerebellum and back to the vestibular nuclei. This control probably consists mainly of a kind of dampening of vestibuloocular transmission.

If one considers the more complex biological situation in which rotation occurs in the presence of vision, the action of the climbing fiber system on to the vestibulocerebellum will be added. Its function may be to inform the cerebellum on the performed or ongoing eye movement. In fact, it has been demonstrated that the presence of vision suppresses vestibular nystagmus, and that removal of the vestibulocerebellum abolished visual suppression of nystagmus (Cohen, Takemori, 1973). Visual activation of the climbing fiber system to the flocculus may in part be responsible for this effect.

It has been pointed out that the vestibuloocular system lacks a straightforward negative feedback from the eye to the labyrinth (Ito, 1970) that would signal to the labyrinth a performed compensatory eye movement. The circuit illustrated in Fig. 3 is reminiscent of a feed-forward control system in which the cerebellum may be viewed as a computer with some learning capabilities receiving information from the labyrinth (head movement) and eye (eye movement) which is being used for the optimalization of compensatory eye movements.

It appears that this system offers very promising aspects in the search for cerebellar function in general as well as for a better understanding of climbing and mossy fiber function. The circuitry is well elaborated, quantifiable vestibular and visual stimuli can be applied and the eye movements can be measured with great precision. Several groups are presently working in this area and it may well be that important discoveries will occur within the next few years. Similarly, in other fields of cerebellar research, in particular the spinocerebellar system, considerable progress has been made which promises interesting new concepts in the near future (Eccles, 1973).

Concluding remarks: The above brief notes on cerebellar morphology and physiology have demonstrated that even though we know a large amount of detailed anatomy and physiology we are still unable to formulate a clear concept regarding overall function of the cerebellum. Certainly the classical notion that the cerebellum is implicated in the coordination of movement is still valid although it may also be involved in functions other than motor. The problem, however, of how this coordination is being performed still exists. Also, there are a number of examples in which a large discrepancy exists between the size of the cerebellum and motor abilities of an animal, e.g. in certain electric fish. Such findings suggest that the term motor-coordination may be too limited to describe the cerebellar function.

As for the general mode of operation of the cerebellum in motor control, it has been suggested that it may serve as an error-correcting device. The cerebellum could achieve this goal as part of a closed loop system, i.e. it would correct for errors that have already been produced and that were transmitted to the cerebellum by the proprioceptive system (Eccles et al., 1967). Such type of control mechanism may be suited for the control of slow movements. For fast movements, however, a closed loop system would be too slow since it would start functioning only after the movement had been initiated. Therefore, an open-loop correcting system would be required for rapid movements. In such a system errors in the motor programs destined for the final common path would be corrected prior to motor execution. Even such common movements as handwriting and voluntary eye movements are of the ballistic type. Recently in several laboratories studies have been begun in which cerebellar neuron activity is recorded simultaneously with particular movements. Even though this work is far from being completed, it can be stated that often changes in Purkinje cell firing precede the onset of movements. This is exactly what one would have to postulate if the cerebellum were to function as a ballistic error-correcting device.

References

Baker, R., Precht, W. and Llinás, R.: Cerebellar modulatory action on the vestibulo-trochlear pathway in the cat, Exp. Brain Res. 15, 364-385 (1972).

Braitenberg, V.: Is the cerebellar cortex a biological clock in the millisecond range, Progr. Brain Res. 25, 334-346 (1967).

Braitenberg, V. and Onesto, N.: The cerebellar cortex as a timing organ: Discussion of a hypothesis, Atti 1° Congr. Int. Med. Cibernetica, Giannini, Naples.

Chambers, W.W. and Sprague, J.M.: Functional localization in the cerebellum. I: Organization in longitudinal corticonuclear zones and their contribution to the control of posture, both extrapyramidal and pyramidal, J. comp. Neurol. 103, 105-129 (1955a).

Chambers, W.W. and Sprague, J.M.: Functional localization in the cerebellum. II: Somatotopic organization in cortex and nuclei, Arch. Neurol. and Psychiat. 74, 653-680 (1955b).

Cohen, B. and Takemori, S.: Visual inhibition of nystagmus by the flocculus, Fed. Proc. 32, 339 (1973).

Dow, R.S. and Moruzzi, G.: The Physiology and Pathology of the Cerebellum, The University of Minnesota Press, Minneapolis, 1958.

Eccles, J.C.: The cerebellum as a computer: patterns in space and time, J. Physiol. 229, 1-32 (1973).

Eccles, J.C., Ito, M. and Szentágothai, J.: The Cerebellum as a Neuronal Machine, Springer-Verlag, Berlin, 1967.

Eccles, J.C., Llinás, R. and Sasaki, K.: The excitatory synaptic action of climbing fibers on the Purkinje cells of the cerebellum, J. Physiol. 182, 268-296 (1966a).

Fox, C.A. and Barnard, J.W.: A quantitative study of the Purkinje cell dendritic branchlets and their relationship to afferent fibers, J. Anat. 91, 299-313 (1957).

Freeman, J.A.: The cerebellum as a timing device: An experimental study in the frog. In: Neurobiology of Cerebellar Evolution and Development, ed. by R. Llinás, pp. 397-420. American Medical Association, Education and Research Foundation, Chicago, 1969.

Fukuda, J., Highstein, S.M. and Ito, M.: Cerebellar inhibitory control of the vestibuloocular reflex investigated in rabbit IIIrd nucleus, Exp. Brain Res. 14, 511-526 (1972).

Ito, M.: Neurophysiological aspects of the cerebellar motor control system, Intern. J. of Neurol. 7, 162-176 (1970).

Jansen, J. and Brodal, A.: Aspects of Cerebellar Anatomy, Johan Grundt Tanum, Oslo, 1954.

Larsell, O.: Comparative Anatomy and Histology of the Cerebellum, 1: 1-291, ed. by J. Jansen, University of Minnesota Press, Minneapolis, 1967.

Llinás, R. (ed.): Neurobiology of Cerebellar Evolution and Development, Amer. Med. Ass. Educ. and Res. Fed., Chicago, 1969.

Llinás, R.: Functional aspects of interneuronal evolution in the cerebellar cortex. In: The Interneuron, ed. by M.A.B. Brazier, Univ. Calif. Press, Los Angeles, 1969b.

Llinás, R.: Neuronal operations in cerebellar transaction. In: The Neurosciences: Second Study Program, ed. by F.O. Schmitt, Rockefeller Univ. Press, New York, 1970.

Llinás, R.: Frog cerebellum: Biological basis for a computer model, Math. Biosciences 11, 137-151 (1971).

Llinás, R. and Nicholson, C.: Electrophysiological analysis of alligator cerebellum: A study on dendritic spikes. In: Neurobiology of Cerebellar Evolution and Development, ed. by R. Llinás, pp. 431-465. American Medical Association, Education and Research Foundation, Chicago, 1969.

Llinás, R., Precht, W. and Clarke, M.: Cerebellar Purkinje cell responses to physiological stimulation of the vestibular system in the frog, Exp. Brain Res. 13, 408-431 (1971).

Marr, D.: A theory of cerebellar cortex, J. Physiol. (London) 202, 437-470 (1969).

Maekawa, K. and Simpson, J.I.: Climbing fiber responses evoked in vestibulocerebellum of rabbit from visual system, J. Neurophysiol. 36, 649-666 (1973).

Oscarsson, O.: The sagittal organization of the cerebellar anterior lobe as revealed by the projection patterns of the climbing fiber system. In: Neurobiology of Cerebellar Evolution and Development, ed. by R. Llinás, pp. 525-537. American Medical Association, Education and Research Foundation, Chicago, 1969.

Palkovits, M., Magyar, P. and Szentágothai, J.: Quantitative histological analysis of the cerebellar cortex in the cat. I: Number and arrangement in space of the Purkinje cells, Brain Res. 32, 1-13 (1971a).

Palkovits, M, Magyar, P. and Szentágothai, J.: Quantitative histological analysis of the cerebellar cortex in the cat. II: Cell numbers and densities in the granular layer, Brain Res. 32, 15-30, (1971b).

Palkovits, M., Magyar, P. and Szentágothai, J.: Quantitative histological analysis of the cerebellar cortex in the cat. III: Structural organization of the molecular layer, Brain Res. 34, 1-18 (1971c).

Palkovits, M., Magyar, P. and Szentágothai, J.: Quantitative histological analysis of the cerebellar cortex in the cat. IV: Mossy fiber - Purkinje cell numerical transfer, Brain Res. 45, 15-29 (1972).

Precht, W. and Llinás, R.: Functional organization of the vestibular afferents to the cerebellar cortex of frog and cat, Exp. Brain Res. 9, 30-52 (1969).

CAN NEURONS DEVELOP RESPONSES TO OBJECTS?

C.R. Legéndy*
Institute for Information Sciences
University of Tübingen, Germany

1. Introduction

If the brain had to start from a completely naive state, could it still organize the sensory environment into "objects", or is that in some way logically impossible? In answer, I shall outline a scheme whereby it might do so, and discuss its plausibility. To place it: it is in the same category as the scheme of Hebb (1949), in calling neither for an occasional "now print!" command (Livingston, 1968; Legéndy, 1967; Marr, 1969) nor for reinforcement (Rosenblatt, 1960). Reinforcement, it is assumed, gains prominence later in ontogenesis, when the brain is already organized to be responsive to objects of the world, thus there are fewer independent variables left to determine by trial and error. (To achieve organization through reinforcement is to achieve it by trial and error; and in a truly naive brain that would take well over 10^{10} years!)

The present scheme (stated in a useful special form) differs from Hebb's in that the lasting change is not caused by frequent co-activity between a presynaptic and a postsynaptic neuron but by frequent co-activity between several presynaptic ones, or several ones that would become presynaptic to a common neuron after the change. The change makes the postsynaptic neuron responsive to the co-activity. Stated in a more general form, the scheme is to detect certain kinds of statistical correlation between the signals coming into a region of neuropil, and channel the correlated activity appropriately to influence the firing of some neuron there. The present paper discusses the consequences of certain assumptions along these lines, and argues that the assumptions are realistic.

One thing required of the correlations in question is that they be non-constant in time; in other words, sometimes observable, sometimes not. This increases the likelihood that their cause lies outside

* This research was supported in part by the Alexander von Humboldt-Stiftung and in part by the Consiglio Nationale delle Ricerche (Italy).

the neural network. Conversely, constantly present correlations are expected to be caused by network connectivity. For instance, Noda and Adey (1970) found that pairs of cortical neurons lying close together often showed very strongly correlated firing in deep sleep; some pairs showed it in the waking state as well. Such correlation, according to the present scheme, does not cause plastic change. (However, in the schemes of Hebb (1949) and Marr (1970), it would.)

Of special interest in the sequel will be events representing dramatic evidence of correlation, where some combination of neuron signals, relatively elaborate and essentially "never" expected to occur, in fact occurs. For instance, when a combination is expected, as calculated from the occurrence rates of its components, to come up about once in 10^6 years, and comes up instead on one occasion several times in a few minutes.

For picturesqueness I shall call such extreme events "local miracles". (Questions surrounding their detection, and a possible physical mechanism, are discussed in Sections 5 and 6.) As will be appreciated, there is nothing miraculous about these; they only appear to be miracles from the vantage point of local brain tissue. An observer aware of the global relations could readily see the reason for each coincidence, or, as will be said, the "object" behind it. (The use of the word object here is not as in "objects scattered on the table" but as in "object of the discussion". This usage is maintained throughout the paper. The present scheme will be regarded as defining, for the present purposes, the notion of "object"; see Sec. 4.) It will be argued that local miracles are a common occurrence in neuronal networks, and that it is possible to interconnect a network in such a way that the objects in the outside world cause an abundance of such occurrences inside the net.

The inference from a local miracle to the existence of a reason (an object) may be illustrated from another angle, by an example involving behavior. Let us consider the common occurrence where we speak of a long-gone and long-forgotten friend, then go down into the street and run into him, are quite struck by the event, and, if possessing any "sense of drama" at all, conclude that telepathy is at work. The question is, why the wild conclusion? The answer: because an event occurred whose occurrence frequency we unconsciously (and maybe quite incorrectly) estimated to be once in aeons. Our immediate reaction has been to infer that a reason must exist, and postulate one, though vaguely. The same inference at other times yields less aesotheric results; for instance when in research we find a correlation and infer that something causes it. But in no case does the inference require direct contact with the reason!

Hence the usefulness of the present scheme in establishing the unplanned contacts: the inference penetrates the barrier of ignorance surrounding any small piece of undeveloped neuropil (the barrier consisting of the dependence, for inputs, on a network largely unplanned). Accordingly, neurons can develop the (superficial) appearance of having a "direct line" to the outside world. Researchers of the visual apparatus have long been impressed by an analogous feat: the apparent ability of brains to ignore the retinal image in certain ways, and see beyond it; the incorrectness of introspectionism (Köhler, 1929).

Sections 2 and 3 discuss the simplest phenomena resulting from the scheme, and the simplest model that derives from it, where neurons can only respond or not respond to their inputs, and whenever they do, their response is to an object. A few of the consequences are: visual recognition is relatively insensitive to missing features; it is dependent on context; and it is not hierarchical. Section 4 shows the link between correlations and "reality", points out that in Hebb's scheme such a link cannot be established, and sketches a way in which goal-directed motor activity can be formulated in the present framework. In Section 5 certain known brain structures are reinterpreted in the light of this scheme, namely _glia_ cells, the early transitory structures like Retzius-Cajal cells, and _glomerular_ synaptic elaborations; and a possible physical mechanism of correlation detection for the special form of correlation consisting of co-activity in the literal sense is outlined. In Section 6 some calculations and numerical estimates are made, giving a sufficient condition for detectability of correlations.

2. A Simple Model

This section will serve to show that many qualitative characteristics of what may be called the "data structure" in the human brain can already be explained by postulating that (i) all useful neuron outputs are responses to objects and (ii) neurons can acquire responsiveness to new objects by way of the "correlation scheme" outlined in Sections 4, 5 and 6.

It will be found that restatements of (i) and (ii) in the light of cortical anatomical and physiological data, or deductions from them trivial enough to require no mathematics, translate into known macroscopic traits. The local-to-global translation utilizes the assumption, discussed in Section 3, that the organism as a whole will confuse two objects with one another if all its neurons respond to them identically; and can respond to an object (other conditions being appropriate) if some of its individual neurons can develop responsiveness to the object.

The examples and illustrations in this section are taken from vision; but the usage of the word "object" remains as stated above. Thus, occasionally I shall have "object" where the most frequently used term would be "feature". Some of the properties listed separately are not truly distinct, and are separated for emphasis.

Objects in terms of objects. From Postulate (i), neurons respond to objects. At the same time, we know from physiology that they respond to their inputs. From Postulate (i), again, the inputs themselves are also responses to objects. Since these input responses suffice to dictate a response, a corresponding connection is expected to exist between the underlying objects. We may say therefore that Postulate (i) implies that neurons define, in an elementary way, each object in terms of other objects (its "component" objects, as they will be called below). A similar notion was enunciated by Hubel and Wiesel (1962) when they assumed that complex cells received their inputs from simple cells. Since different neurons may define the same object from different components, clear separation into components is not automatically expected to exist macroscopically. However, in recognition experiments certain "cues" are often identifiable.

Tolerance to imperfection. From the threshold property of the neuronal membrane it follows that synapse sets able to cause a response can often be decreased without loss of the response. From what was just said it therefore follows that a neuron often responds to an object though it does not receive signals from all its components. This is in agreement with the common fact of vision that an object is often recognized even though parts of it are invisible, for example obscured by other objects.

Blind spot. Another expected macroscopic consequence is occasional inability to detect absence. An instance of such inability exists in vision: our retina has a blind spot, yet we do not see black spots, or the like, in our visual field. In general, presence is known to be easier detected than absence.

Complex objects. If a neuron A sends synapses to neuron B, and that to neuron C, there may exist an object to which B responds, which is component of a response object of C and of which a response object of A is component. (When one does not exist, it is because one neuron can have many response objects, each with a different set of components.) Since an object can often be considered more "complex" than one of its components (not always, as will be seen), it follows that at each synaptic junction of a neuron chain there is potential for augmentation of the complexity of neuronal response objects. Since in the cerebral cortex long chains of neurons have been traced (Lorente de Nó, 1939) within single cortical areas, single cortical areas can potentially augment greatly the complexity of response objects; such augmentation does not require a long chain of cortical areas. It is reasonable that this should be so, as the succession from simple to complex to hypercomplex cells (Hubel and Wiesel 1965, 1969) appears to proceed much too slowly to reach complex patterns in the few areas left beyond area 19. Recently, Gross et al. (1972) found an inferotemporal lobe neuron in a monkey which apparently responded to the visual picture of a whole paw. It may be noted that, if it were not for the U-fibers (long and short) and intracortical fibers, a single cortical area could not augment the object complexity by much.

Acquired responses built upon acquired components. Because of Postulate (ii), responses to complex objects (or to their components, etc.) need not be genetically determined; they can develop by themselves under the influence of the sensory input. For the "correlation scheme", as described below, is a strategy for recruiting components to would-be response objects (chosen in the recruiting), and is insensitive to the complexity of the components. Thus a model of self-evolving complexity need describe it only once, like mathematical induction. The ability of advanced brains to learn complex recognitions is beyond dispute.

"Rank" ill-defined. It follows from the insensitivity of the strategy to the complexity of would-be components that the self-evolving scale of responses cannot correctly be called a hierarchy. Response objects in general end up defined in terms of components of mixed complexity; some components may be more complex than the object they help to define. Though in macroscopic examples care must be exercised in speaking of components, it is easy to construct examples of a similar tendency in pattern vision. For instance: the hour hand of a watch is, by any reasonable heirarchy of complexities, lower than the whole watch and of the same level as the minute hand. Yet, both the watch as a whole and the minute hand are used in visual recognition of the hour hand. Placed alone on a table it would at most be recognizable as a pointer of a watch.

Context dependence. A corollary is that context should in general be helpful in recognition, as indeed it is. The question of perceptual organization, and in particular integration of the visual field, has received much attention, especially from the Gestalt school. Its explanation has been, for instance, a principal goal of Köhler's now-disproven electrical field theory (Köhler and Wallach, 1944; Lashley, 1954).

Analogy. Response of a neuron to an object is by necessity based on a limited collection of components. When a neuron receives the same components from two objects it responds identically to them. Since the nature of component objects is just such as to account for analogies this raises the possibility that what we call analogy between two objects starts from neurons unable to distinguish the two. The suggestion is compatible with the fact that most of our concepts are generic notions, based on analogies between non-identical objects. The analogies make them the same object in certain contexts.

3. Discussion of the Simple Model

Limitations. As is seen from the above list, the brain characteristics immediately explained from postulates (i) and (ii) of the last section are ones related to object learning and identification (as opposed to orienting; Held, 1970) and to the general data structure found in brains. These postulates cannot by themselves explain matters relating to the perception of quantity, such as colour vision, constancy phenomena, locating and orienting. Such matters are expected to depend on detailed consideration of the gradation of neuron responses which is overlooked in the simple model, where neurons either respond or do not respond. The postulates also cannot by themselves explain matters relating to absence, avoidance and figure-ground discrimination, as these are expected to depend on detailed consideration of inhibition, also overlooked in the simple model, where all neuron responses are treated alike.

Survey of the evidence. Direct evidence for Postulates (i) and (ii) is scarce; like the postulates used in atom models, they are probably destined to stand or fall on indirect evidence. Let us note again that in the pattern recognition literature the word "features" is used for some of my "objects".

There is now clear evidence that feature-responsive neurons exist at least in the visual apparatus (Hartline, 1938, 1940; Lettvin et al., 1959; Hubel and Wiesel, 1959, 1962, 1963, 1965; Pettigrew et al., 1968; Gross, et al., 1972; McIlvain, 1972). The evidence is unclear that they exist in the auditory (Goldberg and Brown, 1969) or the olfactory (Lettvin and Gesteland, 1965; Gesteland et al., 1968) apparatus. That feature-responsiveness can be acquired is harder yet to establish, and there is no direct evidence showing that it can; but animals raised selectively deprived of certain visual features were found to have a subnormal number of neurons responding to those features (Blakemore and Cooper, 1970; Hirsch and Spinelli, 1970) and persons born with cataracts and surgically freed of them later have temporary and sometimes permanent deficits in pattern vision (von Senden, 1932). How much of brain function is made up of neuron responses to features of the world is again hard to say at this time, because in unanaesthetized animals cortical single-unit recording gives notoriously uninterpretable results (Hubel, 1959). The inconclusiveness of such results does not necessarily mean that the probed neurons do not respond to features; it may rather mean that the features are more general (i.e., "objects"), and that they partly depend on activity within the subject's brain, rather than only on the subject's sensory environment. Some such assumption would in fact be necessary if we believe in the existence of selective attention and perceptual "set" (Lashley, 1954). In addition, centrifugal excitations are believed by many neurophysiologists to be the reason for the "noisy" appearance of unit activity in the sensory cortex of unanaesthetized animals. The abundant centrifugal fibers in the auditory apparatus have been thoroughly studied (Fex, 1968; Eldredge and Miller, 1971).

The question of neuronal co-operation. It is possible to object to the assumption underlying the local-to-global translation (Section 2), on the basis that the "inputs" and "outputs" of the brain as a whole are all many-neuron signals, and in addition the responses of individual neurons appear unreliable and barely decipherable. It might seem to follow from these that "read-out" of useful information is only possible from a large number of neurons whose outputs are co-ordinated by means of cross-connections (and therefore that the ways in which single neurons respond or distinguish are by themselves of no macroscopic interest).

However, a closer look shows that this does not follow. What follows is that read-out is only possible from a large number of neurons whose outputs are correlated.

Cross-connections are one way of achieving correlated outputs. It has for instance been suggested (Legéndy, 1967, 1970) that neurons form huge groups, by making many connections within groups and fewer connections between them, such that each group is capable of "ignition", an unstructured forest-fire of responses by group members, recognizable to other neurons because the membership of each group is fixed. But it must be emphasized that near the "input end" of the brain the need for cross-connections is not absolute; at least, they are not required for providing correlation between outputs of large neuron collections. Such correlation will be automatic when neurons respond to objects which themselves are interrelated. Recently Elul (1972) presented EEG evidence for partial synchrony among cortical neurons. His data indicate that such synchrony requires the availability of thalamic inputs, which lends support to the latter position.

4. The Correlation Scheme

Local miracles. Let us return to the inference mentioned in the Introduction, from correlations to objects behind them. It is probably fair to say that the inference is backed by an empirical rule which may safely be considered reliable, as it is the main motivation behind all of statistics-based research: that when there is a correlation between nature-made random variables, there is always a reason. Detection of the correlation never reveals the reason, only tells the researcher that a reason exists. The present application hinges on noting that the rule is still valid when it is unknown what the correlated random variables signify, as for correlated trains of firing spikes.

The reader is justified in asking: if behind any non-constant correlation of firing there is an object, why were dramatic events, "local miracles", singled out in the Introduction? The reasons are practical, involving the complexity of the required biological correlation detector: (1) If the correlation is not dramatically exhibited, its detection poses inordinate information storage requirements. (2) Statistical estimation is a difficult and controversial matter, but not when the data are overwhelming and the standard deviations negligible compared to the means.

Another detail needing justification: why the emphasis on correlations that are non-constant? Here, the reason is that constantly present correlations are quite likely to have their reasons inside the neural net (e.g. firing trains may come from a common source).

It may be desirable to give an example of a local miracle, in terms of neurons and impulses. Here is a descriptive example: If cortical neurons fire at a rate of 16 spikes per second (Evarts, 1968), then in any 10 msec interval about 16% of all neurons are expected to give off a spike, thus a fairly large number of neurons. However, for instance,

20 *given* neurons will fire together, within a 10 msec time spread, only about once in 10^6 years. An extreme example of a local miracle would be when 20 neurons, usually firing their independent ways, fired together on one occasion, and were found to do so 10 more times in the next 5 minutes. Any neuron equipped with synapses from the 20 neurons in question, so as to respond when those fired together again, would presumably respond to "something". For then, its sometimes-correlated inputs show their correlated behavior.

Discussion of the "object" concept. The notion of *object* in this general sense is rather vague: strictly speaking we cannot name anything that is not an object, for by naming it we make it one. Yet, to human intuition the notion appears to have substance as was already clear to Aristotle, and is demonstrated well by our spoken language. Everyday discourse is replete with statements like: "this occurrence signifies something", implying that it could alternatively signify nothing.

One may speculate that the general notion of *object* is in fact a logical absurdity, that the notion is "all psychological", and that it arises from the way in which our brain tissue operates. This is the point of view adopted in this paper. Brain tissue develops responses to neural events that deviate from the random and uncorrelated, and the organism functions as if on the assumption, right or wrong, that there is always some real reason for such events. Spoken language is developed, to a large extent, by naive speakers, and is much concerned with the subjective truth, as viewed from the brain. Accordingly if brain function attributes reality to a class of items (namely, the "reasons" behind local miracles), it is to be expected that language shall have expressions dealing with these items, even if the class does not actually exist.

Hebb's scheme. A comparison may be made here between the correlation scheme and the scheme of Hebb (1949). The main difference is that Hebb did not describe a way whereby his postsynaptic neuron could acquire the response habit enabling it to fire in the first place; the present correlation scheme provides that way. Hebb's scheme, at the same time, implicitly assumes that both the presynaptic and the postsynaptic neuron are equipped to respond to *objects* to start with; in other words, genetically. Otherwise the responses never would show correlation (nor would the psychological interpretations make sense). In this light, it is not surprising that computer simulations of Hebbian networks (Rochester et al., 1961; Farley and Clark, 1961) never gave encouraging results, and were eventually abandoned. A point of similarity is that the correlation scheme does, under certain conditions, give rise to ignitable neuron groups, similar to cell assemblies (Sec. 6).

Objects faceless. Motor skills. We may raise a point here, a partly philosophical one, that although correlation detection tells when an object is present, and although when the same signals recur the same object is behind them (within its definition, as was seen), there is no way ever to tell what the object is. It would seem that this should lead to a form of blindness and inability to react effectively to the world outside.

However, two lines of argument indicate that this is not so. First, it may be argued that even from the vantage point of the whole organism it is impossible to tell what anything is. We tend to answer questions of "what" things are by reference to other things with which they correlate. The dictionary defines all words in terms of each other. Second, the ignorance in places of decision-making does not inherently prevent learning of motor skills; in fact the correlation scheme is easily extended to deal with learned goal pursuit. In essence, a neuron becomes part of a skill through detection of a correlation between its own output and the outputs of other neurons both before and after the neuron's own. It amounts to detection that by responding to signals A the neuron can (help to) bring about the occurrence of signals B (Legêndy, 1970). Learning is acquisition of the habit to respond in that way. The object behind B is (part of) a goal attained; the object behind A is the set of circumstances under which B is usually attainable. In other words, A is a hint that the circumstances are right, both inside and outside the brain, for attaining the goal. (For instance, when the goal is to drink water, the outside circumstances might be that a glass is in the visual field, with water in it, and the inside circumstances that a co-operative neuronal effort is underway to reach out for the glass, guaranteeing that the local neuron's own output, weak and undecipherable alone, will be supplemented by properly timed outputs of other neurons.)

5. Possible Mechanism of Correlation Detection

Nature of the strategy. For concreteness, I shall consider one particular way to achieve this end which is, as will be seen, at least not ruled out by the available data. The postulated biological mechanisms arrange that when the signals in some set of axons are found to be correlated, these axons will grow collaterals to make synapses on a common receptive surface. The choice of neuron whose receptive surface is thus utilized is immaterial, and it does not matter if the neuron is already equipped with many other synapse sets and has a separate response habit corresponding to each (Wickelgren, 1969).

Schema of correlation detection. The following is a step-by-step description of a conceivable course of correlation detection. First the logical schema.

Since correlations show themselves through repetition of patterns, their detection requires information storage. The storage requirement is expected to be unmanageably high unless a way is found to recognize good "candidate" patterns without any repetition, and to establish the correlation, with fair likelihood, in not more than one repetition. The latter of these means that the pattern must have a very low expected occurrence frequency (once in several months, for instance); then a stored candidate pattern can either be erased after a few minutes (if it does not recur before then) or pronounced meaningful (if enough of it does recur). The need for convergence, mentioned in the Introduction, arises from the need for low expected occurrence frequency.

For spotting good candidate patterns, an easy method offers itself if the pattern, like in the above extreme example (Sec. 4), simply consists in nearly-simultaneous arrival of responses. For then the detector, subject to a constant multi-channel rainfall of neuron outputs, need only detect when the rainfall is momentarily more intense than the usual. When it is, it might be because, superimposed on the more-or-less constant "noise", it contains a "clump" of responses that belong together and often occur together. (Later, recurrence of the "clump" would set it apart from the "noise".)

Possible involvement of the glia. The available data on glia cells suggests that they may be suitable for the spotting of candidate patterns by such a method. Kuffler and Nicholls (1966) in their review article on glia cells remarked:

> By registering the K^+ concentration in their environment glial cells indicate in a non-specific way the activity of groups of nervous elements, without differentiating the discharges in inhibitory and excitatory neurons. Whether the nervous system makes use of the 'information' is not known.

In addition, Palay (1969) found that in many parts of the brain, lamellar processes from astrocytes cover essentially the whole neuron surface not covered by synapses or myelin. In neuropil, they cover many unmyelinated portions of axons. Although I know of no counts of the number of axons thus in contact with an astrocyte, from the dimensions of astrocytes, and the density of neuropil, it is probably safe to estimate that the number is about 10^2-10^4. Finally, astrocytes are quite numerous; they make up, to order of magnitude, one half of all glia cells, and glia cells in man outnumber neurons by a factor of about 2-4. The factor about triples in ascent from mouse to man (Friede, 1954; Blinkov and Glezer, 1968). Correlation detection could then be achieved, for instance, by the following combination of mechanisms:

Let us suppose first that a mechanism exists whereby a large influx of K^+ ions from axons into an astrocyte temporarily renders the astrocyte surface membrane (relatively) impermeable to K^+ ions everywhere except in places where K^+ ions entered in the course of the large influx. Then, while the impermeability lasts, fluctuations of the potential within the astrocyte are indicators of the extent to which the "candidate" pattern, the geographic pattern of the large influx, was repeated. Let us suppose next that a mechanism exists whereby, if during impermeability the potential swings beyond a critical value, the portions of surface membrane where K^+ entered the second time are marked in some chemical manner. The critical value is high enough to make its attainment by accident unlikely. The marked membrane portions then stake out a once-confirmed candidate pattern; the axons apposed to the marked portions have shown a tendency to fire together.

It then remains (a) to bring these axons into synaptic contact with the receptive surface of a neuron (any nearby neuron will do) and at the same time (b) to confirm the pattern again and again by detecting further recurrence; each time add to the pattern or delete from it, if the recurrences so dictate. Of these, (b) need not be discussed in detail, as the discussion would lead deeper into speculation, and would involve nothing essentially different from the procedure already described.

The occurrence of some step like (a) in nervous systems is suggested by a number of deprivation experiments (Ruiz-Marcos and Valverde, 1969; Coleman and Riesen, 1968; Lund and Lund, 1972) which may be interpreted as linking learning, at least early learning, to synapse formation. That (a) is mediated in particular by the astrocytes is not in contradiction with the known facts. Astrocytes have end-feet on blood-vessels (Cajal, 1911) and thus may have access to nutrients, which they might pass on to the growing portions of fibers (Kuffler and Nicholls, 1966); and Hydén (1968) and his co-workers present evidence that reciprocal chemical changes in fact do occur between neurons and glia cells during various learning tasks, though their data have been severely criticized (Kuffler and Nicholls, 1966). In addition, in tissue culture glial cells appear able to grab and pull (Lumsen and Pomerat, 1961).

Transitory structures and glomeruli. The glia cell is not the only structure that may be suspected of having to do with correlation detection. There are suggestions that in some cases the neurons themselves may be involved in it.

There are a number of examples of greater neuronal elaboration in immature brains than in mature brains. In 3-day old cat cerebral cortex the molecular layer is full of Retzius-Cajal cells (Cajal, 1911). By the 8th day they are gone (Purpura et al., 1964). Other examples are the early extra branches on cerebellar granules and other cells (Cajal, 1911) which disappear later. It is tempting to speculate that the various transitory dendritic structures are feelers, apposed to many axonal fibers and developing the areas of apposition into synapses where signals from the facing fibers correlate with others.

Finally, one may speculate that the glomeruli found in the cerebellum and thalamus (Szentágothai, 1970) as well as similar synaptic elaborations found in the superior colliculus (Lund and Lund, 1972) are meeting places at which, at an earlier time, axons showing correlated signals have been brought together with dendrites.

6. Conditions of Detectability

In all the foregoing discussion of correlation detection, it was taken for granted that the detecting cell, e.g. the glia, often encounters correlations between the signals it monitors. The motivation behind such an assumption is that local miracles are a dime a dozen; that these correlations are as frequent as are objects in the outside world; and that, of course, the outside world is made of "nothing but" objects. However, such reasoning is fallacious, because the arrival of correlated inputs at the sense organs does not guarantee that a correlation detector exists on which the correlated disturbances converge. The requirement of detectability imposes conditions on the correlation, on the number of neurons whose outputs must be correlated, and on the network.

Appropriate correlations. The correlations suitable for detection in the ways described are set-correlations with the property that if they link a set of responses, they also link (many) subsets of the set. They originate, for instance, from sets of objects that "belong together". The correlation considered above, consisting of a tendency by a set of neurons to fire together, belongs in this class. It will be appreciated that if such a correlation extends to enough neurons, some detectors will likely monitor enough of them; and to achieve this, the axons need not be connected to the correlation detectors by any *a priori* individual selection. In a net where correlations of the said type are frequent and link enough neurons, one can choose any set of neurons at random, and expect their outputs to show correlations now and then; therefore it does not matter much what neuron sets the detectors monitor.

In such a net there is no rigid upper bound to the self-developed scale of complexities (Sec. 2) because when a monitored set of neuron outputs shows appropriate correlation, it can be turned into responsiveness whether the components are complicated or simple or mixed. The same neuron may end up equipped for some simpler, some more complex objects. In such a net, also, the output fibers, i.e. the set of efferents leading to a particular other brain structure, need not be carefully selected neurons. Within certain limitations, they too can be chosen at random.

Size of correlated set. Let us next calculate the number C of neurons whose firing must be correlated in order that the correlation be detectable. Let us say that a correlation is not detected unless a detector receives $\tau = 10$ or more correlated signal components, each from a different neuron. (Occurrence frequency considerations yield estimates of τ between 5 and 15). For calculational purposes we assume that we deal with a network of 10^8 neurons, of which each correlation detector monitors 10^3, and take account of the frequent arborization of axons near their cell body by assuming that every neuron has a relatively high probability $\lambda = 10^{-2}$ of being monitored by any given detector "near" it (say a few hundred microns), while it has $\lambda = 10^3/10^8 = 10^{-5}$ of being monitored by any of the rest. The number of correlation detectors is taken to be $N=10^5$ in cases where all the correlated neurons are close together, and $N=10^8$ otherwise. Assuming that all subsets of the correlated set are correlated, the expected number of detectors monitoring τ or more of these neurons is:

$$x = N \sum_{i=\tau}^{C} \binom{C}{\tau} \lambda^i (1 - \lambda)^{C-i} \tag{1}$$

With the given numbers, setting x=1 and solving approximately for C, under conditions when all neurons are close together and close to the correlation detectors, thus when

$$\lambda=10^{-2} \quad \text{and } N = 10^5 ,$$

we find:

$$C \cong 250.$$

When the neurons are arbitrarily scattered, thus when

$$\lambda = 10^{-5} \text{ and } N = 10^{8}$$

we find:

$$C \cong 250{,}000.$$

In other words, no detector is expected to detect the correlation unless 250 or 250,000 neurons, in the two cases, respectively, are correlated, and subsets of the correlated sets also show the correlation. The situation is not as hopeless as may seem from this, as with a fourfold increase of the correlated sets, i.e., at C = 1,000 and C = 1,000,000, respectively, we have

$$\lambda C = \tau \ , \tag{2}$$

at which the majority of the correlation detectors will make a detection. Eq. (2) and the assumptions above (1) constitute a sufficient condition for detectability of a correlation.

The foregoing calculation shows that the correlation must indeed be a set-correlation of the aforementioned class. For, setting C = 10, the probability of detection is essentially nil.

The advantage of mapping. Remarks. Let us note the wide discrepancy between the requirements when the correlated set is concentrated and scattered. The discrepancy shows how a neighborhood-preserving mapping of the retinal image onto the cortex facilitates feature detection. Visual images cause many more correlations between details near one another than between details far apart; thus by mapping, details close together will excite neurons close together, and relatively few need to belong together before their correlation becomes detectable.

It is seen that the metric properties of the retina are unimportant to the process, except in facilitating the hunt for correlations, which checks against observations that vision is not permanently impeded by distorting prisms and the like (Howard and Templeton, 1966). The example also shows one way in which gross neurospecificity (Sec. 5) can go far in aiding object detection.

The condition for ignitible group formation may be obtained by replacing N in Eq. (1) by N', the number of correlation detectors "serving" a neuron (i.e., the number of detectors able to equip a neuron for new patterns), setting $x = EC/\tau$, where E is the number of inputs required for causing neuron response, and solving for C. Fairly insensitive to x, the result will be approximately the second set of figures given above, the figures obtainable from (2).

The figures underneath Eq. (1) show one other thing: that correlation detection is impossible in the total absence of planned contacts; indeed, impossible unless a large number of neurons is equipped for objects to start with, i.e., genetically. Otherwise interrelations between objects in the world will not translate into correlations in the brain.

7. Conclusions

1. It is assumed that not every contact in the human cerebral cortex and cerebellum is "planned", in the sense that its presynaptic and postsynaptic neuron are individually specified genetically. For many contacts, genetic specification extends only to neuron type, synapse type, location and the like.

2. For determining the unplanned contacts in a way that makes neurons responsive to objects of the sensory environment, the signals locally available in the immature neuropil are sufficient; they need not be aided by exogenous reinforcement or "now print!" cues. (The word "object" is used in the general sense.)

3. A scheme in which it may be done involves detecting certain local miracles and equipping neurons to respond to them. A local miracle is an event of dramatic (overwhelmingly convincing) evidence of correlation between signal trains arriving into a neighborhood within the neuropil. Its dramatic nature makes its detection practicable, and its non-constancy increases the likelihood that its underlying reason is outside the neural network.

4. By making neurons responsive to objects, this scheme tends (under certain conditions, believed to exist in the cerebral cortex) to develop responses to complex and simple objects within the same neuron net, acquired responses built on acquired components, responses aided by contextual cues and insensitive to some extent to object imperfections.

5. A certain number of planned contacts, and genetically object-responsive neurons, must exist in the brain; without them, the scheme is

helpless.

6. Speculation is presented describing astroglial cells as mediating the scheme, via their responses to extracellular K^+ and via their end-feet on bloodvessels, with access to nutrients to aid the local growth.

Summary: This paper deals with the "unplanned" contacts in the advanced central nervous system and the manner in which they become established. A contact is considered unplanned when its genetic specification does not single out individually the presynaptic and postsynaptic neuron, but extends only to neuron type, location and the like. A scheme is described which combines the establishing of unplanned contacts with the primitive neural process of organizing the sensory environment into "objects". The scheme involves detection and utilization of a special class of occurrences within the immature neuropil: when a number of signal trains, usually independent, exhibit dramatic evidence of correlation. These occurrences, which are like "local miracles", are often due to causes lying outside the brain, and are useful in bringing neurons under the influence of features of the environment. Questions surrounding their detection are discussed, and a possible physical mechanism sketched, involving astroglial cells.

References

Bell, C.C. and Dow, R.S.: Cerebellar circuitry, Neurosciences Res. Prog. Bull. 5 Nr. 2, A Work Session Report (1966).

Blakemore, C. and Cooper, G.F.: Development of the brain depends on the visual environment, Nature 228, 447-448 (1970).

Blinkov, S.M. and Glezer, I.I.: The Human Brain in Figures and Tables: A Quantitative Handbook, Plenum, New York, 1968.

Braitenberg, V. and Attwood, R.P.: Morphological observations on the cerebellar cortex, J. Comp. Neurol. 109, 1-27 (1958).

Brindley, G.S.: Nerve net models of plausible size that perform many simple learning tasks, Proc. Roy. Soc. London B 174, 173-191 (1969).

Coleman, P.D. and Riesen, A.H.: Environmental effects on cortical dendritic fields, 1. Rearing in the dark, J. Anat. (Lond) 102, 363-374 (1968).

Colonnier, M. and Rossignol, S.: Heterogeneity of the cerebral cortex. In: Basic Mechanisms of the Epilepsies, ed. by H.H. Jasper, A.A. Ward, Jr., and A. Pope, Churchill, London, 1969.

Eldredge, D.H. and Miller, J.D.: Physiology of hearing, Ann. Rev. Physiol. 33, 281-310 (1971).

Elul, R.: Randomness and synchrony in generation of the electroencephalogram. In: Mechanisms of Synchronization in Epileptic Seizures, ed. by H. Petsche and M.E. Brazier, Springer Verlag, Vienna, 1972.

Evarts, E.V.: Unit activity in sleep and wakefulness. In: The Neurosciences: A Study Program, ed. by G.C. Quarton, T. Melnechuk and F.O. Schmitt, pp. 499-515 and 886-886. Rockefeller University Press, New York, 1967.

Farley, B.G. and Clark, W.A.: Activity in networks of neuron-like elements. In: Information Theory (Fourth London Symposium), ed. by C. Cherry, p. 242. Butterworths, London, 1961.

Fex, J.: In: Hearing Mechanisms in Vertebrates, ed. by A.V.S. DeReuck and J. Knight, pp. 169-181. Little, Brown, Boston, 1968.

Fox, C.A. and Barnard, J.W.: A quantitative study of the Purkinje cell dendritic branchlets and their relationship to afferent fibres, J. Anat. 91, 299-313 (1957).

Friede, R.L.: Der quantitative Anteil der Glia an der Cortexentwicklung, Acta Anat. 20, 290-296 (1954).

Gesteland, R.C., Lettvin, J.Y. and Chung, S.H.: A code in the nose. In: Cybernetic Problems in Bionics, ed. by H.L. Oestreicher and D.R. Moore, Gordon and Breach, New York, 1968.

Globus, A. and Scheibel, A.B.: The effect of visual deprivation on cortical neurons: a Golgi-study, Exp. Neurol. 19, 331-345 (1967).

Goldberg, J.M. and Brown, P.B.: Response of binaural neurons of dog superior olivary complex to dichotic tonal stimuli: some physiological mechanisms of sound localization, J. Neurophysiol. 32, 613-636 (1969).

Gross, C.G., Rocha-Miranda, C.E. and Bender, D.B.: Visual properties of neurons in inferotemporal cortex of macaque, J. Neurophysiology 35, 96-111 (1972).

Hartline, H.D.: The response of single optic nerve fibers of the vertebrate eye to illumination of the retina, Amer. J. Physiol. 121, 400-415 (1938).

Hartline, H.K.: The receptive fields of the optic nerve fibers, Amer. J. Physiol. 130, 690-699 (1940).

Hebb, D.O.: The Organization of Behavior, Wiley, New York, 1949.

Held, R.: Two methods of processing spacially distributed visual stimulation. In: The Neurosciences: Second Study Program, ed. by G.C. Quarton, T. Melnechuk and G. Adelman, pp. 317-324. Rockefeller University Press, New York, 1970.

Hirsch, H.V.B. and Spinelli, D.N.: Visual experience modifies distribution of horizontally and vertically oriented receptive fields in cats, Science 168, 869-871 (1970).

Howard, I.P. and Templeton, W.B.: Human Spatial Orientation, Chapter 15, Wiley, New York, 1966.

Hubel, D.H.: Single unit activity in striate cortex of unrestrained cats, J. Physiol. 147, 226-238 (1959).

Hubel, D.H. and Wiesel, T.N.: Receptive fields of single neurones in the cat's striate cortex, J. Physiol. 148, 574-591 (1959).

Hubel, D.H. and Wiesel, T.N.: Receptive fields, binocular interaction and functional architecture in the cat's visual cortex, J. Physiol. 160, 106 (1962).

Hubel, D.H. and Wiesel, T.N.: Receptive fields of cells in striate cortex of very young, visually inexperienced cats, J. Neurophysiol. 26, 994 (1963).

Hubel, D.H. and Wiesel, T.N.: Receptive fields and functional architecture in two nonstriate visual areas (18 and 19) of the cat, J. Neurophysiol. 28, 220 (1965).

Hubel, D.H. and Wiesel, T.N.: Visual area of the lateral suprasylvian gyrus (Clare-Bishop Area) of the cat, J. Physiol. (London) 202, 251-260 (1969).

Hydén, H.: RNA in brain cells. In: The Neurosciences: A Study Program, ed. by G.C. Quarton, T. Melnechuk and F.O. Schmitt, pp. 499-515 and 886-887. Rockefeller University Press, New York, 1967.

Köhler, W.: Gestalt Psychology, New York, 1929; especially the section devoted to critique of introspectionists.

Köhler, W. and Wallach, H.: Figural aftereffects: an investigation of visual processes, Proc. Amer. philosoph. Ass. 88, pp. 269-357 (1944).

Kuffler, S.W. and Nicholls, J.G.: The physiology of neuroglial cells, Ergeb. Physiol. 57, 1-90 (1966).

Lashley, K.S.: Dynamic processes in perception. In: Brain Mechanisms and Consciousness, ed. by J.F. Delafresnaye, pp. 422-437. Blackwell, Oxford, 1954.

Legéndy, C.R.: On the scheme by which the human brain stores information, Math. Biosciences 1, 555-597 (1967).

Legéndy, C.R.: The brain and its information trapping device. In: Progress in Cybernetics, Vol. 1, ed. by J. Rose, pp. 309-338. Gordon and Breach, New York, 1970.

Lettvin, J.Y. and Gesteland, R.C.: Speculations on smell. In: Cold Spring Harbor Symposia on Quanitative Biology, Vol. 30 (1965).

Lettvin, J.Y., Maturana, H.R., McCulloch, W.S. and Pitts, W.H.: What the frog's eye tells the frog's brain, Proc. Inst. Radio Engr. 47, 1940-1951 (1959).

Livingston, R.B.: Brain circuitry relating to complex behavior. In: The Neurosciences: A Study Program, ed. by G.C. Quarton, T. Melnechuk and F.O. Schmitt, pp. 499-515 and 886-887. Rockefeller University Press, New York, 1967.

Lorente de Nó, S.: The cerebral cortex: architecture, intracortical connections and motor projections. In: Physiology of the Nervous System, ed. by Fulton, Oxford, 1939.

Lumsen, C.E. and Pomerat, C.M.: Normal oligodendrocytes in tissue culture, Exp. Cell Res. 2, 103 (1951).

Lund, J.S. and Lund, R.D.: The effects of varying periods of visual deprivation on synaptogenesis in the superior colliculus of the rat, Brain Res. 42, 21-32 (1972).

Marr, D.: A theory of cerebellar cortex, J. Physiol. 202, 437-470 (1969).

Marr, D.: A theory for cerebral neocortex, Proc. Roy. Soc. Lond. B, 161-234 (1970).

McIlwain, J.T.: Central vision: visual cortex and superior colliculus, Ann. Rev. Physiol. 34, 291-314 (1972).

Mirsky, A.E. and Ris, H.: The desoxyribonucleic acid content of animal cells and its evolutionarity significance, J. Gen. Physiol. 34, 451-462 (1951).

Noda and R. Adey: J. Neurophysiology 33, 672 (1970).

Palay, S.L.: Morphology of neuroglial cells. In: Basic Mechanisms of the Epilepsies, ed. by H.H. Jasper, A.A. Ward, Jr., and A. Pope, Churchill, London, 1969.

Palkovits, M., Magyar, P. and Szentágothai: Quantitative histological analysis of the cerebellar cortex in the cat. II. Cell numbers and densities in the granular layer. Brain Res. 32, 15-30 (1971a).

Palkovits, M., Magyar, P., and Szentágothai, J.: Quantitative histological analysis of the cerebellar cortex in the cat, III. Structural organization of the molecular layer, Brain Res. 34, 1-18 (1971b).

Paul, J.: General theory of chromosome structure and gene activation in eukaryotes, Nature 238, 444-446 (1972).

Perkel, D.H. and Bullock, T.H.: Neural Coding. In: Neurosciences Research Symposium Summaries, Vol.III, ed. by F.O. Schmitt, T. Melnechuk, G.C. Quarton and G. Adelman, pp. 405-527. MIT Press, Cambridge, 1969.

Pettigrew, J.D., Nikara, T. and Bishop, P.O.: Responses to moving slits by single units in cat striate cortex, Exp. Brain Res. 6, 373-390 (1968).

Purpura, D.P., Shafer, R.J., Housepian, E.M. and Noback, C.R.: Comparative ontogenesis of structure-function relations in cerebral and cerebellar cortex. In: Progress in Brain Research, Vol. 4, ed. by D.P. Purpura and J.P. Schadé, pp. 187-222. New York, Elsevier, 1964.

Rall, W.: Dendritic neuron theory and dendrodendritic synapses in a simple cortical system. In: The Neurosciences: Second Study Program, ed. by G.C. Quarton, T. Melnechuk and G. Adelman, p. 317. Rockefeller University Press, New York, 1970.

Ramón y Cajal, S.: Histologie du Systeme Nerveux de l'Homme et des Vertébrés, Maloine, Paris, 1911.

Rochester, N., Holland, J.H., Haibt, L.H. and Buda, W.L.: Tests on a cell assembly theory of the action of the brain, using a large digital computer, IRE Transactions on Information Theory, IT-2, pp. 43-53 (1961).

Rosenblatt, F.: Principles of Neurodynamics: Perceptrons and the Theory of the Brain Mechanisms, Spartan, Washington, 1962.

Ruiz-Marcos, A. and Valverde, F.: The temporal evolution of dendritic spines in the visual cortex of normal and dark raised mice, Exp. Brain Res. 8, 284-294 (1969).

Sholl, D.A.: The Organization of the Cerebral Cortex, Methuen, London, 1956.

Szentágothai, J.: Glomerular synapses, complex synaptic arrangements, and their operational significance. In: The Neurosciences: A Second Study Program, ed. by G.C. Quarton, T. Melnechuk and G. Adelman, Rockefeller University Press, New York, 1970.

Vogel, F.: A preliminary estimate of the number of human genes, Nature, 201, 847 (1964).

von Senden, M.: Raum-und Gestaltauffassung bei operierten Blindgeborenen vor und nach der Operation, Barth, Leipzig, 1932.

Wickelgren, W.A.: Learned specification of concept neurons, Bull. Math. Biophysics 31, 123-142 (1969)

Note added in proof. Very recently, Kelly and Van Essen, J. Physiol. (Lond.) 238, 515 (1974), recorded from glial cells identified by Procion yellow in the visual cortex of cat, and found that many of these responded to visual stimuli by slow graded depolarization. Several of the cells so responding showed preference for a stimulus orientation that matched the optimum orientation of some nearby neurons. Thus cortical glia are not unlike others, used above.

A prediction. If a glial cell is responsible for having equipped a neuron for a feature, it is expected to (i) be near the neuron, (ii) respond to the same feature, (iii) have depolarizing apposition with the same axons that make the neuron respond to the feature; and consequently, (iv) its responsiveness to the feature must not be solely a result of the apposition to the neuron, and <u>should survive if the neuron is killed</u>.

ON THE REPRESENTATION OF OBJECTS AND THEIR RELATIONS IN THE BRAIN

V. Braitenberg
Max Planck Institut für biologische Kybernetik
Tübingen, Germany

1. Introduction

Situations within the central nervous system are elusive when we try to render them in anything more abstract than approximate verbal descriptions. A description in terms of ionic movements through cell membranes, for example, might be useful when dealing with information transmission in sense cells but becomes unwieldy if applied to situations involving many neurons of the brain. Attempts at applying the formalism of a logical calculus, inspired by the success of binary algebra in dealing with switching networks, seem to impose a restriction on the working of neurons (e.g. by assuming discrete time) which is not justified by electrophysiology. Occasionally an insight may be gained by reasoning in terms of channel capacity, amount of information and redundancy, but this language again seems more successful in the analysis of peripheral sensory events than in the study of central processes.

Thus the anatomists and physiologists of the brain do not know to what theory their findings contribute. Must they resign themselves, then, to producing results in a merely descriptive form, or is there a sufficiently precise indication of the theories to come for them to formulate already the relevant questions?

A number of statements about the brain are of such a general nature that they will undoubtedly carry over to whatever theory will be developed later. These are principally statements about different types of symmetry of neuroanatomical structures which must correspond to different types of symmetry in the abstract scheme of information handling. Also, metric invariants which can be found in the geometry of some nerve nets in the brain, as opposed to only topological invariants in others, may lead to strong propositions about the possible schemes of information handling in the two types of nerve nets.

An unquestionable acquisition of brain science with strong implications for all theories is the orderly projection of sensory spaces onto certain regions of the brain, in such a way as to give to the internal coordinates of the brain the meaning of external (not necessarily spatial:

e.g. acoustic frequency) coordinates of the sensory space represented. This leads to the idea of the brain as an image of the environment, a small scale replica or a dynamic model of its structure.

But we run into conceptual difficulties when we make our descriptions more detailed, when we talk about single neurons and synapses and ask ourselves, e.g., whether it still makes sense, at this level, to think in terms of an image of the environment. What is the thing in the environment of which a neurons is an image? And what is imaged by a fiber connecting two neurons within the brain? If we had a clear idea of this elementary imaging process, as we analyse more complex situations within the brain, we could synthesize more complex situations of the environment which are "imaged" there.

I shall follow up this idea in order to arrive at an interpretation of the structure of the cerebral cortex.

2. What Is Represented by a Neuron in the Brain

We suppose that a neuron emits a signal through its axon whenever a sufficient density of excitation is produced by the activity of its afferents. Let us suppose that not all the afferents have to be active at the same time for the neuron to emit a signal. It will then be impossible, by looking at the output of the neuron, to determine which particular set of afferents was responsible for it. This information got lost in the neuron in the process of the summation of the incoming excitation. We say that a certain image is a good representation of an object, when the information which is lost in the imaging process was not relevant information as regards that object. What sort of object is it, then, which is represented, given this tolerance in the input-output specification of the neuron? We are driven to the conclusion that it must be an object defined by a collection of properties which are bound by the following relation: any (large enough) subset implies the rest of the properties. In fact, then, if each property is represented by one afferent, it does not matter if the neuron responds to activity in some of its afferents as if all of them had been active. The neuron "hallucinating" the object of which it is an image even if only partial evidence for the presence of this object was provided, makes a reasonable guess if the object has that kind of coherence.

More abstractly, one may define an object thus: a set of properties such that the probability of any of the properties occurring is a monotonically increasing function of the number of the properties belonging to the same set which are known to have already occurred.

$$\frac{\Delta p(E_j)}{\Delta \sum_{i=1}^{n} E_i} > 0 \quad \text{for all } j$$

where $E_1..E_i..E_j..E_n$ are functions describing the presence or absence of the n properties belonging to an object by assuming the values 1 (presence) or 0 (absence). $p(E_j)$ is the probability of E_j occurring.

The probability of the entire object being there is clearly a monotonically increasing function of the number of properties constituting it which are known to have occurred.

If we assume that the output of a neuron is not all-or-none, but graded, being stronger the more afferents have been activated, then the output of the neuron is a measure of the probability of the object of which the neuron is an image having presented itself to the sense organs.

This definition of an object may seem arbitrary and all too obligingly adapted to what we know already about the neuron. On the other hand, the relation between "neuron" and "object" may be a natural and essential one. Terms of such stupendous generality as "objects," "properties," "events" are intrinsically anthropomorphic. They have little meaning in physics, but are rather anchored in some general principle of the working of our brains. A definition of "object" derived from considerations about neurons which represent objects may be the best (most natural) definition possible.

The same description of the function of a neuron can also be interpreted in different terms. For one, if neurons are characterized by the compound logical propositions which describe the conditions for their becoming active in terms of elementary propositions on the state of the sensory cells, as in the McCulloch and Pitts theory, then such a neuron corresponds to a conjunction of disjunctions (or to a disjunction of conjunctions) of elementary logical propositions. In this language the "coherence" of properties which make up an object, is, however, not evident.

Also, this description of a neuron makes sense in terms of information theory. If there are such objects occurring in the environment of the animal, this represents a redundancy of the environment, considered as a source of information, which it pays to take advantage of in the coding of the information. If coherent sets of properties, properties which have a high probability of occurring together are coded in individual neurons, this represents an obvious saving in channel capacity.

3. What is Represented by a Synapse Connecting Two Neurons in the Brain?

An axon of a neuron $\mathcal{N}$ influences another neuron $\mathcal{M}$ with which it is connected synaptically, while we suppose no such influence is exerted in the opposite direction through the same synapse. If N and M are the two events characterized by the activity of $\mathcal{N}$ and $\mathcal{M}$ respectively, the presence of the synapse assures us that in general

$$p\ (NM) \neq p\ (N) \cdot p\ (M)$$

p (N) and p (M) meaning the probability that the corresponding neurons are active in a certain interval of time, p (NM) meaning the probability that they are both active.

This is, however, a statement involving the two neurons symmetrically, while the synapse is, for all we know, an asymmetrical physiological relation between two neurons. One would be tempted to rewrite the above inequality in terms of the asymmetrical conditional probabilities, in order to match the physiological asymmetry of the synapse

$$p\ (M|N) \equiv \frac{p\ (MN)}{p\ (N)} \neq p\ (M)$$

but this is not in general asymmetric, since p (M|N) and p (N|M) are different only if p (N) ≠ p (M). Moreover, p (M|N) > p (M) implies p (N|M) > p (N) and p (M|N) < p (M) implies p (N|M) < p (N), whereas a synapse from $\mathcal{M}$ to $\mathcal{N}$ does not imply in any way a synapse in the opposite direction. Evidently, the conditional probability of the activity of one neuron given the activity of another neuron is not a good abstract description of the synapse connecting the two neurons.

A more complete description of a synaptic connection between two neurons is obtained when the temporal order of their activities is taken into account. If M, Δt, N is the event that the neuron $\mathcal{N}$ is active Δt milliseconds after the neuron $\mathcal{M}$,

$$p\ (M, \Delta t, N) > p\ (M) \cdot p\ (N)$$

describes the situation in which the neuron $\mathcal{M}$ has an excitatory synapse on the neuron $\mathcal{N}$, making it more likely that $\mathcal{N}$ should be active at a time Δt after $\mathcal{M}$ was active than at other times. Similarly

$$p\ (M, \Delta t, N) < p\ (M) \cdot p\ (N)$$

describes an inhibitory synapse.

These are asymmetrical relations between the two neurons, just as the synapse is from a physiological point of view since for positive Δt there is no connection, a priori, between the probability of M, Δt, N and that of N, Δt, M.

These probabilities refer to events characterized by activity of neurons within the brain. We have already noticed that these neurons represent bundles of properties, called objects, of the outside world in the sense that the neurons become active when the corresponding objects present themselves to the sense organs. We have also discussed what kind of objects it should be if the neurons are to be good images of these objects. We may ask now what kind of a relation between the objects it should be that is represented by a synapse between the corresponding neurons if the whole situation, the two neurons connected by a synapse, is to be an image of a situation in the outside world involving the corresponding objects. According to our abstract definition of a synapse, this can only be, for an excitatory synapse, the relation of two objects presenting themselves in succession more frequently than chance, and for an inhibitory synapse, the relation of two objects appearing in succession less frequently than expected a priori. In other words, if neurons are images of objects, the synapses between them are images of the causal laws governing the appearance of these objects in the environment of the animal.

Again, this makes sense in terms of information theory. As the representation of clusters of properties frequently occurring together (the so-called objects) in single neurons constitutes a saving in channel capacity, so does the representation of objects occurring in certain succession in chains of neurons connected by synapses, so that the entire spatio-temporal event of the environment is coded in the spatio-temporal pattern within the brain. The correlation introduced by the synapses reduces the channel capacity of the nerve net in a way that takes advantage of the redundancy of the input. The two obvious kinds of redundancy of the input from the environment, due to clustering of properties into coherent sets on one hand and temporal order of events on the other, may thus be incorporated into the brain in two different forms, dendritic trees of neurons connected to sets of afferents on one hand and connections between neurons within the brain on the other.

We notice at this point that we have surreptitiously introduced an ambiguous interpretation of synapses between neurons. The fibers which we called afferents, connected synaptically to a neuron, we interpreted as implying the object which is represented by that neuron. On the contrary, a fiber of a neuron within the brain, making a synaptic contact with another similar neuron we had interpreted as representing the transition probability between two objects represented, with an accent on the temporal succession which was entirely missing in the first interpretation. We may argue that this fiber too, like many others, is an afferent to that neuron and could be interpreted as implying the higher order object represented by the neuron, just as in the first interpretation. The ambiguity may be resolved as follows.

4. Regions of Uniform Context

If one sort of input reaches a delimited part of the brain, e.g. visual input or somesthetic input or a mixture of the two, or input relayed from some other part of the brain, we may say that the neurons in that region operate in a uniform context. These neurons represent "objects" in the above sense within this context, i.e. correlations between the activities of the afferents to that region. The afferents themselves are the axons of neurons which represent objects in different contexts and, in fact, reside in different regions of the brain. It is not possible in general to define objects between different regions of context. For example, although there is a mixed visual and auditory context, in which phonemes are defined as objects corresponding to letters of the alphabet, there is no mixed linguistic-visual context in which phonemes are correlated with, say, "features" of the peripheral visual input so that it would make sense to define objects in this context. There must be, on the other hand, a linguistic-visual context at a different level, in which correlations exist between words, i.e. complexes of phonemes, and patterns in visual space, so that complex objects, visual forms with their names, can be defined in this context.

If context is what allows objects to be defined, we may also say, the other way around, that a uniform context is defined by the possibility of detecting objects in it. Where there are no correlations there is no context.

The idea of areas of uniform context is related to that of hierarchical levels in perception. Different levels, e.g. that of phonemes and that of words in the perception of language correspond to different contexts. However, the converse is not necessarily true since there may be different contexts for which it cannot be asserted that one is at a "higher" level than the other. The context of musical harmonies is distinct from that of language, although both are fed by the same input. They are not, however, at different hierarchical levels.

If one accepts the idea of separate areas of context in the brain, the two different interpretations of synaptic contacts, as implying objects in one situation and as representing successions of objects in another situation, will seem quite natural. A fiber connecting two neurons within one region can be given a meaning in terms of a relation between the objects which are represented by the neurons. These objects, in fact, belong to the same context. On the contrary, a fiber reaching that region from the outside represents an object which is out of context. It can acquire a meaning in that context only as a member of a set of afferents whose activities are correlated and which therefore will be represented by a neuron within the area of context.

5. A New Look at the Cerebral Cortex

I have been stressing an ambiguity in the interpretation of neurons and synapses as images of objects and relations in the outside world because I believe the two aspects can be used to illuminate the structure of the cerebral cortex, a notoriously obscure, if undoubtedly preeminent portion of the vertebrate (and in particular, human) brain. An interpretation of the cortical anatomy is contained in another paper (J. theor. Biol., in press), I shall only sketch the main point here.

Fig. 1 introduces a familiar arrangement which has frequently been used for discussions on learning in nerve nets.

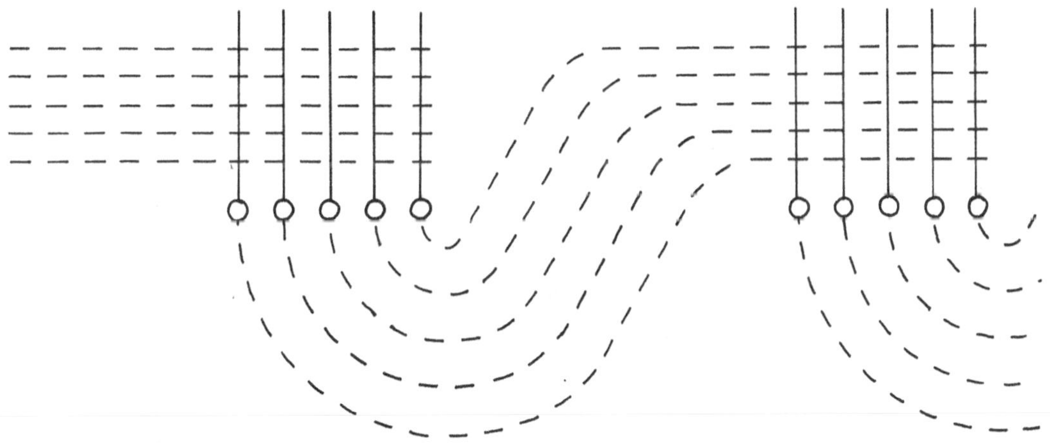

Figure 1

A set of input lines (dotted lines, horizontal) and a set of output lines or dendrites (vertical lines) form a matrix. At the intersection points there are contacts between the lines which are supposed to be so arranged that each dendrite is connected to a set of input lines which tend to be active together. This is the consequence of a memory-law which incorporates the statistics of the input into the connections through a mechanism which we don't want to discuss in detail and which we suppose to be housed in the circles symbolizing the main part of the neurons. Each neuron will become active when the constellation of inputs which it represents becomes active, and it will then emit a signal into its axon. The axons may be again afferents to another such network, so that the mechanism can be cascaded. This scheme corresponds to what we have called the representation of objects in neurons. Anatomical considerations on the cerebral cortex suggest a slightly more complicated scheme (Fig. 2).

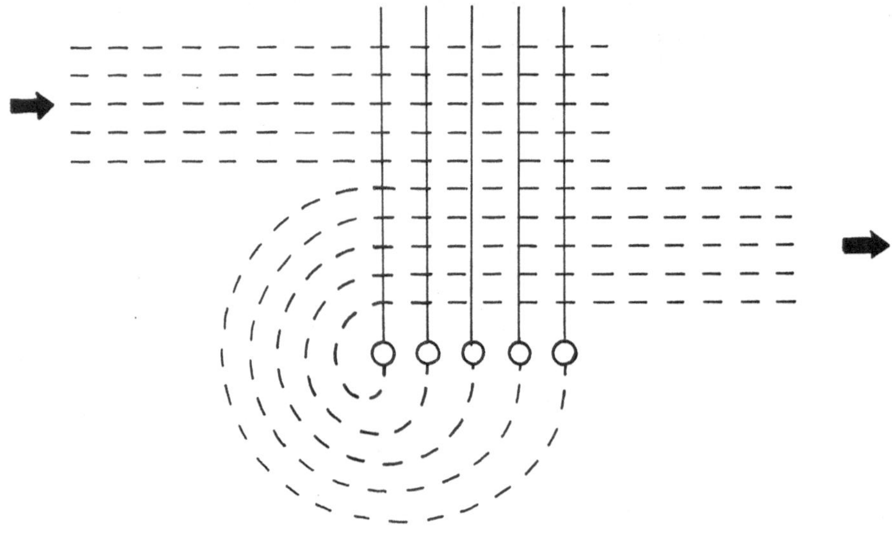

Figure 2

It is characteristic of the main neuron type of the cerebral cortex, the "pyramidal cell" that their dendritic tree, besides being connected to the input, also receives abundant afferents from other similar neurons of the same region through their so-called axon collaterals. The matrix in which the connections are "learned" has an internal as well as an external input (Fig. 2). This means that the "object" learned by each neuron has as essential constituents the presence in the input, a little while before, of "objects" already learned by other neighboring neurons. Neurons will be connected with sets of afferents which have a tendency to be active in succession. They will represent such objects as have a tendency to appear in certain sequences. We may say, they code the environment in such terms as seem to fit into a causal structure.

This is, intuitively, a useful way of representing the world. Not only will the incorporation of a code word into the brain be conditioned by its usefulness in a causal analysis, but the causal relations themselves are incorporated in the brain in the form of the internal connections of the neurons of one region. The internal connections of the brain which activate a set of neurons after another even if no new signals arrive from the outside, will be an image of the laws governing the objects in the physical world. The idea of a representation of the world in the brain acquires a more dynamic aspect. It is not difficult to imagine the usefulness of this in prediction, one of the fundamental attributes of brain function.

To use a paradigm from a different field, we may say that within one area of context in the cerebral cortex, the dictionary and the syntax of incoming messages are stored in a strongly interrelated way. The connections of individual cortical neurons with sets of afferents represent entries in the dictionary. The connections between the neurons within the area represent the rules which govern the successions of these terms, the syntax.

Acknowledgement

I should like to thank Dr. Charles Legéndy for many discussions which I am sure let some of his ideas unwittingly enter this paper.

REPORT ON THE DISCUSSION GROUP ON "LIVING NEURAL NETS"

J.G. Taylor
Department of Mathematics
King's College, London, England

The work of this group was divided into 4 parts, under the headings of synapses, single neurons, small neural nets and large neural nets. The purpose of the discussions was to bring together theorists and experimentalists so as to discuss theoretical questions and their possible experimental answers; the theme running through the several meetings of the group was to find out the 'facts of life' of living nerve nets and how to understand them in terms of theoretical models.

The first heading of synapses, and the main topic of interest was the quantum hypothesis of synaptic transmission described in J.G. Taylor's talk. Prof. Livingston produced beautiful photographs giving graphic proof of this hypothesis, with clear pictures of vesicles in the process of fusion with the presynaptic membrane, and of the conical projections on that membrane and the dense fibrillations on the post-synaptic membrane. Both of these structures were demonstrated to be concerned in synaptic transmission. This leads to the question of modification of structure at a synapse on experience, and Livingston showed how there is a different change in curvature of the presynaptic membrane and of the number of vesicles according as an animal is anaesthetised or not, or on the point of death or not. The possibility of correlated increase of frequency of spontaneous quantal release and mean number of quanta released per action pulse was then described by J.G. Taylor, leading to a local theory of memory; various predictions arising from this were raised.

Dr. Fatt introduced the discussion on neurons with a presentation on the nerve impulse and its ionic basis. A set of questions concerning single neuron activity were then raised by J.G. Taylor and possible answers to them formulated by discussion:

(1) How great is the fluctuation in membrane response? This was thought not to be a very important question, and could be neglected in most questions of neuron activity in a neural net.

(2) What are the types of synapse? There is evidence for the following types: axon/dendrite, axon/soma (cell body), axon/axon, dendrite/ dendrite. Spine synapses are not necessarily modifiable, the spin structure very likely being present to achieve linear response of the post-synaptic neuron.

(3) What is the total single neuron response? This can be obtained by taking proper account of the neuron surface geometry and that there is linear summation of membrane activation, with decrement of this activity by 1/e over 2,000 μ for 100 μ diameter dendrites.

(4) How important are the different types of neuron? It is clear that the various types -- stellate, basket, pyramidal, Purkinje, granular, etc. -- all have their separate functions to perform. To discover the latter it will be necessary to combine the analysis using the answer to the previous question and that involving the single neuron as a unit in the neural net itself.

(5) How important are glial cells? Their relevance to peripheral nerve cell information processing was not clear, but some hints of glial cell involvement, particularly in retinal processing, were raised. Glial cells did not appear to be involved at all in central nervous system information processing.

There is not a clear difference between large and small neural nets, as was evidenced from Prof. Braitenberg's talk on the fly retina. This is very regular, and its activity can apparently be reduced to that of a sub-unit of 7 or so nerve cells. Such units appear to have the ability to multiply inputs, as shown by the optomotor response. Clearly such results indicate that a great deal can be gained by investigation of large nets with suitable regularity to reduce them to small nets. Remarks were also made about progress in analysis of small living nerve nets of various types -- cockroach, worm, and so on. These nets were recognized as important testing grounds for theories of single neurons and of nerve nets, and future activity in this area should only increase.

The final area of large living nets, overlapped greatly with lectures given by Braitenberg, Barlow, Precht, Legendy, Livingston and Taylor. The three main nets under discussion were the cerebral cortex the cerebellum and the reticular system. A description of the neuroanatomy of the cerebral cortex by J.G. Taylor and V. Braitenberg was followed by an outline given by Ch. Legendy of his theory of information processing by 'compacta' of neurons. This was followed by Braitenberg's description of his theory of cerebral activity for the simplified model comprising only pyramidal cells. The Marr theory of cerebral activity was then outlined by J.G. Taylor. These theories each contained their attractive features, but were still at too descriptive a stage to be fully falsifiable.

There was also a more theoretical contribution by Nass on his work with Cooper based on single neurons as linear response mechanisms. This model enabled a wide range of properties to be obtained for large nets, though with no clear predictions by which to test the model. A further contribution was by Shaw, who deduced the existence of persistent states (over one second, for example) on the basis of the quantal transmission of information at a synapse. This was a striking result, yet one which again lacked predictive power.

A description of McCulloch's model of the reticular formation as a parallel processing decision computer was given by J.G. Taylor. Yet again this model possessed no predictive power, though it was based on good neuro-anatomical evidence.

There were clearly very outstanding problems, especially of the cerebellum. Its wiring diagram is reasonably clearly understood, yet it has defied explanation. Both here, in the cerebral cortex, and increasingly for other regions of the brain modelling must be performed which will help in clarifying function and structure. This modelling can only be based on the data of neuro-anatomy, and must have predictive power to be useful.

PART V

Artificial Intelligence and Natural Behavior

COMPLEXITY OF AUTOMATA, BRAINS, AND BEHAVIOR

Hans Bremermann
Department of Mathematics
and
Medical Physics
University of California
Berkeley, California

1. Introduction

Organs other than the central nervous system have functions that are fairly well understood. The heart is a pump, the lung a gas exchanger, the kidney a filter. These functions had to remain mysterious to Hippocrates, Aristotle, and Galen whose times lacked an accurate chemistry, but now most of the mysteries are gone. The brain, however, is still poorly understood, as poorly as the other organs before modern physics and chemistry. The brain's functions are not physical or chemical but cybernetic. The brain is a sensing and control organ. The theoretical framework for its functions is computer science, a discipline that has had much less time to mature than physics and chemistry, and consequently is less well developed. Understanding the brain remains a profound challenge.

In physics and chemistry energy is a fundamental notion, it is a key to physiology. In computer science complexity is emerging as a fundamental notion, possibly as fundamental as the concept of energy is to physics and chemistry. There are, however, still several different approaches. Analogously there were initially various concepts of energy, and the universal principle of energy conservation was not recognized until the middle of the 19th century. Hopefully complexity, in time, will be completely clarified. At the moment there are several concepts of complexity, and we will discuss them in the following as well as make a suggestion for a new complexity notion that seems more appropriate for applications in biology and in artificial intelligence. This paper is essentially limited to finite state automata. In the context of a paper that deals with continuous dynamical systems (Bremermann, 1974) the author addresses the question of how some of these notions can be extended to continuous systems.

2. The Phenomenon of Complexity

Cybernetics, automata theory and artificial intelligence provide the theoretical framework for understanding the central nervous system (CNS). The CNS performs tasks that are necessary for the survival of the organism, or to put it less dramatically: it performs functions that play a role in the course of feeding the organism, protecting it from danger and in the preservation of the species. Among the functions performed in the course of these tasks are: pattern recognition (visual, auditory, tactile, olfactory), motor control, and decisions how to react to and act upon the environment. Cybernetic systems have sensors, effectors and a data switching network (the brain). Automata theory is the theoretical framework for switching nets. Artificial intelligence is the search for algorithms that perform some of the tasks that the brain does. (In the absence of detailed knowledge about the brain circuits, artificial intelligence tries to duplicate at least some of its functions.) We will see that in all these disciplines complexity is an important issue.

Behaviorism has tried to describe CNS functions by stimulus-response associations. Much work has been done on rather simple stimulus-response associations, for example "human learning" has been studied with lists of twelve stimulus-response pairs (nonsense syllables). How does such a stimulus-response list compare with, for example, the responses of a player to the opponent's moves in a card game or a board game such as chess? Intuitively, chess is a rather difficult game; the list of possible moves and countermoves is very large, yet according to the theory of games of Von Neumann and Morgenstern (1953) the optimal moves are completely determined by the rules of the game. How should one compare the "complexity" of chess with the complexity of a nonsense-syllable list? (We use the notion "complexity" here in an intuitive way consistent with ordinary language: complexity = degree of complication (whatever that means)). Even simpler than a nonsense-syllable list is a T-maze. Its "one out of two" decision is the simplest behavior imaginable. Yet it is customary to speak of "learning" in all three cases: learning to run a T-maze, learning a nonsense-syllable list, learning to play chess. Are these situations really comparable? Should one not quantify <u>how much</u> is learned? This requires first of all a notion <u>how complex</u> each of these tasks really is.

Similarly, in cybernetics, systems are being studied that have in common that they make use of feedback, yet intuitively they are very different: one being very simple, another very complex. Among the achievements of cybernetic engineering is the autopilot: it guides an airplane along a predetermined path and counteracts perturbing forces: wind, turbulence, drift, load changes within the airplane (people walking around). It has a small number of feedback loops: rudder, elevator, flaps, and it has a simple goal: to keep the airplane pointing in the right direction and at the right altitude. A thermostat is an even simpler system: one control: on-off, one measured value (temperature), and one target value (desired temperature). In contrast other cybernetic tasks and systems appear to be rather complicated. For example, robot projects at MIT, Stanford, and Stanford Research Institute found that the task of coordinating the "muscles" of an artificial arm-hand to do manipulations such as turning a crank or screwing a nut on a bolt are difficult and require sizeable programming and computing efforts. Should one not be able to measure quantitatively the complexity of cybernetic tasks or systems?

Thirdly, consider pattern recognition. The automatic recognition of printed characters is a commercial reality, recognition of handwritten characters has turned out to be difficult and must be considered as still experimental while speech recognition has proved to be unexpectedly difficult, no method (handling a reasonably large vocabulary and different speakers) has yet been developed. A closer examination reveals that the number of different objects in each of the pattern spaces differs vastly: from a few hundred printed characters, to many billions or trillions for handwritten characters and probably even more for speech. (Compare Bremermann, 1971.)

The mere number of different objects, however, is not a good measure of complexity, neither for pattern recognition nor for games (where one might attempt to measure complexity by the number of possible different board positions and moves). Among the games there are the solvable games, where a simple (or at least manageable) trick makes it possible to play a perfect game (in the sense of Von Neumann) even though the number of possible positions and moves is very large. A well-known example is the game of Nim. Here there can be a large number of possible positions and moves, but a little manipulation of three binary numbers tells each player exactly how to move. (See section 8) Intuitively the game of Nim must be considered as simple in comparison with chess or go or checkers, where no simplifying trick (resulting in a perfect game) has ever been found.

Complexity is not limited to games and behavior but it is a notion that is relevant in many places: circuits, chemical mechanisms, evolution. Thus, for example, if one represents brains by a circuit model, one would like to measure how complex the circuit is. Obviously a leech brain is simpler than a vertebrate brain, and the human brain is presumably the most complex of all.

Complexity also plays a role in mathematics. Numerical tasks are accomplished by algorithms. Each algorithm requires a certain number of arithmetic operations (additions, subtractions, multiplications, divisions). Some algorithms require a great number of arithmetic operations, some require less. Should we count the number of arithmetic operations as the complexity of a problem? Sometimes, however, there are several algorithms. For example, for solving systems of linear equations there is Cramer's rule (determinants) as well as Gaussian elimination (subtracting rows till the matrix becomes triangular). Thus if there are several algorithms for a task one should base complexity upon the best algorithm.

The "best algorithm" need not be known. For example, there is the well-known travelling salesman problem. A salesman wants to visit all cities $A_1, A_2, \ldots, A_n$, pass through each city exactly once (starting from and returning to his home base city A), while minimizing his total mileage.

He can solve this problem by examining the mileage of each of the $n!$ permutations of $A_1, A_2, \ldots, A_n$. By Stirling's formula $n! \approx (\frac{n}{e})^n$, hence $n!$ increases very rapidly. Is there an algorithm that increases with n no faster than some polynomial? The answer to the question is not known. The travelling salesman problem is closely related to many other problems, such that if the answer to the previous question is positive for one it would be positive for all. The problem, however, has so far remained unsolved in spite of vigorous investigation. (For details see Karp, 1972.)

Finally there is also the problem of round-off errors: When a system of equations is "ill conditioned," round-off errors can accumulate and render the result invalid. In such a case double (or multiple) precision arithmetic may be called for, rendering the computation more expensive in computer time (and memory space usage).

How then should complexity be defined? There are at present several approaches. We will sketch them in the following, beginning with arithmetic complexity.

3. Arithmetic Complexity

Karp, Strassen, Schönhage, Winograd (and others) count the number of arithmetic operations (additions/subtractions, multiplications/divisions) that are required in order to carry out an algorithm. Sometimes it is possible to show that a certain algorithm is optimal. (According to Strassen (1972), Ostrowski (1954) has been the first to consider this type of problem. He conjectured the result of the following example.)

Example 1. To compute $\sqrt{2}$ Newton's method may be applied to $x^2 - 2 = 0$, giving the iteration $x_{n+1} = \frac{x_n}{2} + \frac{1}{x_n}$. The algorithm doubles the number of significant digits at each iteration, and the algorithm is optimal (if we don't count division by 2) in the sense that any algorithm can do at most this much per two arithmetic operations per iteration (Karp

Example 2. Evaluation of a polynomial:

$$P(x) = a_n x^n + a_{n-1} x^{n-1} + \ldots + a_1 x + a_0 .$$

It has been shown (Pan, 1966) that Horner's method is optimal. In this method a_n is multiplied with x, then a_{n-1} is added, the sum multiplied with x, a_{n-2} added, etc.:

$$P(x) = (\ldots((a_n x + a_{n-1})x + a_{n-2})x + \ldots a_1) + a_0 .$$

Example 3. Systems of linear equations Ax = b may be solved by various methods, most well known are Cramer's rule and Gaussian elimination. Gaussian elimination requires $n^3 + \ldots$ arithmetic operations for an $n \times n$ system. Cramer's rule requires much more if determinants are computed by the method of developing them by the elements of a row or column (an $n \times n$ determinant in this way requires of the order of n! operations). Prior to 1969 it was generally assumed that Gaussian elimination was the best possible algorithm for solving linear equations. Hence Strassen's result that the Gaussian algorithm is not optimal came as a surprise. Strassen (1969) defined inductively an algorithm which requires only $c \cdot n^{2.71}$ arithmetic operations. (Because of the constant c Strassen's algorithm is more efficient than Gauss only for large n.) It is not known whether this algorithm is optimal, there is little reason to believe that it is. On the other hand it is trivial that any algorithm requires at least n^2 arithmetic operations. Thus for solving systems of linear equations there are upper and lower bounds. Future research, hopefully, will bring these bounds closer together, perhaps all the way.

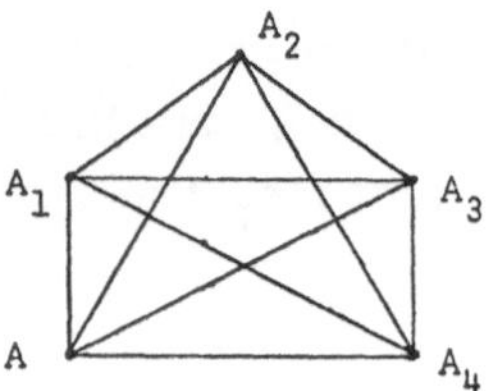

Figure 1

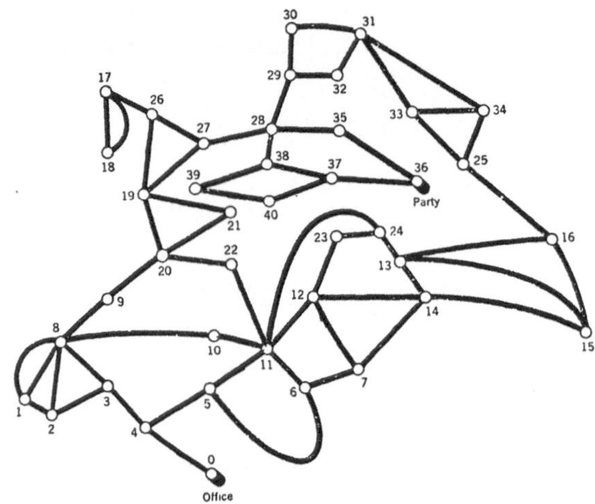

Fig. 2. A map, distance on which is proportional to length of the corresponding portion of the road. The problem--elaborated upon in the next figure--is to determine the shortest path from the office (node 0) to the party (node 36). [Reprinted from Arbib (1972), by permission]

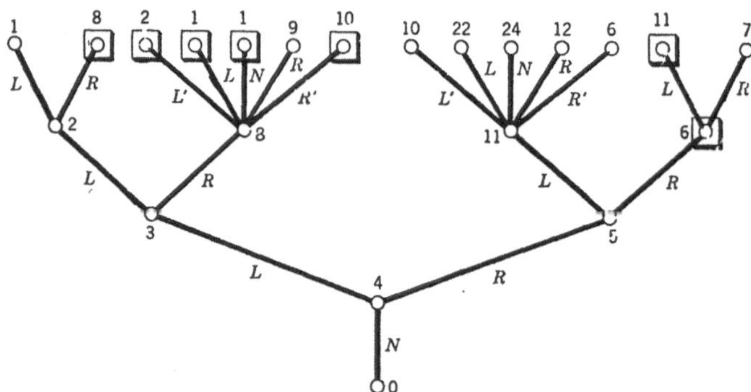

Fig. 3. The first five layers of the "decision tree" for Fig. 2. From each node we grow "branches" to the nodes which can be reached from that node in one "block" using the roads represented in Fig. 2. We "prune" the tree by placing a square around a node as soon as we find a shorter path to a node with the same label. For example, we see two paths to node 8--that via nodes 4 and 3 and the longer one, whose corresponding node in the tree thus has a square about it, via nodes 4, 3, and 2. Note that we refer here to path length on the original map, not to the total branch length of the decision tree, as may be seen by studying the fate of node 6. [Reprinted from Arbib (1972), by permission]

We call arithmetic complexity the minimum number of arithmetic operations required to carry out a given task. (So we have seen: in some cases this number is known, in others upper and lower bounds are known). Should one take the arithmetic complexity as the intrinsic complexity of a given task? The answer is no, because arithmetic operations are special operations and secondly the theory treats them as if they were exact, in other words the theory does not take into account errors that accrue because of truncation. A multiplication can double the number of digits, normally, a computer will discard the least significant digits to keep the word length constant. If the digits are kept, then an algorithm that is optimal with respect to the number of arithmetic operations involved may not be optimal with respect to the number of digit operations.

Also, tasks can be carried out by other than digital operations. For example, the solutions of ordinary differential equations may be computed either by digital computer algorithms (e.g. Runge-Kutta, predictor-corrector) or by means of analog computers. The latter can be very effective for computing solutions of differential equations of moderate accuracy. They are specialized for this task (while digital computers are universal). Analog computers contain non-arithmetic operations (such as integration) and they represent numbers by continuous rather than discrete physical states.

The potential effectiveness of non-arithmetic computation is illustrated by the following example (cf. Fig. 2). Find the shortest path from the office to the party through a net of streets (of an Old World city with irregular and curved streets).

The problem may be described by a problem tree and it may be solved digitally by tree search (cf. Fig. 3).

Alternatively the following analog device also solves the problem in a surprisingly simple way. (We use here the term "analog device" in a general sense, meaning a physical system whose structure bears some resemblance (analogy) to the structure of the abstract problem.) We represent each street by a piece of string whose length corresponds to the length of the street. At intersections the strings are knotted together. The shortest path may be found by picking up the points that represent the office and the party and pulling tight. Those strings that are taut constitute the shortest path!

Analog devices are important. The brain is not strictly a digital computer. For example, in the retina information is transmitted by means of graded potentials rather than nerve impulses and the brain may make use of many other continuous effects. A universal notion of complexity thus should be wide enough to take into account continuous data processing devices. Within the larger class of computing operations--both digital and continuous--arithmetic operations constitute a subclass and thus arithmetic complexity would be an upper bound for a more general complexity measure. Such a more general notion of complexity would involve that we attach a value to all individual operations--in order to be able to compare them. It would also imply comparing digital and continuous operations. We do not solve the latter problem in this paper, though our formalism for discrete operations and some general correspondence principles between discrete and continuous systems show a way how this might be done. We solve the discrete problem by defining a complexity notion for finite state automata. Before doing so, however, we discuss Turing machines, which are an alternative to finite state automata.

4. Turing Machine Complexity

There is a long tradition to model the logical capabilities of the human brain by means of Turing machines. Turing machines were conceived to make precise the concept of algorithm. "There exists an algorithm for a task" can be reformulated: "There exists a Turing machine and a program for it such that the Turing machine will accomplish the task and halt after finitely many operations."

A Turing machine may be described as a finite state automaton (which we define in the following section) which is attached to a tape, a tape drive and a reading and writing head. The tape is unbounded, and this feature gives the Turing machine capabilities that go beyond the capabilities of finite state automata.

There exist Turing machines that are universal, which means that they can compute any function that any other Turing machine (with different internal states and state transitions) can compute. Since a Turing machine computes a function by means of a finite program it would seem reasonable to think that one could minimize over the lengths of all programs that compute a given recursive function. A minimization over program length, however, does not take into account differences in the speed of computation, which may vary from program to program. Optimizing for speed is not possible because of Manuel Blum's (1967) surprising speed-up theorem: there exist recursive functions such that for any program there exist other programs that are exponentially faster (almost everywhere). Another theorem (Paul Young) says: there exist programs that compute the same recursive function but it cannot be proven that they do. Hence a complexity theory of recursive functions runs into difficulties.

In view of these difficulties John Rhodes has suggested abandoning Turing machines and basing complexity theory upon finite state automata. While Turing machines are an interesting mathematical object in their own right, and while they are of great importance for the foundations of mathematics, they are somewhat unrealistic as an actual brain model, or as a model of any earthly computer. The brain is finite (estimates of the number of neurons range from 10^{10} to 10^{12}), its weight is of the order of 10^3g and there are 6×10^{23} atoms or molecules in a mol, hence the brain can contain no more than 6×10^{26} hydrogen atoms and fewer atoms of other kinds. Also, because of quantum effect, the physical quantities that correspond to mathematical operations cannot be measured with infinite accuracy. Hence a bounded, finite computing device is a more realistic model than an unbounded device like a Turing machine. Baer and Martinez (1974) write: "It is an amusing observation that in the development of abstract computational devices, the ultimate device was formulated first (by Turing, 1936), and computationally weaker devices were formulated somewhat later (e.g. the abstract neuronal devices of McCulloch & Pitts (1943) and the nerve nets of Kleene (1956) seem to be the earliest versions of finite automata to appear in the literature)." A rather extensive literature deals with the complexity of computation on Turing machines (compare Hartmanis and Stearns (1965), Hartmanis and Hopcroft (1971)).

A complexity theory based on finite state automata is _not_ a theory of the complexity of recursive functions since such machines cannot compute all recursive functions. Instead it is a complexity theory of _tasks_ and _circuits_. What finite state automata can compute can also be computed by _Turing machines with bounded tapes_. Thus one might try to base the complexity theory of finite state automata upon Turing machines with _bounded_ tapes, taking the _program length_ as a measure of complexity. Again, there is an extensive literature (compare Stearns, Hartmanis & Lewis, 1965; Willis, 1969). Another approach is taken by Schönhage and Strassen (1971). They consider Turing machines with several tapes and count as computational effort the total amount of tape movement and obtain the same kind of results as are valid for "logic nets". All these theories, however, are based upon special kinds of elementary operations. We will in the following consider finite state automata of the most general kind. For these a complexity notion has been defined by Krohn-Rhodes (1968), and Rhodes (1971).

5. Rhodes Complexity

Rhodes has made important contributions to automata theory and he has been the first to develop a complexity theory for finite automata.

A finite state automaton (or machine) is a quintuple

$$M = (X,Y,Q,\lambda,\delta)$$

where

X is a finite set of inputs

Y is a finite set of outputs

Q is a finite set of states

λ: $Q \times X \to Q$ is the next-state function

δ: $Q \times X \to Y$ is the next-output function.

M maps string of input symbols into Y. Strings of input symbols form a semigroup (or monoid) under concatenation (which is associative) with the empty string as identity. This is the free monoid of strings of symbols from X.

The free monoid induces transformations of the state space Q which also form a semigroup. This semigroup is called the semigroup S associated with the machine M. S is finite because Q is finite. Conversely, given a finite semigroup S, a finite state machine may be associated with it that has S as its semigroup (of transformations of Q).

Krohn and Rhodes showed that semigroups have a (loop free) decomposition into simple groups and a special kind of semigroup U_3 which is the semigroup of the flip-flop machine. The decomposition is called the wreath product decomposition. To the wreath product of the groups corresponds the cascade composition of machines:

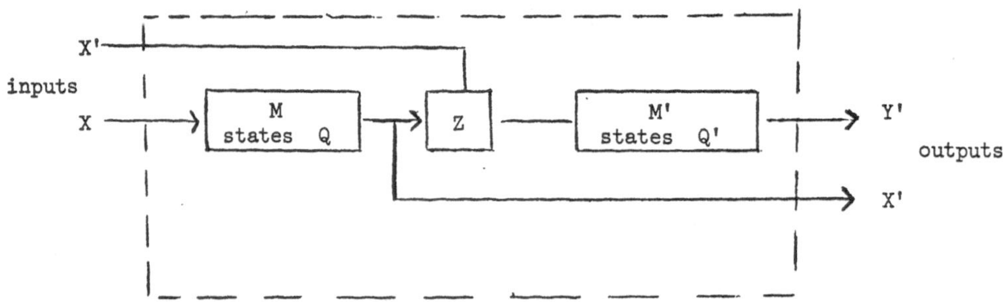

Cascade composition

This diagram describes the cascade composition of machines M and M'. Here Z is a memoryless encoder that converts the outputs of M into symbols suitable as inputs to M'.

Rhodes defines as the complexity of a machine M the minimal number of times simple groups and U_3 alternate in any wreath product decomposition of the semigroup of M.

This complexity function has the following properties:

1) If machine M_1 divides machine M_2 (which means that the semigroup of M divides the semigroup of M_2), then $C(M_1)<C(M_2)$.

2) If M_1 is composed in series with M_2 (see the following diagram), then

$$C(M_1 \text{ series } M_2) \leqslant C(M_1) + C(M_2)$$

3) If M_1 is composed in parallel with M_2, then $C_1(M_1 \text{ parallel } M_2) \leqslant \max(C(M_1), C(M_2))$.

4) C(flip-flop) = 0
C(group machines) = 1.

5) C is the maximum of all integer-valued functions that satisfy conditions 1)-4).

$X_1 \longrightarrow$ [M_1] $\longrightarrow$ [M_2] $\longrightarrow Y_2$

series composition

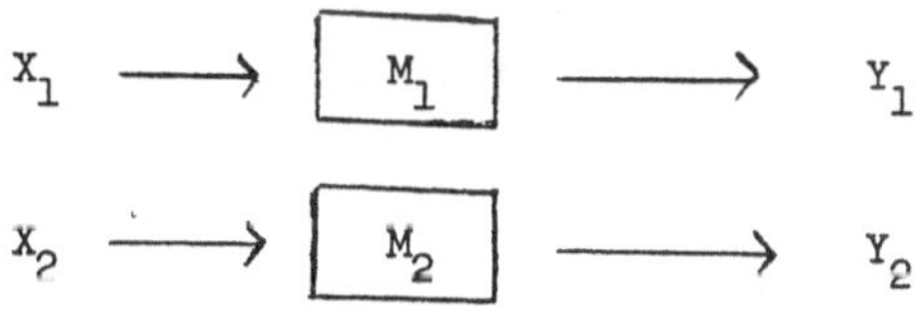

$X_1 \longrightarrow$ [M_1] $\longrightarrow Y_1$

$X_2 \longrightarrow$ [M_2] $\longrightarrow Y_2$

parallel composition

In many ways simple groups are for group theory what prime numbers are for number theory. In fact the finite commutative simple groups are exactly the groups Z^p, p prime, that is the integers mod p, where p is a prime number. The simple non Abelian groups (SNAGS) are partially classified. They play a role in coding theory and other applications and they must be considered as fundamental mathematical objects. The surprising Krohn-Rhodes theorem is a deep result; and Rhodes complexity, which is based upon it, is an interesting aspect of automata theory. It measures the depth of dependence (length of the "chain of command") in a cascade decomposition. A decomposition into group machines and flip-flops is at the same time a decomposition into logically reversible and irreversible machines. We will elaborate this aspect in the following section. (A good exposition of the Krohn-Rhodes theory can be found in Kalman, Falb, and Arbib, 1969.)

6. Logical and Physical Reversibility

Computers are devices that map the abstractly defined states and state transitions of an automaton onto real world physical states and state transitions. Keyes and Landauer (1970) write: "A central question in this area has been, Is there a minimal energy dissipation associated with the nonlinear processes that carry out the typical logic in a computer? The association of an amount of 1/2kT of random thermal energy with a degree of freedom has always made it plausible that the intentional logic signals must be associated with a comparable energy. This was understood by von Neumann, apparently as early as 1949. (Von Neumann, 1966, p. 67 and 1958). Von Neumann indeed suggested that an energy kT is dissipated "...per elementary act of information, that is per elementary decision of a two way alternative and per elementary transmissal of 1 unit of information." A more exact understanding of the reason for the amount of dissipation was provided by one of the authors (Landauer, 1961) who pointed out that general purpose digital computers require the capability of throwing away information, and that it is these information reduction processes which, in turn, require an energy dissipation of the order kT per logical step. In particular, for example, in an operation that requires a bit to be reset to ONE, regardless of its initial value, and in which the data flow initially gives ZERO and ONE equal probabilities, the minimal energy dissipation is $kT\log_e 2$.

Recently Bennett (1973) has introduced the concept of <u>logically reversible computer</u> and he has announced a forthcoming paper in which <u>he will show that any logically reversible computer can be driven with arbitrarily small dissipation of energy</u>. (Bennett, 1973, includes a preliminary argument.) Bennett shows not only that logically reversible computers exist, but that any computation on a Turing machine can be emulated by a reversible Turing machine that saves a copy of the input. Bennett's surprising result answers a problem raised by Landauer. Bennett writes: "Landauer (1961; Keyes and Laundauer, 1970) has posed the question whether logical irreversibility is an avoidable feature of useful computers, arguing that it is and has demonstrated the physical and philosophical importance of this question by showing that whenever a physical computer throws away information about its previous state it must generate a corresponding amount of entropy."

Bennett (1972, p. 525) defines logical irreversibility as follows: "The usual digital computer program frequently performs operations that seem to throw away information about the computer's history, leaving the machine in a state whose immediate predecessor is ambiguous. Such operations include erasure or overwriting of data, and entry into a portion of the program addressed by several different transfer functions. In other words, the typical computer is logically irreversible--its transition function (the partial function that maps each whole-machine state into its successor, if the state has a successor) lacks a single-valued inverse."

Bennett's machines are Turing machines that make autonomous state transitions: given the input (which may be considered as starting the machine in a specific initial state), the machine keeps making state transitions till it stops (or it keeps on going forever). In other words, Bennett's machines are off-line, they receive their input, then make state transitions, possibly for a long time, till they get their result and halt. Rhodes' machines are on-line, they may receive continuing strings of input and produce strings of output. (A real brain has to do this rather fast when it performs cybernetic functions.) Finite state automata contain Bennett's case as a special case: the empty input can be considered as an element of X and can have associated with it state transitions. We generalize Bennett's definition of logical reversibility to _state-output_ machines, that is machines where the output equals the state.

Definition: A state-output machine is called _logically reversible_ exactly if for every $x \epsilon X$ the next state transfer function $\lambda(x,q)$ is a one-to-one map of Q onto Q. A machine that is not logically reversible is called _logically irreversible_.

Proposition: _A group machine is logically reversible_.

Let x_o be any element in X. Suppose that the associated transform T_{x_o} on Q maps two states, say q_i and q_j, into the same state, that is $T_{x_o}q_i = T_{x_o}q_j$. Since the transformation form a group, there exists an inverse transform S (associated with some input string) such that ST_{x_o} is the identity. Now $ST_{x_o}q_i = ST_{x_o}q_j$. Hence, since ST_{x_o} is the identity, $q_i = q_j$. Hence $T_{x_o}q = \lambda(x_o,q)$ is one-one, hence the machine is logically reversible.

Note that in a group machine the effect of any input string can be undone: for each input string there exists an inverse such that the combined string induces the identity transformation on Q. That means that by applying the inverse string the original state can be recovered, whatever it has been. In the flip-flop machine this is not the case.

Proposition: The flip-flop machine is logically irreversible.

The flip-flop machine has a state space that consists of two states q_0 and q_1. Its semigroup, U_3, consists of three transformations: E which leaves the state the same, T_0 which sets it to q_0, irrespective of what it has been before and T_1 which sets the state to q_1, irrespective of the previous state. Its action upon Q is described by the following table:

U_3	q_0	q_1
E	q_0	q_1
T_0	q_0	q_0
T_1	q_1	q_1

This is a semigroup but not a group. For example,

$$T_0T = E$$

has no solution. If the initial state is q_0, then in order to have $TT_0q_0 = Eq_0 = q_0$ we must have $T = T_0$ or $T = E$. However, in order to have $TT_0q_0 = Eq_1 = q_1$ we must have $T = T_1$. These two conditions are contradictory. Hence there are strings of input that cannot be undone. In other words: knowledge of the initial state has been irreversibly lost.

The information loss in a flip-flop amounts to one bit. Thus if we donote by I the number of flip-flops in a machine decomposition, then I is a measure of the logical irreversability of the machine. In a physical realization of a machine IkTlog2 is the minimal entropy increase.

Unfortunately the Rhodes complexity does not measure the number of flip-flops, I, except, perhaps, in special cases. The Krohn-Rhodes theory upon which it is based, however, is important since it implies that any finite state machine is decomposable into purely reversible machines, the group machines, and into the most elementary irreversible machine, the flip-flop.

This author has previously pointed out (Bremermann, 1962, 1967a, 1967b) that computing processes may be dissipative and that, since there are limits to the amount of heat that can be dissipated, there are thermodynamics barriers to computation. This author, however, was not too clear about how the number of bits and kT's were to be identified. With a hedge (1967a, pp. 18-19) he tended to a one-for-one identification. Conversations with Charles Bennett in August, 1972 convinced him that there are "reversible computers" that can make many state transitions while dissipating negligible amounts of heat. Only now (March, 1974) it dawned on him that Rhodes' complexity is based upon a decomposition into logically reversible and irreversible machines.

Incidentally, this author has argued (1962, 1967a, 1967b) that there exist not only thermodynamic barriers to computation but quantum barriers as well which affect the speed of computation. His arguments have received little attention with one exception: Ross Ashby has referred to them in many of his later writings. Recently, René Thom has made, independently, an equivalent assertion (Thom, 1972, p. 143). This subject matter, however, would go beyond the frame of this paper.

Returning to the subject of complexity we may ask, Is Rhodes' complexity a generalization of the arithmetic complexity that we considered before? Obviously it is not. It is not concerned with the sheer number of elementary operations or with the sheer amount of circuitry. There ought to be such a notion, analogous to the concept of arithmetic complexity, only based on a much larger class of elementary operations. In the following we try to define such a notion. We let ourselves be guided by the problem of defining the complexity of games and tables.

Rhodes had to resort to an artifice in defining the complexity of games. (Games are an important part of artificial intelligence. They may also be considered as models of behavior. "One does not have to be a Levi-Strauss to realize that chess is a model of ritualized warfare" (Rhodes, 1971)). For a two person game with complete information there is, according to von Neumann, an optimal move for each board position. An automaton that plays a perfect game can simply consist of a list that has board positions as entries and the associated optimal move as output. Such a table can take on horrendous proportions. Chess, for example has $> 2^{142}$ possible board positions.* (This compares with only 10^{55}g of matter in the entire universe, according to Eddington.) Yet a table is simply a one-state automaton, its semigroup is trivial and it has consequently complexity zero. To get a reasonable theory Rhodes considers

*See note at end of paper.

the complexities of strategies rather than move tables and in this way he obtains numbers that are in agreement with our intuitive understanding of the difficulty of games.

It seems reasonable to search for an alternate measure of complexity that measures the complexity of a table in a reasonable way and that does not assign to all tables complexity zero, no matter how many entries there are. It turns out that a notion of complexity for function tables has been formulated and investigated by Vitushkin (1961). Vitushkin's theory is based upon techniques that were developed by Kolmogorov for the solution of Hilbert's 13th problem. Actually Vitushkin's notion is trivial for function defined on <u>finite</u> sets. It is designed for the approximate tabulation of functions defined on <u>infinite</u> sets. Vitushkin and Kolmogorov's theory, however, contain important ideas that are applicable to function tables of finite set functions and to finite state automata.

7. Vitushkin's Complexity of Tabulation

In 1959 Vitushkin published (in Russian) a book "Complexity of Tabulation Problems" which appeared in 1961 in English under the title, "Theory of Transmission and Processing of Information" (a title which is much less descriptive of its contents than the original title). In his book Vitushkin elaborates and extends Kolmogorov's notion of ε-entropy which Kolmogorov had developed as a tool for his solution of Hilbert's 13th problem.

Hilbert's 13th problem is concerned with the question whether functions of several variables can be expressed by means of functions of fewer variables and through substitution and addition operations. Hilbert conjectured that there exist analytic functions of three variables that cannot be so expressed by means of analytic functions of two variables. Later Hilbert proved this conjecture and he also conjectured that the same would be true more generally. This is not the case. In 1956 Kolmogorov (1956) and Arnold proved that any continuous function of n variables can be represented in the form of the superposition of continuous functions of a single variable: for any $n>1$ and $s>0$ there exist s times differentiable functions of n variables that cannot be represented as superpositions of p times differentiable functions of m variables if $\frac{m}{p} < \frac{n}{s}$, while such a representation is always possible if $\frac{m}{p} \geq \frac{n}{s}$. Also, any continuous function of n variables can be expressed in terms of continuous functions of one variable (Kolmogorov, 1956; Vitushkin, 1961; Kolmogorov-Tichovirov, 1960).

The essential technique used by Kolmogorov is the volume of a function table that represents a function up to an accuracy of ε (in sup norm). The volume is measured in terms of the number of bits (binary digits) that are contained in the table. The tables are organized as economically as possible, in other words no more bits are placed in the table than is absolutely necessary. Kolmogorov then obtains asymptotic formulas for the minimum volume (ε-entropy) for $\varepsilon \to 0$. Different function spaces (analytic functions, continuous functions, s times differentiable functions of n variables, etc.) all have their characteristic asymptotic ε-entropy behavior.

Vitushkin expands upon Kolmogorov and he speaks of the complexity of a table (Vitushkin, p. xiii): "The complexity of a table is characterized by two factors: (a) its volume (the total number of binary digits required to write down all the parameters of the table) (b) the complexity of the decoding algorithm of the table."

The decoding algorithm is a procedure that constructs from the numbers recorded in the table the actual function (or rather another function that differs from the desired function by no more than ε). Vitushkin restricts his decoding algorithms, however, to polynomials of the recorded parameters and to a few other elementary operations and the complexity of his decoding algorithms does not contribute to his asymptotic formulas. Vitushkin's results are all for functions whose domains are infinite sets. The basic idea, however, to measure the complexity of a function by means of the volume of a tabulation and the complexity of a decoding algorithm can be carried to functions on finite sets. There arises, however, the following difficulty: In the finite case we cannot avoid defining the complexity of the decoding algorithm and there is no sense in restricting the nature of this algorithm. Thus in order to define the complexity of a function we must define the complexity of algorithm. A function, however, is an algorithm of a special kind (we elaborate this remark in the following section). Hence we seem to have replaced our problem only by a more difficult one. In the following section we present our own definition of complexity which resolves this difficulty.

8. Tabular Complexity of Finite State Automata

A finite state automaton is specified by two finite functions: the next-state function $\lambda: X \times Q \to Q$ and the next-output function $\delta: X \times Q \to Y$. If Q consists of only one state, then λ becomes trivial and δ is a function of the inputs only. Let us consider initially this special case. If X has n elements and Y has k elements, then δ can be represented by a table that contains $n \lceil \log k \rceil$ bits ($\lceil \log k \rceil$ denotes the smallest integer larger than log k). This number, however, is not a good measure of the complexity of δ since there may exist ways to compute δ.

To illustrate the importance of this possibility consider a two person game with complete information and no chance moves. As already mentioned, there exists for each board position (or more generally for each state of the game) one (or several) optimal moves. A tabulation of the optimal moves is all that is needed to play a perfect game. Such a table, however, can be horrendous. We mentioned already that chess has $> 2^{142}$ board positions, and hence the table would have to have $> 2^{142}$ entries. No algorithm is known to compute the chess function except a complete search through all possible games. Such a complete search is so enormous that it transcends the capabilities of any physical computer (Bremermann, 1967b). For some games, however, the moves can be computed by a simple algorithm. Such games are known as solvable games. A well known example of a solvable game is the game of Nim.

Nim actually refers to a class of games. There are three piles of "matches," consisting initially of n_1, n_2, n_3 matches. Each player removes in turn from one to all matches of a pile. The player who takes the last match wins. The number of states of the game is $n_1 \times n_2 \times n_3$ and each move can be characterized by the number of matches taken and by the pile number from which they are taken. Hence $\max_{i=1,3} \{\lceil \log n_i \rceil\} + 2$ bits will suffice to describe the move. Thus for example if $n_1 = n_2 = n_3 = 1000$, then a complete tabulation would require about $10^9 (\lceil \log 1000 \rceil + 2) \approx 10^{10}$ bits. The optimal move, however, can be computed by a little algorithm: represent the remaining number of matches in each pile by a binary number.

Arrange the numbers in a scheme where the 2^0 digits form a column, the 2^1 digits another column, etc:

$$\begin{array}{lr} n_1: & 101101 \\ n_2: & 10111 \\ n_3: & \underline{10} \\ & 000eee \end{array}$$

When the number of ones in each column is even, then the scheme is called even, otherwise odd. (The scheme in the example is odd.) When a player is faced with an odd scheme, he can make it even. (In our example he removes 11000 matches from pile one, leaving 10101 matches). However, when a player is faced with an even scheme, any move will make it odd. The last (winning) move leads to all zeros, which is an even scheme. Hence each player moves such as to confront the opponent with an even scheme--if possible. By this simple rule optimal moves can be computed--in fact a short computer program can be written that implements this rule. Obviously the complexity of the Nim game should not be measured by the volume of its table of optimal moves but by the complexity of the "best" algorithm that computes it.

Any finite algorithm can be implemented by a (loop free) finite state automaton. Such an automaton is characterized by the next-state function λ and the next output function δ:

λ	$q_1, q_2 \cdots q_p$
x_1	
x_2	next state q
$\vdots$	
x_n	

δ	$q_1, q_2 \cdots q_p$
x_1	
x_2	next output y
$\vdots$	
x_n	

The table for λ has a volume $n\,p\lceil \log p \rceil$, and the table for δ has a volume $n\,p\lceil \log m \rceil$, where n is the number of input elements, m the number of output elements and p the number of states. As we have demonstrated for the case of game tables (where $p = 1$) we should not take the table volumes as the complexity of the automaton, instead we should allow for the possibility that δ and λ can be computed by simpler algorithms. Such simpler algorithms can be implemented by loop free finite state automata which may be composed of subautomata. (The cascade, series and parallel composition of section 5 were examples of automata compositions.)

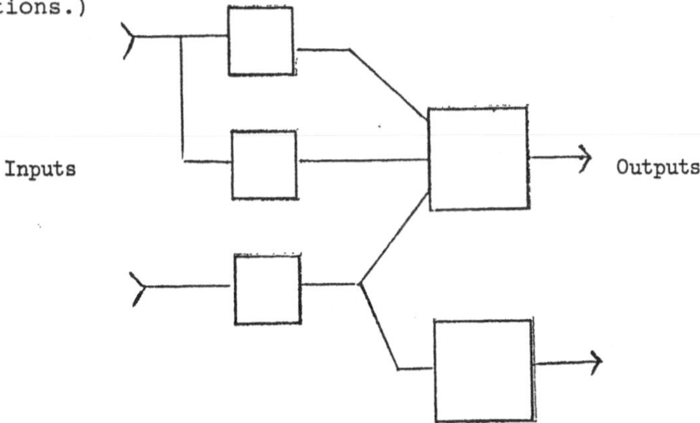

Example of an automaton composed of subautomata

Each subautomaton is characterized by its next-state and transfer functions. In addition there is the wiring diagram. The wiring can be described by a "wiring function" that associates with each output either an ultimate output of the automaton or one or several subautomata inputs. We define as the volume of a decomposition the sum of the volumes of all tables of the transfer functions of the subautomata and of the "wiring function."

We further allow decomposition of the subautomata. A decomposition of a decomposition into further subautomata is also a decomposition of the original automaton. Thus we might attempt to define the complexity of the original automaton as the minimum of the volumes of all decompositions. Such a definition, however, leaves out the possibility that the wiring functions could in turn be computed by automat. To be consistent we must allow wiring functions to be replaced by wiring automata, which in turn may be decomposed. We call such decompositions, where the wiring functions are also decomposed <u>total decompositions</u>.

<u>Definition</u>: <u>The tabular complexity of a finite state automaton is the minimum of the volumes of all total decompositions</u>.

From the definition follows immediately:

<u>Proposition</u>: <u>The tabular complexity of a finite state automaton is a positive integer less or equal to the sum of the volumes of the tabulations of its undecomposed transfer functions</u>.

In many cases the exact value of the tabular complexity of a finite state automaton is not known. However, each decomposition of the automaton gives an upper bound.

<u>Definition</u>: <u>The tabular complexity of a game is the minimum of the tabular complexities of all finite state automata that play a perfect game (in the sense of von Neumann)</u>.

<u>Example</u>: Tabular complexity of binary addition and multiplication. Binary addition is an operation that associates with each pair of n bit binary numbers an $n + 1$ bit number. The operation can thus be represented by a table that has 2^{2n} entries and $n + 1$ bits per entry, that is a volume of $(n+1)2^{2n}$ bits. For binary multiplication the result is a $2n$ bit number, hence its table has a volume of $n \times 2^{2n+1}$ bits. The actual complexities are, of course, much lower. Both operations can be reduced to bit-wise operations. Addition can be achieved by the circuit on the following page. Here $a_1,\dots,a_n$ are the bits of the first summand, $b_1,\dots,b_n$ the bits of the second summand. The adding units have two bit outputs, one of which is a <u>carry</u>, except for the last

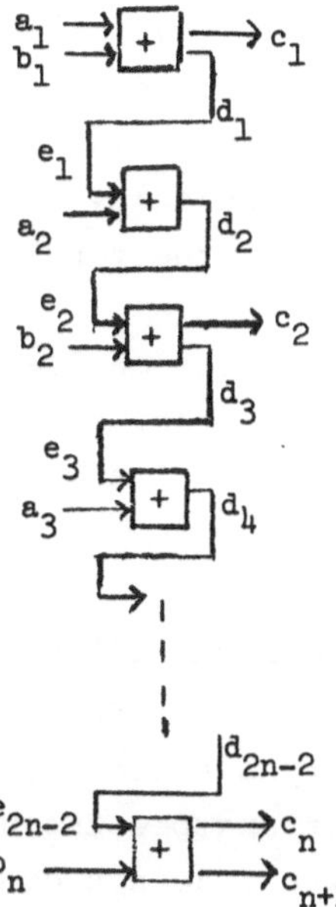

unit where the carry bit becomes output.

Thus $2n - 1$ elementary binary adding units are required. Each can be represented by a 4 bit table (taking into account symmetry). The wiring diagram is described by the wiring function:

c_1	out
c_2	out
$\vdots$	
c_n	out
c_{n+1}	out

d_1	e_1
d_2	e_2
$\vdots$	
d_{2n-2}	e_{2n-2}

Of course this function could be more economically described by the rule: $e_i = d_i$.

Similarly binary multiplication can be reduced to bit-wise operations. In the standard method which corresponds to the scheme:

```
1011 X 1101
-----------
  00 0 1011
  00 0 0000
  01 0 1100      n
  10 1 1000
-----------
1000   1111

    2n
```

These are of the order of n^2 bitwise binary arithmetic operations. Recently Schönhage and Strassen (1971) have described a different method that requires (asymptotically) only c n logn log logn arithmetic operations. Earlier, Karacuba and Ofman (1962) found the estimate $c \cdot n^{1.50}$. The exact bitwise arithmetic complexity of binary addition and multiplication is unknown. Analogously, the tabular complexity of these operations is unknown.

Conclusion: There remains the challange of computing the tabular complexity for all sorts of important tasks (that can be implemented by finite state automata). Each actual algorithm will contribute to an estimate of the complexity.

9. Applications

The brain can be modelled in different ways: as a Turing machine, as a finite state automton or as a continuous device. There are interesting continuous models, for example the hologram theory of memory. Most continuous models, however, do not come to grips with the computer aspects of the brain: the brain is a meat computer. A mature brain is full of "hard-wired" subroutines and it is this algorithmic structure that must be understood. The experience in artificial intelligence has taught us that many of the tasks that the brain performs require highly sophisticated algorithms. In fact, some of the functions of the brain, especially in the pattern and speech recognition area, have turned out to be impossible to duplicate by any means so far. The functions that the brain performs are _cybernetic_. A brain must render its services _in real time and for a purpose_. This is a point emphasized by John Rhodes (1972). Rhodes writes (p. 126): "...I subconsciously owe very much in the following to that excellent early book, _Design for a Brain_ by W. Ross Ashby (1952) which I recommend to the reader for a valuable viewpoint. It is that viewpoint coupled with powerful algebraic techniques which we present in the following." (Ross Ashby, incidentally, shared the author's concern about the existence of fundamental limits to computation.) In contrast, the study of Turing machines does not spring from cybernetics. Turing machines were first introduced in the context of mathematical logic to clarify certain problems in the foundations of mathematics. They are half way between pure, Platonic abstractions and real, earthly computers.

For earthly computers, and especially for biological computers like the brain, questions of the amount of effort required and of speed of computation are all important. Thus we have introduced the _volume_ of a circuit or machine and since any actual computer program can be converted into an automaton, this notion can also be applied to computer programs. While it may be tedious to determine the volume of a computer program by hand, this determination can be carried out by the computer itself. In this way it becomes possible to compare different computer programs and different machines. (Running times are, of course, roughly related but are a less accurate measure of program size). We can then compare the volume of a program that carries out a task with the intrinsic difficulty of a task for which the tabular complexity is a measure. We can then determine how efficient a particular program or circuit is. Thus we can ask questions like: "How efficient is this nervous system in doing its (cybernetic) tasks?" "Has the evolution of the nervous system produced circuits that are about as efficient as they can be?" These are, of course, long term projects, little is known at present.

An interesting application of Rhodes complexity is the following: J. Yost (1973) has applied Rhodes complexity theory to the nerve impulse after modelling the various phases of the impulse and various inputs (sub-threshold, superthreshold, hyperpolarizing, etc.) by a finite state machine. The model is a 10 state machine and it turns out to have complexity 9. If we apply now the propositions of section 6 which relate Rhodes complexity to physical implementations and estimating that the number of flip-flops also is about 9, we conclude each nerve impulse must dissipate heat in the amount of $9kT \approx 10kT$ per impulse. Assuming that there are 10^{12} neurons firing in the average $\approx$ once per second we have a dissipation of 10^{13} kT per second. At a temperature of $37^{\circ}C = 310^{\circ}K$ we have $kT = 1.38 \times 310 \times 10^{-16} erg \approx 4.3 \times 10^{-14}$. Hence the brain would have to dissipate .43 erg per second; say about one erg per second = 10^{-7} watts. The actual dissipations is of the order of 10 watts (von Neumann, 1958).

This discrepancy of a factor of 10^{8} is puzzling. Much of it can be explained, however. First of all Yost's 10 state model is only an approximation of the exact nerve impulse. More states are needed for closer approximation. Yost (1973) has shown: <u>For an n state model of the nerve impulse the Rhodes complexity remains maximal, having the value $n - 1$</u>. Thus if more states are used in the model to get a close approximation, perhaps a factor ten can be explained. Secondly, the average firing rate could be higher: ten firings per second would explain another factor of ten. That still leaves a factor of 10^{6}.

This factor of 10^{6} can be explained if we consider the nerve impulse as a local event. In fact, neurons have numerous dendrites and synapses. For example, Purkinje cells in the cerebellum make as many as 200,000 contacts with other fibres. Arbib (1972) estimates that the average neuron makes as many as 10^4 synaptic connections with other neurons. Each synapse is a unit that has to override thermal noise. If we thus treat each synapse as a unit that is capable of its own action potential, then we explain another factor of 10^4. This leaves only a factor of 100. This factor could be easily explained as a safety factor, or as a factor that results if segments of each axon and each dendrite must be counted as individual units.

In summary: The mystery of the presumed inefficiency of the brain disappears if we apply Yost's theory and our propositions of section 6 not to entire neurons as units, but to subunits of neurons: synapses and unit segments of axons and dendrites. Thus it appears that the brain, as a neuronal network, is thermodynamically about as efficient as it can be.

Besides applications to the brain, finite state automata have many other uses as models, for example as models of behavior. In some cases behavior is merely a list of stimulus-response associations. The volume of a simple n stimulus m response list is $n\lceil \log m \rceil$. If $m = n$ and if there are no repeated responses, then the number of possible associations is $n!$. The volume of such a table is $\log n! \approx n \log(n/e)$, by Stirling's formula. If there is no simpler way to compute the association, then this is the complexity.

Economic organizations can also be modelled as finite state machines and notions like Rhodes complexity can be applied. (This theme has been elaborated recently by Göttinger (1974)).

Rhodes (1971) has suggested numerous applications of finite state automata theory and of his complexity notion. In particular he has elaborated an application to bacterial metabolism and applications to psychology. In summary: Finite state automata are a very versatile tool for modelling numerous phenomena of the life and social sciences. In any application the notions of complexity and reversibility are bound to be fundamental.

10. Genetic Complexity

Besides complexity and Rhodes complexity a brain has another complexity aspect: it is organized by the genes, and this number is limited (of the order of a million or less in mammals) and the genes have many other tasks besides organizing the brain.

The developing organism may be considered as something analogous to a compiler. A compiler receives a source program written in FORTRAN or Algol or some other programming language. From this program it constructs the actual machine language program. It is conceivable that the source program is simpler than the machine program. For example, a minimal finite state automaton may contain repetitive subautomata. In fact our binary adding and multiplying circuits contained repeated subautomata. If they were minimal circuits the subautomata would have to be counted as often as they occur since our complexity is the sum of the volumes of all tables of all subautomata. A compiler, however, could generate a repeated subautomaton from a simpler instruction. This question is related to the complexity of tesselation automata (cf. Baer and Martinez (1973)) and to von Neumann's self-reproducing automata (von Neumann-Burks (1966)).

The brain is composed of billions of neurons that are more or less alike. There are only a few classes of basically different neurons. Hence it is conceivable that the complexity of the brain is larger than the complexity of the genes. The latter may be considered simply as a table. These are of the order of 4×10^9 nucleotide pairs in human DNA. This would give an upper bound of $8 \times 10^9 \approx 10^{10}$ for the genetic table. (If the effective entries are not nucleotides but genes, then we must consider the number of genes multiplied with the logarithms of their number of alleles. The volume of such gene tables are smaller than 10^{10} (cf. Bremermann (1963)).

In an earlier paper (Bremermann, 1963) this author has attempted to find limits for the complexity of stimulus-response behavior that can be purely genetically determined. In view of the fact that the tabular complexity of the brain--as defined in this article--can be greater than the genetic volume, there are complications. However, it seems that the conclusions of the earlier paper (which is based on Shannon's concept of information-entropy) remain valid if the stimulus-response association is random in the following sense: the stimulus-response table has maximal tabular complexity, that means that its tabular complexity equals its volume. In other words, there is no way to compute the stimulus-response function by means of an algorithm that is simpler than listing the explicit table.

Studies of the complexity of behavior should be of considerable interest. For example one could quantify how much behavior is innate and how much behavior is learned. Also, it would be possible to compare different kinds of behavior quantitatively. Perhaps some new light could be shed on the nature-nurture controversy.

Our definition of a random association as one of maximum complexity is analogous to a definition by Solomonoff (1964) and Kolmogorov (1965 and 1968), except that their definition is based on Turing machine complexity. (Turing machines permit a definition of randomness for infinite sequences as well as finite ones.) Solomonoff and Kolmogorov's definitions initially led to complications that were later resolved in different ways by Kolmogorov (1968) and by Willis (1969). Our definition is very simple. Unfortunately, in individual cases randomness may be hard to prove. Note that our definition of randomness as maximal complexity lends itself to defining degrees of randomness. In other words, it can be used to make randomness a fuzzy notion in the sense of Zadeh (1965).

It would be interesting to pursue this subject further, however, it seems that this paper already exceeds optimum length.

Acknowledgements

The author gratefully acknowledges helpful suggestions and criticisms from John Rhodes, Roy Solomonoff, Laszlo Csanky, David Krumme, Robert Miller, and Stan Zietz.

Bibliography

Arbib, M.A.: The Metaphorical Brain, Wiley-Interscience, New York, 1972.

Ashby, W.R.: Design for a Brain, Chapman and Hall, London, 1952.

Baer, R.M. and Martinez, H.M.: Automata and biology, to appear in Ann. Rev. Of Biophy. a. Bioeng., Ann. Rev., Inc., Palo Alto, California.

Bennet, C.H.: Logical reversibility of computation, IBM Journ. Research and Developm. 17, 525-532 (1973).

Blum, M.: A machine independent theory of the complexity of recursive functions, Journ. Assoc. Comput. Mach. 14 322-336 (1967).

Bremermann, H.J.: Limitations on data processing arising from quantum theory, Part I. in, Optimization through evolution and recombination. In: Self-Organizing Systems 1962, ed. by M.C. Yovits, G.T. Jacobi, and G.D. Goldstein, pp. 93-106. Spartan Books, Washington, D.C., 1962.

Bremermann, H.J.: Limits of genetic control, IEEE Transact. Mil. Electr. Vol. MIL-7, 200-205 (1963).

Bremermann, H.J.: Quantitative aspects of goal-seeking self-organizing systems. In Progress in Theoretical Biology, Vol. 1, ed. by M. Snell, pp. 59-77. Academic Press, New York, 1967.

Bremermann, H.J.: Quantum noise and information, Proceed. Fifth Berkeley Sympos. Math. Stat. and Prob. IV, 15-20, University of California Press, Berkeley, Calif. (1967b).

Bremermann, H.J.: What mathematics can and cannot do for pattern recognition. In: Pattern Recognition in Biological and Technical Systems, ed. by O.J. Grüsser, pp. 31-45. Springer-Verlag, Berlin, 1971.

Bremermann, H.J.: Manifolds, models, automata and computability. To appear in Journal of Math. Biology (1974).

Gottinger, H.W.: Computable organizations, representation by sequential machine theory, Memorandum ERL-M426, March 1974, Electronics Research Lab, University of California, Berkeley, Calif. (1974).

Hartmanis, J. and Hopcroft, J.E.: An overview of the theor of computational complexity, Journal Assoc. Compt. Mach. 18, 444-475 (1971).

Hartmanis, J. and Stearns, R.E.: On the computational complexity of algorithms, Trans. Amer. Math. Soc. 117, 285-306 (1965).

Kalman, R.E., Falb, P.L. and Arbib, M.A.: Topics in Mathematical Systems Theory, McGraw-Hill, New York, 1969.

Karp, R.M.: Lecture Notes, Dept. of Computer Science, University of California, Berkeley, Calif. (1972).

Keracuba, A. and Ofman, J.: Multiplication of many-digit numbers by computer automata (Russian), Dokl. Adad. Nauk, SSSR 145, 293-294 (1962).

Keyes, R.W. and Landauer, R.: Minimal energy dissipation in logic, IBM Journ. Res. and Developm. 14, 152-157 (1970).

Kleene, S.C.: Representations of events in nerve nets and finite automata. In: Automata Studies, ed. by C.E. Shannon and J. McCarthy. Annals of Math. Studies No. 34, Princeton University Press, Princeton, N.J. (1956).

Kolmogorov, A.N.: On the representation of continuous functions of several variables by superposition of continuous functions of a smaller number of variables, Dokl. Akad. Nauk SSSR 108, p. 2 (1956).

Kolmogorov, A.N.: Three approaches for defining the concept of information quantity, Information Transmission 1, 3-11 (1965).

Kolmogorov, A.N.: The logical basis for information theory and probability theory, IEEE Transact. Information Theory, Vol. IT-14, 662-664 (1968).

Kolmogorov, A.N. and Tichomirov, W.M.: ε-Entropie und ε-Kapazität von Mengen in Funktionalräumen, Arbeiten zur Informationstheorie III, VEB Deutscher Verlag der Wissenschaften, Berlin (1960).

Krohn, K. and Rhodes, J.L.: Complexity of finite semigroups, Annals of Math., 88, 128-160 (1968).

Landauer, R.: Irreversibility and heat generation in the computing process, IBM Journ. Research and Developm. 5, 183-191 (1961).

McCulloch, W.S. and Pitts, W.: A logical calculus of the ideas immanent in nervous activity, Bull. Math. Biophysics, 5, 115-133 (1943).

Ostrowski, A.M.: On two problems in abstract algebra connected with Horner's rule, Studies in Mathematics and Mechanics, pp. 40-48, Academic Press, New York (1954).

Pan, V. Ya.: Methods of computing values of polynomials, Russian Math. Surveys, 21, 105-136 (1966).

Rhodes, J.L.: Application of automata theory and algebra via the mathematical theory of complexity, Notes, Dept. Math., University of California, Berkeley, Calif. (1971).

Schönhage, A. and Strassen, V.: Schnelle Multiplikation grosser Zahlen, Computing, 7, 281-292 (1971).

Solomonoff, R.J.: A formal theory of inductive inference, Pt. I, Inform. Control, 7, 1-22, Pt. II, ibid., 224-254 (1964).

Stearns, R.E., Hartmanis, J. and Lewis, P.M.: Hierarchies of memory limited computations, IEEE Conf. Rec. on Switching Circuit Theory and Logical Design, IEEE Pub. 16C13, 179-190 (1965).

Strassen, V.: Gaussian elimination is not optimal, Numerische Mathematik, 13, 354-356 (1969).

Strassen, V.: Evaluation of rational functions. In: Complexity of Computer Computations, ed. by R.E. Miller, J.W. Thatcher and J.D. Bohlinger, pp. 1-10. Plenum Press, New York, 1972.

Thom, R.: Stabilité Structurelle et Morphogénèse, Benjamin, Reading, Mass., 1972.

Turing, A.M.: On computable numbers, with an application to the Entscheidungsproblem, Proceed. London Math. Soc. Ser. 2, Vol. 43, 230-265 (1936).

Vitushkin, A.G.: Theory of the Transmission and Processing of Information, Pergamon Press, New York, 1961.

Von Neumann, J.: The Computer and the Brain, Yale University Press, New Haven, Conn, 1958.

Von Neumann, J. and Burks, A.W.: Theory of Self-reproducing Automata, University of Illinois Press, Urbana, Ill., 1966.

Von Neumann, J. and Morgenstern, O.: Theory of Games and Economic Behavior, Princeton Press, Princeton, N.J., 1953.

Willis, D.G.: Computational complexity and probability constructions, Journ. Assoc. Comp. Mach. 17, 241-259 (1970).

Yost, J.L.: Modelling a Nerve Cell Membrane as a Finite-State Machine, M.A. thesis, Dept. Math., University of California, Berkeley, Calif. (1973).

Zadeh, L.A.: Outline of a new approach to the analysis of complex systems and decision processes, IEEEE Transact. Systems, Man, Cybernetics, Vol. SMC-3, 28-44 (1973).

Note: This is actually the number of ways 32 pieces can be placed on the board, taking into consideration that identical pieces should not be counted differently.

MODEL NEURON BASED ON THE JOSEPHSON EFFECT

Bruce T. Ulrich
Department of Physics and Department of Astronomy
University of Texas, Austin, Texas 78712

A superconducting Josephson junction may be used to model the behavior of a biological neuron, but on a much faster time scale. I first list the properties of the biological neuron to be modeled, then introduce the Josephson effect, and finally describe how the properties of the Josephson junction model neuron correspond to the properties of the biological neuron. An important advantage of the Josephson junction model neuron compared to the biological neuron is its speed. A Josephson junction model neuron functions on a 10^{-9} to 10^{-12} second time scale, whereas the biological neuron functions on a 10^{-3} second time scale.

The Josephson junction model neuron differs from the active transmission line (called a "neuristor") which mimics the behavior of a biological axon. The model neuron receives input pulses from other model neurons, and fires a fixed amplitude output pulse when a threshold is exceeded. The neuristor, active transmission line, discussed by Crane, Scott, Parmentier, Ghausi and others,[1] propagates a fixed amplitude pulse along its length. The objective of the model neuron is to model the behavior of the biological neuron, whereas the neuristor models the behavior of the biological axon. The two are related in the biological systems being represented, in the sense that an "action potential" can propagate inside a neuron, as well as along an axon. However, from a functional viewpoint, the two are distinct: the neuron is a "node" in an information processing network, a node at which inputs from axons which form a synapse with a neuron are processed, and from which an output pulse can be generated. The axon is a "transmission line" in the network and conveys the output pulse from one neuron to other neurons. Although it is possible to arrange neuristor transmission lines so they perform logical operations, it appears that the information processing in biological systems occurs largely at the neurons rather than in the axons. Thus I emphasize the Josephson junction model neuron as an information processing element. Neuristors would be useful to convey output pulses from Josephson junction model neurons to other model neurons.[2]

The following properties[3] of a biological neuron are mimicked by a

Josephson junction model neuron:

1. Each model neuron can receive the output pulses from other model neurons to which it is connected, and is capable of generating its own output pulse. That is, the inputs to each model neuron are the output pulses of other model neurons.
2. The state of a model neuron is determined by an <u>internal parameter</u> which corresponds to the Post Synaptic Potential (P.S.P.) in a biological neuron.
3. Both excitatory and inhibitory inputs are possible. A given input channel to the model neuron can be arranged so the input pulses either raise or lower the level of the internal parameter. This response of the model neuron corresponds to the Excitatory Post Synaptic Potentials, and to the Inhibitory Post Synaptic Potentials observed in biological neurons.
4. The internal parameter of the model neuron decays toward zero exponentially with time, as does the PSP of a biological neuron, so input pulses must be in near time coincidence to be effective.
5. When the internal parameter exceeds a threshold, the model neuron fires a fixed amplitude output pulse.
6. After the model neuron fires, the internal parameter resets to near zero. To fire again, the model neuron must have received new input pulses such that the threshold is exceeded again. This characteristic results in an effective "dead time" for the model neuron.[4]

The properties of the model neuron result from the physical properties of a superconducting Josephson junction; I must now introduce the Josephson effect. The Josephson effect[5] occurs when two superconductors are weakly coupled together. The weak connection between the superconductors is called a Josephson junction. In this case, "weak" means that the <u>total</u> coupling energy between the two superconductors is of the order of a few electron volts. Such weak coupling typically is achieved by placing a thin (10-20Å) oxide barrier between the two superconductors.[6] A supercurrent then can tunnel quantum mechanically through this barrier from one superconductor to the other. The Josephson "effect" describes the fact that the wave function of the supercurrent from one superconductor interferes with the wave function in the other superconductor. The current that can flow between the superconductors is controlled by the difference ϕ between the phases of the wave functions on opposite sides of the barrier by the equation

$$I = I_o \sin\phi \tag{1}$$

The effect is entirely quantum mechanical and cannot be described classically. The above Eq. (1) describes the flow of supercurrent regardless of whether or not there is a voltage difference between the two superconductors. If there is a voltage difference V, then the phase difference ϕ is not constant, but depends on time according to

$$\frac{d\phi}{dt} = \frac{2eV}{\hbar} \tag{2}$$

where $\hbar$ is Planck's constant, e is the charge of the electron, and $(1/2\pi)(2e/\hbar) = 483.6$ MHz/μV. These two equations together are known as the "Josephson equations". Although a measurement of I constitutes a measurement[7] of ϕ the directly observable quantities outside a junction are the current I and voltage V across the junction. The current and voltage must satisfy boundary conditions imposed by whatever electrical circuit is external to the junction. The properties of the model neuron to be described here result from the interaction between a Josephson junction and an external circuit.

The neuron model is based on a Josephson junction placed in the circuit shown in Figure 1. The circuit as drawn is simplified to illustrate the principles of operation,[8] and does not include the bias circuit which supplies energy to replace the energy dissipated in the resistances. The Josephson junction is represented by the X in the circuit. R_p is the impedance of the input circuit, and also the dissipative element which controls the decay time of the parameter corresponding to the Post Synaptic Potential. R_o is the junction internal shunt resistance, in parallel with the output impedance of the model neuron.

I discuss in turn how each of the properties of the neuron model arises from this circuit. I begin the discussion heuristically, and later present the nonlinear differential equation that describes the model exactly.

1. The input to the circuit consists of current pulses from the current source $I_i(t)$. This pulse represents an idealization of the output of other neuron models. We suppose that the duration t_i of each pulse is short compared with the time constant of the inductive circuit,

$$t_i \ll \frac{L}{R_p}$$

Each input pulse produces a voltage $I_{io}R_p$ across the input resistor R_p, as indicated in Fig. 1. During the input pulse, the circulating current through the inductance and the Josephson junction changes by an increment

$$\Delta I_p \approx \int_o^{t_i} I_{io}R_p \frac{1}{L}dt$$

For circulating current $I_p < I_o$, the phase difference ϕ is constant and there is no voltage drop across the junction.

2. <u>The internal parameter corresponding to the Post Synaptic Potential is the circulating current</u> I_p, around the loop consisting of the inductance, the Josephson junction, and the input impedance R_p.
3. Depending on whether the input current pulse I_{io} is positive or negative, I_p will increase, or decrease, corresponding to an excitatory or inhibitory input.
4. The internal parameter I_p decays exponentially with time constant $\tau = L/R_p$, so input pulses must arrive coincidentally within a time τ to be effective.
5. The maximum current that can pass through the Josephson junction is I_o, and this value is the threshold for the internal parameter I_p, beyond which the neuron model fires. When the current through the inductance starts to exceed I_o, then some of the current flows through the resistance R_o and the capacitance C. For such a current to flow, a voltage V develops across the junction, and ϕ starts to increase from the value of $\pi/2$ at which the current $I_o \sin\phi$ achieved its maximum value I_o. As ϕ increases, less current flows through the junction, as determined by the Josephson relationship $I_j = I_o \sin\phi$. The current through the inductance cannot decrease rapidly, and thus the excess current flows through R_o and C, by developing a yet higher voltage V, which in turn further increases the rate of increase of ϕ The voltage pulse is the output of the neuron model. While there is current passing through R_o, the decay time of I_p reduces to $L/(R_p+R_o)$, which may be made much shorter than L/R_p which controlled the decay of the I_p before the model neuron fired. Thus I_p decreases rapidly below I_o. If the capacitance C were not present, then V would drop to zero as soon as $I_p < I_o$ and the model neuron would not stop firing. However, with C present, a voltage will remain across C even with $I_p < I_o$. Thus ϕ will continue to increase (by more than 2π), and

energy will be dissipated in R_o, further decreasing I_p. Depending on the amount of energy stored in C, this process can continue until I_p is reset to near zero.

6. After the model neuron has fired, a new set of input pulses must arrive together within time L/R_p before the model neuron will fire again.

By combining the Josephson equations with the equations describing the circuit external to the junction, a differential equation may be derived to describe the neuron model. It is most convenient to write the differential equation in terms of the phase difference ϕ

$$I_i(t) = \frac{\hbar LC}{2eR_p}\frac{d^3\phi}{dt^3} + \left(\frac{\hbar L}{2eR_pR_o} + \frac{\hbar C}{2e}\right)\frac{d^2\phi}{dt^2} + \left[\frac{LI_o}{R_p}\cos\phi + \left(\frac{1}{R_o} + \frac{1}{R_p}\right)\frac{\hbar}{2e}\right]\frac{d\phi}{dt} + I_o\sin\phi. \tag{3}$$

This equation exhibits the behavior described above. It may be solved numerically,[9] but I do not know a solution derivable by techniques of analysis.

The cycle time of the neuron model is given approximately by $L/(R_p+R_o)$. If the Josephson junctions are made by microcircuit techniques, then L can be in the range 10^{-9}H to 10^{-11}H, and with $R_p+R_o = 1$ Ohm the cycle time is 10^{-9} to 10^{-11} second. This cycle time should be compared to that of the biological neuron which is of the order of 10^{-3} second.

I have measured the output pulses from a single Josephson junction model neuron element subjected to a random input. The external circuit elements were such that the time between pulses was long, and the output pulses were easily observable with a relatively narrow bandwidth amplifier. Note the similarity between the output pulses shown in Fig. 2, and the trains of output pulses observed for biological neurons.

The Josephson junction used for this experiment was formed of a chemically etched niobium point contact that was pressed against a flat surface of tin by a differential screw mechanism that could be adjusted while the junction was immersed in liquid helium at T=2.0 Kelvin. The current to the junction was supplied by a noisy current source I(t), shunted by $R_p = 1$ ohm. The inductance of the leads, approximately 1 meter long, going to the junction served as the inductance L indicated in Figure 1. The shunt resistance R_o was internal shunt resistance of the junction, with R_o = .7 ohm, measured from the asymptotic slope of the junction current-voltage characteristic. Junction critical current I_o = 270.μA. The capacitance C was the internal junction shunt capacitance, and was not measured. The voltage pulses from the junction were observed with a Princeton Applied Research Type B preamplifier connected across the junction in series with a 500μF d.c. blocking capacitor. The type B preamplifier, used separately from the P.A.R. HR-8 mainframe, had a gain of 2000. Because the Type B amplifier has a transformer input, it is the frequency response characteristic of this amplifier that determines the shape of the pulse shown in Figure 1, rather than the pulse shape as determined by the solution of equation (3).

The detailed dynamics of the phase ϕ is important in determining the high speed behavior of the Josephson junction model neuron. This approach, based on Eq. (3), is to be contrasted to the approach where analysis is based on time averaged current-voltage curve which contains a negative resistance region. Suppose for the moment a constant current $I(t) = I > I_o$ impressed upon the junction. Numerical integration[9] of Eq. (3) shows that for some choices of the circuit parameters, the junction can oscillate stably at a frequency that is a subharmonic of the "Josephson frequency"

$$\nu = \frac{2\ e\ V_{avg}}{h} .$$

When the junction oscillates in this mode, the trajectory in the ϕ, $d\phi/dt$ plane is a stable limit cycle which is periodic in ϕ with a period of $2\pi n$ where n is an integer. The voltage across the junction is then periodic in time at the nth subharmonic of the Josephson frequency. In terms of the model neuron, the above statement means that the output pulse frequency can be a submultiple of the frequency that would be given by the Josephson voltage-frequency relation Eq. (2), by using the average voltage across the junction. Evidence for subharmonic oscillation of a Josephson junction and of oscillations at frequencies other than at the Josephson frequency has been obtained in experiments on microwave emission[10], parametric oscillations[11], and frequency mixing with Josephson junctions.[12] For the model neuron, detailed consideration of Eq. (3) becomes important when the output pulse frequency approaches the Josephson frequency. The output pulses from Josephson tunnel junctions, as opposed to Josephson point contact junctions, have been discussed by Vernon and Pedersen[13] in terms of a phenomenological model which does not take into account the detailed dynamics of the

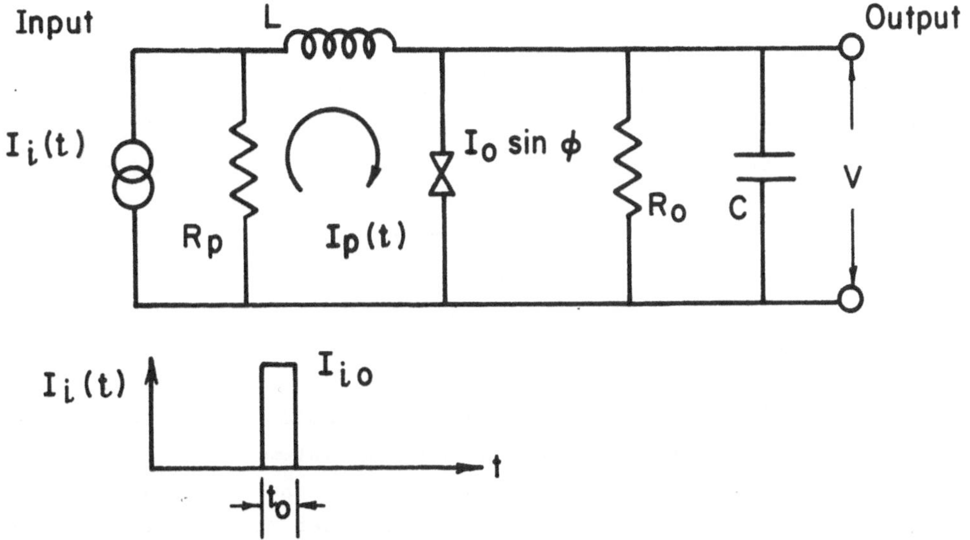

Fig. 1. Circuit for Josephson junction model neuron. The Josephson junction is indicated by the X. The model neuron contains an internal parameter corresponding to the Post Synaptic Potential of the biological neuron. The internal parameter is the circulating current I_p through the Josephson junction when the junction is in the zero-voltage state. The model neuron fires a fixed amplitude output pulse when I_p exceeds the junction critical current, I_o. To simplify the discussion of circuit operation, the input pulses from other model neurons have been represented as rectangular pulses. The short input pulses from other model neurons affects the internal parameter additively.

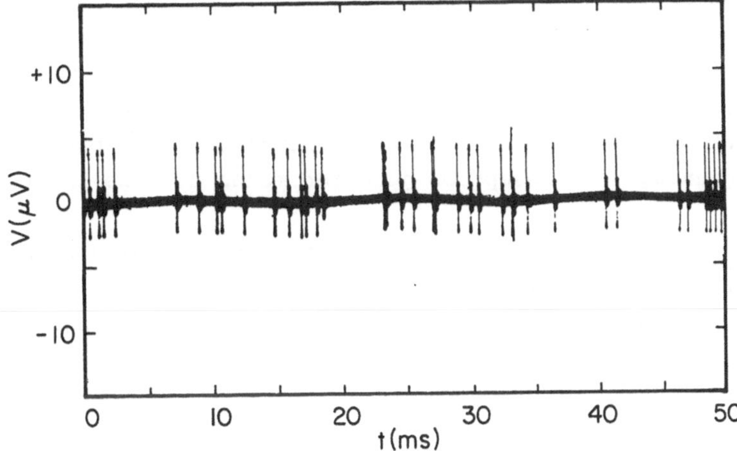

Fig. 2. Output pulses observed from a Josephson junction model neuron subjected to a random input $I_i(t)$. The width of the pulses displayed is determined by amplifier bandwidth rather than by the intrinsic cycle time of the Josephson junction model neuron. The amplifier bandwith also reduces the amplitude of the pulses. Note the similarity between the output pulses, and the trains of output pulses observed for biological neurons.

phase ϕ. For the purposes of a model neuron, the point contact junction, or the microbridge junction is more appropriate because of its lower shunt capacitance, which makes possible a neuron like voltage pulse, rather than the voltage pulse observed by Vernon and Pederson.

Possible uses for the Josephson junction neuron element are the following:

1. The first application is a biological one, in which the Josephson junction model neuron can be used to deduce properties of interconnected biological neurons. Analog models of arrays of interconnected neurons can be constructed from arrays of model neurons, and the behavior of the arrays can be studied, and compared to the behavior of interconnected biological neurons. In this sense, the model neuron can be used to further understanding of biological neural systems.

2. A second application is toward computing and artificial intelligence. It may be possible to construct a computer using Josephson junction model neurons, and based on the same principles that underlie biological cognitive systems. Of course it would be helpful to understand how a biological system works, before trying to construct a computer modeled after a biological cognitive system. Nevertheless, I make this suggestion in the hope that it will be possible to do both simultaneously, by a kind of bootstrap process, starting with simple systems of interconnected neurons. At each of the successive stages, understanding of a given system of interconnected neurons would suggest the next to study. This task may prove feasible with Josephson junction model neurons because of their advantages, mentioned later in this article. Ultimate objectives for a computer based on model neurons would be "associative" parallel processing of data in a manner similar to that performed by neurons in the brain. Tasks might include pattern recognition, decision making and control based on properties of the patterns.

The advantages of Josephson junction model neurons are the following:

1. The time scale of the Josephson junction model neurons ultimately can be much faster than in neurons in the brain. The Josephson junction model neurons can cycle on a 10^{-9} to 10^{-12} second time scale, whereas biological neurons cycle on time scales of 10^{-3} second.

2. The signal processing properties of the Josephson junction model neurons are similar to the signal processing properties of biological neurons, and thus it may be possible to perform decision making tasks by "association" in the manner done by the brain.

3. By microcircuit techniques, Josephson junctions can be made with dimensions of a few micrometers, and thus it is possible to pack large

arrays into small volumes.

4. Josephson junction model neurons operate in the low noise environment of temperatures of a few Kelvin, and thus may be intrinsically lower noise devices than are biological neurons.

5. Power dissipation is low: on the order of 10^{-9} watts per junction on the average.

References

1. The neuristor active transmission line was proposed by H.D. Crane, Proc. IRE 50, 2048 (1962). For references to the extensive literature on this subject, see the articles by L.S. Hoel, W.H. Keller, J.E. Nordman and A.C. Scott, Solid State Electron, 15, 1167 (1972), and by N.J. Elias and M.S. Ghausi, J. Franklin Institute, 293, 421 (1972).
2. The neuristor has been generalized by Hoel et al.[1] from one dimension to two dimensions, and a model built to illustrate the modes of oscillation possible in the two dimensional system. Although discrete Josephson junctions were used in this model, the emphasis was on modelling the multimode oscillatory behavior of a continuous two dimensional neuristor active transmission line, rather than on modeling a neuron network in which each of the Josephson junctions modeled the behavior of a biological neuron.
3. Eccles, J.C.: The Physiology of the Synapses, Academic Press, New York and London, 1964.
   Griffith, J.S.: Mathematical Neurobiology, Academic Press, New York and London, 1971.
   Hubbard, J.I., Llinas, R., Quastel, D.M.J.: Electrophysiological Analysis of Synaptic Transmission, Monograph Number 19 of the Physiological Society, ed. by H. Davson, A.D.M. Greenfield, R. Whittam, and G.S. Brindley. Williams and Wilkins, Baltimore, 1969.
   Katz, B.: Nerve, Muscle, and Synapse, McGraw-Hill, New York, 1966.
4. The model neuron differs, however, from the biological neuron in the following respect: during the dead time of a biological neuron, inputs that would have caused the PSP to exceed the threshold do not cause the biological neuron to fire.
5. B.D. Josephson, Phys. Letters 1, 251 (1962), Rev. Mod. Phys. 36, 216 (1964), Advan. Phys. 14, 419 (1965). For further references, see L. Solymar, Superconductive Tunnelling and Applications, Chapman and Hall, London, 1972.

6. Other techniques may be used to weakly connect two superconductors. These techniques include point contacts, crossed wires, thin film microbridges, proximity effect weakened microbridges, pressed superconducting powders. The technique most suitable for initial studies of the neuron model probably is the point contact technique, because of the suitable values of the inductance and capacitance presented to the junction.
7. An observation of the current between the two superconductors constitutes a measurement, using macroscopic apparatus, of a purely quantum mechanical quantity, the difference ϕ between the phases of two wave functions. This measurement is possible only because of the long range (meters) over which the wave function is coherent in a superconductor, and because a large number of current carrying particles are all described by this same wave function. (The typical distance over which a wave function is coherent in a metal is 1 Å to a few thousand Å.)
8. As mentioned earlier, it may be convenient to interconnect the model neurons with active transmission lines ("neuristors") which transmit fixed amplitude pulses, while resupplying the energy lost during transmission, in a manner similar to that of the biological axon.
9. Numerical experiments with this equation are in progress, and will be reported in another publication: B.T. Ulrich, in preparation.
10. B.T. Ulrich, Physics Letters 42A, 119 (1972); B.T. Ulrich and E.O. Kluth, Proc. IEEE 61, 51 (1973); B.T. Ulrich and E.O. Kluth, Proc. Applied Superconductivity Conf. May 1972, Annapolis, IEEE Conference Record, IEEE Catalog Number 21 CHO 682-5 TABSC.
11. B.T. Ulrich, Proc. of 13th International Conference on Low Temperature Physics, Boulder, Colorado, August, 1972.
12. B.T. Ulrich, H.-A. Combet and G. Talalaeff, in preparation
13. F.L. Vernon, Jr. and R.J. Pedersen, J. Appl. Phys. 39, 2661 (1968).

ADEQUATE LOCOMOTION STRATEGIES FOR AN ABSTRACT ORGANISM IN AN ABSTRACT ENVIRONMENT - A RELATIONAL APPROACH TO BRAIN FUNCTION

O.E. Rössler
Division of Theoretical Chemistry, University of Tübingen, Tübingen, Federal Republic of Germany

Contents

1. Introduction

The idea to be presented in the following - namely a relational approach to brain function - belongs to the class of so-called teleonomic (Pittendrigh, 1958) approaches in biology in which certain features and organizations are predicted as (more or less) necessary implications of a particular function which is known to be performed by the system to be analyzed.

The "relational" approach was invented, so to speak, by Nicolas Rashevsky (1954, 1965). Following the model of d'Arcy Thompson (1917), Rashevsky had always toyed with the idea that the form and organization of biological systems obeys certain mathematical optimality principles, finally casting it into the postulated "principle of adequate design" (Rashevsky, 1961, 1973). It means adequate design with respect to the performance of a distinct function, even if this function is not yet known. Apart from this postulate, Rashevsky (1954) also formulated the "principle of biotopological mapping" (or, equivalently, of "relational epimorphism"). It stresses that analogous parts of different organisms can be mapped upon each other (like the pulsating vesicle of the unicellular flagellate Euglena viridis can on the kidney of a vertebrate, or its eye-spot can on a real eye, etc.). These two different ideas (principle of adequate design, and principle of biotopological mapping) were finally brought together by Robert Rosen (1959), when he added his own abstract theory of metabolic systems (i.e. (M,R)-systems) which is based on the second principle. The name of relational biology became both broadened and sharpened in the following (Rosen, 1972a). Rosen points to the rather analogous situation in theoretical physics where also, besides structure-oriented descriptions, purely relational principles (like that of least action, and other optimality principles) could be formulated (Rosen, 1967).

The "relational" approach to the analysis of systems may be characterized best by showing that it is "<u>dual</u>" to the conventional approach, employing a complementary method. In the usual physico-chemical and physiological approaches, the system is analysed "upwards", so to speak: starting at a certain level of phenomenological description, the higher-level properties of the system are derived from those observed at the lower level. (Hence also the name of "reductionism", namely reduction of higher-level phenomena to lower-level ones.) A relational approach, on the contrary, goes "downwards", starting typically with the highest-level phenomena. Hereby it is not so much the original system itself which is analysed, but "analogs" of it, namely a whole equivalence class of systems, all defined according to their sharing a single aspect (e.g. a functional performance) with the original system. The strategy of descending downwards, then, consists in the stepwise narrowing of the at first very large model class. After deriving the characteristic features of some major subclasses, they are checked, as hypothetical predictions, against the presence or absence of corresponding properties in the original system; and so forth (Mittelstaedt, 1961). If the predicted properties are indeed found again in the original system, it turns out usually that the fundamental role they play in the functioning of the original system could hardly have been detected by any other method, even though the properties themselves may have been well-known.

The indicated procedure is the more appropriate the more complicated a given system is. The obvious non-uniqueness of the approach at its starting level is complementary to the non-uniqueness of the reductionist method at its own starting level, because (as Rosen, 1972b, put it) not only can one function always be realized in many physical ways, but also must one physical system always realize many functions. Hence the problem of finding the "essential" function of a system which is known by its physical properties only (i.e. when starting from below), is just as great as the dual problem of finding the "actual" physical structure of a system which is known by its function only (i.e. when starting from above).

Once the theoretical possibility of the method is granted, along with its asset of optimal applicability to the most complicated systems, its liabilities must also be considered. Building a model (i.e. class of analogs) of a given system may appear much less troublesome to some people than going into the phenomenological details of the same system. However, at the same time the great risk is incurred that the whole model will be unavailing. This is due to the fact that a model can have non-trivial implications for the understanding of the original system only when the aspect of the original system determining the choice of the model had been a "constitutive" (rather than "accidental") aspect; and sometimes the only way to find out to which type an aspect belongs is, to build a model ... [Systems which are not "designed", i.e. tailored to a particular performance, may not possess a constitutive function at all. This possibility is excluded, in the case of biological systems, by the principle of adequate design. For several strategies for proving this principle, see Rössler (1972).]

Therefore, the prospects for finding a relational approach to the brain which is not unavailing in the described sense are very poor. So far, just one attempt to define a possible constitutive function of the brain has become known: Ashby's (1952) remarkable idea of "ultra-stability". Unfortunately, the implications of the corresponding model system, the so-called homeostat, proved not too closely related to observable properties of real brains.

2. A Possible Constitutive Function of Brains

There exists a relational theory of biological systems in general, based on the hierarchically highest constitutive function of biological systems which is known so far. It is, simply, the capability of autonomously increasing (and maintaining) population biomass in a given environment. Hence every environment (and every succession of environments) defines its own, specific constitutive function. The corresponding equivalence classes of systems comprise analogs of certain species and groups of species (whereby the biological notion of analogy is extended). The common properties of the members of these classes allow for a verification of the principle of adequate design in a number of instances (Rössler, 1972). One such instance is our present problem.

One particular succession of environments (and hence of constitutive functions) is characterized by the gradual increase in restrictivity of a single factor: the degree of accessibility, in space, of certain growth-essential chemical and physical factors (and, in parallel, the degree of accession of growth-impeding factors like predators). If the metabolic needs of a species living in this succession are assumed to be immutable, the problem of biomass-increase and maintenance becomes one of adequate locomotion of individuals. This problem can be formalized.

The conjecture that the postulated constitutive function of brains is comprised by this problem can be tested by screening the corresponding equivalence class of systems.

Our way of proceeding is now determined. First, our "hypothetical constitutive function" will be formalized by introducing a number of admissible idealizations. Then, the set of simplest realizations of the constitutive function, depending on one parameter, will be considered. (The accent will hereby lie on the formulation of abstract realizations, named "strategies".) Finally, some of the results will be confronted with biological reality.

3. The Formalized Problem

A. Definition of the Problem

The following relatively simple problem can be derived, by way of minor idealizations, from the above mentioned, hypothetical constitutive function for the brains of locomotive animals:

A single dot ("abstract organism") is to move among a set of stationary points ("sources"). The sources belong to several (n) types and are scattered uniformly (in both directions) in a two-dimensional space without borders. The regimen of locomotion is subject to the constraint that the time interval between two encounters of the same type (i.e. encounters with a source of the same type) never exceeds a certain upper bound T_i $(i = 1,\dots,n)$. The speed of locomotion (v) is also bounded (v_{max}).

More specifically, $\underline{v}$ is assumed to be constant. The $\underline{T}_1$ (each of which is envisionable as the "depletion-time" of a "store" of type $\underline{i}$ inside the abstract organism) are assumed to be equal (= $\underline{T}$). All sources have the same (unit) capacity, and the sources of all types have the same mean density $\underline{n}_{t_i}$, and hence the same "regular mean distance" $\underline{d}_i \equiv 1/\sqrt{n_{t_i}}$. [The regular mean distance $\underline{d}$ is the side-length of the regular simplices which are formed under a regular dense packing of the same amount of sources; $\underline{d}$ constitutes one possible generalization of one-dimensional mean distance to higher dimensions.] In addition, every individual source is to disappear, following one encounter, whilst another source of the same type is appearing somewhere else; and each source is to move very slowly in a random fashion. (The last two assumptions serve the exclusion of pattern-specific strategies.) Fig. 1 summarizes the main assumptions.

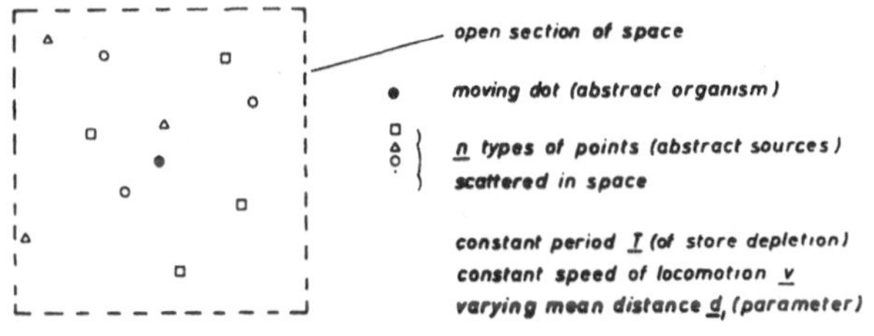

Fig. 1 The problem.

Of course, all these idealizations can be relaxed. In addition "negative" sources, with the special case of obstacles, can be admitted, and a spontaneous mobility of sources (following different strategies of their own) can be introduced. At last, the sources might be given spatial extension and other properties, requiring different sub-strategies in order to be exploited. However, most of these generalizations will be neglected in the following. (In fact, the majority of them, except for the admission of regularities, seem to have no major effect on the solution).

The problem posed is to be solved for different values of the regular mean distance $\underline{d}_i$, ranging from values close to zero up to values approaching vT, the maximum allowed regular mean distance between sources of the same type. It is obvious that very simple strategies will be sufficient in the first case, whereas rather complex ones will be required in the second.

B. Comparison of the Problem with that of the "Traveling Salesman"

The problem just posed can be considered as a modification of the well-known traveling salesman problem (cf. Bellmore and Nemhauser, 1968). There, too, a number of points in space ("towns") have to be visited by a moving dot ("salesman") under the observation of a certain constraint (e.g. minimum milage). [Accordingly, our present problem could be given a more picturesque name, too, like (e.g.) the "rallying truck problem": the moving dot is represented by a truck rallying (among others) on a salt-lake (or in a dense road-network), the sources consist of scattered supply depots (for fuel, water, oil, etc.) providing instantaneous service, the corresponding "tanks" of the truck all run out continuously at a fixed depletion rate, and the rules of the "rally" demand or forbid nothing except that any driving "on reserve" is put under a heavy fine.]

There are a number of minor and major differences between both problems: the pattern of the points to be visited is non-stationary now; not all points have to be visited; the cost is not a continuous function of time, being zero for most of the time, but suddenly rising to very high values when a violation of the rules occurs; the paths are not confined to direct connections between sources; etc.

C. A Supra-optimal Solution

The traveling salesman problem is usually solved numerically by a more or less effective heuristic search strategy (cf. Nilsson, 1971). An analogous procedure can be applied in the present case. [Mount a camera on the top of a truck, take, at a certain moment of rest, a picture of all surrounding sources within a certain maximum distance (whereby the diameter of the circle depends on the assumed velocity of change of the overall pattern). Then rectify the picture and identify the types and positions of all sources shown. Now read off all meters gauging the actual depletion states of the tanks. Finally, feed all this information into a computer which is connected with the truck's steering apparatus and requires only a negligible computation time. (The computer-program may be an adapted version of a traveling salesman program. As an optimality criterion for the computer program, the largest depletion of an individual tank occuring during a simulated path may be chosen.) Then follow the obtained best sub-path until either the outer periphery of the circle or a corresponding timelimit is reached. Now repeat the whole procedure, and so forth.] The resulting overall path will be a near-optimal one, being composed of optimal sub-paths.

By this effective reduction of our problem to another, better known one, we have solved it, in principle, for all possible values of d, up to the maximum possible one. However, we have solved a task different from the originally posed one, by our treating the problem as an optimality problem rather than as the adequacy problem that it actually is.

In the original problem, the optimality criterion does not distinguish between almost total depletion and less than total depletion of a store, so that any strategy avoiding total depletion of all stores is equally optimal or, synonymously, "adequate". Of course, any optimal (or near-optimal) solution is also adequate. But we are interested in all adequate solutions in dependence on d, and especially in the "simplest" ones (whatever this means) in dependence on d. The near-optimal solution, obtained in the above-described way, is a "simplest" solution (namely the only one left) only for extremal values of d. Hence the majority of simplest solutions, namely for nonmaximal d, still have to be found.

D. The Problem of "Simplicity"

There are many different ways to simplify the above described strategy, if the regular mean distance d is sufficiently small. So the range covered by the camera could be diminished, the information on the actual depletion state could be cut, and the complexity of the computer program and the computer hardware could be reduced, and all this could be done in many different ways. The number of different possibilities is decreasing again, however, when the lower end of the scale (with very small values of d) is approached: Then the camera can be disposed of completely, and so can the gauges of the tanks, and the computer can be replaced by a simple random number-generator which forces the truck through a random path. (Even a straight path would be sufficient if the sources were penetrable.)

Therefore, there are two fixed-points, so to speak, from which successions of increasingly or decreasingly simple strategies can be derived: the "last" strategy (for maximum d), and the "null" strategy (for negligible d). It seems that the latter is more appropriate as a starting point for the identification of a "natural" set of simplest strategies.

The set of strategies to be discussed in the following meet the following 4 simplicity constraints:

a) Every element, except for the last one, is a "single-step" strategy (or synonymously, a "breadth-only" strategy; this is the extremal case of a "breadth-first" strategy as defined by Nilsson, 1971), meaning that only possible next steps (but not possible steps after next) are entering the weight functional. b) The strategies are "linear" in the sense that the contribution of a particular source to the weight-functional does not depend on the presence or absence of other sources (superposition principle). (A minor exception will be provided by Eq. 16 below.) c) Each strategy covers a considerable range of $\underline{d}$-values, so that the whole set is rather small. d) The whole set is partially ordered, and every element can be derived from the next-lower one by way of minor modifications ("continuity principle").

The possibility that an even "simpler" set can be found (observing a catalog of even more canonic constraints) cannot be ruled out completely.

4. The Set of "Simplest" Strategies

At first, some additional notation is introduced: (1) Since all strategies refer to a situation with a probabilistic structure (namely a uniform distribution of sources in space), an effectivity of hundred per cent is excluded by definition. The allowed mean frequency of failures per depletion period $\underline{T}$ is denoted by ε (where $\varepsilon \ll 1$, for example $\varepsilon = 10^{-5}$). (2) For part of the strategies considered, a finite diameter of the sources, $\underline{1}$, is assumed. (3) For convenience, $\underline{T} = 1$ time-unit, $\underline{v} = 1$ unit length/time-unit, so that $\underline{vT} = 1$ unit length.

A. First Strategy: Random Locomotion

The random strategy consists either in a straight line (if the sources are assumed to be penetrable) or in a series of straight line-segments between collisions. The angles show hereby a uniform angular distribution in the simplest case (cf. Fig. 2). Quite formally one could

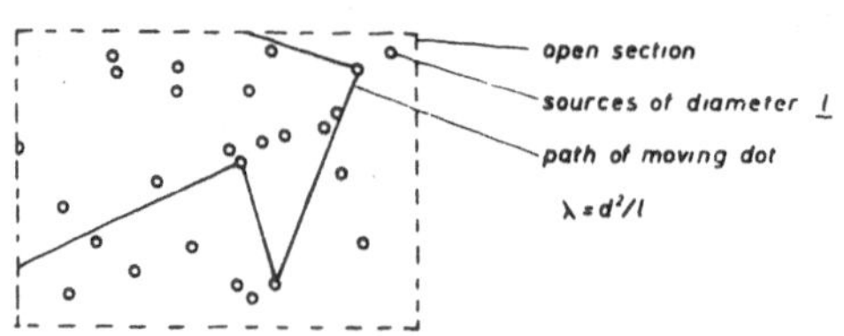

Fig. 2 Display of the first strategy ("random locomotion").

say in this case that a weight functional has been assigned locally to the set of polar directions, such that the optimum changes its place following every encounter, according to a uniform polar distribution.

The situation is mathematically identical with the well-known 2-dimensional collision problem of statistical physics (cf. Schpolski, 1970). The mean free path is $\lambda = \frac{1}{n_t l}$, and the probability for a particular path to have free length $\alpha\lambda$ is $P_\alpha = e^{-\alpha}$. [The probability for a free path of length x, P(x), follows from the evident formula $dP/P = -n_t l dx$, where $n_t l dx$ is the fraction of the unit length blocked out by the targets in the thin strip of space of thickness dx. This differential equation yields $P(x) = e^{-n_t l x}$, if P(0) = 1. Normalization of the integral ($\int_0^\infty P(x)dx = 1$) leads to the normalized probability density $P_{n.}(x) = n_t l e^{-n_t l x}$. $P_{n.}(x)dx$ is the probability that a free path of length $\geq$ x will end in the interval x, x + dx). Hence the probability for a particular path to possess length > x is $\int_x^\infty P_{n.}(x)dx$, i.e. $e^{-n_t l x}$. Setting $x = \alpha\lambda = \frac{\alpha}{n_t l}$ yields P_α. λ, the mean free path length, is obtained as the integral over all products $xP_{n.}(x)dx$, i.e. $\lambda = \int_0^\infty x n_t l e^{-n_t l x} dx = \frac{1}{n_t l}$.]

Therefore, the effectiveness of the strategy, when measured in terms of the maximum regular mean distance d beyond which it ceases to be sufficient (adequate), is easy to calculate. The probability for a particular free path (streight segment of the whole path) to exceed unit length, so that the strategy fails at this step, has to be less than ε, i.e. $e^{-n_t l} \equiv e^{-1/\lambda} < \varepsilon$, so that $\lambda < -1/\ln \varepsilon \approx -(2{,}30 \lg \varepsilon)^{-1}$. For example, if $\varepsilon = 10^{-5}$, $\lambda < 0.09$. Since the fraction of the unit area occupied by sources ($n_t \pi l^2 = \pi l/\lambda$, with $n_t = (\lambda l)^{-1}$, should be less than 1/3 (in order for the above used statistical formulas to be applicable), l has an upper bound, depending on λ: $l_{max} = \frac{\lambda}{3\pi}$. This yields, using $d = 1/\sqrt{n_t} = \sqrt{\lambda l}$,

$$d_{max} = \lambda/\sqrt{3\pi} \approx \frac{-1}{7 \lg \varepsilon}. \qquad (1)$$

Thus,

$$d_{max} = 0.03$$

if $\varepsilon = 10^{-5}$. Accordingly, $n_{t_{min}} = 1/d_{max}^{2} = 1000$, if $\varepsilon = 10^{-5}$. This result has been derived for $n = 1$, but holds true independently for every i ($i = 1,\ldots,n;\ n > 1$). However, since $\underline{l}$ has to be reduced with $1/n_t$, in order to keep the occupied area small, in effect

$$d_{i_{max}} = d_{max}(n{=}1) \cdot \frac{1}{\sqrt{n}}. \qquad (2)$$

To conclude, the present strategy is, for $n = 1$, adequate up to values of 3 per cent of the theoretical upper limit of the regular mean distance (which is, approximately, $d = 1$). This is an unexpected high proportion.

A precondition for the preceding result has been that the contact-times with the sources can be neglected as against the traveling-times. In case that considerable refueling-times have to be taken into account, two modifications of the strategy will enhance the range of adequacy once more: random locomotion of "unspecifically arrestable" type, and random locomotion of "specifically arrestable" type. In the former case, an automatic slowing-down in the immediate neighborhood of every source is provided for, thus enhancing the contact time, whereas in the latter case the halt occurs only near those sources which belong to the same type as the presently most depleted store.

B. Second Strategy: Controlled Locomotion, "Nonselective Type"

If the mean distance d_i surpasses the three per cent threshold of the preceding strategy, the provision of a capability of controlled, goal-directed locomotion toward the nearest respective source should yield a considerably more effective strategy than having the moving dot run in arbitrary directions (former strategy). Of course, mechanisms allowing for direction detection and distance estimation of neighboring sources are now presupposed, as well as a steering mechanism for locomotion. (Even more restrictive requirements of this sort will be encountered in all subsequent strategies.) The strategy is illustrated in Fig. 3.

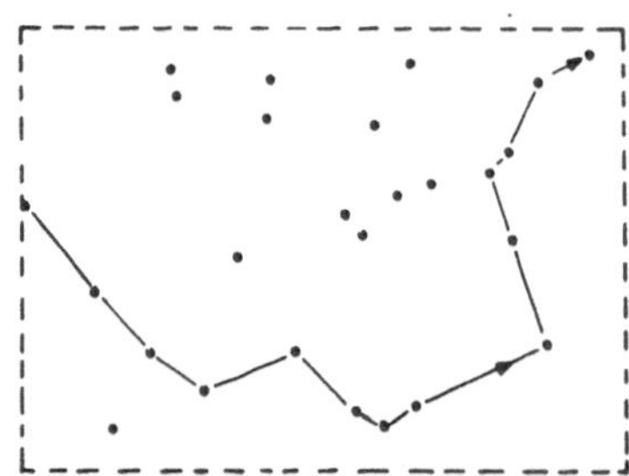

Fig. 3 Display of the second strategy ("nonselective approach", being based on distance alone).

Again, a "priority functional" $\underline{w}$ can be assigned formally to the set of neighboring sources (or rather their directions):

$$w(\beta) = f(x_\beta), \tag{3}$$

where β = angular direction, x_β = distance of the nearest source in this direction. If f decreases in a strictly monotonous way with growing $\underline{x}$, Eq. (3) yields $\beta(w_{max})$ as the optimum direction, as is illustrated in Fig. 4. The functional may be evaluated either just once per trip

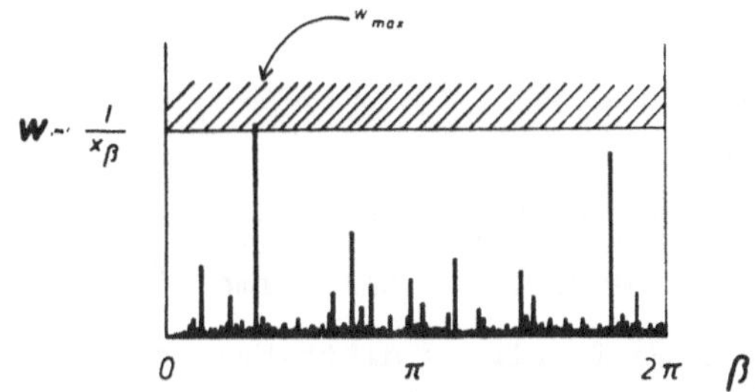

Fig. 4 A possible way to determine the optimum local direction in strategy 2 (using a functional and a variable threshold).

(namely in the very moment before leaving the last source, just as in the preceding strategy) or continuously.

The effectivity of the present strategy is to be determined in three steps. First, the probability of <u>not</u> finding the next source at a certain distance $\underline{x}$ has to be known. It is $p_o = e^{-n_t \pi x^2}$. [In uniform isotropic space-distributions with mean density n_t, the probability $\underline{p}$ for finding $\underline{m}$ points in any domain of area $\underline{V}$ and mean density n_t is described by the Poisson distribution $p(m; Vn_t) = \frac{(Vn_t)^m}{m!} e^{-Vn_t}$ (Feller, 1966). Setting $m = 0$ and $V = \pi x^2$ yields p_o.]

Then, the trivial case (n = 1) is considered. An adequate strategy is obtained if p_o can be shown to stay below ε during every $\underline{T}$ (i.e. for all $x \leq 1$). Therefore, $d_{max} = 0.52$ and, equivalently, $n_{t_{min}} = 3.7$, if $\varepsilon = 10^{-5}$. [From the requirement $p_o(x{=}1) \equiv e^{-\pi n_t} \equiv e^{-\pi/d^2} \leq \varepsilon$ follows $d_{max} = \frac{-\pi}{\ln \varepsilon} \approx \frac{-1.36}{\lg \varepsilon}$, yielding the result. Note that the result is independent of $\underline{l}$, so that l may also be zero in the present strategy.] This result means that the present strategy is adequate for regular mean distances about 17 times greater, and mean source densities three hundred times smaller, than the former strategy, if n = 1. This is even the best result which can be obtained with any single-step strategy, if n = 1.

Finally, the nontrivial case (n > 1) is considered. The average distance between two hits, λ_o, depends on $\sqrt{n_t}$ now (rather than on $\underline{n_t}$, as in the preceding strategy): $\lambda_o = (\pi\sqrt{n_t})^{-1} = d/\pi$. Therefore, $\underline{d}$ has $\overline{\text{to}}$ decrease with $\underline{n}$, if $\lambda_o n$ is to be kept constant (so that the average frequency of type-specific hits remains unchanged with growing n). Using $d_i = d\sqrt{n}$, one obtains

$$d_{i_{max}} = d_{max}(n=1) \cdot \frac{1}{\sqrt{n}}. \qquad (2)$$

This equation is the same as in the preceding strategy; however, $\underline{1}$ does not play any role in its derivation this time. [λ_o, the mean shortest distance in a 2-dimensional uniform distribution, is obtained, in analogy to λ, from the expression $\lambda_o = \int_0^\infty x p_{o_{n.}} dx$, where $p_{o_{n.}}$ is the normalized probability density p_o: $p_{o_{n.}} = p_o / \int_0^\infty p_o dx$.

Since $\int_0^\infty p_o dx = \int_0^\infty e^{-\pi n_t x^2} dx = \frac{1}{\sqrt{\pi n_t}} \int_0^\infty e^{-t^2} dt = \frac{1}{\sqrt{\pi n_t}} \frac{\sqrt{\pi}}{2} \operatorname{erf} \infty =$ $= \frac{1}{2\sqrt{n_t}}$, one obtains $p_{o_{n.}} = 2\sqrt{n_t}\, e^{-\pi n_t x^2}$. Insertion into the expression for λ_o yields, through integration by parts,

$$\lambda_o = 2\sqrt{n_t} \frac{-1}{2\pi n_t} e^{-\pi n_t x^2} \Big|_0^\infty = \frac{1}{\pi\sqrt{n_t}} = d/\pi.]$$

Eq. (2) is about 2 times too favorable for the present strategy, since the fact that the type of a hit source is selected at random has not yet been taken into account. [This is done by requiring that the probability for the total absence of a particular source type in a series of q hits, p_(q), times the probability for the occurence of q hits per unit traveling distance, p_{q_1} (being normalized), is less than ε:

$$p\_(q_1) \cdot p_{q_1} \leq \varepsilon \text{ for all } q_1, \qquad (4a)$$

where the index 1 refers to the given time interval. (The probabilities for the absence of 2 and more source types in the series which should be added to the left-hand side of Eq. 4a are small enough to be neglected.) If the integral over all $q_1 \geq q_1^*$ would be 1, i.e. $\int_{q_1^*}^{\infty} p_{q_1}\, dq_1 = 1$, it would suffice to consider the equation

$$p_-(q_1^*) \leq \varepsilon, \tag{4b}$$

i.e. $(1-\frac{1}{2})^{q_1^*} \leq \varepsilon$, yielding $q_1^* = \frac{\lg \varepsilon}{\lg\frac{n-1}{n}} \approx 12n$, if $\varepsilon = 10^{-5}$. The result (q_1^*) would not be changed, if the probability density of $q_1 < q_1^*$ would be somewhat greater than zero, but smaller than

$$\tilde{p}_{q_1} = \frac{\varepsilon}{(1-\frac{1}{2})^{q_1}} = \varepsilon e^{q_1 \ln\frac{n}{n-1}} \approx \varepsilon e^{q_1/n} \tag{4c}$$

for all $0 \leq q_1 < q_1^*$. Therefore, that particular p_{q_1} (d) which just stays below $\tilde{p}_{q_1}$ has to be chosen, yielding d_{max}.

p_{q_1} cannot be calculated directly. (It is a Markov process governed by a set of q_1 linear functional differential equations; the damping term which characterizes the process -- in analogy to a Poisson process -- is $z = 2\pi n_t t_m$, where $t_m = 0$ up to the moment of the mth hit, T_m, and $t_m = t - T_m$ afterwards; z is obtained from the solution of the first (ordinary) differential equation of the set, $\dot{p}_o(t) = (p_o$, as described above)$^{\cdot}$; $^{\cdot} = d/dt$; see Feller, 1966). However, it is reasonable to assume that p_{q_1} has approximately the form of a Poisson distribution, with mean and variance $1/\lambda_o$ (as is true for $q_1^* \to \infty$). This distribution, in turn, can be approximated by a normal distribution possessing the same mean and variance, so that

$$p_{q_1} \approx p(q_1;\, 1/\lambda_o) \approx \frac{1}{\sqrt{2\pi/\lambda_o}}\, e^{-\frac{1}{2}\frac{(q-1/\lambda_o)^2}{1/\lambda_o}}. \tag{4d}$$

This distribution will stay below $\tilde{p}_{q_1}$ for all mean values $1/\lambda_o$ more than two standard deviations $(\sqrt{1}/\lambda_o)$ to the right of q_1^* and, at the same time, more than 4 standard deviations to the right of $q_1 = 0$ (as follows readily from a comparison of the decay of a normal distribution (Eq. 4d) with increasing deviation from the mean, and the decay of the exponential function (Eq. 4c) to the left of q_1^*). The preceding requirement is approximately equivalent to the following, which can be calculated without the necessity of iterations:

$$1/\lambda_o = \max(q_1^* + 2\sqrt{q_1^*},\ 4\sqrt{2q_1^*}),$$

that is (since $q_1^*(n=2) \stackrel{\sim}{\sim} 24$ already),

$$1/\lambda_o = q_1^* + 2\sqrt{q_1^*}. \qquad (4e)$$

Insertion of λ_o and q_1^* finally yields $\frac{\pi}{d_{max}} = 12n + 2\sqrt{12n}$, so that $d_{max} = \frac{\pi}{12} \quad \frac{1}{n+\sqrt{n}/3}$, and, using $d_i = d\sqrt{n}$,

$$d_{i_{max}} = \frac{\pi}{12}\frac{1}{\sqrt{n} + \sqrt{1/3}} = \frac{0.26}{\sqrt{n} + 0.6}.] \qquad (2')$$

Therefore

$$d_{i_{max}} = 0.13 \quad (n = 2)$$
$$= 0.11 \quad (n = 3)$$
$$= 0.07 \quad (n = 10).$$

The present strategy is about 6 times more effective than the former, random strategy, when measured in terms of regular mean distance (and about 36 times in terms of mean densities).

C. Third Strategy: Controlled Locomotion, "Sequential Nonflexible Type"

In the last strategy, the "priority degree" $w(\beta)$ of a certain direction depended only on the distance of the nearest source in this direction, x_β (see Eq. 3). Now, the particular type i of this source is also of influence. Some mechanism enabling the discrimination of the type of a detected source is hereby presupposed. The strategy is displayed in Fig. 5.

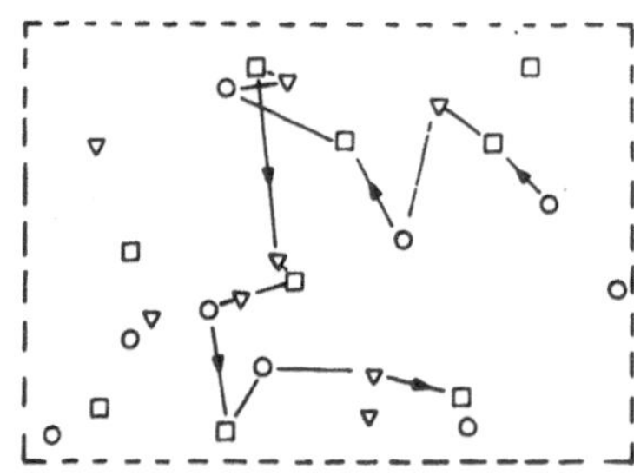

Fig. 5 Display of the third, and fourth, strategy ("selective approach", depending on source-type, time, and distance). 3 source-types are depicted.

In the simplest case (present strategy), the priority weight depends on t_i alone. From all surrounding sources, the nearest of the selected type is approached. Thus, two functionals are presupposed, first the source-type selector,

$$w_1(i) = f_1(t_i), \tag{5a}$$

yielding $i_{opt} = i(w_{1_{max}})$, and second the source-selector

$$w_2(\beta, i_{opt}) = f_2(x_{\beta_{i_{opt}}}), \tag{5b}$$

which is equivalent, for each i_{opt}, with Eq. (3) of the last strategy. Both functionals have to be evaluated immediately after each hit, but may as well be evaluated continuously. The simplest possibilities for f_1 and f_2 are $f_1 = t_i$, and $f_2 = 1 - x_{\beta_{i_{opt}}}$ (where $x_{\beta_{i_{opt}}} \leq 1$).

For n = 1, the strategy is identical with the former, yielding $d_{max} = 0.52$ if $\varepsilon = 10^{-5}$ (see above). If n > 1, the nth part of the unit length has to suffice for finding a source of this type with probability $1 - \varepsilon/n$. This leads to

$$d_{i_{max}} \approx d_{max}(n{=}1) \cdot \frac{1}{2}. \tag{6}$$

[From the requirement $p_o(x{=}1/n) \equiv e^{-\pi/(nd_i)^2} \leq \varepsilon/n$ follows

$$d_{i_{max}} = \frac{1}{2}\,\frac{-\pi}{\ln \varepsilon/n} = \frac{1}{2}\,\frac{-1.36}{\lg \varepsilon - \lg n} \approx \frac{1}{2}\, d_{max}(n{=}1), \text{ if } n < 100.]$$

Specifically,

$$\begin{aligned} d_{i_{max}} &= 0.26 \quad (n = 2) \\ &= 0.17 \quad (n = 3) \\ &= 0.05 \quad (n = 10) \\ &= 0.004 \quad (n = 100). \end{aligned}$$

The strategy is superior to the former for n < 6. (For n > 300, it is even inferior to the first, random strategy.)

D. Fourth Strategy: Controlled Locomotion, "Sequential Flexible Type"

It is now assumed that the priority degree of a source-specific direction depends on its three determinants (i.e. distance $\underline{x}$, type $\underline{i}$, and store depletion state $\underline{t_i}$) in a more flexible way than in the preceding two strategies.

Formally, n+1 functionals have to be determined now: $\underline{n}$ times the source-type specific selector of the nearest source

$$w_i(\beta) = f(x_{\beta_i}) \qquad (i = 1,\dots,n), \tag{7}$$

yielding $\beta_i = \beta(w_{i_{max}})$ as the optimum local direction of $\underline{i}$ and x_i as the distance of the corresponding source; and once the source-type selector

$$w_{n+1}(i) = f(t_i, x_i), \tag{8}$$

yielding $i_{opt} = i(w_{n+1_{max}})$ and, along with it, $\beta_{opt} = \beta_{i_{opt}}$ as the locally optimal direction, to be connected to the control system of locomotion. Both equations have to be evaluated continuously now.

The former two strategies are obtained, if f in Eq. (8) is set equal to $1-x_i$ (yielding the second strategy) or equal to t_i (yielding the third). This situation is illustrated in Fig. 6a and $\underline{b.}$ (Note that any strategy fails for sources outside the range

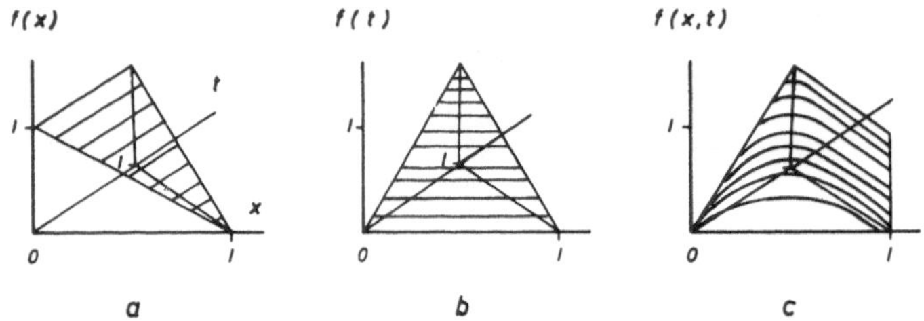

Fig. 6 Comparison of the functionals of the second (a), third (b), and fourth (c) strategy. (Confer Eqs. 3, 5a, and 8.)

$x_i + t_i = 1$, so that this region is excluded in the Figure, just as negative values of $\underline{x_i}$ and $\underline{t_i}$ are.) The determination of an <u>optimum</u> $f(t_i, x_i)$, however, as symbolically sketched in Fig. 6c, constitutes a <u>variational problem</u>.

The optimum form of this function depends on both $\underline{n}$ and $\underline{d_i}$. For every given $\underline{n}$, that particular optimum $\underline{f}$ has to be selected which allows for the maximum $\underline{d_i}$.

Obviously, effective numerical optimization procedures are required for obtaining the exact solution of our problem (for example, the "hierarchical" procedure of Rechenberg, 1973, may be used). However, the qualitative form of f can be derived in a more straightforward way. The result can then be checked and improved by means of ordinary simulation.

The importance of a nearest source of type i increases with its closeness, as long as the corresponding t_i is small (but not zero), since a detour is the more justifiable the smaller it is; and it increases with its distance when t_i approaches one. The highest possible value should be attained if its position comes close to the line $x_i + t_i = 1$ (cf. Fig. 6c).

Quantitatively, the "gain", due to an "unnecessarily early" refilling of one tank, is proportional to the obtained increase of the average filling state above the minimum possible mean (of half-filling), supposed this tank would always be refilled at this level. The gain is, therefore,

$$g = \frac{1}{2}(t_i - t_i^2), \qquad (9a)$$

as illustrated in Fig. 7a (upper half). In the same vein, the "loss",

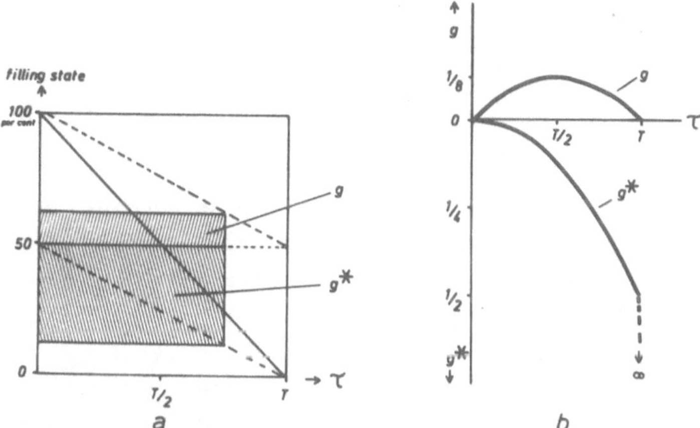

Fig. 7 The two "temporal" weights of either filling (g) or not filling (g^*) a store.

a: graphical construction
b: corresponding graphs

due to not filling up (supposed it is just possible) at a particular moment $t_i < 1$, is proportional to the obtained decrease of the average filling state below the maximum possible mean (of total filling), supposed that a refilling would always be renounced of at this level. This negative gain is, therefore,

$$g^* = \frac{1}{2} t_i^2 \qquad (9b)$$

(see lower half of Fig.7a). These "basic weights" have to be "modulated" differently by the spatial factor: in the first case, the modulation has to decrease with distance,

$$h_1 = \frac{k_1}{k_1+x_i}, \qquad (10a)$$

in the second it has to increase with distance (up to the maximum possible distance of $1-t_i$):

$$h_2 = \frac{k_2}{k_2+(1-t_i-x_i)}, \qquad (10b)$$

so that

$$f(t_i,x_i) \approx gh_1 + g^*h_2. \qquad (11)$$

k_1 and k_2 (from Eqs. 10a and b) are constants which have to be adapted numerically (for example, $k_1 = 1/20$ and $k_2 = 1/10$). Of course, Eqs. (10a,b) each stand for a whole class of similarly behaving functions.

The double-sum structure of Eq. (11) can be justified theoretically by a game-theoretic consideration (maximization of the own chances and minimization of the risks, i.e. the chances of the imaginary opponent; Rössler, in preparation).

The actual rise in effectivity of the present strategy (i.e. Eqs. 7 and 8 with 11), as compared to the preceding two strategies, is hard to assess. The strategy certainly avoids some "mistakes" which the former two cannot (namely the sometimes fatal "distractability" of the strategy of Fig. 6a, and its contrary, the sometimes fatal "stubbornness" characterizing the strategy of Fig. 6b). Having its own blindnesses, however, the strategy yields a net factor which probably does not exceed 2 or 3.

With this strategy, obviously the "knee toward the asymptotic branch" has been surpassed, so that ever more sophisticated means will be required from now on in order to cope with ever smaller increments of the environmental parameter (namely the regular mean distance of sources).

An interesting, self-evident modification of the present strategy applies if the radius of source-detection, r, is assumed to be small as compared with vT: $r \ll 1$. Then a random-locomotion substrategy has to be built in, being switched-on as soon as the maximum radius $(1-t_i)$ within which some source of a particular type has to be found, is going to approach r from outwards.

The next strategy of the series takes advantage of the fact that one and the same direction may offer several reasons for following it.

E. Fifth Strategy: Controlled Locomotion, "Simultaneous Type"

The weight of each direction is evaluated not so much any more with respect to the question whether or not it points toward the presently most important single source, but rather whether or not it points to the presently most important cluster of sources.

Every direction, when chosen, changes the absolute distances to many sources. Specifically, every direction in space pointing away less than 90 degrees from a particular source will effectively contribute to an approach of this source through vector-addition. Therefore, every nearest source of each type automatically determines, as far as the distance factor alone is concerned, a radial vector field, emanating at the local position of the moving dot (Fig. 8).

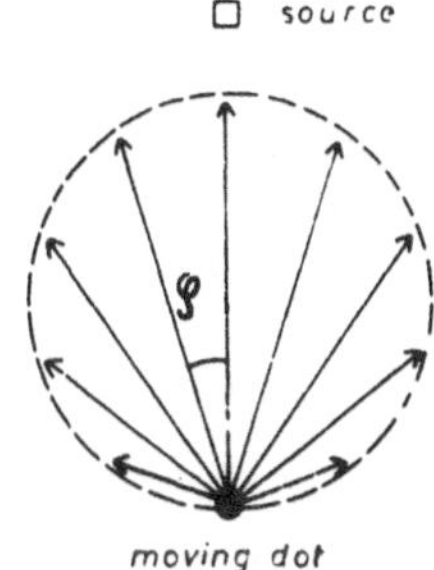

Fig. 8 Relative contributions of different directions to the approach of one source.

The circumference is a Thales circle. Its equation is (in polar coordinates and with normalized diameter)

$$\rho = \begin{cases} \cos\phi \\ 0 \end{cases} \text{if } \phi \; \begin{matrix} \leq \\ > \end{matrix} \; \pi/2 \tag{12}$$

If several sources of the same type, but different distances, are surrounding the dot, then each possesses a vector field of the same form, but differing diameter (decreasing in proportion with the distance of the source), so that each source becomes the "nearest" in a certain range of directions (whereby the individual fields do not add). The resulting, polycyclically shaped type-specific vector field is displayed in Fig. 9. It is evident that no more than about 3 "effective nearest sources" exist for each source type at the same time.

After modulation with their actual "temporal" importance (depending on t_i), the individual type-specific vector-fields may be added, specifying a locally optimal sum-direction. There is no danger that the

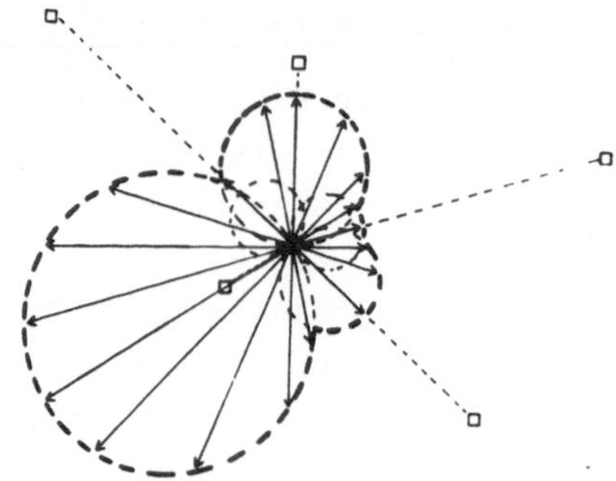

Fig. 9 Relative contributions of different directions to the approach of some source.

resulting intermediate course (which does not directly aim at any particular source) will miss all sources; rather the automatic change of weights occuring during the voyage of the abstract organism determines an ever increasing attraction in the neighborhood of first one and then another source. (The involved "unnecessary" curves cost much less time than is gained through the simultaneous attraction by the members of the "cluster" which, then, are readily exploited in succession.)

Formally, the strategy presupposes once more $\underline{n}$ functionals (yielding the directions of the 3 effective nearest sources of each source-type, and also the separating directions between the regions of influence of these sources). Since this is hard to formalize (and perhaps also to realize), the specification of a single optimum direction per source type is assumed here for the sake of simplicity (implying a small drop of the overall effectivity), so that Eq. (7) applies once more. It has to be evaluated continually.

In addition, a single function (rather than functional) must be present (comprising the above-named additions):

$$w_{+}(\beta) = \sum_{i=1}^{n} \rho_i \cdot f(t_i, x_i). \tag{13a}$$

Again the problem of the optimum $\underline{f}$ is encountered.

Before turning to the solution of this problem, we observe that an alternative strategy exists which is just as reasonable as the preceding one. Its basis is illustrated in Fig. 10, and the equation of the "negative" vector field of a source, being dual to Eq. (12), is

$$\rho^{*} = 1 - \cos\phi, \tag{14}$$

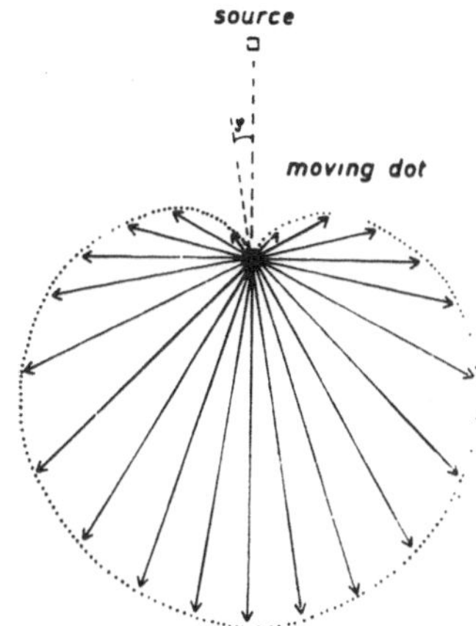

Fig. 10 Relative contributions of different directions to the non-approach of one source.

which is the equation of a cardioid (see Fig. 10).This vector field describes the participation of all directions in the non-approach of a nearest source (corresponding, again, to the interests of the imaginary opponent in the game). It leads (if again the selection of only 1 effective nearest source is assumed for the sake of simplicity) to an alternative function to Eq. (13a), namely

$$w_{-}(\beta) = -\sum_{i=1}^{n} \rho_i^* \cdot f^*(t_i, x_i). \tag{13b}$$

The negative sign is necessary if the maximum of $w_{-}(\beta)$, being least negative, is to be chosen for direction-determination.

Both equations (Eq. 13a and b) are reasonably added, yielding

$$w(\beta) = w_{+}(\beta) + w_{-}(\beta). \tag{13c}$$

As in the preceding strategy it is again found that f and f^* are different with

$$f = g\,h \tag{14a}$$

and

$$f^* = g^* h^*, \tag{14b}$$

where h and h^* are the "spatial" effects (h^* depending also on t_i), and g and g^* are from Eq. (9a and b).

h and h^* are somewhat different from h_1 and h_2 (Eq. 10a,b), since arbitrarily large ratios between different attracting forces have to be admitted in the function of Eq. (13c) (which no longer is a selector switch (functional) in which the influences of all forces except for one are automatically suppressed). h remains the same as h_1:

$$h = \frac{k}{k + x_i}, \qquad (15a)$$

whereas h^* has the form

$$h^* = \left(\frac{1-t_i}{(1-t_i) - x_i}\right)^{\gamma} - 1. \qquad (15b)$$

Again, k and γ have to be determined by numerical simulation (for example, $k = 1/100$ and $\gamma = 3$).

The complete strategy is described, apart from Eq. (7), by

$$w(\beta) = \alpha \sum_{i=1}^{n} \rho_i g h - \sum_{i=1}^{n} \rho_i^* g^* h^*, \qquad (13)$$

where $\alpha = 1$; ρ and ρ^* are from Eqs. (12 and 14); g and g^* from Eqs. (9a and b); and h and h^* from Eqs. (15a and b). The equation yields the locally optimal direction, $\beta_{opt} = \beta(w_{max})$.

Eq. (13) can be interpreted as a force-field. It is possible, therefore, to indicate the corresponding (locally existing) potential possessing this force-field as its gradient. (Concerning the question of realizability, the unbound values potentially occuring in Eq. (13) are probably most easy to realize as gradients.)

A mathematical proof that Eq. (13) has the optimal form can, possibly, be derived from game theory (Rössler, in preparation). As a final improvement, an inhibition of w_+ by large values of w_- (rendering the whole strategy "nonlinear" in the sense of section 3d), can be provided:

$$\alpha = \frac{1}{a + w_-}, \qquad (16)$$

where $a < 1$ (e.g. 0.25). This is reasonable since the relative directional sharpness of w_-, which is much greater than that of w_+, is impaired even by relatively small values of w_+.

The structure of Eq. (13) is interesting from a biological point of view (see Discussion).

The effectiveness of the present strategy will be greater than that of the preceding strategy by a factor of less than 2 or 3. A similar factor can be gained once more with the last strategy.

The strategy to be considered next is no longer a breadth-only (i.e. single-step) strategy, as all former strategies were, even though it is a modification of the last described strategy. At the same time its requirements concerning information about the surrounding space are more modest: a simultaneous record of all neighboring sources during every moment is no longer necessary (a sequential scanning of small sections of the environment being sufficient).

F. Sixth Strategy: Controlled Locomotion, Simultaneous Type, "Supra-local" Version

The last-described breadth-only strategy yields an optimal direction from the standpoint of a single point in space. A considerably greater effectiveness is obtained if, instead of a single standpoint, several standpoints are evaluated before the optimum direction (which now has become a whole path segment) is chosen.

One possibility how to realize technically such a supralocal strategy has been discussed above in the context of an ideal strategy (Section 3C). The solution offered there has been based on the assumption of a serially executed heuristic search program (see Nilsson, 1971). There is no doubt that parallel processing computers and programs would provide a great advantage. Of the large set of abstractly conceivable solutions of this last-named type, one is closely related to the preceding strategy (constituting an extended version), so that the "continuity of design principle" (Section 3D) is fulfilled. Due to the more primitive strategy contained, the new strategy can never act less efficiently than the former, even if the time available for a decision is greatly reduced.

Although a mathematical formulation of the new, modified strategy is not to be attempted here, one possibility how to realize it may be mentioned. Three assumptions are made: a) All environmental inputs are processed through a set of (continuously graded) delay channels, and the actual directional decisions are fed back in such a way that an immovable representation of the surrounding environment (and especially the actual force-field contained) is generated, along with a representation of the actual position of the moving dot inside this landscape ("generalized reafference principle"; see von Holst & Mittelstaedt, 1950). b) The presence of rapidly decaying "virtual forces",corresponding to fictitious sources which appear in arbitrary directions for a short while (like will-o'-the-wisp's), determines the pursuit of short stretches of (from the local standpoint) "nonsense directions" inside the landscape. c) Every time such a nonsense path happens to lead to a better point (possessing a higher sum gradient than the former), the intermediate path (formerly bridged by the rapidly decaying virtual gradient) is "filled up" with a longer-lasting, "real" gradient in the landscape. -- All 3 assumptions together determine an iterative process which automatically converges to a new locally optimal direction. (In case that the conditions deteriorate more rapidly than the whole process converges, the iteration is automatically interrupted in favor of a new relatively optimal direction.)

The proposed organization has one obvious disadvantage: all paths have to be pursued at least twice in order to find the best one which is then actually (and "seriously") followed. This liability is automatically made up for by the following, intrinsic property of the assumed reafference type organization: locomotive impulse which, by some reason, are prevented from actual execution allow for a very rapid "internal" screening and smoothing of the landscape, requiring hereby only a few actually acquired landmarks. This "short-circuit" mode of operation allows the whole system to detect new optimal paths without their being carried out actually. The system is thus equivalent, from outside, to any other system using a different heuristic search program (like that mentioned in Section 3C), unless specific tests are applied.

The preceding engineering-type description still requires some formalization before it can be fed into a simulatory program. A mathematical proof that no less complex solution to the problem exists (once the principle of parallel processing is granted) is, again, hard to formulate. It even is not clear whether or not game theory will be of any help.

Thus, the proposed solution remains just one possibility at the moment. However, some of its properties are nontrivial enough to render its further investigation (and even artificial realization) a rewarding task: the system performs compromises, shows anticipatory behavior, and possesses self-teaching properties (within a certain, "shortest term" memory range). It thus invites comparison with other "adaptive" systems (see Holland, 1970).

It is finally noted that, depending on the assumed temporal width of the 4-dimensional "window" of the landscape, any desired approximation to the optimal solution (remaining the only possible solution at maximum d) can be achieved, without any accompanying, greater reduction in the velocity of performance. (This is a major difference to other heuristic search programs, the computing times of which crucially depend on the complexity of the problem; cf. Nilsson, 1971).

Thus, the upper bound (d_{max} = d(theoretical maximum)) has been reached with this strategy, so that the whole series is complete.

5. Discussion

The outlined approach provides an example of an essentially deductive procedure in theoretical biology, deriving the minimal properties that any system, performing a function which is comprised (as an apparently trivial part) in the unknown set of functions carried out by real brains, must possess. Methodologically, the procedure is thus related to Darwin's approach who also asked for the minimal properties that any system, performing an (apparently trivial) performance of biological systems (namely self-maintenance in a varying environment), must possess.

The result will, or will not, possess a greater impact on biology, in dependence on the (unknown) relative importance of the minimum performance chosen. It turns out that the strategies, obtained as "simplest" solutions in dependence on the parameter "average distance between sources", all can be interpreted as "motivational" systems (if motivation means preferential choice in dependence on environmental factors and endogeneous state), as the described functionals are proving (Eqs. 3; 5a,b; 7,8; 7,13; last strategy). They thus form a partially ordered series of artificial motivational systems of greatly varying complexity. Even if the elements of this set will prove to have little in common with biological motivational systems, their very existence may suffice to stimulate further research on the real class. At least the specific predictions they allow should be easy to confirm or disprove.

Thus the following 4 heuristic hypotheses may be raised: (1) The "motivational" system of certain unicellular organisms follows strategy 1. (2) The motivational system of certain insects can be interpreted as possessing strategy 4 as its prototype. (3) The motivational system of certain very low vertebrates can be interpreted as a realization of strategy 5. (4) The motivational system of somewhat higher vertebrates can be interpreted as a realization of strategy 6.

The first hypothesis is in accordance with a view expressed by Lorenz (1973). The two modifications mentioned at the end of Section 3a (unspecifically and specifically arrestable type of random locomotion) are well-known in biology under the name of "kinesis" (Fraenkel & Gunn, 1961), whereby only 1 type of "arresting" information (acidity) has been identified so far. (In order to avoid misunderstandings it may be added at this place that the case of amoebae -- which at the first sight fit rather into strategy 5 -- is not covered by the preceding series of strategies, since their behavior presupposes the presence of actual gradient fields in the environment, i.e. of nonpunctiform positive and negative sources.)

A discussion of the other 3 hypotheses is beyond the scope of this note. Only the global observation is made that they cannot be rejected out of hand, i.e. without a careful and specific experimental examination being carried out in each case. This fact supports the view that the present type of approach (of which only the first iteration step, being based on the most idealized assumptions, has been carried out so far) will yield even more interesting results at later stages. For example, it can be expected that in an environment containing certain spatio-temporal regularities (like re-occuring source patterns), the pertaining set of simplest strategies will involve a different type of learning each.

Apart from their possible implications for experimental psychology, ethology, brain theory, and bionics (question: do the preceding strategies, when being realized by electronics' hardware or software, constitute "bionic systems"?), the results of the present approach may have some interest for two branches of mathematics, namely artificial intelligence (including game theory) and logic theory.

One possible further application is to be considered shortly as a final point. Zeeman (1965) wrote on possible topological models of the functioning of the brain, and later (1971) showed that Thom's (1970) catastrophe theory is applicable to the sudden behavioral transitions occurring in certain animals in response to a continuously varied parameter (slow approach of a potential aggressor; Hediger, 1955). From Eq. (13) (fifth strategy), very similar results may be derived, even though "negative" and "moving" sources (like a potential aggressor) have not been included in the set of environmental assumptions made so far. It seems that all 5 types of 3-dimensional elementary catastrophes described by Thom (1970) can be generated in a 2-dimensional environment, including the "cusp" type considered by Zeeman (1971).

The fact that catastrophe theory plays a role in both types of astract approaches to the brain which are possible (namely the "upwards relational" approach of continuous, and discrete, neural nets -- see Rössler, 1974a,b --; and the "downwards relational" approach considered here) may be taken as evidence that a mathematical understanding of the qualitative functioning of real brains is no longer out of reach.

Summary: The method of analyzing a complicated biological system from the top to the bottom (starting from the constitutive performance of the whole system) is applied to the control system of locomotion.

The minimum and, perhaps, constitutive performance of a brain can be seen in the generation of an adequate locomotion strategy in a spatial environment containing "sources" of several types which are distributed in space-time in a more or less complicated manner. In the simplest formalization, only positive essential sources, being uniformly distributed as points in 2-dimensional space, are considered along with a steerable punctiform organism moving at constant speed and requiring a source-type specific "hit" at least every unit time interval. Then different minimum-complexity strategies are adequate in dependence on source density (random locomotion; directed approach toward the nearest neighboring source; toward the most important neighboring source; toward the best neighboring source; toward the best neighboring cluster of sources; pursuit of the best path in a whole neighborhood).

Despite the simplifications made, some of the strategies show a certain "convergence" to biological reality (motivational behavior of insects and vertebrates). The degree of analogy remains to be tested. More realistic assumptions will lead to modified strategies each involving a specific type of "learning". The described first generation of artificial brains can be investigated further both by means of computer simulations and hardware realization (highly parallel type).

References

Ashby, W.R.: Design for a Brain, Chapman & Hall, London, 1952.

Bellmore, M. and Nemhauser, G.: The Traveling Salesman Problem: a Survey, Operations Res. 16, 538-558 (1968).

D'Arcy Thompson, W.: On Growth and Form. Rev. ed. Macmillan, New York, 1945.

Darwin, C.: On the Origin of Species by Means of Natural Selection, Joha Murray, London (1859).

Feller, W.: An Introduction to Probability Theory and its Applications, Wiley, New York, 3rd ed., 2 Vols., I 14, 41; II 159, 448-450 (1859).

Fraenkel, G.S. & D.S. Gunn: The Orientation of Animals, Claredon Press, Oxford (1961).

Hediger, H.: Studies of the Psychology and Behavior of Captive Animals in Zoos and Circuses, Butterworth, London (1955).

Holland, J.H.: Hierarchical Descriptions, Universal Spaces and Adaptive Systems. In: Essays on Cellular Automata, ed. by A.W. Burks, Urbana, Chicago, London, University of Illinois Press, 320-353 (1970).

Lorenz, K.: Die Rückseite des Spiegels (The Other Side of the Mirror), Piper-Verlag, München, 1973.

Mittelstaedt, H.: Control Theory as a Methodic Tool in Behavior Analysis (in German), Naturwissenschaften 48, 246-254, p. 248 (1961).

Nilsson, N.J.: Problem-solving Methods in Artificial Intelligence, McGraw-Hill, New York, 1971.

Pittendrigh, C.S.: Perspective in the Study of Biological Clocks. In: Perspectives in Marine Biology, La Jolla, Calif., Scripps Institute of Oceanography, 1958.

Rashevsky, N.: Topology and Life: In Search of General Mathematical Principles in Biology and Sociology, Bull. Math. Biophys. 16, 317-348 (1954).

Rashevsky, N.: Mathematical Principles in Biology and their Applications, Thomas Publ., Springfield, Ill., 1961.

Rashevsky, N.: Models and Mathematical Principles in Biology. In: Theoretical and Mathematical Biology, ed. by T.H. Waterman & H.J. Morowitz, Blackwell Publ. Co., New York, 36-53 (1965).

Rashevsky, N.: The Principle of Adequate Design. In: Foundations of Mathematical Biology, ed. by R. Rosen, Academic Press, New York, Vol. 3, 143-175 (1973).

Rechenberg, I.: Evolutionsstrategie, Optimierung technischer Systeme nach Prinzipien der biologischen Evolution, Friedrich Frommann-Verlag (Günter Holzboog K.G.), Stuttgart-Bad Cannstatt, 1973.

Rosen, R.: A Relational Theory of Biological Systems II. Bull, Math. Biophys. 21, 109-128 (1959).

Rosen, R.: Optimality Principles in Biology, Butterworth, London, 1967.

Rosen, R.: RelationaleBiology. Fortschritte der experimentellen und theoretischen Biophysik, VEB Georg Thieme, Leipzig, Vol. 15, 1972a.

Rosen, R.: Relations between Structural and Functional Descriptions of Biological Systems, Intern. J. Neurosci. 3, 107-112 (1972b).

Rössler, O.E.: Design for Autonomous Chemical Growth under Different Environmental Constraints, Progr. Theor. Biol. 2, 167-211 (1972).

Rössler, O.E.: A Synthetic Approach to Exotic Kinetics, with Examples, These Proceedings (1974a).

Rössler, O.E.: Chemical Automata in Homogeneous and Reaction-diffusion Kinetics, These Proceedings (1974b).

Schpolski, E.W.: Atomphysik, VEB Verlag der Wissenschaften, Berlin, Vol. 1 (Section 23 and Appendix), 1970.

Thom, R.: Topological Models in Biology. In: Towards a Theoretical Biology, ed. by C.H. Waddington, Edingburgh University Press, Edinburgh, Vol. 3, 89-116 (1970).

von Holst, E. and Mittelstaedt, H.: Das Reafferenzprinzip (The Reafference Principle), Naturwissenschaften 37, 464-479 (1950).

Zeeman, E.C.: Topology of the Brain (Mathematics and Computer Sciences in Biology and Medicine), Medical Research Council, London, 1965.

Zeeman, E.C.: The Geometry of Catastrophe, Times Literary Supplement, 10th December 1971, 1556-1558 (1971).

THE ARTIFICIAL INTELLIGENCE/PSYCHOLOGY APPROACH TO THE STUDY OF THE BRAIN AND NERVOUS SYSTEM

B.D. Josephson
Department of Physics
University of Cambridge
Cambridge, England

A number of different approaches to the study of the brain and nervous system are reviewed, with particular reference to those of artificial intelligence, psychology and psychiatry.

1. Introduction

In this talk I shall discuss the interrelationships between a number of different approaches to the study of the brain and behaviour, and deal in particular with two which have not been dealt with elsewhere in this summer school, namely those of artificial intelligence and psychology. Both artificial intelligence and psychology attempt to tackle the problem of the brain by finding models for behaviour.

The models dealt with involve concepts such as planning, achieving goals, recognition, interpretation, concept formation and storage and recall of information. These models are thus at a different level to those considered elsewhere in this summer school, whose basic elements are neurons, synapses and nerve impulses. Because the concepts of the AI/psychology approach are ones of which one has intuitive knowledge, it is likely that progress in this direction will be more rapid than by the approach advocated by some neurophysiologists, which is to study the structure of the nervous system and the behaviour of nerve impulses in ever greater detail, in the hope that thereby one day illumination will suddenly dawn. I doubt if we shall be in a position to use such detailed knowledge until much more information as to the principles on which the brain works has been obtained by other means. On the other hand, certain general types of information may be of great use, for example knowledge that in the nervous system information transfer takes place by nerve impulses, processing in neurons and storage at synapses, and that there exist neuronal feature detectors and areas of the brain with specialized functions of various kinds. Again, it seems unlikely that automata theory or information theory approaches will be very helpful, as the brain is a very special type of automaton rather than a general one, and the information it handles (about the real world) has a special type of redundancy not considered at all in conventional information theory (as shown, for example, in the way natural language offers a far more efficient way to transmit information relevant to the actual environment than mathematical language or any type of code yet devised by the communications engineer).

2. Artificial Intelligence

The aim of artificial intelligence research is to study specifically intelligent behaviour. Generally the theories proposed are checked by writing appropriate programs for a digital computer. What the results obtained so far suggest is that, with the possible exception of processes involving a high degree of creativity, any process involving the exercise of intelligence, such as natural language comprehension, or scene analysis, can be reduced to a purely mechanical operation which can be simulated on a computer, if enough time and effort is put into writing the computer program. Of course, we do not yet have a computer program with the capability of a human being, but it has been found that each particular aspect of intelligent behaviour which has been studied so far has yielded to this type of attack.

It is appropriate at this point to consider the connection between the AI concept of "carrying out a program" and the concepts of the psychologist. It appears that at the most fundamental level all that is required is an associative or sequential memory which can execute the program steps in the right sequence and at the right times, and a filter which can select the appropriate step when a choice is possible (this corresponds to the psychologists' concept of discrimination learning). It is comforting to note that for all these processes there are simple models in neural net theory.

Besides the execution of a program it is necessary to consider the planning operation which precedes it. Here we are concerned with sequences of mental images, which probably involve the same operations in the part of the brain concerned as when the plan eventually formulated is actually executed. In this case, however, (the planning mode) they are decoupled from the outside world, which must therefore be represented by some internal world-model. This world model is probably used also in the execution mode, since any discrepancy between its predictions and the actual state of the world is generally a sign of the need for appropriately directed action. Again, pyschological evidence suggests that the world-model plays an important role in perception.

Let us now consider some criticisms which can be made of the current direction of artificial intelligence research. One feature of essentially all AI research to date has been the concentration, for historical reasons, on serial processing of information. However, the brain is not a serial processor but a complex interacting system, and it is well known that such systems can possess characteristic properties such as separation into domains with sharply defined boundaries. Such behaviour could provide simple explanations for such observed phenomena as separation of the visual field into objects and background, and channelling of information. It is also likely that operations involving parallel processing, essential to the correct functioning of the brain, are taking place during periods of "restful alertness", when the brain is active but not attending to any problem in particular. While it is true that an interacting system of arbitrary complexity can always be simulated by a sequentially operating digital computer, this is perhaps not the best way to approach the problem in practice.

Secondly, it may be asked whether current AI research is tackling the right problems or not. The main criterion fcr success in AI research so far has been the degree of intelligence shown by the resulting computer program. As a result, the subject of the acquisition of intelligence has been neglected, and it is probably a fair comment to make that in all cases to date more intelligence was required to write the computer program than was displayed by the program when run. It is not clear that this problem can ever be overcome without studying the question of the acquisition of intelligence.

3. Psychology and Psychiatry

The subject of acquisition of intelligence is not easy to tackle by the methods normally used in AI research, since anyone intelligent enough to do AI research cannot easily imagine what it is like not to be intelligent. The problem can, however, be tackled experimentally. In one type of experiment the psychologist sets up an environment in which one aspect of behaviour such as learning, motivation or memory can be studied in isolation as far as is possible. The response to a given type of task is then analyzed quantitatively. Another type of experiment is less formalised, and can consist, for example, of observation of the gradual process of acquisition of knowledge, skills of various types and intelligence in a child as it grows up.

Yet another approach is that of psychiatry, pioneered by Freud. In the initial stages this method involves the study of the sequences of mental images and emotions experienced by the investigator himself. Gradually patterns are discovered in the sequence of events, and models which account for them are tested against the experiences of patients and other investigators. In this way a far more detailed picture of the dynamics of mental processes has been built up than could be obtained by any other method. In the early days of psychiatry there was a tendency to make generalisations on the basis of inadequate evidence, but in the present state of maturity of the subject the basic ideas as given, for example, in the book by Berne, may reasonably be assumed to be valid. In studying the subject it is important to realise that the theories are <u>models</u> for what is observed, and that at the moment we have very little idea of what the connections are between the processes occurring in these models and neuronal events.

One important result obtained from the psychiatric approach is that "normal" behaviour is only the surface manifestation of much more subtle processes, which are normally observed directly only during dreaming or in cases of mental illness. Similarly, memory is found to be organised in such a way that at any given moment only a minute fraction of the stored information can be retrieved directly.

4. The Philosophy of Wittgenstein

Wittgenstein began his philosophical researches by following the traditional line taken by such people as Russell. After some years he decided that this approach was not a fruitful one, and turned to a line of his own, which is summarised in his book <u>Philosophical Investigations</u>. The ideas in this book are quite close to those discovered later by workers in artificial intelligence, and deserve close study. In this connection, it is interesting to speculate how many man-years would have been saved by workers on machine understanding of natural language if they had been familiar with Wittgenstein's work.

5. The Validity of the Description of Mental Processes in Conventional Physical Terms

It is worth considering here the question of whether a description in conventional terms (which could be either deterministic or quantum-mechanical) is adequate to discuss everything that goes on in the brain. There is indeed one significant omission in the description of the world given by quantum mechanics. The way in which one's description of a system changes when the system is <u>not</u> being observed is well prescribed, being given by Schrödinger's equation. On the other hand, the description of the effect of an <u>observation</u> on a system (involving spectral decomposition of the state vector) requires the specification of the operators corresponding to the quantities being observed. When one actually uses quantum mechanics, the appropriate operators are determined by heuristic arguments, and there is no way to derive them from the theory. The fact that the gap in the theory is connected with the question of observation is probably of particular relevance to the study of living systems, since observation is a basic property of living systems. It is likely that this problem is only the tip of the iceberg, and that only slightly deeper lie questions such as whether living systems can to some extent <u>control</u> the state of the world, for example by forming intentions. No experiment yet performed has any bearing at all on this question, and if we are interested in finding out about living systems it may well be a mistake to regard such questions as falling outside the realms of scientific enquiry.

6. Concluding Remarks

The work reported above constitutes the results of an investigation into the different ways to approach the brain problem, in which a striking feature observed on many occasions was the way in which research in one area appeared to be blocked by ignorance of or an antagonistic attitude towards research in a different area on the part of those concerned. In the course of this investigation a number of interdisciplinary links were discovered, some of which have been referred to above. These links might well offer promising grounds for future research.

Acknowledgements

I should like to express my thanks to Dr. B.J. Nicholson and Mr. J.K. O'Regan for stimulating discussions during the early stages of this work.

Reading List

Artificial Intelligence

Schank, R.C. and Colby, K.M. (eds.): Computer Models of Thought and Language, Freeman, San Francisco.

Psychology

Borger, R. and Seaborne, A.E.M.: The Psychology of Learning, Penguin, London.

Miller, G.A.: Psychology, the Science of Mental Life, Penguin, London.

Sandström, C.I.: The Psychology of Childhood and Adolescence, Penguin, London.

Psychiatry

Berne, E.: A Layman's Guide to Psychiatry and Psychoanalysis, Penguin, London.

Philosophy

Wittgenstein, L.: Philosophical Investigations, Blackwell, Oxford.

PART VI

Molecular and Modifiable Automata

CELLULAR AUTOMATA

(An Introduction)

Wolfgang Merzenich
Department of Computer Science
University of Dortmund, Fed. Rep. Germany

1. Introduction

The mathematical theory of cellular automata (CA) was first used by J. v. Neumann[1] who showed that complex abstract machines (mathematical structures) have the property of reproducing themselves. So the self-reproducing property is not only specific for biological systems but turns out to be a property of complex structures.

One of the main features of CA that may provide an application of the theory to a theory of nets of neurons is the following: a CA can be looked upon as very large net (of automata) that may have more than two dimensions (or only one dimension). We will see later how these nets can be defined formally. Such a net consists of "points" that are connected to other points of the nets by "lines". The points are interpreted as finite automata and the lines as the input or output wires of these automata. A single such automaton will be called a cell. A cell can be in one of a finite number of internal states at any instant of time and the information about the state of a cell is transmitted to other cells by means of the output wires. On the other hand a cell changes to a new state according to a certain rule, using the information it receives about the states of the cells that are connected with its input wires. Each cell of the net does so and in this way a configuration of states in the net will change to a new configuration at each time step. (We assume discrete time (i.e. time steps) and a synchronization of the cells.) This change of the configurations will be referred to as the global behaviour whereas the behaviour of a single cell will be called the local behaviour. A very important property of CA is that the global behaviour is completely determined by the wiring scheme and the local behaviour of the cells. An example at the end of this paper will show that a simple net with simple cells can produce complex global behaviour. For some reasons that will be clear from the text, the model of the CA as a mathematical structure is not assumed to be a model that is well adapted to the biological problems. But this model may be taken as a first approximation to a more biologically adequate one. The main reason for an introduction of this model is to show that a net of more or less simple functional units (cells) may produce global behaviour of the net that is rather complex (e.g. such a net can simulate a Turing machine).

2. A Formal Definition of Cellular Automata

First we need a description of the net. There is a mathematical structure that can express any sort of intuitive net. This structure is studied in the theory of graphs. The theory of cellular automata restricts the class of possible nets to certain graphs that have quite homogeneous structures. These graphs are called arrays or tesselations and can be viewed as a kind of "geometry". An array is given by an abelian group D together with a neighborhood relation (NB-relation) ν $D \times D$. The NB-relation must have the following properties:

i) For each $x \varepsilon D$ the set $\{y | y \; \nu \; x\}$ has k elements (k positive integer)

ii) $x \; \nu \; y \rightarrow (x + z) \; \nu \; (y + z)$

We give a more intuitive description of these properties, to show that this geometry is quite homogeneous and therefore may not be adequate for a description of nerve nets that have an extremely inhomogeneous geometry. But there may be simple nerve nets with a more homogeneous structure, too.

To illustrate the properties of our net-geometry we take $\mathbb{Z}^2$ as abelian group D (i.e. the group of pairs of integers with componentwise addition). The elements of this group may be represented by the points of the two-dimensional plane which have integral coordinates (Fig. 1).

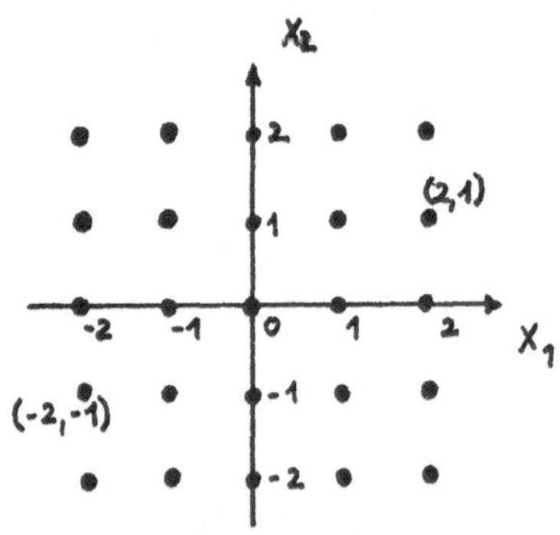

Figure 1

The first property of the NB-relation is that each point has exactly k neighbors and the second property means that the relative position of the neighbors of a point is the same for all points in this array. So the NB-relation is completely determined by the neighbors of one point. For example, if the point (0, 0) has the neighbors (-1, 1), (1, 1) and (0, -1) then the point (2, -1) has the neighbors (1, 0), (3, 0) and (2, -2) (Fig. 2). So the NB-relation may be defined by the k-tuple of neighbors N = ((-1, 1), (1, 1), (0, -1)). This shows the homogeneous structure of an array.

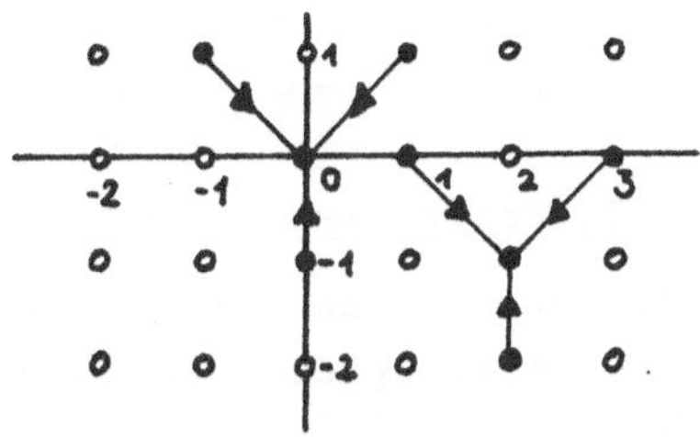

Figure 2

Now we think of the group elements $x \in D$ as addresses for cells (finite automata), i.e. each point represents a cell and the arrows of the NB-relation that lead to a point represent the input wires of the cell it represents. As a further simplification we assume that all cells have the same function or that all cells are copies of one cell, and this cell behaves according to a given function τ that assigns a state to a k-tuple of states of the k neighbors of the cell.

As in the case of finite automata we let S be a finite nonempty set and call it the state-set and define the operation of the cell by giving a function

$$\tau : S^k \to S$$

that we call <u>local</u> <u>transition</u> <u>function</u>.

We want to explain this function and the way it determines the global behaviour of the CA. By $c_t(x)$ we denote the state of the cell with address x at time t, so $c_t(x) \in S$. The neighbors of cell x are the cells $\{x + a_i \mid i = 1,\dots,k\}$ if $N = (a_1,\dots,a_k)$ is the k-tuple of neighbors of o. At time t we have therefore a function $c_t : D \to S$ and call it the <u>configuration</u> of the CA at time t. If c_t is the configuration at time t we can compute the next state of cell x:

$$c_{t+1}(x) = \tau\,(c_t(x + a_i) \mid i = 1,\dots,k).$$

In this way a configuration c_t is transformed into a configuration c_{t+1}. This transformation of configurations expresses the global behaviour of the CA and is called the <u>global transformation</u>. Let A : = S^D the set of all functions c : D → S (i.e. A is the set of all possible configurations). Then the global transformation is a function H : A → A. So we see that a CA behaves in a deterministic way, i.e. if we know the configuration at time t, we can predict the sequence of configurations that follow c_t if we know the neighborhood of cell o and the local transition function.

<u>Def</u>: A cellular automaton (CA) is a quadruple M = (D, N, S, τ) where D is an abelian group, $N \in D^k$ $(k \in \mathbb{N})$ and $\tau : S^k \to S$.

3. Properties of CA

Without going into the details and without giving mathematical proofs we will explain some of the properties of CA in this paragraph. As was mentioned in the introduction J. v. Neumann[1] first introduced the model of CA in a special form for a special purpose. He wanted to show that there exist nontrivial mathematical structures that can reproduce themselves in a certain sense. To avoid triviality he investigated structures with certain computational capabilities, i.e. he investigated structures that can simulate the operation of a Turing machine and in this way are computational universal. As abelian group D he used $\mathbb{Z}^2$ which we have already shown as an example in the previous paragraph, and as neighbors of the cell with address (O, O) he defined the 5-tuple ((O, O), (-1, O), (O, 1), (1, O), (O, -1)) (Fig. 3).

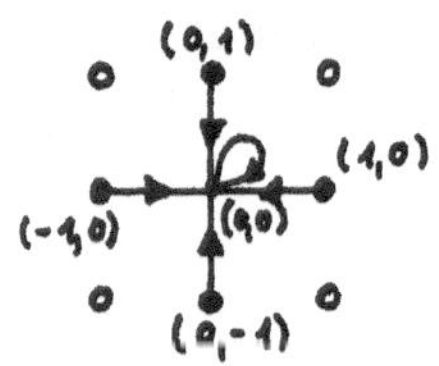

Figure 3

The state set consists of 29 states and the local transition function is quite complex. Such a CA can hold a configuration c, where all but a finite number of cells are in a special state, the quiescent state. This configuration may simulate the operation of a Turing machine and besides that ability it can produce a copy of itself in an area of the net that had been quiescent before. For details the reader is referred to [2, 3].

Another result is that there may exist configurations that never can be reached by the global transformation. These configurations can only exist at time t = 0. The configurations (if they exist) are called Garden-of-Eden configurations. Moore[4] and Myhill[5] gave necessary and sufficient conditions for the existence of Garden-of-Eden configurations. We put as a question whether this may be a property of nerve-nets, too. An interesting mathematical result that is due to A.R. Smith III[6] is that the number of neighbors may be drastically reduced (in the case of $\mathbb{Z}^2$ to 3) at the cost of an increase of the number of states. This result states that each CA for a special group can be simulated by a CA with a normed NB-index. But the cell structure would be complex when the simulated CA has a complex NB-index. So the complexity of the structure of a CA is given by the complexity of the NB-index and that of the cell. One can reduce one of them but has to increase the other in order to perform the same behaviour. This result may be interpreted for nerve nets. Either the neuron is a very simple functional unit and the complexity of behaviour of the nerve net is achieved by a complex wiring scheme or the wiring scheme is very simple and the neuron is a complex unit or reality is somewhere between these extremes.

To show how a simple cell and a simple wiring scheme can establish a CA with a variety of interesting phenomena we want to give an example that is taken from K. Zuse[7]. Again we use $\mathbb{Z}^2$ as the group for the CA and the NB-index is that of Fig. 3. We will define the state-set and the local transition function informally by a graphical representation (Fig. 4a-e).

A cell may carry a pulse in one of the four directions up, down, left and right. Such a pulse will be represented by an arrow starting from the cell (Fig. 4a); call this situation a single pulse. A single pulse will (by an appropriate definition of the local transition function) propagate in the direction the arrow is pointing (Fig. 4b); it moves from one cell to the next at each time step. When two pulses happen to meet at one cell we have three situations:

a) The two pulses have the same direction. In this case only one pulse of that direction shall propagate (4c).
b) The two pulses have opposite directions. In this case they shall annihilate each other. So no pulse will be emitted from the cell (4d).
c) The two pulses are perpendicular. In this case each pulse will be shifted one cell in the direction of the other pulse (4e).

If two pulses propagate along the same line in opposite directions it depends on their phase difference whether they annihilate each other or whether they cross without interference (5a, 5b). If two pulses meet at a cell in perpendicular directions, a new, complex pulse propagates

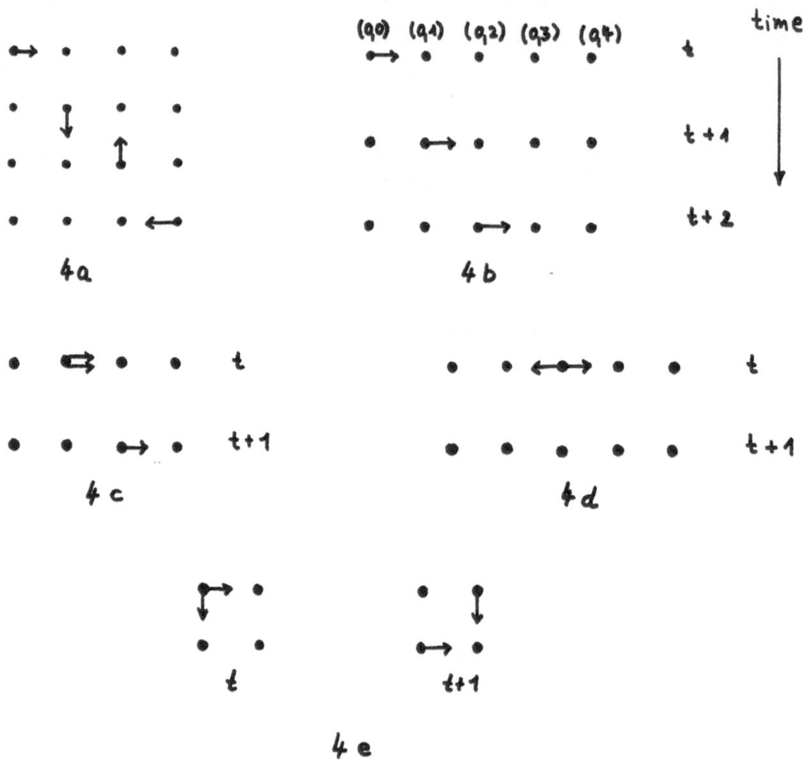

Figure 4

in diagonal direction 5c. In this figure the different positions and forms of the complex pulse are drawn in the same plane, i.e. at time t there is only the configuration of the two perpendicular pulses at one cell and at time t + 1 there is a configuration of two single pulses and so on. What happens if two such complex pulses propagate along the same diagonal line in opposite directions? (Fig. 5d) In this case propagation stops and a two-phase complex pulse will remain at the same position, but will change alternatively from state I to state II. This is an oscillator. Can such an oscillator be destroyed? This could happen if a pulse (single or complex) moves into the position of the oscillator. For a single pulse there are two possible phase conditions to hit the oscillator (Fig. 6a, 6b). In Fig. 6b the pulse travels through the oscillator without interference and in Fig. 6a the pulse is absorbed by the oscillator. Figures 7a and 7b show the situation for a complex pulse that is absorbed in one case and that travels through the oscillator

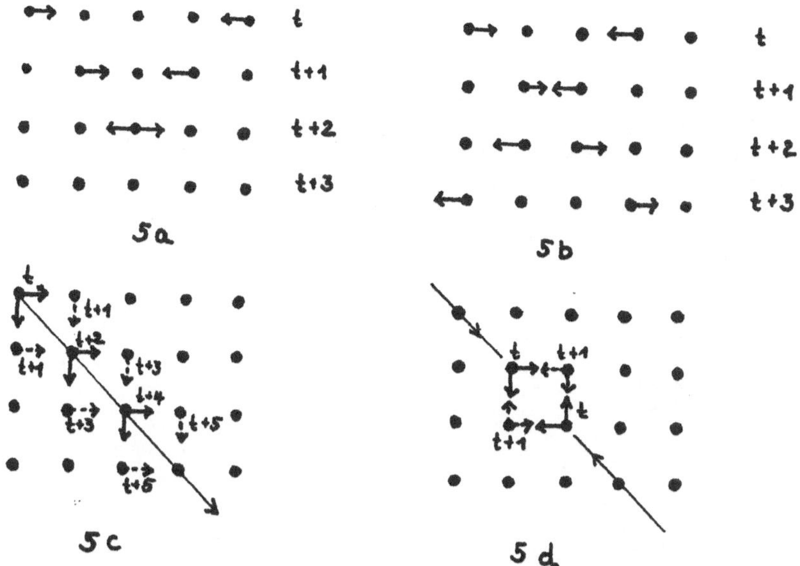

Figure 5

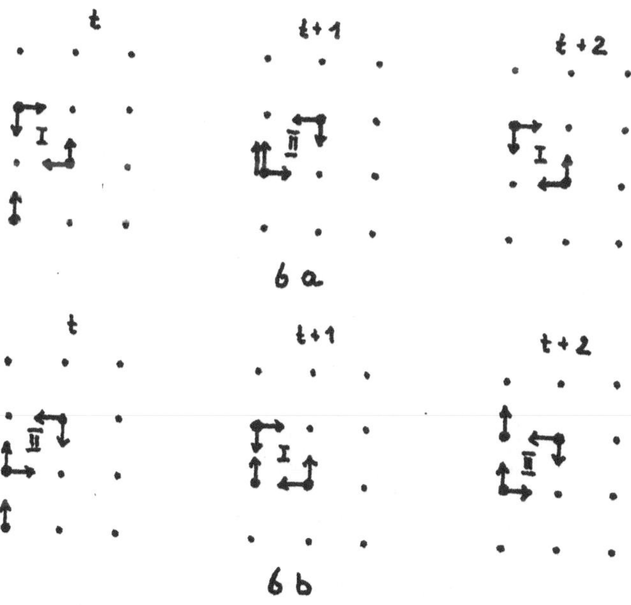

Figure 6

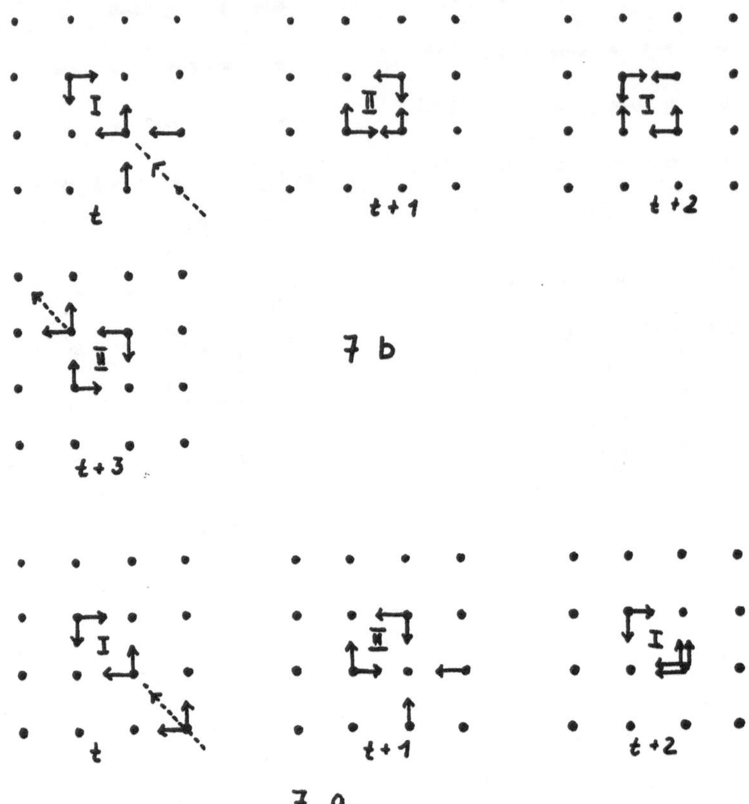

Figure 7

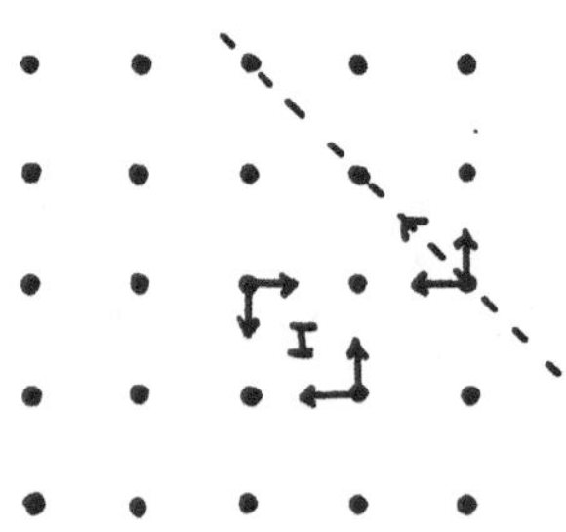

Figure 8

like the single pulse in the second case.

The reader may show that the complex pulse of Fig. 8 will destroy the oscillator completely and after collision two single pulses will propagate. Another example of a CA of complex behaviour is the game "life"[8] that also has a very simple local transition function. These examples show that a great complexity of behaviour can be achieved by a large net of simple functional units. The theory of CA is a first approach to a mathematical study of the relation between the local transition function and the global behaviour of a special kind of nets of similar automata.

References

1. v. Neumann, J.: Theory of Self-Reproducing Automata, ed. by A.W. Burks, University of Illinois Press, Urbana, 1966.
2. Burks, A.W. (Ed.): Essays on Cellular Automata, University of Illinois Press, Urbana, 1970.
3. Codd, E.F.: Cellular Automata, ACM-Monograph Series, Academic Press, New York, 1968.
4. Moore, E.F.: Machine models of self-reproduction, Proc. of Symp. in Appl. Math., vol. 14, AMS, 1962.
5. Myhill, J.: The converse of Moore's Garden-of-Eden theorem, Proc. of the Amer. Math. Soc. 14, 685-686 (1963).
6. Smith III, A.R.: Cellular Automata Theory, Techn. Rep. No. 2, Stanford University, Dec. 1969.
7. Zuse, K.: Rechnender Raum, Schriften zur Datenverarbeitung, vol. 1 Vieweg & Sohn, Braunschweig, 1969.
8. Gardner, M.: On cellular automata, self-reproduction, the Garden-of-Eden and the game life, Sci. Amer. 224, 112-117 (1971).

ON TURING MACHINES WITH MODIFICATIONS*

Roland Vollmar
Institut für Mathematische Maschinen und Datenverarbeitung (III)
(Codirector: Prof. Dr. W. Händler)
Universität Erlangen-Nürnberg
Erlangen, Martensstr. 1
Fed. Rep. Germany

After some informal definitions of Turing machines and their applications the importance of this concept for biological information processing will be mentioned. We shall see that Turing machines are very rough models. Therefore we shall try to build into this model "forgetfulness"--one of the phenomena observed in nature.

We will consider some modifications:

1) Turing machines for which the control unit becomes defective if their states have not been reached within a certain period of time and
   a) for which information on the tape is forgotten if it remains "untouched" for a long time
   b) which can reach information on its tape only if this information is already written on this tape a certain period of time before.
2) Turing machines with a modified control unit (as in 1) which operate on a two dimensional storage with failures
3) Turing machines which are imbedded in cellular automata with a forgetful structure.

It will be shown that all these modifications are equivalent (in a certain sense) to the basic model if certain restrictions on the functions which describe the forgetfulness and (in 2) on the distribution of the failure in the storage are fulfilled.

Introductory Remarks about Turing Machines, Their Definitions and Their Capabilities

First we will discuss the motivation which led to the model of Turing machines and then we shall describe its definitions in such a way that the following concepts can also be understood by nonmathematicians.

The introduction of Turing machines is motivated by searching for an exact meaning of the idea of "computability". In 1936 A.M. TURING proposed a simple formalism, later called a Turing machine (abbreviated as TM), as a model for computational work. We briefly introduce TM. We choose a specific version, the so-called on-line 1-tape TM. It is possible to prove (mathematically) that there are a great number of versions which are all equivalent (in a certain well-defined sense) to each other.

A TM consists of a finite automaton, called finite control unit,

*Some of the results mentioned in this paper are based on research supported by the Bundesminister für Forschung und Technologie (No. NT - 187)

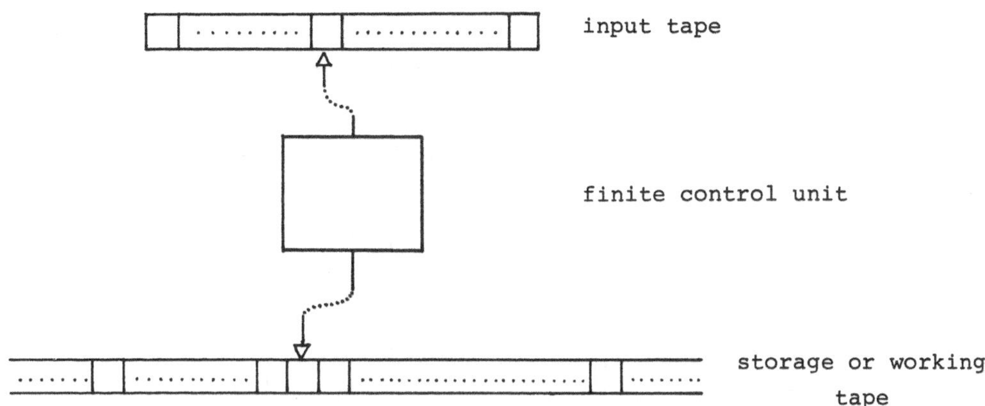

which can reach any state out of a finite nonempty state set, and two tapes which are chains of squares. One of the tapes is infinite in both directions. The finite control unit has access to each tape by means of heads, to the input tape by a read-only head and to the storage tape by a read-write head. The heads are placed on one of the squares of the two tapes at any given time. The TM works on discrete time $t_o, t_1, \ldots$ It is started in a designated start-state. An input word out of the set of all words of finite length over a finite alphabet is written on the input tape. The head on this tape initially scans the leftmost symbol of the input word; this read-only head may only move to the right. Initially the storage tape contains only special "blank" symbols.

The transition function, which is incorporated in the finite control unit, determines the behaviour of the transition of the TM - the next state, the "new" symbol to replace the symbol under the head of the working tape (it may be the same) and the movement of each head - depending on the present state and the symbols under each head.

In the set of states a subset is designated as the set of final states. If the TM reaches a configuration for which the transition function is not defined, then it stops.

Example for the working process of a TM: For a given integer $n \geq 0$ $2 \cdot n$ must be computed. (For this simple function the capabilities of the TM have not been utilized but it is too lengthy to discuss real examples!) The input (n) is encoded (on the input tape) in the following form: $\underbrace{II \ldots I}_{n}*$. The result should be on the storage tape in the same form as the input (2n strokes plus *).

Transition function:

conditions			consequences			
actual state	read symbol on input tape	storage tape	symbol to be written on the storage tape	move of storage tape head	input tape head	next state
q_1	I	b	I	R	R	q_2
q_1	*	b	*	R	R	q_3
q_2	I	b	I	R	N	q_1
q_2	*	b	I	R	N	q_1

q_1: start state b: "blank" symbol R: move to the right

N: no move

The importance of the formalism proposed by TURING for the foundations of logic and mathematics cannot be discussed here in detail, but the following will give a rough idea:

TURING'S thesis: Everything which is computable in the intuitive sense is also computable in the precise sense (by TM) and conversely.

or: The sketched definition of computability is an adequate explication of the intuitive concept of computability.

It is worthwhile to notice that this is not a mathematical theorem which is to be proved or to be disproved. Rather in this sentence there are words which have not been given a precise meaning before. Nevertheless today almost all mathematicians accept this thesis - and there are many reasons in favor of it. Perhaps it is interesting to note that TURING had the psychology of computing of man in his mind.

Motivation for concepts of modified TM: We shall interpret TM as formalisms for the mapping of strings of symbols into strings of symbols. For example the "computation" of the product of the two integers "11" and "234" is a transformation of the string "11·234" into the string "2574". Furthermore, the recognition of patterns is a transformation into the values "yes" and "no".

Example: The recognition of symmetry in a chain of symbols

uvuuvu → yes

vvvuvvuv → no

In my opinion pattern recognition is very important for the struggle of life and therefore, I think TM can be considered not only as models (very simple and rough ones) of technical data processing systems, but also of biological ones. One possibility for constructing more realistic models of the process of biological pattern recognition will be the introduction of phenomena observed in nature. Therefore we have defined TM with special modifications and we have compared the capabilities of these machines with those of the "classical" TM.

In the following I shall only try to develop the concepts and to cite the results and shall make only very few remarks about the proofs, which in general are rather tedious. All results involve comparisons between the TM and the modified TM with respect to the capabilities for recognizing sets.

"To recognize a set" means the following. The device "knows" the property defining the set; e.g. a set containing only strings of symbols of a certain alphabet which consist of three identical parts:

uvuuuvuuuvuu → yes
uvuu → no
uuu → yes

This knowledge is incorporated in the finite control unit. Elements (or words) whose membership in the set it must decide are written on the input tape. The machine reads the input tape and in general uses its storage tape to note some intermediate information. A word is said to be recognized (by a TM M) if after the complete reading of the input tape, M stops and is in an "accept"- or "reject"-state, i.e. in a final or in a nonfinal state, respectively.

The set of all words recognized by M is called the recognized set. We shall consider here only such TM which will stop for all words written on the input tape.

1. Turing Machines with a Modified Finite Control Unit and a Forgetful Storage Tape

In this first section we shall consider TM with the following modifications:

- States of the finite control unit are blocked if they are not reached during a certain period.
  If a state is blocked then the structure of the finite control unit is in general destroyed.
- Information on the working tape disappears if it is not touched, i.e. rewritten or reread, during a certain period of time.

It is worthwhile to notice that the information on the input tape is not changed (as is the case in all the other sections of this paper); it is unaffected during the whole working process. This assumption is made because the input comes from the "outside world". The assumption about the forgetfulness of the working tape is motivated by a principle of economy: Information which is not used during a long period of time is probably not relevant.

Some informal definitions:

Values of the function t are determined by the number of state transitions that have been performed so far.

Squares of the working tape are numbered beginning with the square on which the head is initially set and are designated by s_i $(i=0,1,\ldots)$.

A square of the working tape is touched by the corresponding head if the head writes on this square or reads from this square.

$\sigma(s_i,t)$ determines the moment $\leqslant t$ at which the square s_i was last touched by the head.

The value of the function $\tau(z,t)$ is defined as the time of the last occurrence of state z before time $t+1$.

We define $\tau(z,t_i):=\infty$ for all $t_i \leqslant t$ if state z is not reached until t. Analogously σ is defined. (Here and in the following states are denoted by z.)

1.1 A Turing machine with a variable structure of type 1 (TMVS1) is a TM described above with two additional functions μ,ν:

$\mu(t,\phi_1)$ defines the set of states blocked at moment t. After t the blocked states are no longer reachable.

$\nu(t,\phi_2)$ determines the set of squares of the working tape "erased" at moment t.

(These two functions are applied immediately after state transition and do not affect the time function t).

The recognizing of a set is defined in an analogous manner as before. If we define

$$\mu(t,\phi_1) = \{z \;/\; (t-\tau(z,t)) > \phi_1(t)\}$$
$$\nu(t,\phi_2) = \{s_i \;/\; (t-\sigma(s_i,t)) > \phi_2(t)\}$$

then we obtain the following result:

Let M be a TM. Let $\phi_1(t)$ be a positive function and $\phi_1(t) \geqslant c_1$ holds for all t where c_1 is a constant depending on the number of states of M. Let $\phi_2(t)$ be a positive, monotonically increasing, unbounded function. Then there exists a TMVS1 M_v such that M and M_v recognize the same set.

We will give some comments concerning the proof. Starting with M we shall construct M_v in such a manner that all the relevant states have been reached in a certain rhythm and that the information at the tape is "renewed" currently. For that purpose the number of tape symbols is increased in M_V with respect to M. By the rhythmic reaching of the states it is sufficient to use the function $\phi_1(t)$ with values which are greater than a constant depending on M. But it is not possible to transfer any information (in the usual way!) from one section of the tape to another section (because the control unit enters any state only for a fixed amount of time).

In this case we have no need for further knowledge about the function $\phi_2(t)$ - except what is supposed - because M_V checks at each addition of a square whether $\phi_2(t)$ is "sufficiently great". For this M_V writes "test information" on the tape first and tries to read it at a later time.

1.2 A modification of type 1 is obtained if we assume that information written on the storage tape is blocked during a certain period. (The aging of states happens in the same way as above.) This assumption is motivated by the fact that old people often remember things which happened a long time ago but not events which took place only a short time ago.

Instead of the function σ we shall now consider $\rho(s_i,t)$. This function determines the last moment $\leqslant t$ at which a symbol was written into the square s_i, but with the provision that it is not the same as that contained in the square immediately before.

A Turing machine with a variable structure of type 2 (TMVS2) is a TM defined above with two additional functions μ,λ:

μ is the same as in TMVS1 and $\lambda(t,\phi_3)$ determines the set of squares of the storage tape which cannot be read at the moment t. These squares are marked by a special symbol. (The blocking and deblocking do not affect the time function t.)

If we define

$$\mu(t,\phi_1) = \{z \,/\, (t - \tau(z,t)) > \phi_1(t)\}$$

$$\lambda(t,\phi_3) = \{s_i \,/\, (t - \rho(s_i,t)) \leqslant \phi_3(t)\}$$

then we obtain the following result:

Let M be a TM. Let $\phi_1(t)$ be a positive function and for all t let hold $\phi_1(t) \geqslant c_1'$ where c_1' is a constant depending on M. Let $\phi_3(t)$ be a positive function and for all t may hold $\phi(t) < c_3 t$ where $c_3 < 1$. Then there exists a TMVS2 M_V such that M and M_V recognize the same set.

The idea of the (constructive) proof is similar to that of the last result: the states are reached in a rhythmic manner. If a tape square must be changed, first the information about what is to be done--namely the next state, moves of the heads of the input and the storage tape and the symbol which must be written on the tape--is inscribed in this square. Then the head stays on this square - at that time the states are reached rhythmically - until the information written is "sufficiently old" and can be read in its correct form again.

Remarks:

1) The modified versions of TM introduced in section 1 can be understood as devices with a specified occurrence of faults. The defects arrive in a deterministic (but not in a probabilistic) manner and they are dependent on the processing time of the TM. It is possible to interpret the principles applied in the proofs of the theorems above as strategies for the self-repair of TM, which work "incorrectly" in the described manner.
2) For a special sort of the TMVS defined above, it has been shown that there is no possibility of recognizing some sets with the same speed as the unaltered TM.

2. Turing Machines with a Modified Finite Control Unit and a Two-dimensional Storage with Defects

As a starting point the TM is considered again; but instead of the one-dimensional working tape it is equipped with a two-dimensional storage divided into squares. The access to this storage is made by one read-write head. (It is known that this type of TM has the same recognizing capabilities as the TM defined above.)

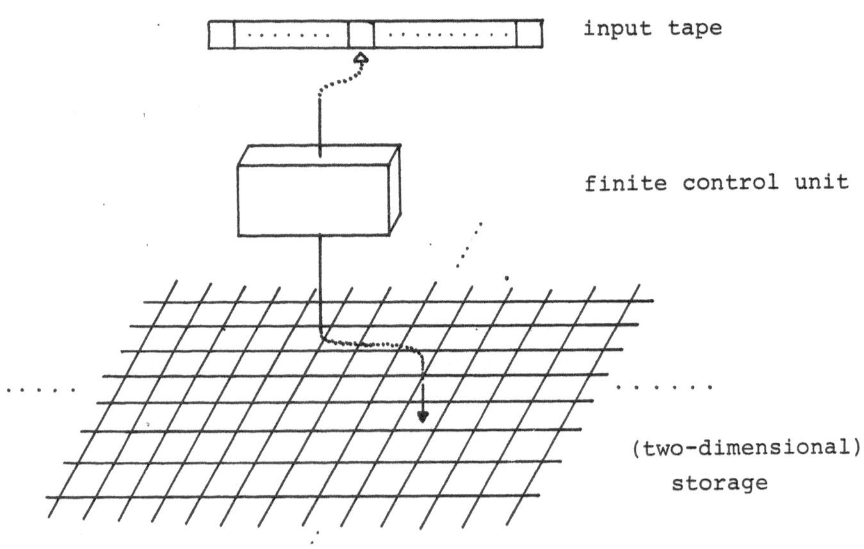

A Turing machine with a defective storage (TMDS) is a Turing machine with a two-dimensional storage sketched above with two additional constraints:

- States in the finite control unit which are not reached during a certain period are blocked and thus the structure of the finite control unit is destroyed.
- A set S defines those squares on the storage which are "defective", i.e. which contain a special symbol. (This symbol is unalterable.) S is called crossable if an infinite number of nondefective squares is reachable beginning with the square the head was on at the start of the process.

The definition of this type of Turing machine is motivated by the search for models for working in a defective environment. If we define the function μ as above we get the following result:

Let M be a TM. Let $\phi_1(t)$ be a positive function and for all t $\phi_1(t) \geq \bar{c}_1$ may hold, where $\bar{c}_1$ is a constant depending on M. Then for all crossable sets there exists a TMDS M_D such that M and M_D recognize the same set.

In the proof we make use of the fact that for each TM there exists such a device M' with a one-side bounded working tape which performs the same task.

M_D is constructed in the following way. The storage of the TMDS is considered as a maze. We apply the well-known algorithm of TREMAUX (in a slightly modified version) to find the next nondefective square - and there are such squares according to our assumption about crossable sets - if M' needs a "new" square at its storage tape. The alphabet of the storage is increased to such an extent that it is possible to store information about the connections between squares. (This information is independent of the size of the storage "used".) Then the control unit works in this part of the storage which is marked correspondingly. Simultaneously all the relevant states are reached in a rhythmic manner.

3. Turing Machines in Cellular Automata with a "Forgetful" Structure

In the last years there has been an increasing interest in the so-called cellular automata. On the one hand these are models for array computers and on the other they may be considered as a first approach towards describing neural networks by algebraic methods, and also the development of some simple sorts of organisms (in the form of so-called LINDENMAYER systems).

We shall briefly sketch the concept of cellular automata (abbreviated as ca) and of our modified version. In every lattice point of a d-dimensional space there are attached finite automata of the same structure. Each of these automata is connected with a finite number of other automata (determined by the template). These connections are homogeneous in the whole net. A ca is given 1) if the automata change their states synchronously at discrete steps depending on their own state and on the states of the automata connected with them (and possibly on an input from the "outside world") and 2) if there exists a state, the so-called quiescent state, determined by the fact that those automata which are in this state change into the same state when all of their neighbors are in this state as well. A subset of the set of states is designated again as the set of final states. At each time only a finite number of automata are allowed to be in the non-quiescent state. (This is only one possible definition but not the most general one.)

We shall also use in this section the functions τ and μ; but now we must consider these functions separately for all the automata in the ca and therefore we shall note them as $\bar{\tau}(\underline{x},z,t)$ and $\bar{\mu}(\underline{x},t,\phi_1)$ respectively, where $\underline{x}$ designates the automaton in the lattice point $\underline{x}$. In analogy to the former sections we define $\bar{\mu}(\underline{x},t,\phi_1) = \{z \;/\; (t - \bar{\tau}(\underline{x},z,t)) > \phi_1(t)\}$.

A forgetful cellular automaton (fca) is a cellular automaton sketched above with a transition function (for each automaton $\underline{x}$) which is modified in the following way: An automaton $\underline{x}$ reaches (at time t) a state (controlled by the transition function) only if this state is not contained in $\bar{\mu}(\underline{x},t,\phi_1)$; otherwise $\underline{x}$ will enter the quiescent state. (The structure of the finite automaton $\underline{x}$ is then (in general) destroyed.) Therefore the quiescent state is the only state which does not "age".

Again we want to recognize sets which are presented from the outside world. Therefore we shall consider some sort of more specialized ca (sometimes called iterative automata). An example of such a type is given in the following picture:

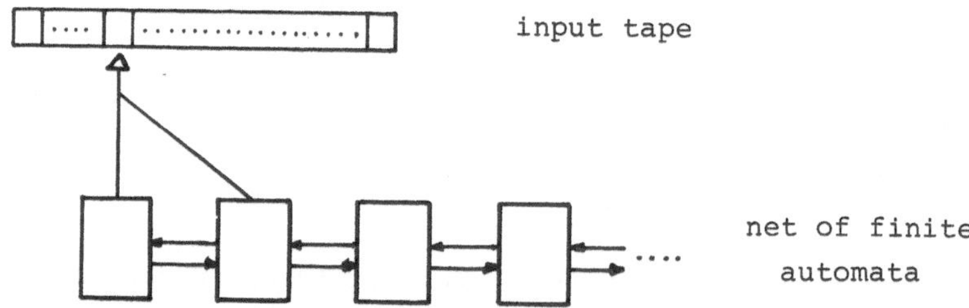

The automata in this one-dimensional one-side bounded net exchange information at each step only with the immediately neighboring automata (in the ordinary meaning). The input is read only by the first automaton and the automaton (automata) connected with it directly; such automata we shall denote as input automata.

Changing the usual definition we define: A ca (fca) recognizes a word if all the input automata are in the same final state after having read the whole input word (normally after a certain period). Then we obtain the following result:

> Let M be a TM. Let $\phi_1(t)$ be a positive function such that for all t $\phi_1(t) \geqslant \chi$ holds where χ is a constant depending on M. Then there exists a fca ζ such that M and ζ recognize the same set.

To prove this we construct a ca which recognizes the same set as M first (this is possible with the neighborhood sketched in the figure). Then we construct using a procedure of HÖLLERER the fca: the ca is "inflated" (at the same time the neighborhood must be increased).

During the working phase of the fca the automata are set in three different modes in a regular way: The automata in the so-called C-mode do the actual "work". The automata in the R-mode reach all their states in a rhythmic manner. Each pair of automata in either the R- and in the C-mode or in the C- and in the R-mode is separated from such other pairs by automata which are in the Q-mode. These automata stay in the quiescent state. After equal periods the automata in the C-mode and the automata in the R-mode change their modes. This change guarantees that no state of the automata in the C- and in the R-mode has been "forgotten". The automata in the Q-mode are needed to identify the other automata. If a word is to be accepted, the input automata (independent of their modes) reach the same final state.

Concluding Remarks

In this paper we considered some possibilities for simulating online Turing machines in devices with defects which occur in a deterministic manner. The simulations are possible if certain assumptions about the "fault functions" are fulfilled. The sketched devices must not be understood as biological models. This is rather an approach - as mentioned earlier - to introduce phenomena observed in nature (greatly simplified) into the behavior of the "classical" Turing machine. The intention was to sketch the concepts, and therefore the descriptions of definitions and of results are quite simplified. One can find details in the papers referred to below and in some forthcoming papers.

References

(The following list should not be taken as a complete collection of all the literature used.)

The following two books contain a lot of material about the theory of computability and Turing machines:

Davis, M.: Computability and Unsolvability, McGraw-Hill, New York, 1958.

Minsky, M.: Computation: Finite and Infinite Machines, Prentice-Hall, Englewood Cliffs, New Jersey, 1967.

In

Arbib, M.A.: Theories of Abstract Automata, Prentice-Hall, Englewood Cliffs, New Jersey, 1969.

you can find a survey of automata theory and the theory of computability as well as references to the biological relevance of some of the results of these topics.

There exist some connections between the material presented in sections 1 - 3 and the papers of

Lofgren, L.: Self-repair as a computability concept in the theory of automata, Proc. of the Symp. on Math. Theory of Automata, Polytechnic Press, Brooklyn, N.Y., 1963, 205-222.

Salomaa, A.: On finite automata with a time-variant structure, Information and Control 13, 85-98 (1968).

A part of the material presented in section 1 is contained in a weaker form (but the proof method doesn't work for the results cited above) in

Vollmar, R.: On Turing machines with a variable structure, Arbeitsberichte des interdisziplinären Modells DORA, Nr. D1-3, Erlangen, 1972.

Other modifications of Turing machines (without a change of the finite control unit) are considered in

Vollmar, R.: Über Turing-Maschinen mit variablem Speicher, Elektronische Informationsverarbeitung und Kybernetik 9, 3-13 (1973).

Material about cellular automata is contained e.g. in

Burks, A.W. (Ed.): Essays on Cellular Automata, Univ. of Illinois Press, Urbana, 1970.

Results about the simulation of Turing machines in cellular automata are in, e.g.

Smith III, A.R.: Simple computation-universal cellular spaces, Journal of the ACM 18, 339-353 (1971).

The algorithm of TREMAUX is described in e.g.

Müller, H.: Stackautomaten in Labyrinthen, Archiv für Math. Logik und Grundlagenforschung 14, 127-134 (1971).

The cited procedure of Höllerer will appear in

Höllerer, W.O.: Über zellulare Netze mit variabler Struktur, Arbeitsberichte des interdisziplinären Modells DORA

CHEMICAL AUTOMATA IN HOMOGENEOUS AND REACTION-DIFFUSION KINETICS

Otto E. Rössler
Division of Theoretical Chemistry, University of Tübingen
Tübingen, Federal Republic of Germany

Contents

1. Introduction

A finite automaton or, synonymously, a finite state machine is in the simplest case a triple (X,I,λ), whereby X is a finite set of states, I is a finite set of inputs, and λ is the next-state mapping, such that $\lambda : X \times I \rightarrow X$ (cf. Arbib, 1969). For example, if x_1 and x_2 are two state variables each possessing two possible states, the whole automaton has four possible states, and λ specifies the transitions between these states in dependence on a given input. If the input is constant, one speaks of an autonomous automaton, otherwise of a nonautonomous automaton.

In contrast, the dynamics of a chemical reaction system is usually described by a differential system, i.e. a dynamical system. A dynamical system is a pair (X,π) whereby X is a continuous set with a topology defined on it, and π a mapping such that $\pi : X \times R \rightarrow X$, whereby R, the real axis, corresponds to time (cf. Bhatia & Szegö, 1970). Through each point in state space a unique trajectory is passing. The state space contains a finite number of basins separated by separatrices. Every basin contains an attractor (limit set), which in the simplest case (steady state) is a fixed-point of the mapping of X onto itself.

Specifically, three types of chemical reaction systems (all subject to the constraints of being isothermal and of observing an appropriate concentration range) will be considered in the following, all being naturally described in terms of a differential system. They are:

i) Homogeneous reaction systems. They are described by ordinary differential systems of the form (in vector notation; $\cdot$ = time derivative):

$$\dot{x} = f(x), \tag{1}$$

whereby $\underline{f}$ is the rate function of the system. In the majority of chemical systems of this type, the mass-action law applies, so that the $\underline{f}_i$ are polynomials of less than the fourth degree in the substances x_i. In the special case that only steady states are considered, or that the so-called steady-state approximation is valid (see Heineken, et al., 1967), rational algebraic functions of the concentrations may also be present on the right-hand side as summands (for example Michaelis-Menten terms).

ii) Compartmental reaction systems. In the simplest case the different compartments (or, synonymously, "cells") all contain the same reaction system, whereby corresponding substances are coupled by diffusion-terms D_i. In the case of just two compartments, the rate equations have the form (in vector notation; the prime refers to the second compartment which obeys an analogous equation):

$$\dot{x} = f(x) + D(x' - \alpha x). \tag{2}$$

$\alpha \neq 1$ refers to the presence of active transport between the cells.

iii) Reaction diffusion systems. These non-stirred reaction systems without convection are described by parabolic partial differential systems:

$$\partial_t x = f(x) + D\nabla^2 x. \tag{3}$$

Eq. (2) with $\alpha = 1$ can be considered as the first discrete approximation (Taylor-expansion) of Eq. (3).

The question to be posed is whether or not, and under which conditions, reaction systems of these 3 types can be described also by automata theory. The more general question is, of course, whether or not dynamical systems exist which admit a (in a sense) canonically equivalent description in the language of finite automata theory (yielding a "short-hand description"). This question is not to be confounded with the other, much more general problem, recently raised in general systems theory (see Kalman, et al., 1969), whether or not a common language can be defined for all types of mathematical systems representations, including finite automata and dynamical systems.

A first "correspondence principle" (in the restricted sense we are interested in here) has been formulated by Rosen(1968): all states belonging to the same basin of a dynamical system are considered equivalent to one state of a corresponding automaton; the problem of state-transitions is hereby circumvented by the postulate of sudden changes of initial conditions in the dynamical case. This type of "dynamical automata" (as they may be called), involving pulse-shaped inputs, will be referred to as "AC type" dynamical automata in the following. This notion suggests that "DC type" dynamical automata are also possible (see below).

In what follows, a heuristic procedure will be adopted. We will try to arrive at a first description of chemical automata by simply looking for homogeneous reaction systems which are mathematically equivalent to other physical realizations of finite automata (like mechanical automata and electronic computers). The results will then be generalized to the compartmental and the reaction-diffusion case. A discussion of the mathematical existence problem, however, (which is necessarily the same for all kinds of physico-chemical realizations of finite automata) will not be attempted.

2. Homogeneous Chemical Automata

Almost any physically realized finite automaton belongs to a small subclass of finite automata, the so-called $\underline{Z}_2$-networks (cf. Hotz & Walter, 1969). Since any finite automaton can be simulated by a $\underline{Z}_2$-network, no restriction of generality is implied by the consideration of this subclass.

$\underline{Z}_2$-networks consist of three major classes of elements:

(a) flip-flops,

(b) logical elements, and

(c) delay elements.

The logical elements, (b), may, for example, comprise the three logical functions NON, AND, and OR. Since this is a "logically universal" set of logical functions, the so-defined class of $\underline{Z}_2$-networks is universal (cf. Shannon, 1938).

If these 3 types of elements can be shown to exist (and to allow for the appropriate wiring) in homogeneous kinetics, the existence of homogeneous chemical automata will be established up to analogy with other physico-chemically realized finite automata.

A. Homogeneous Flip-flops

A flip-flop has two possible states and two possible inputs (in the simplest case of an RS flip-flop to be considered here), one input "re-setting" the flip-flop and the other "setting" it (Hurley, 1961). So the next-state is described by

$$x_{t+1} = S \vee (\overline{R} \wedge x_t), \tag{4}$$

whereby the specific restriction applies that both inputs are never "on" simultaneously, i.e.

$$S \wedge R \neq 1. \tag{4a}$$

Accordingly, the system remains in the "upper" state, once it is set, even if S becomes 0 again, and it remains in the lower state, after R has become 1, even if R is 0 again, and so forth.

This type of behavior is an example of so-called hysteresis behavior: Since R and S are disjoint, they can be considered as a single parameter P (P = 0 if both R = 0 and S = 0; P = 1 if S = 1 and R = 0; and P = -1 if R = 1 and S = 0). With respect to recurrent up-and-down variations of this parameter, the state of the whole system runs through a hysteresis loop.

Indeed any physical realization of an RS flip-flop makes use of this fact. Both the mechanical "toggle" (as contained in a toggle-switch, for example), and the electronic toggle form examples. The best-known electronic toggle is the Eccles-Jordan trigger (Eccles and Jordan, 1919) which is still used as an active storage element in modern digital computers (Fig. 1). This system is, like any physico-chemically realized hysteresis system(Rashevsky, 1929), a bistable system (cf., for example, Andronov, et al., 1966). Now bistable systems are rather well-known in reaction kinetics. In fact, they form one of the major paradigms in exotic kinetics.

As a first example of a homogeneous bistable reaction system, the

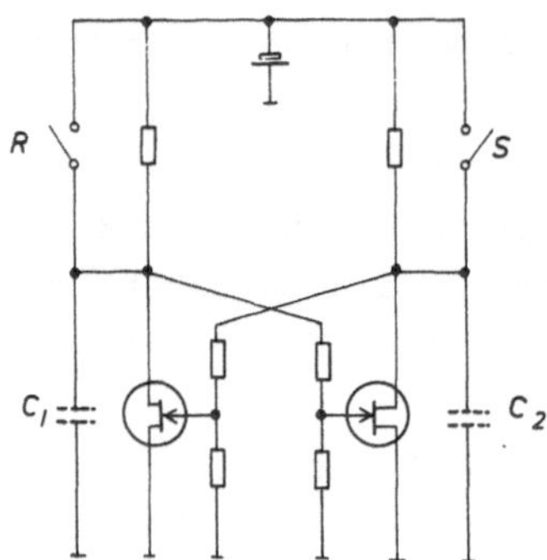

Fig.1. The electronic Eccles-Jordan trigger (using field-effect transistors).

"chemical Eccles-Jordan trigger" may be mentioned: see Fig. 6 and Eq. (6) of the preceding communication (Rössler, 1974c).

The <u>simplest</u> homogeneous bistable system, from the viewpoint of realizability, seems to be formed by an autocatalytic reaction (or reaction sequence, respectively) which is prevented from starting autonomous growth by the presence of either a second-order reaction taking place with some exogeneously supplied inhibitor I (Eq. 5),

$$\begin{array}{l} \longrightarrow X \longrightarrow \\ \quad\; X + I \longrightarrow \\ \longrightarrow I \longrightarrow \end{array} \qquad (5)$$

or a wall-reaction. The latter is dynamically identical to the presence of a Michaelis-Menten catalyst I of constant concentration (Eq. 6).

$$\longrightarrow X \longrightarrow \quad (\text{catalysed by } I) \qquad (6)$$

In both cases, an additional, second-order outflux reaction of <u>X</u> (as it is always present, both due to the postulate of reversibility of all chemical reactions and due to the fact that the reaction takes place in a solution) imposes an end to the growth process, and a small leakage influx to X allows the process to start, if $\underline{I}$ is sufficiently small. The system thus can be set by reducing $\underline{I}$ or its influx, respectively, and be reset by increasing it.

The simple differential equation corresponding to Eq. (6) is

$$\dot{x} = j + x - x^2 - I \frac{x}{x+K} , \tag{7}$$

where $\underline{I}$ is the resetting-setting parameter, and j = 0.01, and K = 0.05 (e.g.). The resulting hysteresis loop is determined by a numeral-2-shaped steady state curve (Rössler, 1972b).

By means of a minor modification (mass conservation between influx and outflux; Rössler & Hoffmann, 1972b), an almost perfect z-shaped curve of steady states can also be obtained, from the following equation

$$\dot{x} = j \frac{(C-x)}{(C-x)+K_1} + x \frac{(C-x)}{(C-x)+K_2} - I \frac{x}{x+K_3} \tag{8}$$

(where C = const., K_1, K_2, $K_3 \to 0$). This equation is a single-variable analog to the equation of the chemical Eccles Jordan trigger which also yields a potentially ideal hysteresis curve (Eq. 6 and Figs. 8 and 9 of the preceding note).

Asymmetrically-built homogeneous chemical hysteresis systems need not to involve an autocatalytic reaction (see, for example, O'Neill et al., 1971, and Lutz, 1973), although a self-activating cycle can always be shown to be present in a steady state-analysis, by contracting the number of variables.

Finally an example for a concrete reaction network possessing the structure of Eq. (5) is pointed out (Fig. 2). It is a partial system contained in the well-known Belousov-Zhabotinsky reaction (see Vavilin & Zhabotinsky, 1969). Curiously enough, an experiment proving the prediction of hysteresis behavior of this system has not yet been performed.

The preceding examples provide sufficient evidence for the existence of (both abstract and actual) homogeneous <u>RS</u> flip-flops.

B. Homogeneous Realizations of the Logical Functions NON, AND, and OR

The function NON is automatically realized by any more or less symmetrically-built <u>RS</u> flip-flop (like that of Eq. 5). Here the second variable always shows the opposite behavior to that of the first. In the case of asymmetrically-built <u>RS</u> flip-flops (like that of Eq. 6), an additional reaction leading to the same result can always be introduced.

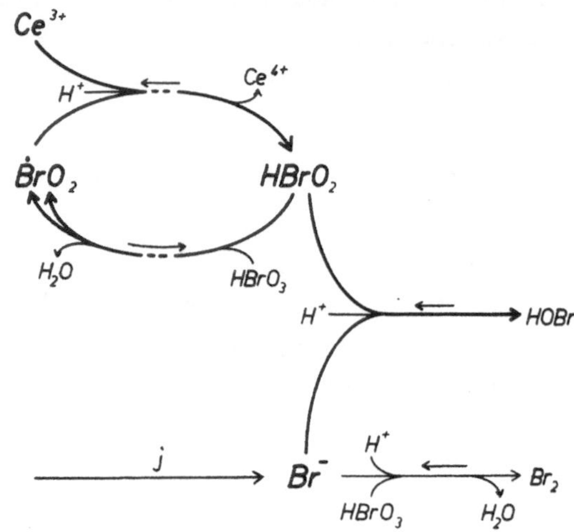

Fig.2. A homogeneous chemical hysteresis system. j = exogeneous influx = hysteresis eliciting parameter, Ce^{3+} and $HBrO_3$ are assumed to be pools.

The functions AND and OR are trivially realized by algebraic addition, in the presence of a threshold. Now any flip-flop is a threshold device. Hence any pair of "convergent" reactions (leading to the same product) realizes these functions in a straightforward manner. If both influxes are required for the setting (or resetting, respectively) of the flip-flop, AND is realized, and if either is sufficient, OR.

In electronics, a threshold logic of the here described type has been called "direct coupled logic" (Ware, 1963).

C. Homogeneous Delay Elements

The presence of a threshold renders any chain of consecutive reactions (of one or more elements, and with or without additional substances participating) an appropriate delay unit.

However, this sort of delay is only applicable if the next-following flip-flop is to be switched in the DC manner, i.e. by a slow change of the input variable (i.e. the "hysteresis eliciting parameter" P).

If rather a fast, reversible change of the parameter is preferred (AC type switching), either a whole chain of flip-flops has to be used, or a single DC delay has to be combined with a consecutive flip-flop.

It may be added in parenthesis at this point that a third (combined AC and DC) way of switching a bistable system exists (von Neumann, 1956; Wigington, 1959; Keyes & Landauer, 1970): to render the bistable system globally stable for a while, using an appropriate parametric change (AC pulse) of sufficient duration, and to provide, in addition, a small DC type perturbation just when the system is in the susceptible state (i.e. before bistability is going to be re-established). Adopting Zeeman's "cusp" terminology (see Fig.16 of Zeeman, 1972), this type of switching consists in the soft circumnavigation of the edge of the cusp.

The contents of the preceding three subsections allow the following conclusions to be drawn:

1) Homogeneous chemical Z_2-networks of logical universality and arbitrary complexity can be constructed in principle.

2) Any hard catastrophe-type homogeneous chemical reaction system can be considered as a homogeneous DC type dynamical automaton. (This result, which will be specified in the next section, is the direct correlate to the above mentioned result of Rosen, 1968, concerning AC type dynamical automata).

3) DC type homogeneous automata, involving both a direct-coupled logic and DC type delay elements, lead to the least-complicated reaction systems and are, therefore, more straightforward to realize.

D. The Simplest Homogeneous Automata

Now that the possible existence of logically universal homogeneous automata of arbitrary complexity has been established (up to analogy with other physically realized finite automata),a look at the simplest examples is allowed.

The three simplest finite automata are (apart from the RS flip-flop itself): the so-called astable flip-flop, the monostable flip-flop, and the T flip-flop. The first is an autonomous automaton, flipping back and forth between either (quasi)stable state. The second falls back spontaneously after a while, but has to be pushed out of the first state by an exogeneous input. The third switches back and forth in response to the same (T) input (T,because this letter follows R and S). All three automata can be realized on the basis of a single RS flip-flop and a single DC type delay element (the "next" flip-flop is hereby identical with the first).The three simplest reactions designed after this rule are shown in Fig.25 of the preceding paper (Rössler,1974c), where also a simulation result is found (Fig.15). The respective flip-flop is set and reset in dependence on its own former state.

The corresponding equations are obtained by adding a single differential equation, involving the hysteresis-eliciting parameter I of the RS flip-flop which is hereby rendered a slowly changing variable ($\mu<10^{-2}$):

$$\dot{I} = \mu(x - aI + b). \qquad (9)$$

By shifting the slope of the nullcline determined by Eq. (9), either

3 or 1 intersection points with the nullcline (or, more generally, the "slow manifold"; Zeeman, 1972) of the flip-flop can be obtained. The bistable situation (3 intersection points) can be changed both into the astable situation (only the unstable intermediate steady state being left) and into the monostable situation (one of the stable steady states being left). The former shift yields T flip-flop behavior when being reversed soon enough.

It is finally mentioned that relatively simple homogeneous T flip-flops can also be obtained on the basis of the above described von Neumann switching principle (see the example described by Seelig & Rössler, 1972).

E. A Concrete Example

Although the Belousov-Zhabotinsky reaction network is very complicated (see Noyes, et al., 1972, and Field, et al., 1972), so that it appears almost impossible to classify all its possible behaviors, depending on temperature and initial conditions, one mode of action of the system can, probably, be related to that of the composed system obtained by the coupling together of Eqs. (7) and (9). A definite proof will of course have to await an experimental "pruning" of the system (such that only the RS flip-flop of Fig. 2 and a delay network are left over). Nevertheless, there is one strong argument speaking in favor of our "relational" interpretation (in the sense of putting the system into a larger equivalence class of systems, namely that of astable flip-flops): the system can also be run, with just slightly changed parameters (addition of bromide), as a monostable system (Winfree, 1972, 1974). An appropriate parameter for the elicitation of the third behavioral type (T flip-flop behavior) has not yet been looked for.

F. Prebiological Evolution Viewed as a Growing Automaton

One can conceive of a more or less homogeneous stage of prebiological evolution in which an increasing number of ever more complicated "nonresettable" flip-flops of autocatalytic type have been present, whereby those being set earlier are both producing and (after a while) triggering the new ones; the former effect being due to their providing new pools (i.e. new substances maintained at a high concentration level) and the latter due to the slow filling up of delay-channels (i.e. networks of consecutive reactions) which finally trigger another one of the flip-flops. A system of this sort constitutes a primitive "growing automaton" (Burks, 1961).

A growing automaton of this type possesses an interesting feature: it may itself, as a whole, be analogous to each of its parts. Depending on the preassumed wiring pattern (that is, its richness of connection), the whole system itself may constitute a triggerable system (i.e. non-resettable flip-flop). This result can be obtained most simply on a level which is even more abstract than that of automata theory, namely with the theory of growing graphs (Erdös & Rényi, 1960).

The obtained "domino theory" of earliest prebiological evolution (Rössler, 1971) has the advantage that it allows one to view the whole evolutionary process as being governed by a single principle. The possibility to rely more and more on the production of chemically <u>related</u> flip-flops, being automatically exploited after a while by the growing automaton, appears as the greatest qualitative transition ever occurring during the whole process. Later, at biological stages, where the flip-flops have, as "species", acquired more and more complicated properties, the readiness of transformability (such that ever harder to exploit growth-conditions, "niches", can be tolerated by a successor system) still constitutes the fastest-developing property ("recursivity principle of biological evolution", Rössler, 1972c; "self-optimization of evolutionary optimization", Rechenberg, 1973).

Of course, the preceding functional-reductionist view of biological evolution constitutes only a "minimal theory" of the whole process (Rössler, 1972c).

Another growing chemical automaton will be encountered below in the context of morphogenesis (Section 3a).

<u>G. An Unsolved Mathematical Problem</u>

The understanding of synchronous homogeneous automata does not involve any deeper problem, once an exogeneous "clock" is presupposed (so that all flip-flops switch only when this input, being ANDed to the other inputs, is one). In fact, the dependent switching systems need not be real flip-flops (i.e. bistable systems), but may as well be devices possessing (more or less sharp) sigmoidal characteristics.

The situation is different, however, for asynchronous chemical (and physical) automata. The problem exists equally in AC and DC type automata, but it is easier to demonstrate for the latter case. It can be comprised in the tentative notion of a "continuous perturbation".

If a flip-flop is switched from one basin to the other by means of a parametric perturbation, the problem is easy in the case of discontinuous parametric perturbations, since the powerful theory of structural stability of dynamical systems is applicable (cf. Andronov, et al., 1971), allowing one to compare the qualitative structure of the system (including attractors and basins) before and after the parametric jump. The essential notion is that of hard bifurcation, referring to the sudden extrusion of the representative state into a new basin (after the former attended one has disappeared like a submerged island), so that the state has to drop toward a new attractor. However, if the system is to be switched by a continuous parametric perturbation, so to speak, i.e. by a slow continuous change of the same bifurcation parameter, one cannot speak of the "disappearance" of an attractor and basin any longer, since there exists just one dynamical system, possessing (at least in the case of the simplest astable flip-flop, namely Liénard's equation; Rössler, 1973) one global attractor only, namely the limit cycle itself. The "disappearing basins" therefore become, at best, "sub-basins" within the single global basin present. This notion, however, is just as undefined as that of continuous perturbation, or that of quasi bifurcation (which also appears reasonable at first sight).

The basic mathematical question arising is whether or not the notion of structural stability can be extended in such a way that different, but adjacent parts of state space of the <u>same</u> dynamical system can be compared (just as, so far, the same part of state space of <u>different</u> but adjacent dynamical systems can be compared).

If the problem cannot be solved at this general level, one may try to solve it at least for a very simple special case, namely for the equation

$$\dot{x} = - x^2 + P,$$

P = slowly decreasing parameter, starting at positive values. Locally, any case of hard quasi bifurcation will look like the right hand half of Fig. 3, when compared to the case of genuine hard bifurcation (left

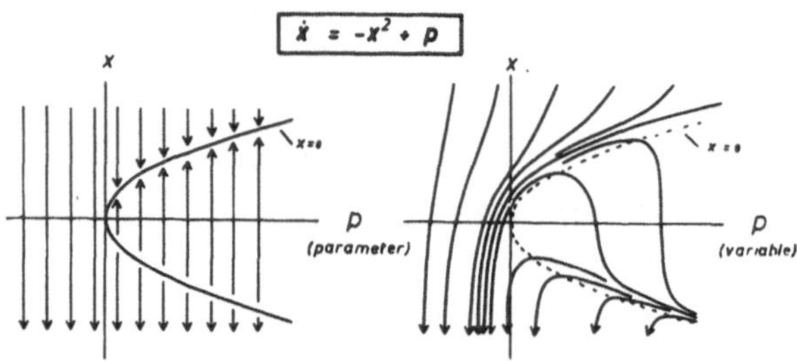

Fig. 3. Prototypes of hard bifurcation (left) and hard quasi bifurcation.

hand side). Explicit computation of the solutions, leading to Hankel functions (R. Wais, personal communication), does not help. The existence of a single distinguished trajectory, starting at a saddle-point at infinity, also is of no help, since it is not distinguished locally and, of course, neither disappears nor suddenly loses its attracting properties near P = 0. Equally, the computation of the trajectorial divergence in the system dx/dP also does not solve the problem. The suggestion to look at the potential of the orthogonal trajectories (isohypses), defining a <u>curved valley</u> which, as a sub-basin, indeed runs into a main basin, finally leaves one with the problem how to define the bottom of a curved valley. Clearly there is no distinguished place for a single trajectory possessing similar attracting properties as a river in a real valley. However, the possibility that a distinguished "bottom region", possessing some of the expected properties, can be defined still appears as a real one. In the positive case, the result would be a generalization of the notion of attractor, obtained at the expense of losing several of the major properties of invariant sets (see Bhatia & Szegö, 1970).

Needless to mention that the same problem (of "hard quasi bifurcation") is equally important in other, more complicated sorts of catastrophe-type systems (for instance those considered by Thom, 1972).

The same problem of DC type switching is also essential for "synergetics", as an interdisciplinary field dealing with co-operative phenomena in terms of a common language has been called (Haken, 1973).

<u>H. Chemical Systems Optimization and Homogeneous Automata</u>

In any Z_2-network, a building-block principle applies, as we have seen. In addition, the particular building-block "flip-flop" can be realized by a number of alternative circuits. Hence an unusually great freedom characterizes the design of homogeneous automata. It appears possible, therefore, that this class of systems will play a predominant role in an era of computer-aided chemical systems design to come.

The same asset can also be derived the other way around. In any dynamical system, especially if it is nonlinear, the number of pertaining parameters grows in a nonlinear manner with the number of variables. At the same time, the number of "regions" in parameter space (separated by bifurcation surfaces) grows even faster, and so does the complexity of their intermingling. As a consequence, any systematic technique of systems optimization (i.e. one based on the pursuit of more or less continuous paths in parameter space) becomes inapplicable very soon.

In this apparently hopeless situtation, homogeneous automata form a singular exception, because the very liability just described has been turned into an asset in this class of systems: the unavoidable bifurcation thresholds in parameter space have been "tamed" and even play an essential role in the dynamics of the whole system.

More generally speaking, one could say that among the whole set of nonlinear (multibasin) dynamical systems, only dynamical automata are structurally stable against ordinary ("medium-size") perturbations. (An analogous reasoning led Thom, 1970, to propose his special subclass of distributed dynamical automata, to use the present terminology, as models of a biological process which necessarily is subject to exogeneous perturbation, namely morphogenesis.)

3. Compartmental Chemical Automata

Two types of chemical automata of this class can be distinguished. In the first sort, homogeneous automata are poured into the well-stirred "cells". In the second case, no individual subsystem inside a cell is, when isolated, a chemical automaton. This more complicated case is considered first.

A. Nontrivial Compartmental Automata

The prototype system is the simplest Rashevsky-Turing system (see Fig. 32 of the preceding note, Rössler, 1974c). The individual subsystems are globally stable (showing either damped oscillations or aperiodic behavior). The selective coupling D_b renders the whole system bistable (i.e. a morphogenetical RS flip-flop) beyond a critical value of D_b, as outlined loco citato.

The same system is also a von Neumann-type T flip-flop (Rössler & Seelig, 1972): a short-lived, reversible reduction of D_b switches the system between either stable steady state, i.e. between the "positive" and the "negative" differentiation pattern. The system behaves in just the same manner as the first (homogeneous) example of a chemical T flip-flop based on von Neumann's principle (Seelig & Rössler, 1972).

The pattern formation observed in larger arrays of cells (either one- or two- or three-dimensional) was investigated for the original Turing system by Turing (1952) and Othmer & Scriven (1971). Similar equations have been simulated also (as successful "models" of hydra morphogenesis) by Gierer & Meinhardt (1972). Investigations on the response of cellular arrays to shortlived changes of an evocation parameter (generalization of T flip-flop behavior) have not been performed so far, neither abstractly nor in actual experiments.

Turing's results show that the overall outcome corresponds to a small discrete number of possible patterns (multistable flip-flop). Therefore, the whole system can be considered as an automaton being characterized (a) by the possession of a very restricted number of states, and (b) by the fact that each state is highly organized in space.

It is worth stressing that both the number of states of this "flip-flop" and their form in space depend on the boundary conditions, whereas the number of cells involved plays virtually no role (if their size is decreased in proportion to their number, and membrane permeability is increased in proportion to their radial density). The situation is very similar to the buckling of plates made up of rigid parts and elastic couplings, where an analogous smooth limiting case applies.

A continuous change of the boundaries (for example through growth) or, equivalently, a continuous change of the properties of all cells, may result in the sudden transition toward a new pattern (state), as numerical experiments of Meinhardt & Gierer (1973), performed on a similar system, suggest. This behavior corresponds to that of a growing automaton (stepwise increase of the number of possible states).

B. Compartmental Automata Sensu Stricto (Cellular Chemical Automata)

The theory of cellular automata is a recognized subdiscipline of automata theory (see Ulam, 1970). It was invented by Ulam and von Neumann in the late 40's (cf. Ulam, 1952). The observed patterns are determined by two factors: a) the particular sort of automaton contained within the cells, and b) the particular connection pattern. More than one cell-type is also possible.

Chemical reaction kinetics is well suited both for the abstract and the concrete realization of cellular automata. The well-stirred compartments (cells) are coupled by either selective or nonselective diffusion (through membranes), and the connection pattern is either simple (next neighbor connection) or complicated. In the following, only the three simplest cases are considered: RS flip-flops coupled through diffusion either in the anti-parallel or the parallel way, and monoflops coupled by diffusion.

The antiparallel coupling of RS flip-flops (such that the opposite switching state of the neighbor is facilitated) is, in the case of diffusion-coupling, bound to the intercalation of a variable which (a) depends on the switching state of the flip-flops (being either activated or inhibited) and (b) has an opposite effect on the flip-flop itself (inhibiting the state by which it is activated, or facilitating the state by which it is inhibited, respectively). This structure is, however, just that of a multivibrator which acts as an <u>RS</u> flip-flop with reduced thresholds.(The latter activity forms, besides astable, monostable, and <u>T</u> flip-flop behavior, one of the four basic behavioral capabilities of a chemical multivibrating reaction scheme; Rössler, 1974c.) Hence any of the three prototype reaction schemes of Fig. 25 (Rössler, 1974c) is appropriate. As a third requirement (c), the coupling has to selectively prefer the intercalated variable.

A cellular automaton of this type can, again, be considered as a single flip-flop. In the simplest case, the flip-flop has only two states (namely that of a checkerboard pattern and that of the corresponding negative). In general, the flip-flop is multistable, however, since the coupling through diffusion has an integrating effect, connecting each cell not only with its direct neighbors, but through them with a whole neighborhood, so that several surrounding patterns may yield the same response of an individual cell. This effect becomes greater with decreasing cell size (and parallel increase of local permeability). Thus, the situation encountered above with Rashevsky-Turing cells applies once more.

It may be mentioned that it is possible to trace back all Rashevsky-Turing systems to this class of systems as a prototype. Indeed the "push-pull" structure between the cells allows one stable state inside the individual cells to be deleted; for example, if in the lowest scheme of Fig. 25 the second-order outflux is omitted, this reaction system becomes immediately identical with a subsystem of the Rashevsky-Turing system of Fig. 32 (loco citato). At the same time, the schemes of Figs. 26 and 27 (ibidem) which also rely on nonresettable flip-flops, become equally appropriate. This explains why the Prigogine-Lavenda oscillator which is an analog of the upper scheme of the named Fig. 27 (Rössler, 1972a) could be employed successfully in computer-studies on chemical morphogenesis (Herschkowitz-Kaufman & Nicolis, 1972; Martinez, 1972). The same classification (as potential single-sweep monoflops; Fig. 27, as cited) applies to all those reaction schemes indicated by Turing (1952) and Gierer & Meinhardt (1972) which do not show the "typical" structure of the scheme of Fig. 32 (ibidem).

We now come to the (simpler) class of <u>parallel</u> (i.e. directly and nonselectively) coupled, identical flip-flops. Again, the whole array is a single bistable flip-flop, possessing, however, two spatially uniform states now. At the same time, the flip-flop shows a preferential state, since (depending on the relative readiness of the two states to be triggered by an analogous neighboring state; a one-to-one ratio can hardly be realized), border lines between two regions of different state tend to move in such a way that the easier-to-trigger state finally covers the whole area (Franck, 1973).

Another interesting feature of this bistable flip-flop is the fact that its two stable steady states are separated by a great number of unstable states. This is because small islands of the "stronger" state (in which the cells possess a greater triggering power towards their neighbors) either grow or disappear, depending on their size, whereby an unstable intermediate situation exists. Since several islands can, theoretically, be simultaneously present (forming a piecemeal pattern), the number of possible unstable states is very large. This result, again, applies in independence of cell-size (when permeability is increased with the radial density).

In automata-theoretic terms, the "unstable" patterns are so-called "Garden-of-Eden" states which are never attained again once they are left (cf. Moore, 1970).

So far, exogeneous inputs were assumed to be absent. Corresponding to the nature of the flip-flops, two sorts of such inputs are possible: (a) biasing inputs comparable to those from a neighboring flip-flop (leading, if they are strong enough, to a hard switching), and (b) evocating inputs (leading to the non-bistable condition of the reaction system in a soft manner). By applying a sufficiently strong input of the first type (a) to a region, the region may both be triggered to the other state and be <u>maintained</u> in that state, even if it happens to be the weaker state (otherwise, the locally triggered state would propagate). This situation is automatically achieved at a certain point in space,if a "gradient" of the type (a) parameter (being relatively smooth if the cell density is large) is assumed to be present. The resulting picture (stable border line) can be called a fold. If at the same time a gradient of a type (b) parameter is assumed to be present along another direction in two-dimensional space, the result will be a whole cusp (see Thom, 1970, and Zeeman, 1972).

Thus, typical parts of Thom's (1970) theory can be reproduced in the present automata-theoretic setting. Furthermore, since chemical reaction systems realizing this type of automaton can be devised (most simply for the limiting reaction-diffusion case of infinite cell density), Thom's predictions can also be realized experimentally.

Instead of Eq. (7) or any other bistable flip-flop, also the following equation, describing a tristable chemical flip-flop, may be used:

$$\dot{x} = j + x^2 - \frac{x}{x + x^2 + K} - x^3 - x \qquad (10)$$

It is, just like the system of Eq. (7) (cf. also Eq. 5), abstractly realizable by resolving the involved steady-state approximations through the introduction of additional chemical variables (Higgins, 1967). This tristable system can show an additional behavioral type under the application of an appropriate, graded input: two stable steady states can be caused to merge into a single unstable one. The result is a new type of bifurcation (which may be termed "antisoft" bifurcation, since simply the stability properties of the soft picture — Fig. 31, Rössler, 1974c — are reversed). Although a specific name has not been proposed for this third basic type of bifurcation so far, it plays a decisive part both in the swallow's tail and the butterfly singularity. Indeed all 4 catastrophes of co-rank 1 (i.e. involving a single essential variable), namely the fold, the cusp, the swallow's tail, and the butterfly (Thom, 1972) will be found again in a cellular automaton of type (Eq. 10), if it is realized in 3 dimensions and subject to 3 appropriate inputs graded in space and a fourth graded in time. The problem caused by the fact that not all locally emerging state-changes will be nonpropagating has not yet been taken into account. The same applies to the problem of co-rank two chemical systems.

Finally, cellular chemical automata based on monoflops are to be considered briefly. The relation to neural networks has been mentioned in the preceding communication (Rössler, 1974c). From the historical point of view it is interesting to note that part of the discontinuous idealizations made by McCulloch & Pitts (1943) in their attempt to describe a neural network (which gave rise to automata theory) can now be omitted, without losing any desired result. The reverberator, or synonymously, dynamical flip-flop is a stable structure not only in discrete cellular automata (see the following example)

(--- = initial excitation and simultaneous block of propagation in one direction), but also in continuously realized cellular automata, and so are analogous behavioral modes. They are all based on the presence of a self-reproducing "leading free edge" (forming an at most one-dimensional surface) which is characterized by the mutual adjacency in space of an excited, an excitable, and a nonexcitable region (cf.

Gul'ko & Petrov, 1972). The nonexcitable ("core", Winfree, 1972) region shows a passive, phase-wave-like response. Thus, a region of non-automata-like behavior side-links the "lower-state" and the "upper-state" regions, which are directly connected by the moving fold (that is, the actively propagated "trigger wave" front, Winfree, 1974).

4. Reaction-diffusion Automata

Part of these systems are obtained as the limiting case of cellular automata involving nonselective diffusion, when turning to a smooth reticulation. Examples involving bistable and tristable flip-flops and monoflops have already been considered (Subsection 3b and c). Actual chemical RS flip-flops and monoflops are available. The possibility of considering the whole reaction volume as a single automaton possessing several complicated stable regimes has been discussed.

If one turns to selective diffusion or to diffusion-mediated transport (Rashevsky, 1940), Rashevsky-Turing type systems are obtained as one possibility. They can, as outlined in Subsection 3a, also be considered as multiple-basin dynamical systems and hence as multistable, triggerable flip-flops, i.e. simple automata. Additional classes of reaction-diffusion type automata (e.g. pattern-reversing systems, Rössler, 1974 c) exist.

5. Conclusions

1) Homogeneous chemical automata exist, up to analogy with other, physically realized finite automata.
2) Simple DC type homogeneous automata (RS, astable, and monostable flip-flops) even exist in chemical reality.
3) Prebiological evolution can be interpreted as an autonomously growing Z_2-network.
4) Cellular Rashevsky-Turing systems can be considered both as a kind of cellular automata and as multistable flip-flops.
5) Cellular automata consisting of diffusion-coupled bi-or tristable compartments exhibit morphogenetical catastrophes.
6) Cellular automata based on diffusion-coupled chemical monoflops are continuous analogs of simple neural nets.
7) Reaction-diffusion systems which form the limiting case of cellular automata based on diffusion-coupled RS (or monostable, or astable) flip-flops, can be realized.

Summary: Chemical automata of the homogeneous, the compartmental, and the reaction-diffusion type are described.

Homogeneous automata are, in the simplest case, asynchronous Z_2-networks of DC type, consisting of bistable partial systems as RS flip-flops, and intercalated reactions as logic and delay elements. An actual example is the either astable or monostable Belousov-Zhabotinsky reaction.

Compartmental automata comprise both chemically realized cellular automata and morphogenetic multi-cellular systems of DC type. Thom's catastrophe theory can be applied to part of these systems.

Reaction-diffusion automata correspondingly include AC and DC type morphogenetic systems. The theory of the former class is still in an undeveloped state.

Concerning applications, homogeneous Z_2-networks form the most easy to realize class of homogeneous reaction systems showing a prescribed behavior; and cellular chemical automata may be important for the development of a unified view of morphogenesis and brain-tissue function.

References

Note: For additional references, see Rössler (1974c).

Andronov, A.A., Leontovich, E.A., Gordon, I.I., and Maier, A.G.: Theory of Bifurcations of Dynamic Systems on a Plane, Israel Program for Scientific Translations, Jerusalem, 1971 (First appeared in Russian, 1967).

Arbib, M.A.: Theories of Abstract Automata, Prentice Hall, Englewood Cliffs, 1969.

Bhatia, N.P. and Szegö, G.P.: Stability Theory of Dynamical Systems, Springer Verlag, Berlin-Heidelberg-New York, 1970.

Burks, A.W.: Computation, behavior, and structure in fixed and growing automata, Behav. Sci. 6 , 5-22 (1961).

Erdös, P. and Rénji, A.: On the evolution of random graphs, Publications Math. Inst. Hungarian Acad. Sci., 5, 17-95 (1960).

Haken, H.: Introduction to synergetics. In: Synergetics, Cooperative Phenomena in Multi-component Systems, ed. by H. Haken, pp. 1-19. B.G. Teubner Verlag, Stuttgart, 1973.

Heineken, F.G., Tsuchiya, H.M., and Aris, R.: On the mathematical status of the pseudo-steady state hypothesis of biochemical kinetics, Math. Biosci. 1, 95-113 (1967).

Herschkowitz-Kaufman, M. and Nicolis, G.: Localized spatial structures and chemical waves in dissipative systems, J. Chem. Phys. 36, 1890-1895 (1972).

Higgins, J.: The theory of oscillating reactions, Indl. Engrg. Sci. 59, 19-62 (1967).

Hotz, G. and Walter, H.: Automatentheorie und Formale Sprachen II, Endliche Automaten, Bibliographisches Institut, Mannheim-Vienna-Zürich, 1969.

Kalman, R.E., Falb, P.L., and Arbib, M.A.: Topics in Mathematical System Theory, McGraw-Hill, New York, 1969.

Keyes, R.W. and Landauer, R.: Minimum energy dissipation in logic, IBM J. Res. Develop. 14, 152 (1970).

Landauer, R. and Woo, J.W.F.: In: Synergetics: Cooperative Phenomena in Multi-Component Systems, ed. by H. Haken. Teubner, Stuttgart,1973.

Lutz, R.S.: Oscillations and hysteresis behavior in enzymatic systems involving inhibited inhibition, under varying concentrations of inhibitor (in German), Diploma Work, University of Tübingen (1973).

Martinez, H.M.: Morphogenesis and chemical dissipative structures, a computer simulated case study, J. theor. Biol.36, 479-501 (1972).

McCulloch, W.S. and Pitts, W.: A logical calculus of the ideas immanent in nervous activity, Bull. math. Biophysics, 11, 115-133 (1943).

Meinhardt, H. and Gierer, A.: Computer simulation of morphogenetic equations, Lecture held at the Max Planck Institute, Tübingen, November 1973.

Moore, E.F.: In Essays on Cellular Automata, ed. by A.W. Burks, University of Illinois Press, Chicago, 1970.

Othmer, H.G. and Scriven, L.E.: Instability and dynamic pattern in cellular networks, J. theor. Biol. 32, 507-537 (1971).

Rechenberg, I.: Evolutionsstrategie, Optimierung technischer Systeme nach Prinzipien der biologischen Evolution, Friedrich Frommann-Verlag (Günter Holzboog K.G.), Stuttgart-Bad Cannstatt, 1973.

Rosen, R.: Discrete and continuous representations of metabolic models. In: Quant. Biol. Metabolism, ed. by A. Locker, pp. 24-32. Springer Verlag, Berlin-Heidelberg-New York, 1968.

Rössler, O.E.: A system-theoretic model of biogenesis (in German), Z. Naturforsch 26b, 741-746 (1971).

Rössler, O.E.: Design for autonomous chemical growth under different environmental constraints, Progr. Theor. Biol., 2, 167-211 (1972c).

Rössler, O.E.: A synthetic approach to exotic kinetics, with examples, These Proceedings (1974c).

Rössler, O.E. and Hoffmann, D.: A chemical universal circuit, Fourth Intern. Congr. Biophysics, Moscow, August 1972, Abstracts 4, p. 49 (1972b).

Seelig, F.F. and Rössler, O.E.: A chemical reaction flip-flop with one unique switching input, Z. Naturforsch 27b, 1441-1444 (1972).

Shannon, C.E.: A symbolic analysis of relay and switching circuits, Trans. AIEE 57, 713-723 (1938).

Ulam, S.M.: Random processes and transformations. In: Proc. Int. Congr. Mathematicians 1950, Vol. 2, pp. 264-275. Amer. Math. Society, Providence, Rhode Island, 1952.

Ulam, S.M.: In Essays in Cellular Automata, ed. by A.W. Burks, University of Illinois Press, Chicago, 1970.

Von Neumann, J.: Nonlinear capacitance or inductance switching, amplifying and memory organs, U.S. Patent No. 2, 815, 488, issued December 3, 1957.

Ware, W.H.: Digital Computer Technology and Design, Vol.II: Circuits and Machine Design, p. 9:60. Wiley, New York, 1963.

Wigington, R.L.: A new concept in computing, Proc. IRE 47, 516 (1959) (Cited after Landauer & Woo, 1973).

MOLECULAR AUTOMATA

Michael Conrad
Institute for Information Sciences
University of Tübingen, Germany

1. Introduction

In this seminar I want to characterize the information processing properties of systems in which individual molecules play a key role. I will call such systems molecular automata since I would like to describe them as nearly as possible in terms of discrete (automata) concepts which derive from our experience with conventional computers. This is the simplest approach and also the most useful from the standpoint of comparing conventional information processing concepts and those which are suitable for describing molecular biological systems.

2. Basic Definitions

We begin by defining molecular automata and molecular computers in a framework which is sufficiently general to include conventional computers as well. In section 3 we add some fundamental properties to our definition of molecular automata.

2.1 Elemental and complex inputs

The inputs to a system may be elemental in the sense that they are individual elements. The inputs may also be complex in the sense that they are the sequences of elements themselves (or, more generally, spatial arrangements of individual elements). More formally:
Definition. Let $\Theta = \{\theta_u | u=1,\dots,k\}$ be a set of elements and let $\Theta_i = \{\theta_{iu} | u=1,\dots,m\}$ be the set of inputs associated with system i. Also, let $\Theta_i^* = \{\theta_{iv}^* | v=1,\dots,n\}$ be a set of sequences of θ_u associated with this system. The inputs will be called complex if $\Theta_i \equiv \Theta_i^*$ and the sequences are not just the individual elements. Otherwise they will be called elemental. If in certain situations $\theta_u \in \Theta_i^*$ $(\equiv \Theta_i)$ the inputs will be called decomposable. If this is never possible they will be called indecomposable.

The inputs to a system may also be classified on the basis of their source. In what follows we distinguish three types of inputs on this basis:

1. genome inputs ($\Theta = G$, $\theta_u = g_u$, etc.);
2. tape inputs ($\Theta = X$, $\theta_u = x_u$, etc.);
3. environmental inputs ($\Theta = E$, $\theta_u = e_u$, etc.).

The genome inputs are the states of the genome. This may be thought of as a memory space, but with the provision that its states are complex (i.e. $g_{iv}^* \varepsilon G_i$). The tape inputs are also inputs from some memory space, except that they are elemental. The environmental inputs are states of the external environment. These are also elemental.

The distinction between elemental and complex inputs is actually a bit subtle. For example, a folded sequence of amino acids is a complex input and also indecomposable. The DNA sequence which codes for this is therefore also a complex input, but this is obscured by the fact that it is potentially decomposable into elemental inputs (e.g. in transcription).

The following notation will be useful later and also illustrates the distinction between elemental and complex inputs. Suppose that $G = \{A^+, A^-, T^+, T^-, G^+, G^-, C^+, C^-\}$ and that we form some sequence $g_{iv}^* = A^+T^+G^+G^-G^-A^-$. The inputs would be elemental if $g_u \varepsilon G_i$, where $g_1 = A^+$, $g_2 = T^-$, and so forth. The input is complex if $g_{iv}^* \varepsilon G_i$, i.e. if the input is the sequence as a whole.

The symbols in the above example are clearly chosen to correspond to the bases in DNA, with plus signs indicating that the base is accessed (or actively being used in translation to amino acid sequence) and minus signs indicating that it is not accessed (or not being used in translation to amino acid sequence). In this case the sequence is, in effect, a complex input because of the folding property of the proteins. Naturally we can think of certain subsequences of elements which are always "+" or "-" together as genes, or of hierarchies of such elements as operons, etc.

2.2 Molecular automata

Definition. A molecular automaton, M_i, is a quintuple

$$M_i = [G_i, P_i, E_i, \delta_i, \lambda_i] \quad i \varepsilon I \text{ (index set)}$$

where

G_i = finite set of genome states (as previously defined)

P_i = finite set of phenome states ($p_{ik} \varepsilon P_i$)

E_i = finite set of environmental states

δ_i: $P_i \times G_i \times E_i \rightarrow P_i$ (next phenome state function)

λ_i: $P_i \times G_i \times E_i \rightarrow G_i$ (next genome state function)

We assume, for simplicity, that M_i operates on a discrete time scale $(t=1,2,3,\ldots)$. Also, for simplicity we ignore the fact that M_i might influence the state transitions of the environment.

The program (or transition functions) of a molecular automaton is conveniently expressed in terms of a set of quintuples

$$\{p_{iu} g^*_{ir} e_{iu} p_{iv} g^*_{is}\}$$

where g^*_{ir} and p_{iu} are the initial genome and phenome state, g^*_{is} and p_{iv} are the final genome and phenome states, and e_{iu} is the initial environmental state.

2.3 Molecular computer

A general molecular automaton GM_i is a molecular automaton with tape inputs and outputs:

$$GM_i = [G_i, P_i, X_i, E_i, Y, \delta_i, \lambda_i, \omega_i]$$

where X_i is the finite set of tape inputs (as previously defined), Y_i $(Y_i \subset X_i)$ is the finite set of tape outputs, and the transition functions are now given by

$$\delta_i: P_i \times G_i \times E_i \times X_i \to P_i$$
$$\lambda_i: P_i \times G_i \times E_i \times X_i \to G_i$$
$$\omega_i: P_i \times G_i \times E_i \times X_i \to Y_i \quad \text{(output function).}$$

A molecular computer, GMC_i, is a general molecular automaton along with the work (or memory) space:

$$GMC_i = \langle GM_i, T_i \rangle$$

where T_i is the work space associated with GM_i. The simplest model of a molecular computer is a molecular Turing machine. In this case the T_i is a linear tape, with each tape square being in one of a number of states. These states are the tape inputs (or the x_{iu}). The tape outputs change (rewrite) the states and also move the tape, i.e. $Y_i = \{x_{i1},\ldots,x_{in},N,R,L\}$, where N is no move, R is a move one square to the right, and L is a move one square to the left. The program of the Turing machine is given by a set of septuples

$$\{p_{iu} g^*_{iv} x_{ia} e_{ib} y_{ic} p_{iv} g^*_{is}\}$$

where $y_{ic} \varepsilon Y_i$. The behavior of the system is specified by specifying the initial values of g^*_{iv}, p_{iu}, the initial location on the tape, and the initial tape sequence x^*_{iv}.

2.4 Conventional automata and computers

A conventional automaton (with output) is a general molecular automaton with no genome, i.e. a GM_i in which $G_i = \{\Lambda\}$ (the empty word), from which it follows that λ_i disappears also. In this case what we have called the phenome states are the only states of the system (frequently denoted by Q).

A conventional computer, C_i, is a conventional automaton along with a work space which it can manipulate:

$$C_i = \langle A_i, T_i \rangle$$

where A_i is a conventional automaton. For example, a conventional Turing machine is the same as our molecular Turing machine, except that there are no genome states.

Notice that we have described conventional systems as a subclass of molecular systems.

2.5 Multiplication law

Definition. (Functional Notation) Suppose that GM_i is in states p_{iu} and g^*_{iv} at time t. Then its phenome and genome states at time t+1 are given by $\delta_i(p_{iu}, g^*_{ir}, x_{ia}, e_{ib})$ and $\lambda_i(p_{iu}, g^*_{ir}, x_{ia}, e_{ib})$. Also, the tape output at time t+1 is given by $\omega_i(p_{iu}, g^*_{ir}, x_{ia}, e_{ib})$. (If G is empty we omit g^*_{ir} and likewise for X_i and x_{ia}.)

Now we can define the multiplication law which extends the applicability of δ_i and λ_i to (temporal) sequences of elemental inputs.

Definition. The multiplication laws determining the state and output after repeated inputs are given by:

1. $\delta_i(p_{iu}, g^*_{ir}, x_{ia}x_{ib}, e_{ic}e_{id}) = \delta_i(\delta_i(p_{iu}, g^*_{ir}, x_{ia}, e_{ic}), \lambda_i(p_{iu}, g^*_{ir}, x_{ia}, e_{ic}), x_{ib}, e_{id})$

2. $\omega_i(p_{iu}, g^*_{ir}, x_{ia}x_{ib}, e_{ic}e_{id}) = \omega_i(\delta_i(p_{iu}, g^*_{ir}, x_{ia}, e_{ic}), \lambda_i(p_{iu}, g^*_{ir}, x_{ia}, e_{ic}), x_{ib}, e_{id})$

and similarly for the next genome state function. In a particular case the first of these may be rewritten:

$$\delta_i(p_{iu}, A^+T^+G^+G^-G^-A^-, x_{ia}x_{ib}\ldots x_{im}, e_{ic}e_{id}\ldots e_{im}) =$$
$$\delta_i(\delta_i(p_{iu}, A^+T^+G^+G^-G^-A^-, x_{ia}, e_{ic}), \lambda_i(p_{iu}, A^+T^+G^+G^-G^-A^-, x_{ia}, e_{ic}), x_{ib}\ldots x_{im}, e_{ic}\ldots e_{im})$$

where $A^+T^+G^+G^-G^-A^- = g^*_{ir}$. Notice that the multiplication law does not extend to the sequence of elements in a complex input (such as g^*_{ir}). The impossibility of such an extension (or of making a decomposition which allows it) is the fundamental distinction between complex and elemental sequences.

2.6 Populations

Definition. A population of molecular automata (M_I) is a tuple

$$M_I = \{M_i, \alpha, \delta_{rs}, \lambda_{rs} | r,s,i \in I\}$$

where

$\alpha = \{p(G_s|G_1,\ldots,G_n)\}$ (genome transformation probabilities)

$\delta_{rs}: P_s \times G_r \times E_i \to P_t$ (mixed next phenome state function)

$\lambda_{rs}: P_s \times G_r \times E_i \to G_t$ (mixed next genome state function)

α is the set of transformation probabilities which determine the interconversion of genome sequences. We will not here specify the structure of this set except that we may assume that it results from random changes in the internal input sequence or changes which follow the rules of genetics. The mixed transition functions (which will be specified later) are necessary because the transformed genome state must interact with the old phenome states (but notice that, e.g. $\delta_{ss} = \delta_s$).

Naturally we can easily generalize our definition of M_I to include populations of general molecular automata (GM_I) and molecular computers (GMC_I).

3. Fundamental Properties

Now we can formalize the essential properties of molecular biological systems.

Property 1. (Darwinian Property). The λ_i are subject to the following

1. if A^+ changes it can only change to A^- and conversely;
2. if T^+ changes it can only change to T^- and conversely;
3. if G^+ changes it can only change to G^- and conversely;
4. if C^+ changes it can only change to C^- and conversely.

According to property 1 the phenome states affect the genome state only by affecting the pattern of accessing and not by changing the nucleotide sequence. This is accurate even though such changes in sequence may result from cytoplasmic (e.g. enzymatic) processes as long as such changes are essentially random (and therefore included in the transformation function). Property 1 is sometimes called the strong principle of inheritance.

Property 2. (Genetic Code Property). For all r and s the mixed transition functions of a molecular automaton are given by:

$\delta_{rs}: P_s \times G_r \times E_i \to P_r$

$\lambda_{rs}: P_s \times G_r \times E_i \to G_r$.

(These are also the mixed transition functions of a molecular computer except that we include X in this case.)

According to property 2 the genome dominates the phenome. If the genome changes (through some statistical process) from G_s to G_r, the phenome will be driven into the state set associated with system r. Also, none of the genome states are driven out of G_r by phenome states in P_s. This is the genetic code property because the genetic code assures that the sequence of nucleotides determines the sequence of amino acids in proteins (or the sequence of bases in certain phenomic nucleic acids), which in turn determines the properties of the cell. This assumption is realistic, though perhaps not perfect since certain properties of the initial phenotype might in fact be propagated. Also, we must assume that the transformation of the genome does not affect the codes itself. This is also reasonable, however, since such mutations are almost always lethal.

Property 3. (Genome Isolation Property). The next genome state function of a molecular automaton can always be written

$$\lambda_i: P_i \times G_i \to G_i$$

or in the case of a molecular computer as:

$$\lambda_i: P_i \times G_i \times X_i \to G_i \ .$$

The output function of a molecular computer can always be written:

$$\omega_i: P_i \times E_i \times X_i \to Y_i \ .$$

According to property 3 the genome transitions are not directly affected by the external environment, but rather only by the effects of the external environment on the phenome. This is necessarily so since we regard the state as characterizing everything which is within the boundaries of the system aside from the genome, which we suppose has no direct contact with the external environment. We also suppose that any effects which the automaton has on some tape must be mediated by the phenome. These restrictions are not true for the phenome transitions, which may be affected by the genome state, the tape input, and the external environment state.

Definition (Similarity).

(i) If g^*_{iv} differs from g^*_{jv} in only one or a few places we will say that they are similar and will write $g^*_{iv} \sim g^*_{jv}$ (or $G_i \sim G_j$).

(ii) If p_{iu} differs from p_{ju} only slightly we will say that they are similar and will write $p_{iu} \sim p_{ju}$. (What "slightly" means depends on the nature of the states; we will specify this later in a particular case.)

(iii) If x_{ia} differs from x_{ja} only slightly we will say that they are similar and will write $x_{ia} \sim x_{ja}$. (For example, suppose x_{ia} and x_{ja} are coded into O's and 1's. They are similar if they differ by, say, Hamming distance one.)

(iv) If e_{ib} differs from e_{jb} only slightly we will say that they are similar and will write $e_{ib} \sim e_{jb}$. (Again suppose e_{ib} and e_{jb} are coded into O's and 1's and that they are separated by a small Hamming distance.)

Notice that the similarity relation is symmetric, reflexive, but not transitive. In fact, what we have in mind is the same as the tolerance concept described by M. Dal Cin in his contribution to this workgroup (Dal Cin, 1974; see also Zeeman, 1968), except that we will not be using the full machinery of tolerance geometry in this note.

Property 4. (Molecular Adaptability). If $G_s \sim G_r$ there is a "good" chance that $P_s \sim P_r$. (By a good chance we mean one that is itself dependent on the sequence of bases in the genome and therefore adjustable to the particular circumstances.)

Property 4 expresses the significance of molecular hierarchy. The genome (or sequence of bases in DNA) is a complex input because, when translated to the sequence of amino acids in protein it becomes an indecomposable complex sequence, essentially because it folds to form a three dimensional structure on the basis of the weak interactions among the amino acids. Furthermore, the degree to which the properties of these folded molecules change with change in primary sequence depends on the redundancy of weak bonding; for as this increases the probability increases that slight transformations in the sequence have a smaller effect on certain essential features of the shape (cf. Conrad, 1972a).

Finally we introduce two more definitions.

Definition. (Broken Homomorphism) Two general molecular automata, GM_i and GM_j, are quasi-homomorphic if there are maps

$$h_1: P_i \to P_j; \quad h_2: G_i \to G_j; \quad h_3: X_i \to X_j; \quad h_4: E_i \to E_j$$

such that for all u, r, a, b

(i) $h_1[\delta_i(p_{iu}, g^*_{ir}, x_{ia}, e_{ib})] = \delta_j[h_1(p_{iu}), h_2(g^*_{ir}), h_3(x_{ia}), h_4(e_{ib})]$

(ii) $h_2[\lambda_i(p_{iu}, g^*_{ir}, x_{ia}, e_{ib})] = \lambda_j[h_1(p_{iu}), h_2(g^*_{ir}), h_3(x_{ia}), h_4(e_{ib})]$

The two systems are quasi-isomorphic if the h_i are all one-one and onto.

Note that our definition does not extend to the output function and is therefore weaker than the ordinary definition of homomorphism. This is because the output is strictly speaking a property of the phenome state. Thus, what our notion of broken homomorphism means is that the two systems obey more or less the same rule of development (i.e. have more or less the same pattern of accessing the genome), but that the corresponding phenome states may respond differently to the tape or environmental inputs.

Definition (Similarity again). Two general molecular automata, GM_i and GM_j, will be said to be similar if they are quasi-homomorphic and if $G_i \sim G_j$. A family of general molecular automata will be said to be connected by similarity if they are connected by quasi-homomorphism and each can be reached through a sequence of genomes connected by similarity.

4. Neural Representation

Now we are in a position to describe the information processing capabilities of molecular automata. However, instead of doing this in the most abstract way all we have to do is find a particular representation of molecular automata. Certainly molecular automata will at least have the capabilities of this representation.

The simplest representation is a neuron in which the nerve impulse is controlled by individual molecules. This is the enzymatic neuron discussed in our morning lecture (Conrad, 1974b; cf. also Conrad, 1974a). Its equation is:

$$y_k(t+1) = \Phi\left[\sum_{j=1}^{n} \delta(\xi_{jk} - \sum_{i=1}^{m} w_{ijk} x_{ik}(t))\right],$$

where

$$\delta(u) = \begin{cases} 1 & \text{if } u=0 \\ 0 & \text{otherwise} \end{cases}$$

$$\Phi(v) = \begin{cases} 1 & \text{if } v>0 \\ 0 & \text{otherwise} \end{cases}$$

and t = time, x_{ik}(0 or 1) is the dendritic input to neuron k, j is the position of (excitase) enzyme E_j, ξ_{jk} is the electric field which causes this enzyme to catalyze events leading to impulse formation, and w_{ijk} is the weighted value of dendritic input i at position j of neuron k, n is the number of excitase species, and m is the number of dendritic inputs. Each of the E_j has some primary structure of amino acids $a_1 \ldots a_p$ which, though not strictly its genome, is encoded in the sequence of bases in some gene. The position assumed by the excitase in the neuron is determined by its properties, i.e. $j = j(a_1 \ldots a_p)$. Also, w_{ijk} may depend on the input pattern, i.e. $w_{ijk} = w_{ijk}(x_{1k}, \ldots, x_{mk})$. In this case the neuron fires when the input pattern causes sufficient electrical excitation at location j on the neuron surface. Notice that we have omitted the tolerances, which we could include, e.g. by introducing more j values. (Also, recall that we can give an alternative, formally more convenient interpretation in which $\xi_{jk} = \xi_{jk}(a_1 \ldots a_p)$.)

Theorem. The enzymatic neuron is a general molecular automaton if the excitases are such that they can be made to assume all possible values of j by making step by step alterations in the $(a_1 \ldots a_p)$.

Proof: G_k is the set of states of the sequence of bases in the genome of neuron k, including the particular subsequences which code for the excitases (i.e. for the $(a_1 \ldots a_k)$). P_k is the set of possible patterns of states (active or inactive) of all the excitase species in the neuron. Also we say that $P_k \sim P_j$ if P_k differs from P_j by the presence or absence of only one excitase species. The E_i or X_i are the $\{(x_{1k},\ldots,x_{mk})\}$ and Y = {0,1}. The next phenome state "function" is

$$p_k(t+1) = [\delta(\xi_{1k} - \sum_{i=1}^{m} w_{ijk}x_{ij}(t)),\ldots,\delta(\xi_{nk} - \sum_{i=1}^{m} w_{ijk}x_{ij}(t))]$$

where $p_k(t+1)$ is the phenome state at time t+1. The next genome state function is arbitrary and the output function, ω_k, is given by the neuronal equation itself. The equation makes no reference to G_k except through the functional relation $j = j(a_1 \ldots a_p) = j(g^*_{kv})$. Therefore there is no problem in satisfying properties 1-3 of molecular automata. Moreover, by assumption it satisfies property 4, since step by step alterations in the $(a_1 \ldots a_p)$ mean step by step alterations in the g^*_{kv} and also single changes in the presence or absence of excitases at particular locations.

Networks of enzymatic neurons are also general molecular automata. Thus we know that the class of molecular automata have all the capabilities of enzymatic neurons or networks of enzymatic neurons, i.e. that they can simulate any conventional neural network, that they can simulate any finite automata and with quite high efficiency, that they have particularly powerful pattern recognition and generation capabilities, and that the set of patterns generated or recognized by such systems can be varied by making gradual changes in the genome. These capabilities have been discussed elsewhere (Conrad, 1974b). Therefore, we here just restrict ourselves to making one general statement about the class of molecular automata.

Theorem. The class of general molecular automata includes members which can compute all logical functions of their inputs and which are at the same time connected by similarity.

Proof: It is only necessary to show that the enzymatic neuron has this property. Choose the $w_{ijk} = w_{ijk}(x_{1k},\ldots,x_{mk})$ so that each different input pattern gives rise to a different locus of excitation. The inputs are automatically similar since h_3 and h_4 are identity maps. h_1 and h_2 can be chosen arbitrarily since G_i and G_j are disjoint, P_i and P_j are are disjoint, and broken homomorphism places no requirement on Y_i and Y_j (which are not disjoint). Thus all we require is that similar genome states are associated with similar phenome states and that we can produce all patterns of excitase location in a step by step way, which is, of course, the assumption.

The above result is practically trivial once we have concocted the enzymatic neuron. However, it admits of an interesting twist if we add the further requirement that similar input patterns give rise to neighboring loci of excitation. In this case, we can make a slight change in the nature of the inputs (e.g. e_{ia} to e_{ib}) and therefore a slight change in how these are encoded into 0's and 1's (i.e. in the $(x_{1k},\ldots,x_{mk})$). Then two similar enzymatic neurons, differing only by one or a few genetic changes, can react in the same way to these changed inputs. <u>But this ability to remain essentially the same despite changes in the environment is none other than the fundamental biological property of homoeostasis</u>.

5. Programmability

The g^*_{kv} is not the program of the molecular automaton or molecular computer; for it is a complex input and therefore cannot possibly encode the transition functions (e.g. the set of septuples $(\{p_{iu}g^*_{ir}x_{ia}e_{ib}y_{ic}p_{iv}g^*_{is}\})$. What this means is simply that it is an abuse of the concept of programmability to say that biological systems are programmable at the genetic level (cf. Conrad, 1972b). But, of course, this does not mean that the δ_i, λ_i, and ω_i are not such that the states are settable by the sequence of inputs (i.e. the e^*_{iv} or the x^*_{iv}) or that they cannot interpret transition functions encoded in these sequences. Thus molecular computers potentially have the same capabilities as conventional computers; but of course the reverse is not the case since $C_i = \langle A_i, T_i \rangle$ can only be changed by changing the e^*_{iv} or x^*_{iv} (since $G_i = \{\Lambda\}$ in this case).

The above consideration is important because it means that the homeostatic (or, more broadly, adaptability) capabilities of molecular systems are unattainable by their conventional counterparts (cf. Conrad, 1974c).

6. Further Remarks

Our motive idea is that systems which are amenable to gradual transformation of function with gradual transformation of structure are especially amenable to learning; and therefore that the theory of such systems should provide a particularly useful key for unraveling the relationship between structure and function in the brain. The simplest system with this amenability is the protein molecule. Here the weak bonding of many amino acids makes it possible for slight changes in amino acid sequence to be coordinated to only slight changes in three dimensional shape and therefore the function concomitant to this shape. This trick, which is unique to systems whose function is controlled by individual molecules, makes it possible to build families of such systems which, roughly speaking, are different physical realizations of a single abstract system constructed according to some rule; but which, as different physical realizations, can interact differently with the external environment.

In this note what we have concentrated on is the formalization of this notion of alternate physical realizations and the molecular trickery on which it is based, and also on the exhibition of a particular neural representation. This notion suggests an especially powerful mode of learning, one in which each realization is a particular stage of the learning process. This type of evolutionary learning is impossible in conventional computers; for one could hardly program them to produce alternative physical realizations. But of course this does not mean that each of our particular realizations does not itself have the capability of an ordinary computer, assuming that we interphase it with an appropriately structured memory space.

The neural representation is clearly what is of most interest in this summer school. The natural questions in this case revolve around the class of functions which the system is capable of computing, e.g. when interphased with a particular memory space, and also the gradualness with which they are able to transform these functions. However, molecular automata theory has another, equally natural direction. This is the problem of development per se. Here the idea is that related organisms are different realizations of essentially the same developmental rule and therefore that development and biogenesis should be analyzed by imposing the constraint of broken homomorphism on systems with bona fide genome transition functions.

Acknowledgment

I would like to thank M. Dal Cin for many discussions and suggestions.

References

Conrad, M.: Information processing in molecular systems, Currents in Modern Biology (BioSystems) 5, 1-14 (1972a).

Conrad, M.: The importance of molecular hierarchy in information processing. In: Towards a Theoretical Biology, vol. 4, ed. by C.H. Waddington, pp. 222-228. Edinburgh University Press, Edinburgh, 1972b.

Conrad, M.: Evolutionary learning circuits, J. theoret. Biol. 46, 167-188 (1974a).

Conrad, M.: Molecular information processing in the central nervous system, this volume, 1974b.

Conrad, M.: The limits of biological simulation, J. theoret. Biol. 45, 585-590 (1974c).

Dal Cin, M.: Modifiable automata with tolerance: a model of learning, this volume, 1974.

Zeeman, E.C.: Tolerance spaces and the brain. In: Towards a Theoretical Biology, vol. 1, ed. by C.H. Waddington, pp. 140-151. Edinburgh University Press, Edinburgh, 1968.

STRUCTURAL AND DYNAMICAL REDUNDANCY

Elmar Dilger
Institute for Information Sciences
University of Tübingen, Germany

1. Introduction

The problem of reliability arises in biology in so far as it is concerned with models of neural nets, and likewise in automata theory. In automata theory we abstract using formal neurons, e.g. in the sense of McCulloch and Pitts (1943), instead of neurons. These modules or cells will actually work with error. In a network consisting of these cells the faults propagate as long as there are no mechanisms for limiting errors, and hence, the network will become unreliable. But in spite of this abstraction the designs which guarantee reliability in organisms consisting of unreliable components can be the same in both cases. Therefore we shall present here some strategies for reliability in automata and switching theory.

The first contribution to this problem came from John von Neumann (1956), who enlarged the number of net components. Later on Elias (1958) made other contributions to this theme, and Winograd and Cowan (1963) developed a theory which includes the results of von Neumann as a special case. They not only enlarged the number of components, but also augmented the complexity of elements. On the other hand Nillson (1965), Pierce (1966), Uttley (1970), Morishita (1970) and others have discussed certain adaptive organs. The structure of networks consisting of those organs is not fixed, but modifiable and, using this modifiability, adaptive organs can serve to limit errors. We will study here some of these methods and we will show how to combine them to attain reliability without excessive redundancy.

The class of networks to which these methods can be applied is very large. We consider networks consisting of a finite number of cells or modules, with each cell or module able to compute a boolean function, that is a function $f: \{0,1\}^n \rightarrow \{0,1\}$ for some integer $n \geq 0$. For instance a module can be a logical element such as "and" or "not", a finite memoryless automaton or, with some modifications, a formal neuron in the sense of McCulloch and Pitts (1943).

2. Faults in Modules and Connections

It is evident that actual networks cannot work without error in the long run. As mentioned above there will be faults, faults in the modules and faults in the connections of modules. These faults make an automaton or a network in general more and more unreliable. The aim is to design for a given automaton A a redundant version $\bar{A}$, that is an automaton which has the same input-output behaviour as the (undisturbed) automaton A and which contains mechanisms for limiting or correcting errors due to faults.

Following Cowan (1966), let us split the possible faults into two classes, the "malfunctions" which produce transient errors and the "failures" which produce stationary errors. Corresponding to these two classes of faults there are two different definitions of reliability. In the case of transient errors we have as a measure the relation of the probability of system malfunction to the probability of modular and connection malfunction; in the case of stationary errors we have as a measure the relation of lifetime at failure of modules and connections to lifetime at failure of the system.

3. Reliability in the Case of Transient Errors

The simplest (heuristic) method is due to von Neumann (1956):

> "Instead of running the incoming data into a single machine, the same information is simultaneously fed into a number of identical machines, and the result that comes out of a majority of these machines is assumed to be true."

A second method, also due to von Neumann (1956) is the method of "multiplexing". That means, roughly speaking, if in the modular network there is a connection between the output of module M_i and one of the inputs of module M_j, the modules M_i and M_j are replaced by sets of N identical modules $M_{i,1},\ldots,M_{i,N}$ and $M_{j,1},\ldots,M_{j,N}$, respectively, where N is some integer (in general $N \geq 1000$). For some permutation π of the numbers $1,2,\ldots,N$, the sets of modules are to be connected in the following way. Output of module $M_{i,s}$ $(1 \leq s \leq N)$ is connected with the corresponding input of module $M_{j,\pi(s)}$. Von Neumann was able to show that if $p \leq 0.005$ is the probability of modular malfunction and N is chosen to be 1000, one obtains a probability of network malfunction $P \leq 0.0027$, where the probability of network malfunction is understood to be the probability of network malfunction apart from malfunction of the output modules. Obviously errors in the output modules cannot be controlled by any internal procedure.

A third method leads us to the "anastomotic" modular networks of Winograd and Cowan (1963). This method can be especially applied to the problem of computing many "similar" tasks simultaneously. Basic to this method is the conception of noisy channels and the use of (n,k)-error correcting codes. So in anastomotic modular networks each single computation is done in many different places and each module will do parts of many other computations. The method is based on Theorem 8.1 of Winograd and Cowan (1963) which is formulated in a slightly different way in Appendix 1.

There is a redundancy problem in applying the Winograd-Cowan method, because we do not construct a redundant version of A, the given automaton, but of A^k, the product of k identical (or similar copies of A. If we do not use the full computing capacity of the redundant version of A^k, especially if we are only interested in a redundant version of A, we have as modular redundancy (that is the relation between the number of modules in the redundant automaton $\overline{A}$ and the number of modules in A):

$$R: = \frac{(N - u)n + uk}{N}$$

where N is the number of modules in A, u is the number of output modules in A and n and k are code parameters. Thus R is not bounded if n, k $\rightarrow \infty$.

But if we use the full capacity, e.g. if we compute simultaneously k of those tasks which A can compute, we have a modular redundancy:

$$R = \frac{(N - u)n + uk}{Nk} < \frac{n}{k}$$

which is bounded (see Appendix 1).

An application of this theorem can be made by choosing an (n,k)-error correcting code with $k = r \cdot m$ (r,m integers, r odd). The task is to design a reliable network which is able to compute m similar functions. Let $A_1, \ldots, A_m$ be the (irredundant) automata which can compute these m functions. Take r identical copies of $A_1, \ldots, A_m$. Thus we have all together k automata which compute similar tasks. To these k automata we can construct a redundant version (given in the appendix) and we can check the outputs by r-input majority organs in the sense of von Neumann (1956) or by adaptive majority organs, which we shall treat later on.

For application these methods have limitations. We cannot make the multiplicator arbitrarily large and the assumption that we have arbitrarily complex modules available cannot be maintained. Therefore,

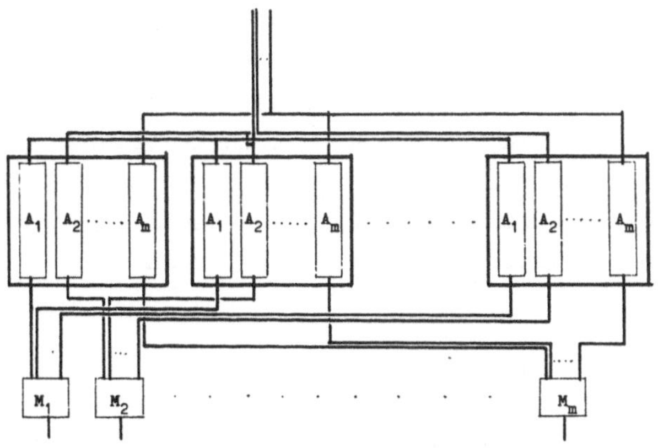

The lines are understood to be bundles of input and output lines

and is understood to be

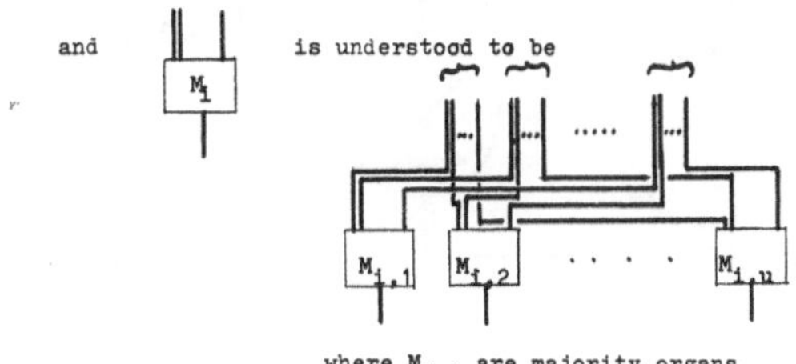

where $M_{i,j}$ are majority organs.

Figure 1

Cowan (1973) suggested: the optimum way to design a reliable network would be to use the techniques of Winograd-Cowan until the limit is reached for complexity, and then to apply von Neumann's multiplexing scheme to the already coded nets.

The redundant networks constructed in these ways are able to mask malfunctions, because in the Winograd-Cowan method, for instance, a malfunction of a module $M'_{i,j}$ is corrected in decoding the outputs of the modules $M'_{i,1},\ldots,M'_{i,n}$ by an error correcting code. It is clear that stationary errors will also be corrected provided they are not too frequent.

4. Correcting Stationary Errors

The redundant networks constructed by these methods are not optimal for correcting stationary errors. For instance it is not possible to switch out a module once and for all which has turned out to be very unreliable. At each computation step a wrong output of such a module has to be masked again. To reach our goal, we should have at our disposal a "module" or an organ which evaluates the reliability of a module and if it has turned out to be unreliable, this organ will switch it out from the circuit. In other words, we must make our networks not fixed but modifiable, in the sense that their organs are able to make decisions.

A simple example of such an adaptive decision element is the adaptive majority organ (Pierce, 1966). This organ computes the function

$$y = \text{sign}\ \left(\sum_{i=1}^{n} a_i x_i \right)$$

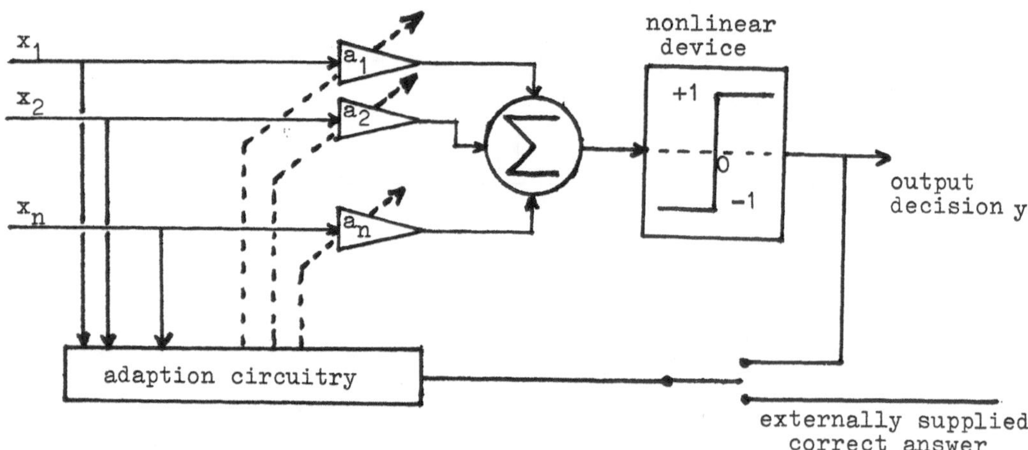

Figure 2

where each of the n inputs $x_1, x_2, \ldots, x_n$ may have the values +1 or -1. If the weights $a_1, a_2, \ldots, a_n$ are chosen to be

$$a_i = \text{ld}\left(\frac{1 - p_i}{p_i}\right)$$

where p_i is the error probability of the input x_i, then it can be shown that this choice of the weights will be the optimal one. But in general these probabilities are not known. Therefore we need an error counting procedure which gives us an approximation of these probabilities. Pierce has shown two ways by which each input is compared either with an externally supplied correct answer or with the output decision of the element. In the first case, if in a cycle of M computations N_i is the number of coincidences of x_i and the externally supplied correct answer, we define

$$a_i = \mathrm{ld}\left(\frac{N_i}{M - N_i}\right)$$

then we have a good approximation provided M is great enough. In the second case the same procedure leads to a good approximation provided suitable limits are placed on the possible values of the a_i's.

A second method which does not need a large storage or an external correct answer is the following. It can serve to find out the most reliable one of the n inputs of a majority organ:

The weights a_i are lowered a certain amount at each time when the input x_i and the organ's output y do not correspond.

In the case n = 3 this organ will switch out with great probability those two of its three inputs x_i, x_j, x_k which have turned out to be the most unreliable ones during some interval of time, by using the following rule:

$$a_i(t+1) = \begin{cases} \delta a_i(t) & \text{if } y(t) \neq x_i(t) \text{ and } \\ & a_i(t) \leq a_j(t) + a_k(t), \\ a_i(t) & \text{else,} \end{cases}$$

with $1/2 < \delta < 1$ fixed. During the time interval which depends on δ and on the error probabilities of the inputs, the organ will work as a majority organ in the sense of von Neumann (1956). After this time the output will be equal to that one of the inputs x_i, x_j, x_k which has turned out to be the most reliable one (see Appendix 2).

Still another method can be used to detect and correct stuck-at-1 or stuck-at-0 (stuck-at-(-1)) faults of a module. The output of such a module will always be +1 or -1, for instance if there is a breakdown in a connection. The rule to lower the weights in this case (n=3) is:

$$a_i(t+1) = \begin{cases} a_i(t) - \delta, & \text{if } y(t) \neq x_i(t) \text{ and } (0 \leq a_i(t) < \\ & a_j(t) + a_k(t) \text{ or } a_j(t) \leq 0 \text{ or } \\ & a_k(t) \leq 0) \\ a_i(t) & \text{else,} \end{cases}$$

with $0 < \delta < 1/2$ fixed. This organ will correct two stationary errors in the three input lines provided these two errors will be both stuck-at-1 or stuck-at-(-1) faults, and these two errors will not enter at the same moment.

Certainly it may not be very economical to replace in a given network all majority elements by adaptive majority elements, because these adaptive majority elements are complex and we cannot assume that they work without faults. However, they can extend the lifetime of networks if they are used at certain critical places (used, for example, as majority organs in the heuristic von Neumann method or in our application of the Winograd-Cowan method).

References

Abramson, N.: Information Theory and Coding, McGraw-Hill, New York, 1963.

Cowan, J.D.: Synthesis of reliable automata from unreliable components. In: Automata Theory, ed. by E.R. Caianiello. Academic Press, New York, 1966.

Cowan, J.D.: The design of reliable systems, preprint, Chicago (1973).

Elias, P.: IBM J. Res. Develop. 3, 346-353 (1958).

McCulloch and Pitts: Bull. Math. Biophys. 5, 115-133 (1943).

Morishita, I.: Analysis of an adaptive threshold logic unit, IEEE Trans. Comp. C-19, 1181-1192 (1970).

Muroga, S.: Rome Air Development Center, New York, Technical Note 60-146 (1960).

v. Neumann, J.: Probabilistic logics and the synthesis of reliable organisms from unreliable components. In: Automata Studies, ed. by C.E. Shannon and J. McCarthy, 43-98. Princeton University Press, 1956.

Nillson, N.J.: Learning Machines, New York, 1965.

Pierce, W.H.: Failure-tolerant Computer Design, Academic Press, New York and London, 1965.

Uttley, A.M.: The informon, J. Theoret. Biol. 27, 1, 31-67 (1970).

Winograd, S. and Cowan, J.D.: Reliable Computation in the Presence of Noise, M.I.T. Press, Cambridge, Mass., 1963.

Appendix 1 (Winograd-Cowan Theorem)

Theorem: Assume modules which can compute any boolean function, all with the same computation capacity C^*. For any set of automata $A_1,\ldots,A_m$ with the same graphs of intermodular connections and which are acyclic, and for all $\delta > 0$ there exists an automaton $\bar{A}$ which has the same input-output behaviour as the collection of automata $A_1,\ldots,A_m,B_1,\ldots,B_s$ (the B_i's can be chosen to be copies of some of the A_j's), and which works with a probability of malfunction less than δ.

In the following P_A is the probability that the automaton A malfunctions. (The computation capacity is the channel capacity of a channel consisting of a noise-free modular network followed by a noisy communication network.)

Proof of the theorem:

Let N be the number of modules of each of the A_i's.
Choose an error correcting (n,k) code such that

$$k > m \quad \text{and} \quad \delta \geq 1 - (1 - 2^{k - nC^*})^N$$

This is possible choosing $k/n < C^*$.
Let A^k be the automaton consisting of the collection $A_1,\ldots,A_m$ and of $k - m$ copies of some of the A_j's, which we denote together $A_1,\ldots,A_k$. The single modules of the automata we enumerate for each of the $A_1,\ldots,A_k$ in the same way (by assumption the automata $A_1,\ldots,A_k$ have all the same graphs of intermodular connections) such that from $i \leq j$ follows that the output of module M_j is not connected with one of the inputs of module M_i. We can do this because the A_j's are acyclic. Let us call the i-th module of the automaton A_s by $M_{i,s}$.
If $e_i(\ldots)_{i=1,\ldots,n}$ are the coding functions of our chosen (n,k) code, we construct $\bar{A}$ from A^k in the following way (beginning with c = 1 and step 1, if $M_{1,1}$ is not an output module, and with step 2, if $M_{1,1}$ is an output module):

step 1: $M_{c,1},\ldots,M_{c,k}$ are not output modules. They may compute the functions $f_{c,1},\ldots,f_{c,k}$, where $M_{c,i}$ may have inputs $x_{1,i},\ldots,x_{s,i}$ where $x_{1,i},\ldots,x_{t,i}$ $(t \leq s)$ are coming from modules $M_{r_1,i},\ldots,M_{r_t,i}$ and $x_{t+1,i},\ldots,x_{s,i}$ may come from outer inputs. Choose decoding functions $d_j(\ldots)_{j=1,\ldots,k}$ which permit the recovery of the message with a probability of incorrect decoding less than or equal to $2^{k - nC^+}$ for the given channel consisting of a noise-free $M_{c,i}$ followed by a noisy communication channel. (Such decoding functions exist by Shannon's second theorem.) Now replace the k modules $M_{c,1}, \ldots, M_{c,k}$ by a set of n modules $M'_{c,1}, \ldots, M'_{c,n}$ which compute the functions $f'_{c,1}, \ldots, f'_{c,n}$ where

$$f'_{c,i} = e_i \begin{cases} f_{c,1}(d_1(f'_{r_1,1}, \ldots, f'_{r_1,n}), \ldots, d_1(f'_{r_t,1}, \ldots, f'_{r_t,n}), \\ \qquad x_{t+1,1}, \ldots, x_{s,1}) \\ \vdots \\ f_{c,k}(d_k(f'_{r_1,1}, \ldots, f'_{r_1,n}), \ldots, d_k(f'_{r_t,1}, \ldots, f'_{r_t,n}), \\ \qquad x_{t+1,k}, \ldots, x_{s,k}) \end{cases}$$

If M_{c+1} is not an output module then go to step 1 with c+1 instead of c.

If M_{c+1} is an output module then go to step 2 with c+1 instead of c.

step 2: Modules $M_{c,1}, \ldots, M_{c,k}$ are output modules. Let the notation be the same as in step 1. Choose decoding functions $d_j(\ldots)_{j=1,\ldots,k}$ with the same conditions as in step 1.
Replace the k modules $M_{c,1}, \ldots, M_{c,k}$ by k modules $M'_{c,1}, \ldots, M'_{c,k}$ which compute the functions $f'_{c,1}, \ldots, f'_{c,k}$ where

$$f'_{c,i} = f_{c,i}(d_i(f'_{r_1,1}, \ldots, f'_{r_1,n}), \ldots, d_i(f'_{r_t,1}, \ldots, f'_{r_t,n}), \quad x_{t+1,i}, \ldots, x_{s,i})$$

If c > N, then go to step 1 if M_{c+1} is not an output module and to step 2 if M_{c+1} is an output module, in both cases with c+1 instead of c.

If c = N we are ready.

In $\bar{A}$ now each rank of n modules $M'_{c,1}, \ldots, M'_{c,n}$ will decode the output of foregoing ranks, thereby it will correct errors, compute the functions $f_{c,1}, \ldots, f_{c,k}$ and encode the results for transmission to subsequent ranks. Thus noise is considered to occur in the noisy channel which follows the noise-free module in the decomposition of a noisy module (Winograd and Cowan (1963), chap. 5). By Shannon's second theorem (c.f. Abramson 1963) there exist (n,k) codes such that the probability of error in decoding the channels output can be made as small as we want provided that $k/n < C^*$. Thus by transition from one of the A_i's to $\bar{A}$ we can recover the message transmitted from rank to rank for large enough n and k with an error probability $P = 2^{k-nC^*}$. Because there is a 1 - 1 relation between modules in one of the A_i's and the ranks in $\bar{A}$, we have as probability that $\bar{A}$ will malfunction:

$$1 - P_{\bar{A}} \geq (1 - P)^N \qquad \text{and therefore}$$

$$1 - P_{\bar{A}} \geq (1 - P)^N = (1 - 2^{k - nC^+})^N \geq 1 - \delta \qquad \text{and}$$

$$P_{\bar{A}} \leq \delta$$

From construction it is clear that A^k and $\bar{A}$ will have the same input-output behaviour in the absence of noise.
It is also possible to extend this procedure to automata which are not acyclic but which have a finite memory.

Appendix 2

Let β_1,β_2,β_3 be the error probabilities of the inputs x_1,x_2,x_3. Without loss of generality let $\beta_1 < \beta_2 \leq \beta_3 < 1/2$. We have to show, that with great probability there is a t for which $a_1(t) > a_2(t) + a_3(t)$. From the error probabilities of the inputs we must go to the probabilities $\varepsilon_1,\varepsilon_2,\varepsilon_3$ with which the organ will alter the weights. (An error in x_1 and x_2 cannot be recognized at the same time, so the organ will record an error in x_3 and alter a_3). A short calculation shows that we also have $\varepsilon_1 < \varepsilon_2 \leq \varepsilon_3 < 1/2$. For j = 2 and 3 the probability, that $a_1(t) > 2a_j(t)$ is

$$P_{1,t} := \sum_n \sum_i \frac{t!}{n!(n+c+i)!(t-2n-c-i)!} \varepsilon_1^{\,n} \varepsilon_j^{\,n+c+i} (1-\varepsilon_1-\varepsilon_j)^{t-2n-c-i}$$

where $c := 1 + \left[\frac{1}{-\mathrm{ld}\ \delta}\right]$.

If we set $P_{2,t}$ to be the probability that

$$1/2\ a_j(t) \leq a_1(t) \leq 2\ a_j(t)$$

and $P_{3,t}$ to be the probability that

$$a_1(t) < 1/2\ a_j(t)$$

then it can be shown that $\lim_{t\to\infty} P_{2,t} = 0$.

As $P_{1,t} + P_{2,t} + P_{3,t} = 1$, we get $\lim_{t\to\infty} (P_{1,t} + P_{3,t}) = 1$.

For any $t > c$ we have the relation: $P_{3,t} \leq (\frac{\varepsilon_1}{\varepsilon_j})^c P_{1,t}$.

From the last two relations we see, if we choose c great enough, that is if we choose δ close enough to one, we have for great t, that $a_1(t) > 2a_2(t)$ and $a_1(t) > 2a_3(t)$ with probability close to one. This means with probability close to one we have $a_1(t) > a_2(t) + a_3(t)$ and therefore $x_1(t) = y(t)$.

MODIFIABLE AUTOMATA WITH TOLERANCE
A MODEL OF LEARNING

M. Dal Cin
Institute for Information Sciences
University of Tübingen, Germany

1. Introduction

We will introduce an abstract mathematical model which mimics the process of learning in certain modifiable systems. In particular, systems are considered which have the capacity to learn to recognize patterns, to be reliable, and to mask errors effectively, and which improve their performance gradually.

We will place our discussion of learning and error masking in the context of abstract automata theory. We present the model in section 2 and touch upon some directions of research dealing with this model in sections 3 and 4. In the Appendix we collect the necessary mathematical notation.

2. A Model of Learning

When we speak about learning we usually think of a goal which has to be reached through the learning process. In general, however, this goal can not be reached unless the learning system changes its state-to-state transition structure. Hence, learning is an adaptive, goal directed process of modifiable systems which is induced by the system's environment and/or experience and which includes recognition as one of its major goals. In order to present our model of modifiable systems we first recall the concept of an automaton (Arbib, 1964; Böhling et al., 1969/70; Kalman et al., 1969).

Let X be a finite set of inputs, Y a finite set of outputs, and Q a set of states of a deterministic or nondeterministic but complete automaton, A, operating sequentially on a discrete time scale. A is described by the quintuple $A = |X,Y,Q,\Delta,\omega|$ where $\Delta \subset Q\times X\times Q$ is the transition structure and $\omega \subset Q\times Y$ the output relation of A (Moore model). Now, a modifiable automaton MA is an automaton with varying transition structure, or more precisely:

Let A = $|X,Y,Q,\Delta,\omega|$ be an automaton and ϕ^A a set of binary relations on Q (called <u>admissible</u> set of state relations) such that $pr_1\phi = Q$ for all $\phi\epsilon\phi^A$. Then the modifiable automaton MA is given by the relational structure

$$MA = |Q,\Delta,\phi^A|$$

and at any time t_n, the performance of MA is that of the (finite or infinite, deterministic or nondeterministic) automaton $MA_n = |X,Y,Q,\Delta_n = \Delta_{n-1} \circ \phi_n,\omega|$ for $n\geq 1$ and MA_n = A for n = O.

The state relation ϕ_n at t_n is specified by the "<u>adaptation circuit</u>" AC of MA executing a map

$$AC: \quad X\times V\times Y\times\phi^A \rightarrow \phi^A \ .$$

V is the set of values of inputs of MA. They are selected by the environment E whose function comprises a specification of inputs SI: $\mathbb{N}\times Y\rightarrow X$ ($\mathbb{N}$ the set of time indices) and a valuation of inputs VI: $X\rightarrow V$ (cf. Fig. 1a,b. Dividing the system into blocks AC, Δ, ϕ^A etc. is arbitrary to a large extent.) Furthermore, let L = $(\phi_1,\phi_2,\phi_3,\ldots)$ be the <u>learning process</u> of MA directed by its AC-part according to E and $\phi_{(n)} = \phi_1\circ\phi_2\circ\ldots\circ\phi_n$, $\phi_i\epsilon L$. Automaton MA_n is called the <u>configuration</u> of MA at time t_n.

Selflearning in MA implies that $|V| = 1$. In specific cases the adaptation circuit AC represents a learning algorithm (algorithmic learning, cf. Appendix 2 and (Tsypkin 1971/73)). Note that MA contains a part, <u>viz</u>. AC, which is not modifiable. That this may be necessary also for adaptive biological systems has been discussed by Arbib (1972). The criterion for structural modifications may be either trial and error evaluation of MA's behavior with respect to some performance measure, reward and punishment, or neural-like reinforcement.

To see whether or not MA acquires specific transition structures we have to define its <u>goal of learning</u>, specifying a set ϕ^L of binary relations on its state space. We say: <u>MA reaches at time t_n the goal ϕ^L under the action E = (VI,SI) of its environment (teacher) iff $\phi_{(n)}\epsilon\,\phi^L$.</u>

If the a priori information required about the learning goal is not completely known, Φ^L may contain "many valued" relations leading to a a nondeterministic or stochastic behavior of MA if it is able to reach its goal. However, even if this is not the case, AC should drive MA as "near" to its goal as possible. A tolerance is then introduced on Q as a suitable performance measure for modifiable automata which gradually and perhaps only approximately reach their goal. It indicates, e.g., similar input-output behavior (see Appendix 1 and Dal Cin, 1973b, 1974). A few concepts of tolerance geometry (Poston, 1971) needed in the following are presented in the Appendix.

The foregoing scheme brings out the features of many discrete learning processes put forward in control theory (Fu, 1971; Menzel, 1970) and biology (Tseltin, 1973; Albus, 1971; Pfaffelhuber, 1974) as well as of fault tolerant computer models (Pierce, 1965).

Since automata with variable transition structures are considered whose state spaces are structured by tolerance relations (Arbib, 1967; Dal Cin, 1973) our model encompasses features of topological (Brauer, 1970; Valk, 1972; Pohl, 1973) and time-variant automata (Salomaa, 1968, 1973; Gill and Flexer, 1967). The languages accepted by finite automata with variable transition structures have been characterized by Salomaa (1968) who showed that languages which are not recursively enumerable can be accepted by time-variant automata with nonperiodic structural changes. In the next section we will discuss some dynamical features of the behavior of modifiable automata and in section 4 we hint at some applications.

3. Learning in Modifiable Automata

It has often been stated that a system is adaptive if it reaches an optimal, terminal state by dint of its interaction with its environment. On the other hand, it may be said that learning occurs only if the system has to modify its transition structure in order to improve its performance. Thus, adaptation takes place on two (sometimes temporarily interchanging) levels: adaptation (a) without and (b) with structural changes. "Animal learning is a case in point. When primates are learning to solve problems, their behavior, though not strictly stationary, remains approximately so: the learning curves can be extrapolated with confidence, and the behavior is predictable. Then, rather suddenly, the creature learns a new concept and subsequently deals with problems in a different way which it sticks to for a further appreciable interval. Once again, the learning curves can be extrapolated and a different kind of behavior becomes predictable" (quoted from G. Pask, 1961, p. 46). Both modes of adaptive behavior of certain modifiable automata have been investigated in [A,B], viz. that of inert modifiable automata which transform their structures gradually and that of modifiable automata with stable configurations. These automata are defined as follows

Let τ be the tolerance of a modifiable automaton, MA. If τ contains all admissible state relations, MA is called inert since in this case all its structural changes take place within tolerance. MA is called fuzzy-state (F-S) if its initial configuration is fuzzy (i.e., if τ is invariant under the semigroup of A) and only fuzrelations are admissible. In this case, no configuration of MA has state-transitions such that small differences in state give rise to large differences in state at later times. This clearly represents a stability property of MA.

In the following let MA = $|Q,\Delta,\Phi^A|$ be a modifiable F-S automaton, ϕ_* an arbitrary fuzrelation and automaton A_* be given as $A_* = |X,Y,Q,\Delta_*= \Delta \circ \phi_*,\omega|$.

Adaptation with structural changes (learning processes).

Suppose now that the objective is that MA learns the transition structure of automaton A_*, but that we are satisfied if after learning the performance of MA is tolerable. This may reduce the cost or time needed for learning, in particular if a large tolerance τ is acceptable. The transition relations of MA_n and A_* are related by τ^τ (i.e. in tolerance) if MA reaches $\phi_*^L := \phi_*(\tau^\tau)$ at time t_n. We then say: MA and A_* are in modification tolerance at t_n. In [C] a useful criterion is given which evaluates whether or not this is the case (cf. Example (a)). Thus we take ϕ_*^L as the goal of learning. Besides knowing the goal, it is, in general, necessary that a quality measure for the learning process L be available, given, for instance, by a function ψ: $Q \to \mathbb{R}$ where $\psi(q)$ is the actual or expected cost for MA in state q (cf. Example (b)). This cost function then also gives rise to a measure of the cost of state transitions and, hence, of L. Adaptation circuit AC should then generate a learning process such that MA inertly modifies its structure in order that MA gradually reaches the goal ϕ_*^L by minimizing the cost of its (state-transition) structure. In [C] it is shown how AC may accomplish this, e.g. by appropriate searching algorithms (Tsypkin, 1971), where the "dynamics" of this learning process is determined by appropriately defined "tangent fields" on Q. In this context, the investigation of the complexity of learning algorithms is an important issue (cf. Bremermann, 1974; Salomaa, 1973), being of great interest also to those who adopt the point of view that the performance of the maturing brain is of an algorithmic nature.

Examples. (a) Generation of a learning goal. Let X = {a,b}, Δ_* be given by Fig. 2 and $\tau = \delta Q \cup \{AB,BA,AC,CA\}$ be a tolerance on the state space

$Q = \{A,B,C,D\}$ of MA. The goal ϕ_*^L induces the structures Δ^i of Fig. 2. The formula which generates Fig. 2 can be found in [C].

(b) Generation of a learning process. The structure Δ, the output relation ω and the tolerance τ of MA with $Q = \{A,B,C,D,E\}$ and $X=Y=\{0,1\}$ is given by Fig. 3 and Table 1, respectively. As cost-function we choose the expectation value $\psi(i,q) = E[SI(i,\omega(q))]$, if MA interacts with a stochastic environment. We valuate input 1(0) as penalty (reward). It follows that $\psi(i,q)$ is equal to the probability $SI(i,\omega(q))$ being 1. Suppose that ψ is time-independent and given by Table 1. If the learning process L is determined by the above mentioned procedure [C] the structure of MA is gradually improving (adapting) according to Fig. 3. At t_i, $i>3$, no new structure appears. Thus, learning terminates after three structural changes.

Adaptation without structural changes (approximate fixed points).

Let us then assume that learning in MA terminated at time t_s. The following theorem [A] implies that MA_s is now potentially adaptive to many tasks (i.e., repetitions of input words generated by a static environment) if its state space is finite and contractible (see Appendix 1). The interaction with a static environment plays a special role in the theory of adaptive behavior because it gives a characterization of an adaptive process by its changing responses to the same situation (Gaines, 1972). A similar mode of adaptation of sensory and nuclear cells to their inputs is discussed in (Bondi, Schmid, 1972).

Theorem*: Suppose that MA is a deterministic modifiable F-S automaton, that its state space is finite and contractible and that ϕ_i is the identity relation ($i \geqslant s$). For any repetitively applied $w \in X^*$ with $qw^\ell = \bar{q}w^m$ (for some $m, \ell \in \mathbb{N}$ and all $(q,\bar{q}) \in Q \times Q$) there is $n_o \in \mathbb{N}$ such that all the states qw^n ($n \geqslant n_o$, $q \in Q$) of MA are mutually within tolerance.

This theorem implies that the performance of MA in its interaction with a static environment becomes eventually either always satisfactory (tolerable) or always unsatisfactory depending on the behavior of MA_s in the state $\bar{q}$ at t_{n_o} (i.e. on $(MA_s)_{\bar{q}}$, see Appendix 1). However, tolerable behavior of MA in state $\bar{q}$ may have been the goal of the learning process taking place before t_s. Subspace $N_w = \{\bar{q}w^i\} i=1,2,\ldots$ of (Q,τ) is a stable set [A] of automata states (approximate fixed points). This notion can be generalized to that of almost stable and attracting state sets [A] . They characterize additional properties of the adaptive

---

*We define: $qw^1 = q\delta_w^{(s)}$ and $qw^{m+1} = (qw^m)\delta_w^{(s)}$ where $\delta_x^{(s)}$ is given by Δs; $x \in X$, $w \in X^*$, $m \in \mathbb{N}$.

behavior of modifiable F-S automata, e.g., forgetting and relearning.

4. Applications of Modifiable Automata

Pattern recognition. Pattern classification by machines and, in particular, by adaptive linear threshold elements (Minsky, Papert, 1969; Arbib, 1972) is an area where modifiable automata provide a useful mathematical model and where the role of an appropriately defined tolerance becomes particularly clear. As may be inferred from the example of Appendix 2, the variety of actions at any learning stage of "classical" perceptrons is rather restricted compared with the possibilities our model provides.*

Pattern recognition and learning in biological systems has been the object of many investigations. It is not our purpose to review the literature, in particular, that on the hierarchical aspect of learning (Fu, 1971) and on learning neurons and neuronal nets, though the hypothesis of Hebb (1949), Brindley (1972) and others that the brain's structure is organized by changing the synaptic weights and thresholds of its neurons (scarcely proved physiologically) proved to be helpful and motivates our investigation of modifiable automata (cf. Appendix 2). Evidently, the ability of machines to perceive their environment is still very limited compared to the ease with which vertebrates perform perceptual tasks and, unlike most artificial learning systems a biological system is able to select its own goals and to determine subgoals (which can be modified in the light of further experience). This capacity is an essential aspect of intelligent behavior.

Language learning. In the context of formal language learning structure Δ_n of a modifiable automaton MA implements the linguistic performance of MA at learning stage n, whereas AC implements the capacity of language learning (faculté de langage, Schnelle, 1965). Sentences (or words) not understood completely at learning stage n may, nevertheless, be partially understood. A tolerance τ can be introduced on the state space of MA which implies the following (Dal Cin, 1974c). If MA is at learning stage n in modification tolerance with an experienced speaker, the partial understanding of sentences gives sufficient information as to how the structure of MA_n is to be changed such that the linguistic performance of MA becomes better at t_{n+1}. Again, we may look for appropriate gradual language learning processes. Evidently, MA will only learn if sentences are presented at t_0 which are, at least, partially understood.

---

*E.g., perceptrons do not answer (at t_n) questions like: Do the patterns appearing at t_n, t_{n-2} and t_{n-7} belong to the same class?

Error masking. Changes in transition structures also occur in nonlearning but unreliable systems (e.g. in certain "noisy" neural nets) due to permanent or temporary state transition errors. Their behavior has been modeled by F-S modifiable automata with tolerance [B] and the phenomenon of error masking and error correction has been studied. From the viewpoint of reliability inert modifiable automata correspond to automata whose structures gracefully degrade under failure conditions of increasing extent. For example, their malfunctioning may be masked to a certain extent by structural or dynamical redundancy (cf. Dilger, 1974; Winograd and Cowan, 1963).

In [B] the notion of l-masking ($l \in \mathbb{N}$) of state-errors has been introduced, where ϕ^A is now interpreted as the set of possible state-errors of MA, MA_i as a faulty version of A and AC as a (random) source of state-transition errors. A decision algorithm has been given which evaluates whether or not a state error is l-masked by MA at t_i. (It is, for instance, shown that no state-error can be l-masked by MA if MA_0 is minimal and $\omega_* \tau^l = \delta Y$.) It was also shown how to generate (reliable) input sequences (controlling experiments) for the faulty automaton MA_i such that these sequences induce state transitions, e.g. resets, in accordance with those of its fault-free reference automaton A. Just as we can use a synchronizing input sequence in order to correct the temporary state-transition errors of an automaton, we can use reliable input sequences in order to neutralize its permanent state-transition errors.

With respect to fault tolerance von Neumann made the differences between biological and artificial systems brilliantly clear: "It's very likely that on the basis of the philosophy that every error has to be caught, explained, and corrected, a system of the complexity of the living organism would not run for a millisecond. Such a system is so well integrated that it can operate across errors. An error in it does not in general indicate a degenerative tendency. The system is sufficiently flexible and well organized that as soon as an error shows up in any part of it, the system automatically senses whether this error matters or not. If it doesn't matter, the system continues to operate without paying any attention to it. ... . This is a completely different philosophy from the philosophy which proclaims that the end of the world is at hand as soon as the first error has occurred." (John von Neumann, 1966, p. 71). Thus the relevance of models of biological system performance can be checked, in part, by determining whether or not the models suffer the same malfunctions as the real system and, in particular, by observing the behavior of these models after an error has occurred, e.g. whether or not the system is able to l-mask certain state-transition errors.

5. Conclusion

Modifiable, time-variant and topological automata are relatively new contributions to learning and systems theory. Their theory is still in a state of development. However, our approach has already led us to formulations and questions which, we think, are relevant to certain aspects of artificial intelligence.

References

Albus, J.S.: A theory of cerebellar function, Math. Biosciences 10, 25-61 (1971).

Arbib, M.A.: Brains, Machines and Mathematics, McGraw-Hill, New York, 1964.

Arbib, M.A.: Tolerance automata, Kybernetika 3, 223-233 (1967).

Arbib, M.A.: Theories of Abstract Automata, Prentice-Hall, Englewood Cliffs, N.J., 1969.

Arbib, M.A.: Organizational principles for theoretical neurophysiology. In: Towards a Theoretical Biology, vol. 4, ed. by C.H. Waddington, pp. 146-168. Edinburgh University Press, Edinburgh, 1972.

Arbib, M.A.: Automata theory in the context of theoretical neurophysiology. In: Foundations of Mathematical Biology, vol. 3, ed. by R. Rosen, pp. 193-282. Academic Press, New York, 1973.

Biondi, E. and Schmid, R.: Mathematical models and prostheses for sense organs. In: Theory and Applications of Variable Structure Systems, ed. by R.R. Mohler and A. Ruberti, pp. 183-212. Academic Press, New York, 1972.

Böhling, K.H., Indermark, K. and Schütt, D.: Endliche Automaten I+II, BI Hochschulskripten, Mannheim/Wien:Zürich (1969/70).

Brauer, W.: Zu den Grundlagen einer Theorie topologischer sequentieller Systeme und Automaten, GMD Bonn, Bericht 31 (1970).

Bremermann, H.: Complexity of Automata, Brains and Behavior, this volume (1974).

Brindley, G.S.: The potentialities of different modifiable synapses, Summaries of the Int. Centre for Theoretical Physics, Miramare Trieste (1972). However, see also: W.L. Kilmer and M.A. Arbib, An automaton framework of neural nets that learn, Int. J. Man-Machine Studies 5, 577-583 (1973).

Dal Cin, M.: Fault Tolerance and Stability of Fuzzy-State Automata, Lecture Notes on Computer Science Vol. 2, pp. 36-44. Springer-Verlag, Berlin, 1973a. (denoted by A)

Dal Cin, M.: Modification-tolerance of fuzzy-state automata, to appear in Int. Journal of Computer and Information Sciences, vol. 4, no. 1 (1973c). (denoted by B)

Dal Cin, M.: Learning in modifiable F-S automata, Proceedings of the Conference on Biologically Motivated Automata Theory, McLean (1974) and Modifiable automata and learning, Proceedings of the Second European Meeting on Cybernetics and Systems Research, Vienna (1974). (denoted by C)

Dal Cin, M.: A Model of Language Learning, preprint, Tübingen (1974c).

Dilger, E.: Structural and dynamical redundancy, this volume (1974).

Fu, K.S.: A critical review of learning control research. In: Pattern Recognition and Machine Learning, ed. by K.S. Fu, pp. 288-296. Plenum Press, New York, 1971.

Gaines, B.G.: Axioms for adaptive behavior, Int. J. Man-Machine Studies 4, 169-199 (1972).

Gill, A. and Flexer, J.R.: Periodic decomposition of sequential machines, J. Assoc. Computing Machinery 14, 666-676 (1967).

Hebb, D.O.: The Organization of Behavior, Wiley, New York, 1949.

Kalman, R.E., Falb, P.L. and Arbib, M.A.: Topics in Mathematical System Theory, McGraw-Hill, New York, 1969.

Menzel, W.: Theorie der Lernsysteme, Springer-Verlag, Berlin, 1970.

Minsky, M. and Papert, S.: Perceptrons, M.I.T. Press, Cambridge, Mass., 1969.

Pask, G.: An Approach to Cybernetics, Hutchinson, London, 1961.

Pfaffelhuber, E.: Information theory and learning in biology, summary of the workgroup sessions, this volume.

Pierce, W.H.: Fault-Tolerant Computer Design, Academic Press, New York, 1965.

Pohl, H.J.: Zur Theorie Topologischer Automaten I, II, EIK 9, Heft 4/5, 217-239, Heft 6 327-339 (1973).

Poston, T.: Fuzzy Geometry, doctoral thesis, University of Warwick, 1971.

Salomaa, A.: On finite automata with time-variant structure, Inform. and Control 13, 85-98 (1968).

Salomaa, A.: Formal Languages, Academic Press, New York, 1973.

Schnelle, H.: Steps towards models of language learning, Kybernetika 1, 365-373 (1965).

Tseltin, M.L.: Automata Theory and Modeling of Biological Systems, Academic Press, New York, 1973.

Tsypkin, Ya.Z: Adaptation and Learning in Automatic Systems, Academic Press, New York, 1971.

Tsypkin, Ya.Z.: Foundations of the Theory of Learning Systems, Academic Press, New York, 1973.

Valk, R.: Topologische Wortmengen, Topologische Automaten, Zustandsendliche, Stetige Abbildungen, GMD Bericht Nr. 19 (1972).

Valk, R.: The use of metric and uniform spaces for the formalization of behavioral proximity of states, Lecture Notes on Computer Science, Vol. 2, pp. 116-122. Springer-Verlag, Berlin, 1973.

von Neumann, J.: Theory of Self-reproducing Automata, ed. by A.W. Burks, University of Illinois Press, Urbana, 1966.

Winograd, S. and Cowan, J.D.: Reliable Computation in the Presence of Noise, M.I.T. Press, Cambridge, Mass., 1963.

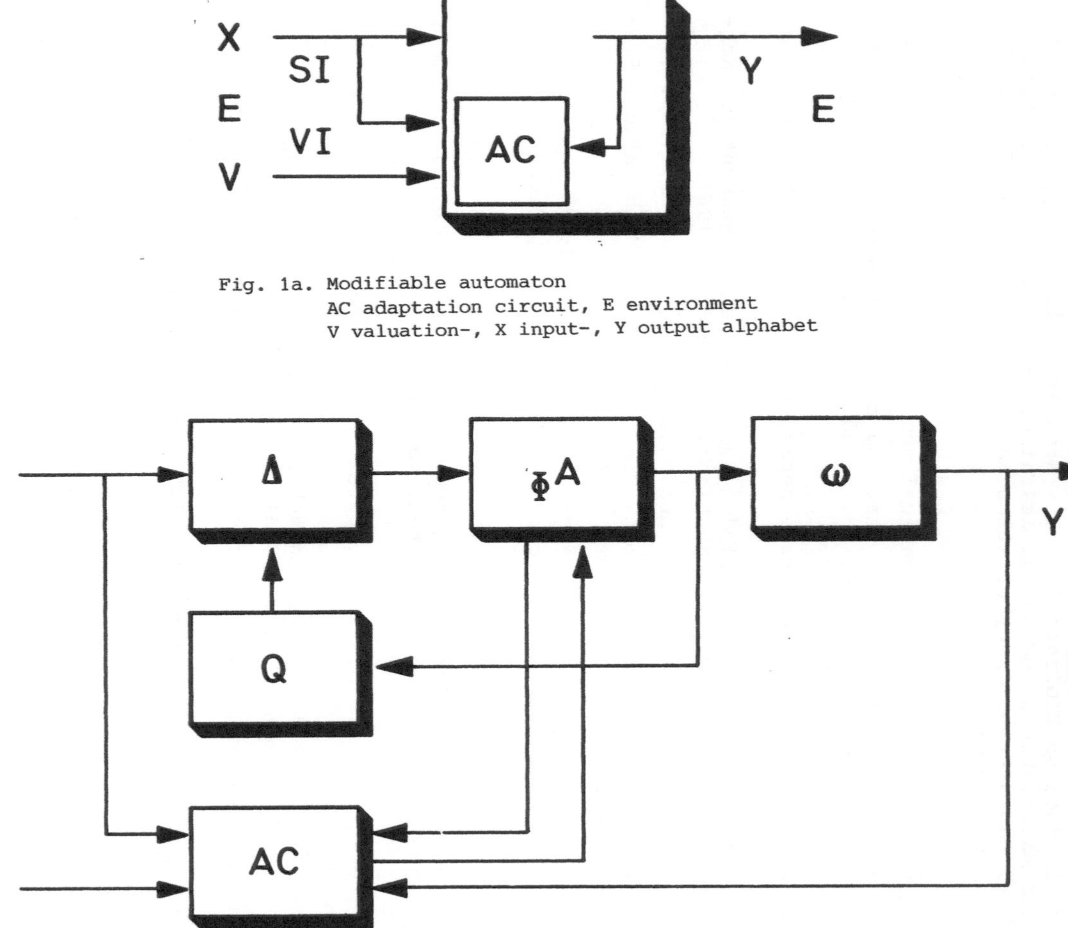

Fig. 1a. Modifiable automaton
AC adaptation circuit, E environment
V valuation-, X input-, Y output alphabet

Fig. 1b. A possible system configuration.

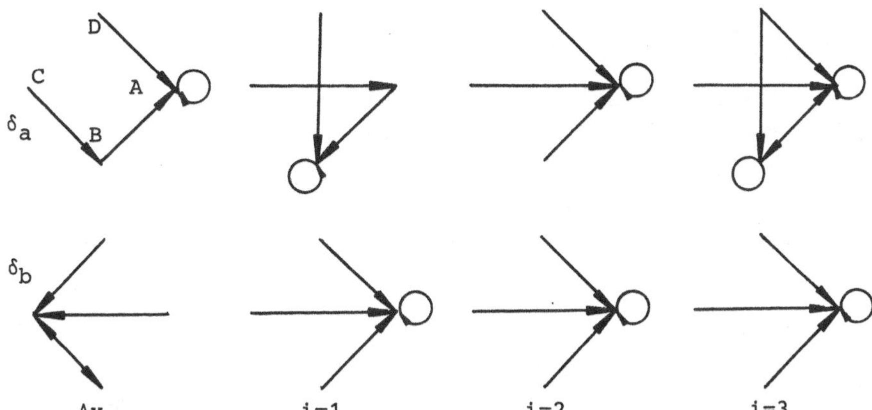

Fig. 2. $\Delta' = \Delta\,\phi'$; $\phi' \varepsilon\, \phi^L_*$

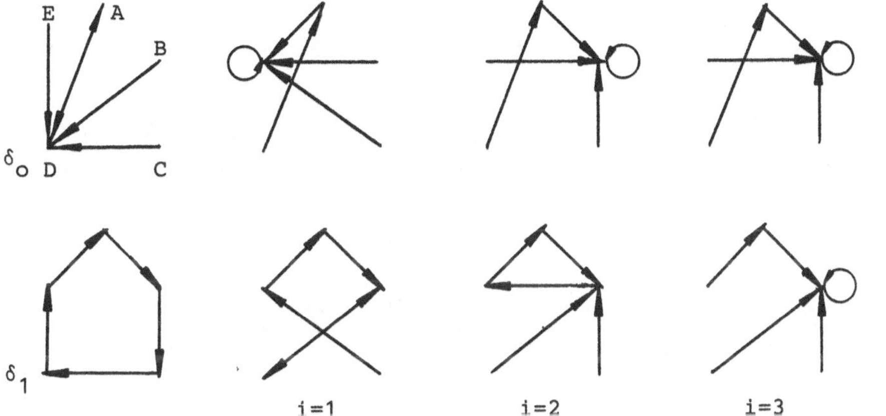

Fig. 3. Δi

Q	A	B	C	D	E
ω	1	0	0	1	0
τ	A	BE	CD	CDE	BDE
ψ	3/5	1/5	1/2	2/5	3/10

Table 1

Appendix 1: Tolerance Spaces

Notation. Let X,Y be arbitrary sets, $\tau_i \subset X\times X$ (i=1,2,...) binary relations on X and $\Delta \subset X\times Y\times X$ a ternary relation between X and Y; τ_i is a tolerance on X iff it is reflexive and symmetric; pr_i: $X\times X \to X$ (i=1,2) are projections. If $(x,\bar{x}) \varepsilon \tau_i$ we also write $x\tau_i\bar{x}$; $\bar{\tau}_i = \{(x,\bar{x}) \varepsilon X\times X \mid (x,\bar{x}) \not\varepsilon \tau_i\}$; $\tau_i^c = \{(\bar{x},x) \mid (x,\bar{x}) \varepsilon \tau_i\}$. Let $S\subset X$ then $S\tau_i := pr_2(\tau_i \cap (S\times X))$. The relative product is given as $\Delta \circ \tau_i := \{(x,y,\bar{x}) \mid \exists z\varepsilon X: (x,y,z) \varepsilon \Delta$ and $(z,\bar{x}) \varepsilon \tau_i\}$ and $\tau_i \circ \tau_j := \{(x,\bar{x}) \mid x\tau_i \cap \bar{x}\tau_j^c \neq \emptyset\}$.

Tolerance spaces. Let $X\neq\emptyset$ be any set and τ a tolerance on X, (X,τ) is called a tolerance space; $\delta X = \{(x,x) \mid x\varepsilon X\}$ is the little tolerance and $iX = X\times X$ the big tolerance on X. If $x\tau x'$ we say x is within tolerance τ of x'. If $f:(X,\tau)\to Y$ is a set theoretic map $f_*\tau := \delta Y \cup f^c \circ \tau \circ f$ is the co-induced tolerance on Y. Let $\rho \subset X\times X$ be a binary relation on (X,τ), ρ is called a fuzrelation or fuzzy iff $\rho_*\tau\subset\tau$. The product of two tolerance spaces (X,τ) and (Y,σ) is given by $(X\times Y, \tau\sigma = pr_1{}^*\tau \cap pr_2{}^*\sigma)$ where $pr_i{}^*\tau := pr_i \circ \tau \circ pr_i^c$. The tolerance $\tau\sigma$ is the biggest tolerance on $X\times Y$ such that the projections are fuzzy. We denote the set of fuzrelations on X by ϕX.

Behavioral tolerances. Let $A = |X,Y,Q,\Delta,\omega|$ be a deterministic automaton, $\delta_x := \{(q,q') \mid (q,x,q') \varepsilon \Delta\}$ and $\delta_w := \delta_{x_1} \circ \delta_{x_2} \circ \ldots \circ \delta_{x_r}$ be the transition relations of A under the action of input sequence $x\varepsilon X$ and $w = x_1x_2\ldots x_r \varepsilon X^*$, respectively; $\delta_\Lambda = \delta Q$ (Λ empty word) and X^* is the free input monoid of A. Suppose that (X,τ_X) and (Y,τ_Y) are tolerance spaces, $\tau_X(\tau_Y)$ expressing similarity of inputs (outputs). Then $\tau_B^{(n)}$, with $q\tau_B^{(n)}q'$ iff $A_q \subset \tau^n A_{q'}$ and $A_{q'} \subset \tau^n A_{q'}$ are natural tolerances* on Q (cf. Also Valk, 1973) where:

(i) $A_q = \{(w,v) \mid q\delta_w \circ \omega = v; \quad v\varepsilon Y\}$,

(ii) $\tau = \tau_{X*}\tau_Y$ and

(iii) $w\tau_{X*}w'$ $(w=x_1x_2\ldots x_r,\ w'=x_1'x_2'\ldots x_s'\varepsilon X^*)$ iff r=s and

$x_i\tau_X x_i'$ (i=1,2,...r),

A_q is the behavior of A in state q and $\tau_B^{(n)}$ relates states in which automaton A behaves similarly where n is a measure of behavioral similarity. (Under certain conditions $(A,\tau_B^{(n)}$ is fuzzy.) On the other hand, let $\hat{\tau}$ be a tolerance on state space Q then $\hat{\tau}$ induces a tolerance, viz. $\omega_*\hat{\tau}$, on output alphabet Y.

---

* $\tau_B^{(n)}$ generalizes the relation (tolerance) which indicates equivalence of automata states.

Homotopic fuzmaps. Two fuzmaps $f, g: (X,\tau)\rightarrow(Y,\sigma)$ are called homotopic ($f\simeq g$) iff there is a bound $m\varepsilon\mathbb{N}$ and a sequence $\{F_i\}$, $i=0,1,2,\ldots,m$, of fuzmaps $F_i:(X,\tau)\rightarrow(Y,\sigma)$ such that $f=F_0$, $g=F_m$ and F_i and F_{i+1} are related by τ^σ, i.e. $F_i^c\circ\tau\circ F_{i+1}\subset\sigma$, $i=0,1,2,\ldots,m-1$. Tolerance space (X,τ) is contractible iff there are fuzmaps $f:(X,\tau)\rightarrow(Y,\sigma)$ and $g:(Y,\sigma)\rightarrow(X,\tau)$ such that $f\circ g\simeq\delta X$, $g\circ f\simeq\delta Y$ and $|Y| = 1$.

Appendix 2: Modifiable Threshold Logic Unit (MTLU)

The area where modifiable automata provide a useful mathematical model includes pattern classification, e.g., by threshold logic devices. The adaptable element discussed in the previous seminar (Dilger, 1974) is such a classifying device which may provide a model of modifiable neurons and which may be represented by the modifiable automaton [C] $MTLU=|W,\Delta,\phi^A|$ acting at time t_n as the automaton $MTLU_n=|X,Y,Q,\Delta_n=\phi_{\underline{c}_n}$, $\omega=\mathrm{sign}(\underline{c}_n\cdot(\underline{x},1))|$

(i) W is the r dimensional weight vector space, $\underline{c}_n\varepsilon W$

(ii) X is the r-1 dimensional feature space, $\underline{x}\varepsilon X$

(iii) $\Delta=\{(\underline{c},\underline{x},\underline{c})\,|\,\underline{c}\varepsilon W,\ \underline{x}\varepsilon X\}$

(iv) $\phi^A:=\{\phi_{\underline{c}}:=W\times\{\underline{c}\}\,|\,\underline{c}\varepsilon W\}$

(v) $Y=\{+1,-1\}$

Examples of learning algorithms executed by AC are:

(a) Selflearning, $|V| = 1$. $AC:X\times Y\times\phi^A\rightarrow\phi^A$ with $(\underline{x}_n,y_n,\phi_{\underline{c}_n})\mapsto\phi_{\underline{c}_{n+1}}:=W\times\{(1-\alpha)\underline{c}_n + \underline{g}(\underline{x}_n,y_n)\}$, $\underline{g}$ a quality criterion, e.g., $\underline{g}(\underline{x}_n,y_n)=\alpha y_n(\underline{x}_n,1)$, $0<\alpha<1$;

(b) Training, $V = \{+1,-1\}$, $AC:X\times V\times Y\times\phi^A\rightarrow\phi^A$ with

$$(\underline{x}_n,e_n,y_n,\phi_{\underline{c}_n})\mapsto\phi_{\underline{c}_{n+1}}:=\begin{cases} W\times\{\underline{c}_{n(\pm)}\alpha(\underline{x}_n,1)\} \\ \text{if } y_n=_{(\pm)}1 \text{ and } e_n=_{(\mp)}1 \\ \phi_{\underline{c}_n} \quad \text{otherwise} \end{cases}$$

In this case, MTLU reaches a terminal state in its learning procedure if $x_i=SI(i,y_i)$, $i=0,1,2,\ldots$, is any training sequence and if the pattern classes which are to be recognized are linearly separable (Arbib, 1973).

Figure 4 shows typical decision surfaces (given by $\underline{c}\cdot(\underline{x},1)=0$); in some cases several surfaces (i.e. MTLU states) separate the two pattern classes equally well. This observation suggests $\tau_s=\{(\underline{c},\underline{c}')\,|\,|\{\underline{x}|\omega(\underline{c},\underline{x})-\omega(\underline{c}',\underline{x}) = 0\}|<s\}$, $s>0$, as appropriate tolerances on W.

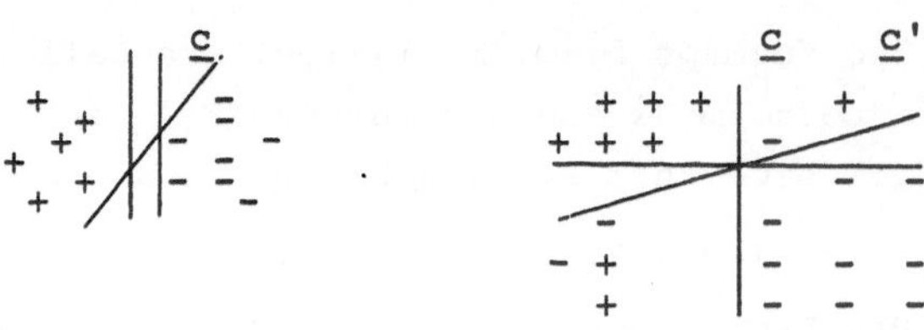

Fig. 4. Decision surfaces.

MTLU's have been composed to give modifiable automata with richer structure and more discrimination power (Minsky, Papert, 1969).

PART VII

<u>The Approach From Information Theory</u>

REDUNDANCY AND PERCEPTION

H.B. Barlow
Psychological Laboratory
Cambridge CB2 3EB

1. Introduction

In the situations that confront communications engineers, it is relatively easy to isolate and identify the component parts to which information theory can be applied, namely a set of messages which are then encoded and transmitted down a communication channel. One can show that the statistical properties of the ensemble of messages have definite calculable effects on the efficiency with which they can be passed down the channel; if the ensemble of messages, after encoding, is highly redundant, much less information can be transmitted. Redundant messages are also less perturbed by errors in transmission but this is a good deal more complicated, and there are problems enough in applying the simpler concepts of information theory to perception.

The first difficulty is to ensure that, if one uses the terminology of information theory, one has agreed about what is the set of messages, what is the encoding process, and what is the communication channel. In this article I get very little farther than the messages and what causes them. These are regarded as arising from the environment of an animal and they constitute the ever-changing sensory scene which it faces. They are, then, the physical stimuli affecting the receptors. There is not much room to doubt the main conclusion that these messages are highly redundant, and that the redundant features of the sensory scene are extremely important for the animal, regardless of any transformations or encoding its nervous system may perform: one function of our perceptual mechanisms must certainly be to identify redundancy in the sensory stimuli.

One would obviously like to use informational concepts inside the nervous system, considering nerves as communication channels for the encoded sensory messages, and higher centres as mechanisms for recoding and receiving these messages, but this is much more difficult because one knows so little about the brain compared either with man-made communication systems, or with the physical processes whereby the environment influences sensory receptors. What is the "cost" of an extra nerve impulse or nerve fibre? How is the environment represented in the brain? What is the nervous system trying to do with all the information it receives? These are the kinds of questions that arise, and because they are unanswered discussion cannot get very far. Nevertheless I think coding to reduce redundancy is an intriguing and perhaps illuminating parallel to some sensory and perceptual neural processes, and in the second part of this article I discuss its strengths and weaknesses, and go on to suggest that redundancy may also be important when it comes to the selective loss or preservation of information. The first part is concerned with the ensemble of messages.

2. Messages from the Environment

In order to decide whether these messages have high or low redundancy examples of each will be compared. Consider first a communication channel that is being used efficiently at nearly its full capacity; that is, the messages arriving along it have low redundancy and the information gained by receipt of a message is, on average, nearly as high as the constraints imposed by the channel allow it to be. This would be the case if the messages appeared to be arriving completely at random, with no regularity or pattern, for under these conditions the prior probability of receiving the message would be 1/(number of alternative messages), and would be unaffected by whatever other messages had just been received. If there was any departure from the apparently random arrival of messages, this would enable one to make some predictions about the messages (possibly only in the form of modified prior probabilities), and this would decrease the average gain of information.

Contrast this with a channel that is being used inefficiently, where the messages are highly redundant. Because of the regularity or pattern of their arrival one can form expectations at any moment, based on the messages received up to that moment and the known patterns, and these reduce the average information gained from each message.

Now if one considers the set of physical messages received by sensory receptors it is immediately obvious that this is a high redundancy situation. There is obviously much regularity or structure: the temperature does not fluctuate randomly between extremes, tactile pressure on the feet occurs at predictable moments in the walking cycle, a gust of wind that disturbs one hair will also disturb its neighbours, and so on.

Another point seems clear as one thinks about this. Subjectively one has a very limited capacity to absorb or take in the random, unexplained, unpredictable elements of one's sensory experience. When the unexpected occurs certainly one becomes aware of it, but in the normal course of events one seems to be dealing with the constraints on the physical stimuli that play on one's senses, rather than with the physical stimuli themselves. Metaphorically, one hears the tune, not the unintelligible background noise. Of course this makes sense because it is the constraints that are caused by objects, and it is knowledge of objects around us that has survival value to us. It is the constraints that enable us to make predictions and find our way about.

Our nervous system handles regularity and pattern in a special way. It may be important to realize that, in informational terms, it is redundancy that is singled out for special treatment, and I think this principle gives unity to an astonishing range of nervous functions reaching up to quite complex mental acts. The search for redundancy may play a role in the informational struggle of intelligent beings comparable with the search for energy rich foodstuffs in the struggle for existence at a lower level. Let me try and justify these claims by pointing out briefly some aspects of the search for constraints on the random arrival of sensory stimuli.

Nature of general constraints. Any departure from complete randomness in the arrival of physical stimuli at an animal's sensory receptors is a form of redundancy, for instance the tendency for light to fall from above, and tactile stimuli to be applied to the animal's lower surfaces. These are obviously important and are reflected in the anatomy of the sensory receptors, their pathways and their connexions, but the main concern here is with more subtle, less universal constraints that enable an animal to find out about its environment. These can be described very inadequately and incompletely as the tendency for physical stimuli to occur together or to follow each other in a sequence. In addition there is a class of associations which must be particularly important, namely those which follow the initiation of a motor act by the animal itself. The brain's model of the world must be built up from its experience of such regularities, but it is extraordinarily difficult to describe the constraints that result from important objects in an animal's world such as a tree blocking its path, or, for us, a traffic sign. The elusiveness of those ill defined constraints is shown by the slow progress of pattern recognition, but there may be some virtue in regarding the various regular and patterned features as informational redundancy, for this is something one can test for and detect on a small scale. There is at least a possibility that it will be easier to understand how an animal knows its environment when we know how it handles structure in sensory stimuli on a small and local level. In other words, if so much hinges on detecting redundancy it will be worth while making a very detailed examination of simple instances of it.

Local constraints. If one considers a very small part of the physical input to the senses, such as the light intensity at a few resolvable elements of the visual field over a few moments in time, the presence of redundancy can be determined by the following test. First find the range of possible intensities, which would be determined by the brightest source in the environment. For simplicity, suppose we can represent the intensity at one element for one moment as one of a finite number of discriminable levels. Then the greatest gain of information would occur if the statistical distribution of observed intensities was rectangular, so that the prior probabilities of all levels were equal, but in practice this distribution would not be followed. This is the simplest form of redundancy, and another type could be detected by observing successive values of light intensity at one point; for optimal information gain two successive values should be statistically independent, but again reflection suggests that one would find a strong tendency for them to be similar, because light levels tend to stay the same or change slowly, or only very exceptionally change widely and rapidly. This would represent a second type of redundancy, and a third type could be found by observing the non-independence of adjacent luminance values resulting from the fact that most luminous surfaces are much larger than a single resolvable element.

It may not yet be evident that much has been gained by observing these regularities, but consider now a fourth type of redundancy. When an object or the eye moves, the intensity at one point on the retina is transferred to a neighbouring point, and one's first idea about detecting this might be to look for instances where the intensity at one point is the same as the intensity at a neighbouring point a moment earlier. Movement is of obvious importance to an animal for a wide variety of reasons, but this would be an ineffective way of detecting it, because of the tendency for neighbouring and successive values to be the same even without any movement. One needs to know the probabilities of the simple events before one can attach any significance to the occurrence of a compound event, and in fact to detect movement one should look for the occurrence of change at one point followed by the same change at a neighbouring point. Since change is comparatively rare, its successive occurrence at neighbouring points is likely to be caused by movement. This is a revealing example of the importance of knowing prior probabilities.

Prior probabilities. The emphasis I am giving to the statistical structure of the sensory stimuli an animal receives is obviously different from the more orthodox way of thinking of stimuli in terms of their survival value. There is a disconcerting circularity in saying that a frog catches bugs because it has bug detectors, while elsewhere (probably in another course in another department) we say it has evolved bug detectors because this enables it to find food; since it is already equipped with bug-detectors, what else could it use? Actually the frog and his ecological niche are so closely mutually adapted that some apparent circularity may be allowable, but there must also be non-circular factors involved in determining what is the best means by which a predator can find its prey, and this is the kind of factor I am drawing attention to. Furthermore these factors are clearly even more interesting and important when an animal uses its senses in a less specialized fashion, as higher animals do. The crucial distinction is that the importance of bugs is often thought of as something that is simply dictated to an animal by its circumstances. On the other view, contour and change are important because of the statistics of visual stimuli, and, if they are made use of, movement and bug detecting come naturally.

<u>Availability and general importance of statistical information</u>. The two environmental forces normally thought to mould the nervous system are the rewards and punishments involved in learning, and the survival or non-survival of an animal that causes genetic selection. The thesis here is that the redundancy of sensory stimuli, or if you prefer it their regularity and pattern, is a third moulding force that is probably more important than conventional trial and error learning.

The first point to make is that there is vastly more information available than is used in most learning situations. A "trial" may involve a lengthy sequence of motor acts and sensory experience, yet it can classically only be right or wrong, yielding one bit of acquired information. For redundancy moulding all that is required is to observe and count how often particular sensory experiences occur; such counting can go on constantly. This suggests that extensive modifications could be brought about this way.

It may be suggested that this may apply to the modifications in the periphery which are involved in sensory adaptation and figural after-effects, but is unlikely to be important in higher functions. A common form of intelligence test requires that one identify some property of a set of figures, numbers or words, that is shared by all of them. The correct identification of this property can then be proved by selecting an additional member of the set from a number of alternatives.

This seems to require observing and counting as before, but here the operation is performed on properties, whereas before we only postulated that it is performed on simple physical stimuli. The trivial intelligence test problem of course has its parallel in finding regularities in the behaviour of physical objects, animals and human beings, and the performance of these tasks constitutes a large part of intelligent behaviour.

3. Sensory Messages Inside the Nervous System

So far we have looked at the physical stimuli reaching sensory receptors, and seen that their redundancy - the regularity, pattern, nonrandomness that they contain - is supremely important because the animal's knowledge of the external world is really contained in this pattern of prior probabilities of its sensory input. Attneave and I have made the suggestion that it is a primary function of sensory centres to recode the incoming messages in such a way as to reduce redundancy while preserving much of the information. I will first very briefly summarize the case for this notion, but I want to spend more time pointing out some difficulties and suggesting a modification. The favourable case can be put under three headings:

A. To devise a redundancy-reducing, information preserving, code requires precisely what we have seen to be important about sensory messages, namely their regularities and the intricate pattern of their prior probabilities. It is one of the paradoxes of such a code that it can, with almost equal legitimacy, be regarded as redundancy preserving, or increasing. If one goes from high redundancy input to low redundancy output, it obviously decreases redundancy, but if one reads backwards and considers what features of the input are specially handled by the code one finds that they are the redundant features, the regular or patterned elements, of the input. So the code captures precisely what the brain needs to know about the environment, and in that sense preserves it. We shall return to this redundancy-conserving viewpoint, for it leads to an interesting suggestion.

B. Finding the structure of the input requires determining the prior probability of some proportion of the possible input patterns. As the sensory input becomes larger the difficulty of this task very rapidly becomes formidable, because the number of possible input patterns increases so rapidly - one encounters the well known problem of the combinatorial number explosion. One of the attractive features of the redundancy reducing proposal is that it shows how the occurrence of quite remote combinations might be discovered. If local redundant elements are first found and coded for, then redundant combinations of these elements can be found, and so on through a hierarchy each stage of which finds structure among previously discovered structural elements. It has not been shown that such a system can do what our brains do, but at least progressive redundancy reduction shows how the number explosion problem might be circumvented - perhaps the only way this can be done.

C. Neurophysiological and psychological evidence seems to fit the notion. I have quite recently reviewed some of this supporting evidence, and although a lot of it fits quite impressively, it is not fully compelling because of difficulties with redundancy reduction that I shall now come to. These difficulties can be conveniently described under five headings:

(1) Need for "display", not transmission. Postulating that sensory centres elaborate a redundancy reducing code for the sensory messages received seems to be saying that their role is simply to transmit sensory information economically to some other position or positions in the central nervous system, but that would hardly be an adequate account of what sensory centres do. Ultimately the information is utilized, and if one is looking for an electronic analogy it might be better to think of the sensory centres as displaying, rather than merely relaying, sensory information. With this in mind, one has to admit that a type of coding which would be appropriate for economizing in the transmission of information might make their interpretation very inconvenient. The essential points here are that the economical use of a number of separate elements demands that every possible combination of elements occurs and has a separate meaning, whereas for convenience one would like a particular element to have some constant significance regardless of what other elements were or were not also in use. Economy and convenience conflict here, as elsewhere in life, and we are not helped in trying to imagine what kind of compromise may have evolved in the brain by our ignorance on the next two matters: what is economical and convenient for the nervous system?

(2) Economy. The purpose of a redundancy-reducing code is to reduce the cost of transmitting messages. It is often sufficient to assume that these costs are simply proportional to the time a channel is occupied, and that coding to minimize this time will also minimize cost. This assumption was, in fact, tacitly made when discussing the redundancy of physical messages reaching sense organs, where time may be important and cost seems totally irrelevant. Inside the nervous system it is a different matter; cost is presumably a matter of selective

disadvantage, and it is not easy to judge the advantages of, say, increasing the number of optic nerve fibres against the disadvantages this would probably cause by increasing the size or decreasing the conduction velocities, or perhaps increasing the susceptibility to damage. Probably any advantages would be marginal.

Similarly an impulse costs the animal a certain (very small) amount in increased metabolic activity of the sodium pump, so there is perhaps a small advantage in reducing impulse frequencies to a minimum. However these "transmission costs" are almost certainly trivial and unimportant, and there would be little to be said for redundancy reduction if fewer nerve fibres and impulses were the only benefit conferred. But I think that the difficulty of utilizing a mass of information depends primarily upon the capacity of the channel that is carrying it, not on the actual information content of the messages. If this is so, redundancy reduction is a desirable first step, and could lead directly to much improved utilization. It is clear that enormous survival benefit can be derived by catching prey better, escaping more predators, or enlarging the viable environment.

The advantage of redundancy reduction here is that it enables the information to be presented in a more compact format; redundancy is wasteful in that it implies the possible occurrence of messages that never actually occur, and demands provision for interpreting such messages. But although preliminary representation in a form that would be agreed to be non redundant (according to some plausible way of assigning costs) is one way of avoiding the wasteful allocation of mechanisms for utilizing sensory information, it is probably not the only way of doing this. We need some clearer ideas about "convenience" as well as economy in information handling.

(3) Convenience. If one is to discuss this one must know what is the aim or purpose of sensory integration. Obviously one must think beyond the physiological problem of getting sensory messages to the brain and consider what they have been got there for; how is sensory information used? A very general answer, presumably, is that it promotes a successful outcome to the current and future motor actions the brain initiates. To some extent this must be achieved by innate mechanisms - nociceptive reflexes, bug-, mate-, and cliff-detectors, and the selective sensory mechanisms for other innate responses, but apart from these the main function must be to represent or display the current sensory scene in a form such that associations can readily be made to it. A particular sensory scene occurs at the same time as salivation induced by an unconditional stimulus, or in conjunction with some

rewarded motor act, and these particular sensory scenes are somehow recorded and given a special status so that, when they recur, they evoke the motor acts. The essential requirement is for the system to report when something "like this" happens again. Redundancy reduction would improve the situation by decreasing the total number of events from among which the actual event was selected, so that events "like this" would have to be selected from a smaller ensemble. It does not, however, say anything about the form of representation that would be convenient for recognising recurrences, nor does it say anything about the similarity structure of the representation. How can scenes be alike without being identical?

(4) Insufficiently prescriptive. The fact that an ensemble of recoded messages has low redundancy is a statistical property, and it is one which is shared by a large number of codes. Suppose, for example, that we have coded the message onto N binary channels; now imagine a further recoding onto another N channels, this being done in a one to one fashion. These will all be exactly the same as the original as regards redundancy, but there are N! such codes and they will differ greatly from each other in important properties such as which outputs are "alike" by various measures of similarity. We certainly need additional criteria to decide between those equally non-redundant codes.

(5) Selection and loss of information. There is another very important aspect of the way that the brain handles sensory information about which the idea of redundancy reduction, in its simplest form, has nothing to say. These codes are reversible, and cause no loss of information. Sensory mechanisms, on the other hand, are highly selective, and an animal can only be thought to respond to or store a small fraction of the information its senses provide. Much of this selection may be of a rather uninteresting and esoteric sort in that it depends on peculiarities of a species, or an individual, or even on haphazard factors. According to the original idea the selective loss of information was thought to be an entirely separate process, but it is possible redundancy is also important here, so I am going to suggest an addition to the hypothesis.

4. Redundancy Criterion for Selective Preservation of Information

It was pointed out earlier that a redundancy reducing code must have special properties which enable it to code more compactly the regular or patterned features of the ensemble of input messages. It must have the means of responding appropriately to these features and in this sense it must recognise them and preserve them in the output. But in the redundancy reducing codes originally proposed and discussed no provision was made for reduction of the output of information by irreversibility, or selective loss of information, so although it is true that regular features of the input were preserved, so was everything else. In view of the importance of the constraints on the input that was emphasized at the beginning of this article one must consider seriously the possibility that lack of redundancy may be a criterion for selective neglect by the sensory coding mechanisms. If sensory events occur which are completely unrelated with any of the constrained, regular, patterned, or redundant features of the input, should not such sensory events be selectively rejected?

Such a policy would make good sense according to the following paradigm. Suppose it is advantageous to represent as much as possible about the physical stimuli playing on an animal's receptors (and hence about his physical environment) and yet we know that the channel, or final display, available for representing the sensory scene has an informational capacity quite inadequate to represent all of it faithfully. The messages are recoded to reduce redundancy, but let us say there is still more information than can be displayed. At this point one must reject something. It is clear that if one rejects sensory messages in which regularity has been found and which have consequently been coded more compactly, one is discarding more of the sensory input than if one rejects unpatterned, apparently random, input messages. Hence it would be sensible to preserve messages which corresponded to redundant (i.e. regular, patterned) inputs, and reject messages which appeared uncorrelated, coincidental or random.

According to that criterion it is right to listen to the tune and reject the noise, but there is an obvious danger in doing so. Noise only remains noise as long as one fails to find any structure in it, and if one rejects all apparent noise one precludes for ever the possibility of finding any structure it may have held. And of course there is the Kantian-inspired fear that, if one listens only for the tune, that is all one will ever hear. These are clearly problems lurking in this area, but I must say they do not seem so deep or insoluble in the current paradigm. What is to stop the system conserving a small, randomly-selected, sample of the uncoded input in order to check constantly that it is not being misled by its own more generally applied codes and rejection-acceptance criteria? Such a system might be slower to detect a change of tune than one otherwise designed, or to put it another way it might have a tendency to interpret a sensory scene in terms of relatively slowly changing concepts and constructs, but one is not forced to believe that these are completely immutable or entirely determined genetically. I believe this extension of the redundancy-reducing paradigm gives a tenable viewpoint on the direction, purpose, or selective advantage of sensory and perceptual processing.

5. Conclusions

Regularity and pattern in sensory stimuli are certainly important. Recognising that this is informational redundancy suggests that it may be detected by distributed, local processes rather than global operations involving much of the sensory input. Much of the influence of the environment on the brain may operate in this way rather than through trial and error learning or the even slower and more drastic method of genetic selection. Redundancy can be exploited to transmit or display information more compactly and it is suggested that it may also be a criterion for preservation or rejection of information. These ideas generate a viewpoint on many perceptual processes, but there are certainly a number of problems which remain untouched.

SENSORY CODING AND ECONOMY OF NERVE IMPULSES*

E. Pfaffelhuber
Institute for Information Sciences
University of Tübingen
Tübingen, Germany

1. Introduction and Summary

"A wing would be a most mystifying structure if one did not know that birds flew" wrote Barlow in one of his papers[1], thus pointing out that in biology (as well as in technology) many structures and processes are often extremely hard to understand and unrewarding to examine in detail unless the trick and purpose behind them is known. This seems true in particular when it comes to the question of how external, environmental constellations are coded in terms of the activity pattern of neurons at a certain depth behind the sensory input layers. An attempt to provide an answer to the problem was made by Barlow who suggested[1-5] that a common principle of sensory coding is to economize the number of nerve impulses, without, of course, degrading too much the information transmitted. Clearly, this is, if an answer at all, only a partial one, since we may still wonder about the purpose behind economizing impulses.

The most naive explanation one could give here, namely that the organism tends to keep the total energy consumption small, is probably in most cases rather out of place. Namely, the human brain is known to have an energy consumption of some ten Joule/sec, so that a spike can be estimated to have a 'cost' of about 10^{-10} Joule if we take the number of neurons in the brain as roughly 10^{10}, each neuron firing with an average rate of 10 impulses per second, and assuming that some ten percent of the brain's total energy consumption is due to spike generation. This amounts to a total cost, due to spike generation, of the order of 0.3 kWh per day which is small as compared to the total daily energy consumption of about 3 to 4 kWh per day. (This is even more true for other animals whose brains' fractional energy consumption is certainly smaller than in man.) A more sophisticated explanation of the economy-of-spikes principle[6] would say that spikes often signal a change in environmental

*Supported by the Deutsche Forschungsgemeinschaft

conditions which may be important for the organism and therefore should be reliably and quickly detectable. Clearly this is easier if the average number of spikes is small, since then the actual occurrence of a spike is a rare event. On an even more abstract level one could finally argue that economy of impulses may have its cause in the possibility that decision processes going on in more central parts of the nervous system may be carried out faster and more easily if the decision mechanisms have to listen only to a few neurons instead of to a whole lot.

Be this as it may, it appears worthwhile to investigate the consequences of the economy-of-spikes principle and try to compare these with experimental facts, similarly as in thermodynamics one studies, in fact with much success, the consequences of the postulate of maximum entropy, though the reason behind the latter principle is still not really clarified[7,8]. We should point out, nevertheless, that sometimes (and possibly quite often) other economy principles appear to be involved in the area of sensory coding, especially the principle of economizing the total number of neural channels necessary to transmit a certain amount of information. Namely, all neurons which possess a non-vanishing spontaneous firing rate (like the Purkinje cells in the cerebellum or the retinal ganglion cells) and thus are able to signal three basic events, namely firing at, below, and above the spontaneous rate (instead of the two possible events, firing and not firing, in the case where there is no spontaneous firing) will therefore reduce the number of channels needed by something like a factor of log 3 / log 2. Clearly, however, under normal circumstances where spontaneous firing is the more frequent event, this economizing procedure goes at the cost of an increased total number of impulses used.

In the following we shall aim at a more precise and quantitative formulation of the economy-of-spikes principle and its consequences, thus putting some of Barlow's[1-5] and Attneaves's[10] qualitative ideas and results on a more mathematical (and hence, hopefully, less ambiguous) basis. In section 2 we will approximate the information transmitted on a neural channel, whose inputs are the possible environmental constellations, by a quantity which is essentially the entropy of the channel's firing rates. In section 3 we will derive the optimal spike frequency distribution which guarantees that (at least) a given amount of information is transmitted along the neural channel while the cost of transmission is minimized. Here we make use of some results on minimum cost encoding of information derived by Blachman[11], the result being that the optimal distribution is a negative exponential of the cost function, so that if the latter is linear in the firing rate, a negative exponential of the spike frequencies results. If the decoding delay resulting from low spike rates contributes to the cost, the optimal probability distribution no longer has its maximum at the lowest, but at intermediate frequencies. The difference between the actual and the minimum average cost for transmitting a certain amount of information (i.e. the cost which actually could have been saved without spoiling information) is a natural extension of the concept of code redundancy in classical information theory (and coincides with the latter if cost is identified with codeword length). In section 4 we shall first consider the case of multichannel transmission of information where optimization with respect to the average cost (or, equivalently, the code redundancy) implies that different nerve fibers should fire independently, and show then that lateral inhibition is a simple device for reducing code redundancy. (Analogous results hold for the phenomenon of adaptation.) If the redundancy reduction is to be effective, lateral inhibition coefficients of the order of some 10% should result. We note that we do not touch here the problem of the existence of feature detectors which obviously has something to do with redundancy reduction in neural codes[1-5], the reason being that the single concept of the cost of coding does not appear sufficient to deal with this phenomenon.

2. The Information Rate of a Neural Channel

Let us start with the following picture. An environmental constellation e out of a set E (which, for simplicity, is assumed to be discrete), with probability $p(e)$, produces a sensory input which in turn determines the corresponding (stationary) firing rate f (the 'codeword' for e) of the single nerve fiber considered. We assume here that the relevant coding characteristic is the spike rate[11-14]. This appears to be true in many cases though most probably a number of exceptions exist where the exact timing of nerve impulses plays the main role[15].

If Δf is the uncertainty in the firing rate of the nerve fiber (Δf is assumed to be roughly independent of which particular environmental event e caused the firing) then we may divide the range F of all possible firing rates into intervals of length Δf and take, say, the center point of each interval as a 'quantized' version of all the firing rates f inside the interval. If F' is the set of all these center points f', then, by construction, to each environmental constellation $e \in E$ there corresponds essentially a unique $f' \in F'$. Consequently the conditional entropy

$$H(f'|e) = \sum_{\substack{e \in E \\ f' \in F'}} p(e)\, p(f'|e) \log(1/p(f'|e))$$

of f' given e is practically zero since the conditional probability $p(f'|e)$ of f' given e (which is, in fact, the probability that, given e, the neuron's actual firing rate f is in the interval of length Δf centered at f') is always close to zero or one. (Note that a similar quantization procedure, in the time domain, can be used to derive the formula of MacKay and McCulloch on the axonal information rate for pulse interval modulation[13].) Since the information T transmitted along the channel whose inputs are the environmental constellations e and whose outputs are the neuron's firing rates f remains, for small Δf, approximately the same if the firing rates f are replaced by their quantized versions $f' \in F'$ (this is a general property of the information rate, cf.[16] p. 9) we have

$$T \approx H(f') - H(f'|e) \qquad (1)$$

where

$$H(f') = \sum_{f' \in F'} p(f') \log (1/p(f'))$$

and

$$p(f') = \sum_{e \in E} p(e)\, p(f'|e)$$

is the unconditional probability of f', i.e. the probability that f lies in an interval of length Δf around f'. We find, therefore, that

$$T \approx H(f') \qquad (2)$$

since the second term in (1) vanishes.

A more formal proof of (2) proceeds as follows. By definition, we have

$$T = \sum_{e \in E} \int_F df\, p(e)\, p(f|e) \log(p(f|e)/p(f)) \qquad (3)$$

where f is the actual (non-quantized) firing rate, $p(f|e)$ is the conditional probability density of f given e, and

$$p(f) = \sum_{e \in E} p(e)\, p(f|e)$$

is the unconditional probability density function of f. By our assumption on the uncertainty of f for given e, we have that f, for given e, may be considered approximately as being uniformly distributed in an interval of length Δf, whence

$$\int_F df \; p(f|e) \log p(f|e) \approx - \log \Delta f.$$

Thus, (3) becomes

$$T \approx \sum_{e \in E} \int_F df \; p(e) \; p(f|e) \; \log(1/(p(f) \; \Delta f)).$$

Hence, by summing over e, we find

$$T \approx \int_F df \; p(f) \; \log(1/(p(f) \; \Delta f)). \qquad (4)$$

Approximating (4) by a Riemann sum and using the fact that, if Δf is small and f is in the interval of length Δf centered at $f' \in F'$, then

$$p(f) \; \Delta f \approx \int_{f'-\Delta f/2}^{f'+f/2} df \; p(f) = p(f')$$

we see that (2) and (4) are the same.

3. Minimum Cost Transmission of Information and Generalized Code Redundancy

Since $H(f')$, and consequently T, from (2) remain unchanged under any permutation of the firing rates $f' \in F'$, we see that by choosing that permutation which makes the probability distribution $p(f')$ a monotonically decreasing function of f', the same amount of information T can be transmitted over the channel. Thus, if the economy-of-spikes principle is to possess any significance at all we expect that usually higher firing rates have smaller probabilities, and that the smallest possible rates occur rather frequently. This result seems plausible, but is not easy to check experimentally, since the probabilities referred to are the overall probabilities of the firing rates, for the whole range of possible environmental stimuli, and thus would have to be measured on an intact animal in natural surroundings.

It should also be noted that it may be necessary to modify this simple result as far as rather low firing rates are concerned. Namely, for these, it takes apparently much time until a reliable estimate about the rate f can be made, so that a large delay in the transmission time of the input information results, which is disadvantageous. Thus, one may expect that too small rates are not preferably used as codewords (at least if the spike rate is the true information carrier), and hence have smaller probabilities, as our above considerations might suggest.

Let us proceed to make the previous, rather qualitative, results more quantitative by introducing explicitly the cost of codewords, i.e. firing rates. It is natural to expect that, for firing rates not too far away from the minimum rate f_{min}, the corresponding cost $C(f)$ will increase linearly with f, since each pulse will contribute a fixed amount to the cost. For larger f-values, however, $C(f)$ will probably increase faster than linearly, so that rates above some hundred pulses per second are practically impossible. We may approximate this behavior by allowing only for values of f below some maximal rate f_{max} (the same would be achieved if we let $C(f)=\infty$ for $f \geqslant f_{max}$) and putting

$$C(f) = \varepsilon \cdot f, \quad f_{min} < f < f_{max} \tag{5}$$

Obviously, $\varepsilon = C(f+1) - C(f)$ is the 'cost' of one spike. A common 'zero point' cost (describing, e.g., the cost of keeping the fiber at normal working conditions even without spike generation) could have been added to (5); as our results will be independent thereof it is omitted.

As noted in section 1, the cost $C(f)$ need not measure a neuron's actual energy consumption when firing with a rate f, but could be, e.g., related to the chance of wrong decisions about the actions to be taken by the organism if the nerve fiber under consideration fires with rate f.

Let us proceed to calculate the minimum $\bar{C}_{min}$ of the average cost

$$\overline{C(f)} = \int_{f_{min}}^{f_{max}} df\, p(f)\, C(f) \tag{6}$$

necessary to transmit (at least) a given amount T of information. We shall make use here of equation (4) for T, as it will yield the optimal probability density $p^{opt}(f)$ of the firing rates f in the whole interval between f_{min} and f_{max} (whereas equation (2) would yield the optimal probability distribution only for the quantized firing rates $f' \in F'$). Thus we have to minimize (6) subject to the boundary conditions (4) and

$$\int_{f_{min}}^{f_{max}} df\ p(f) = 1 \tag{7}$$

This variational problem has, in the discrete case, already been treated in [10]. The continuous case goes through analogously. Putting the variation of

$$\int_{f_{min}}^{f_{max}} df\ p(f)\ C(f) - \beta^{-1}\left[\int_{f_{min}}^{f_{max}} df\ p(f)\ \log(1/(p(f)\Delta f)) - T\right] - \lambda\left[\int_{f_{min}}^{f_{max}} df\ p(f) - 1\right]$$

with respect to p(f) equal to zero (β^{-1} and λ are Lagrange parameters) we arrive at the result

$$p^{opt}(f) = \exp(-\beta C(f))/Z(\beta) \tag{8}$$

where

$$Z(\beta) = \int_{f_{min}}^{f_{max}} df\ \exp(-\beta C(f))$$

and β has to be determined from the boundary condition

$$\int_{f_{min}}^{f_{max}} df\ p^{opt}(f)\ \log(1/(p^{opt}(f)\Delta f)) \equiv -\beta^2 \frac{d}{d\beta}(\beta^{-1} \log(Z(\beta)/\Delta f)) = T\ .$$

In addition we have

$$\bar{C}_{min} = \int_{f_{min}}^{f_{max}} df\ p^{opt}(f)\ C(f) \equiv -\frac{d}{d\beta} \log Z(\beta)\ .$$

To determine the sign of β we consider a probability density p, different from (8), for which the average cost is given by (6) and which satisfies (4) and (7). Then

$$0 = \int_{f_{min}}^{f_{max}} df\ p(f)\ \log(1/(p(f)\Delta f)) - \int_{f_{min}}^{f_{max}} df\ p^{opt}(f)\ \log(1/(p^{opt}(f)\Delta f)) =$$

$$\int_{f_{min}}^{f_{max}} df\ (p(f) - p^{opt}(f))\ \log(1/(p^{opt}(f)\Delta f)) -$$

$$\int_{f_{min}}^{f_{max}} df\ p(f)\ \log(p(f)/p^{opt}(f))$$

Since the second integral on the right-hand side is always positive ([17], p. 14), we find

$$0 < \int_{f_{min}}^{f_{max}} df\ (p(f) - p^{opt}(f))\ \log(1/(p^{opt}(f)\Delta f)) =$$

$$\int_{f_{min}}^{f_{max}} df\ (p(f) - p^{opt}(f))\ (\beta C(f) + \log(Z(\beta)/\Delta f) =$$

$$\beta \int_{f_{min}}^{f_{max}} df\ (p(f) - p^{opt}(f))\ C(f) =$$

$$\beta(\overline{C(f)} - \bar{C}_{min})$$

whence we see that β has to be positive.

(8) is intuitively quite meaningful as it says that the optimal code has to be such that costly codewords are to have a small probability, or, more precisely, that the optimal probability density is an exponentially decreasing function of the cost $C(f)$. From (8) we find that the quantity $\log(1/(p^{opt}(f)\Delta f))$, which represents essentially the information content of the event that the firing rate is in an interval of length Δf around the point f, is a linearly increasing function of the cost $C(f)$. Thus we may also say that the optimal code has to be such that costly codewords carry a large amount of information.

So far, our results are independent of the special form of the cost function $C(f)$. Taking now equation (5) seriously, we arrive at the result that the optimal probability density of the firing rate is a negative exponential of f,

$$p^{opt}(f) = \text{const.} \quad \exp(-\gamma f) \tag{9}$$

the constant being determined by the normalization condition (7), and $\gamma = \varepsilon \cdot \beta$. Again, a firing rate probability density close to the theoretical result (9) does not seem unrealistic, but here, too, the same remarks apply which were already made in connection with our previous, more qualitative finding that the economy-of-spikes principle favours a monotonically decreasing firing rate probability distribution.

In particular, the remark concerning the disadvantage of low spike rates requiring large decoding times can be made more precise as follows. It seems reasonable to assume that the decoding time is proportional to the inverse of the firing rate f, and that a term proportional to the decoding time enters into the cost function, when $C(f)$ is given by

$$C(f) = \varepsilon f + \varepsilon'/f \quad , \quad 0 \leq f < \infty \tag{10}$$

where $\varepsilon' > 0$ is a constant and we assumed for simplicity that all non-negative firing rates can occur. In this case we find from (8) that

$$p^{opt}(f) = \text{const.} \quad \exp(-\gamma(\varepsilon f + \varepsilon'/f))$$

whence p^{opt} is no longer maximal at the lowest possible frequencies, but at $f = (\varepsilon'/\varepsilon)^{1/2}$, so that medium firing frequencies are preferred.

In the classical theory of source coding ([18], p. 39) the code redundancy is defined as the difference between the actual average codeword length and its minimum possible value (which is essentially the source entropy), the idea being that the length of a codeword is the basic cost factor to be considered. Since we have been led to admit for more general cost functions $C(f)$ associated with the elements of the coding alphabet, i.e. the various spike rates f, it is natural to introduce a generalized code redundancy R as the difference between the actual and the minimum possible average cost for transmitting (at least) a given amount T of information,

$$R = \overline{C(f)} - \bar{C}_{min} \tag{11}$$

Thus minimizing the average cost $C(f)$ is identical to minimizing the code redundancy (11), and the principle of economizing nerve impulses is equivalent to inventing good codes in the sense of small redundancy R.

4. More Neural Channels

We have been concerned so far with one particular nerve fiber only, although our considerations can as well be applied to a whole set of parallel fibers carrying information about environmental constellations e. Attaching a label j (j out of a set J) to the different fibers, the codeword for some $e \in E$ is now represented by the collection $\tilde{f}$ of all f_j, $j \in J$, where f_j is the spike rate of the jth fiber. Similar calculations as in section 3 show that the optimal probability density $p^{opt}(\tilde{f})$, minimizing the total average cost $\overline{C(\tilde{f})}$ (measuring the overall cost of the firing rates f_j, $j \in J$) for a given information T which is contained in the collection $\tilde{f}$ about the environment inputs e, is given by a negative exponential of $C(\tilde{f})$, apart from a constant factor which is to be fixed by normalization.

Making now the rather natural assumption that the total cost $C(\tilde{f})$ is made up additively from the costs $C_j(f_j)$ of the individual spike rates (these costs may depend in general upon the fiber label j),

$$C(\tilde{f}) = \sum_{j \in J} C_j(f_j) \tag{12}$$

we find that, no matter what the form of the cost functions $C_j(f_j)$ is,

$$p^{opt}(\tilde{f}) = \text{const.} \quad \exp(-\beta C(\tilde{f}))$$

is factored into the probability densities

$$p_j^{opt}(f_j) = k_j \quad \exp(-\beta C_j(f_j))$$

(the factors k_j being determined by normalization) so that in the optimal case different fibers should fire independently of each other. Again this result seems quite plausible and in accordance with experimental facts. Namely, it is to be expected that, along the neural pathway from the sensory input to the deeper brain centers, the cost or redundancy reducing capabilities of each re-coding station (which can be roughly identified with the anatomically known stations like retina, lateral geniculate body, area 17, 18, and 19, in the case of the visual system) are not too overwhelming, since each recoding process has to be achieved solely by a limited number of neurons and synapses, which is a rather serious boundary condition (and why else should there be half a dozen of such stations instead of just one or two?). Thus we may assume that at each recoding station the code is only slightly improved, i.e. its redundancy is slightly decreased, so that only in the deeper brain centers, but certainly not in the periphery, can we expect a code rather close to the optimal one. Thus our considerations imply that different (and hence also nearby) fibers should fire more and more independently of each other (or, at least, the spatial correlation range of their firing should decrease more and more) the more central they are.

Apart from the cost functions (5), (10), applied to each individual fiber, another choice of C(f) from (12) seems interesting, namely that which interprets costs as the number of fibers firing above a certain threshold rate f_{th} (which could, of course, be zero),

$$C_j(f_j) = \begin{cases} 1 \text{ for } f_j > f_{th} \\ 0 \text{ for } f_j < f_{th} \end{cases}$$

In this case the optimal firing rate density turns out to be constant for rates above as well as below the threshold rate (but with different levels, of course), and it is obvious that very good codewords for environmental inputs are those with only one (or a few) neuron(s) firing above threshold rate. The existence of such codewords has been made plausible in [19] to explain certain psychophysical phenomena.

Let us finally show that lateral inhibition[20,21] provides a simple means of reducing the average cost of codewords, or, equivalently, the code redundancy, while the amount of information transmitted is kept approximately constant under certain conditions on the lateral inhibition coefficients. If $\tilde{f} = (f_j)_{j\in J}$ and $\tilde{g} = (g_j)_{j\in J}$ denote the collections of firing rates of the individual fibers before and after the lateral inhibition layer, respectively, we have

$$g_j = \left[f_j - \sum_{i\in J} \alpha_{ji} f_i \right]_+ , \quad j \in J \tag{13}$$

in the case of forward inhibition which we consider for simplicity. Here α_{ji} is a non-negative coefficient describing the inhibiton which fiber i exerts upon fiber j, and $[x]_+$ is defined as

$$[x]_+ = \begin{cases} x, & x>0 \\ 0, & x\leq 0 \end{cases} \tag{14}$$

(The use of this function in (13) is necessary because firing rates are, by definition, non-negative quantities.) If the labels $j\in J$ are now chosen to represent the (two-component) spatial coordinates of the corresponding fibers in the plane perpendicular to the direction of the bundle of nerve fibers considered, then the inhibition coefficients α_{ji} will in general depend, in good approximation, only upon the relative positions of the two fibers involved, i.e. upon j-i,

$$\alpha_{ji} = \alpha_{j-i} \tag{15}$$

Because of (14) it is plausible that the linear approximation to (13), namely

$$g_j = f_j - \sum_{i\in J} \alpha_{j-i} f_i, \quad j\in J \tag{16}$$

(where use has been made of (15)) is good if the argument of the $[.]_+$ function in (13) is positive in the average, which means that

$$\bar{f} - \sum_{i\in J} \alpha_{j-i}\bar{f} > 0$$

or, assuming that the range of j-values with $\alpha_j\neq 0$ is small as compared to the total range J and neglecting edge effects, that

$$\sum_{j\in J} \alpha_j < 1 \quad . \tag{17}$$

Here we assumed that the average firing rate $\overline{f_j} = \bar{f}$ of fiber j is, at least approximately, independent of its position characterized by j. At the same time condition (17) implies that the recoding procedure mapping $\tilde{f}$ into $\tilde{g}$ according to (16) possesses an inverse and hence is reversible.

Thus, the information about the environmental constellation $e\in E$, carried by the fibers after lateral inhibition layer, viz. by the codewords $\tilde{g}$, is the same as that contained in the fibers before lateral inhibition took place, whose codewords are the collections $\tilde{f}$. (This is another general property of information rates, cf.[16], p. 11.) Hence, if condition (17) is met, then lateral inhibition keeps the amount of information transmitted at a constant level, in the approximation considered. On the other hand, it is easy to see that it reduces cost, and hence redundancy. Namely, by (12), (16) and assuming an equation analogous to (5) for each fiber separately, we find that the average cost $\overline{C(\tilde{g})}$, after lateral inhibition has taken place, is given by

$$\overline{C(\tilde{g})} = \sum_{j\in J} \varepsilon(\bar{f} - \sum_{i\in J} \alpha_{j-i}\bar{f}) = \sum_{j\in J} \varepsilon\bar{f}(1 - \sum_{i\in J} \alpha_i)$$

$$= \overline{C(\tilde{f})}\ (1 - \sum_{j\in J} \alpha_j)$$

where $\overline{C(\tilde{f})}$ is the average cost of the codewords $\tilde{f}$. Thus the ratio of the decrease in the average cost, due to lateral inhibition, to the initial average cost $\overline{C(\tilde{f})}$, is given by

$$(\overline{C(\tilde{f})} - \overline{C(\tilde{g})})\ /\ \overline{C(\tilde{f})} = \sum_{j\in J} \alpha_j \tag{18}$$

It follows that, if an effective cost saving is to be achieved, then (18) should be large, without violating the restriction (17). Remembering that lateral inhibition actually takes place only between nearby fibers, so that α_j vanishes except for the few smallest possible values of $|j|$, we expect therefore values of the non-vanishing lateral inhibition coefficients of the order of 10%. This is, in fact, what has also been found in some experiments[20].

Incidentally, we note that one of the effects of the nonlinearity in (13) is to reduce the relative cost savings (18) as compared to that following from the linearized version (16). This can be seen from the fact that, using (13) instead of (16) we have, by (5)

$$C(\tilde{g}) = \sum_j{}' \varepsilon(f_j - \sum_{i\in J} \alpha_{j-i}f_i)$$

$$\geq \sum_{j\in J} \varepsilon(f_j - \sum_{i\in J} \alpha_{j-i}f_i)$$

$$= \sum_{j\in J} (C(f_j) - \sum_{i\in J} \alpha_{j-i}C(f_i)) \tag{19}$$

where $\sum_j{}'$ involves a summation over those $j\in J$ for which $f_j - \sum_{i\in J} \alpha_{j-i}f_i$ is nonnegative. Taking expectations in (19) yields

$$\overline{(C(\tilde{f})} - \overline{C(\tilde{g}))} \;/\; \overline{C(\tilde{f})} \leq \sum_{j \in J} \alpha_j$$

Let us finally note that analogous considerations can be carried through for the phenomenon of adaptation (or a combination of lateral inhibition and adaptation) where each fiber inhibits, so to speak, its own future firing, so that the label j would mean here a temporal coordinate.

Acknowledgments

The author is most grateful to Dr. C. Legéndy for many stimulating discussions, and to Mr. R. Heim for helpful comments.

References

1. Barlow, H.B.: Possible principles underlying the transformations of sensory messages. In: Sensory Communication, W.A. Rosenblith, ed. M.I.T. Press, Cambridge, Mass., 1961.
2. Barlow, H.B.: Trigger features, adaptation and economy of impulses. In: Information Processing in the Nervous System, K.N. Leibovic, ed. Springer-Verlag, New York, 1969.
3. Barlow, H.B.: The coding of sensory messages. In: Current Problems in Animal Behaviour, ed. by W.H. Thorpe and O.L. Zangwill. Cambridge University Press, Cambridge, England, 1961.
4. Barlow, H.B.: Sensory mechanisms, the reduction of redundancy, and intelligence, Symp. Mechanization of Thought Processes, Natl. Phys. Lab., London, 1959.
5. Barlow, H.B.: Redundancy and perception, this volume.
6. Adrian, E.D.: The Basis of Sensation, Christophers, London, 1928.
7. Grandy, W.T.: Fundamentals of statistical mechanics, University of Wyoming preprint, 1970.
8. Schlögl, F.: Informationstheorie und Thermodynamik irreversibler Prozesse, Arbeitsgemeinschaft für Forschung des Landes Nordrhein-Westfalen, Heft 181, Westdeutscher Verlag, 1968.
9. Attneave, F.: Informational aspects of visual perception, Psychol. Rev. 61, 183 (1954).
10. Blachman, N.M.: Minimum-cost encoding of information, IRE Trans. Information Theory PGIT-3, 139 (1954).
11. Grüsser, O.J.: Die Informationskapazität einzelner Nervenzellen für die Signalübermittlung im Zentralnervensystem, Kybernetik 1, 209 (1962).
12. Grüsser, O.J., Hellner, K.A. and Grüsser-Cornehls, U.: Die Informationsübertragung im afferenten visuellen System, Kybernetik 1, 175 (1962).
13. Färber, G.: Berechnung und Messung des Informationsflusses der Nervenfaser, Kybernetik 5, 17 (1968).

14. McKean, T.A. et. al.: The biologically relevant parameter in nerve impulse trains, Kybernetik 6, 168 (1970).

15. Barlow, H.B.: The information capacity of nervous transmission, Kybernetik 2, 1 (1963).

16. Pinsker, M.S.: Information and Information Stability of Random Variables and Processes, Holden-Day, New York, 1964.

17. Kullback, S.: Information Theory and Statistics, Dover, New York, 1968.

18. Pfaffelhuber, E. and Güttinger, W.: Kybernetik: Informations-, System- und Nachrichtentheorie, University of Munich Lecture Notes, 1967/68.

19. Legendy, C.: Can the data of Campbell and Robson be explained without assuming Fourier analysis?, Kybernetik, in press (1974).

20. Hartline, H.K. and Ratliff, F.: Inhibitory interaction in the retina of Limulus. In: Handbook of Sensory Physiology, vol. 7/2, Physiology of Photoreceptor Organs, M.G.F. Fuortes, ed. Springer-Verlag, New York, 1972.

21. Levick, W.R.: Receptive fields of retinal ganglion cells. In: Handbook of Sensory Physiology 7/2, Physiology of Photoreceptor Organs, M.G.F. Fuortes, ed. Springer-Verlag, New York, 1972.

AN ALGORITHMIC APPROACH TO INFORMATION THEORY*

Roland Heim
Institute for Information Sciences
University of Tübingen, Federal Republic of Germany

1. Introduction

Classical probability theory is based on the well known axioms of Kolmogoroff. A characteristic difficulty of this measure-theoretic approach is the physical interpretation of probability: we can observe only the frequency of events and the order in which they occur, not however the probability in the axiomatic sense. Attempts to formulate a frequency theory of probability are quite old, but it was not until the fundamental work of C.P. Schnorr (1971) that there existed a complete canonical theory of probability and randomness based on the concept of an effective (that is, computable) procedure to detect possible regularities in a sequence of events.

The difficulty of interpreting the axiomatic concept of probability is a critical point in classical information theory too, since the information content of an event is a simple function of its probability. A further important step to overcome the drawback of the purely measure-oriented definition of information has been achieved by Kolmogoroff (1964). He introduced the concept of program complexity: the length of a program by means of which a recursive algorithm computes a finite sequence serves as a measure of regularity of this sequence with respect to the algorithm used and hence provides a measure of information conveyed by the sequence.

Indeed we will show that the program complexity yields not only an equivalent concept of randomness but also the classical results about coding and transmission of messages. The important aspect of computational complexity enters in a natural way in this algorithmic approach which is of great importance in real systems. Recognition, learning, understanding are characteristic abilities of biological systems and are possible only in an environment whose complexity does not exceed the computational resources of the system.

---

* Supported by the Deutsche Forschungsgemeinschaft

2. Information and Program Complexity

The simplest case of an information source is the generation of a binary sequence. For instance we can generate a string of zeros and ones by tossing a coin, by binary encoding of a written message or by representing a (television) picture in digital form. Therefore we may without loss of generality discuss in the sequel only binary sequences as we can represent our environment by zeros and ones with any prescribed exactness.

The approach to define the information content of a message emitted by a source is the well known axiomatic one in the classical information theory. In the case of infinite binary sequences considered as possible outputs of an information source, one defines as a measure of the average uncertainty about the source symbols $x \in X = \{0,1\}$ or, equivalently, as a measure for the average information content per source symbol the entropy

$$H := \lim_{n\to\infty} H(X_n | X_0 X_1 \ldots X_{n-1})$$

or equivalently

$$H := \lim_{n\to\infty} \frac{H(X_0 X_1 \ldots X_{n-1})}{n}$$

with $(\mathrm{ld} = \log_2)$

$$H(X_n | X_0 X_1 \ldots X_{n-1}) = \sum_{\substack{x \in X_n \\ w \in X^n}} p(wx)\, \mathrm{ld} \frac{1}{p(x|w)}$$

$$H(X_0 X_1 \ldots X_{n-1}) = \sum_{w \in X^n} p(w)\, \mathrm{ld} \frac{1}{p(w)}$$

under the assumption of stationarity of the source. H is a simple logarithmic function of the probabilities of events as suggested by some natural requirements about a reasonable uncertainty measure. See for example Ash (1965) or Gallager (1968) for detailed discussion.

Suppose now we have a stationary information source with entropy H. Then by the noiseless coding theorem we can encode each sequence produced by the source by assigning to each block $w \in X^k$ a code word q_w in such a way that the resulting sequence is uniquely decipherable and the average code word length $\bar{n}$ satisfies the condition

$$H \leq \frac{\bar{n}}{k} < H + \frac{1}{k}$$

Therefore, we can represent the output of a source with entropy H by a sequence $q_1 q_2 q_3 \ldots$ which is shorter than the given one by a factor of H on the average and in the limit of infinite block length.

The noiseless coding theorem yields a natural interpretation of the entropy H: For small values of H we can represent the source output by relatively short code sequences, meaning that the information content per source letter is low too. Conversely, if we need relatively long code sequences, the information content is high and this is expressed by a high (close to one) value of H.

This is the intuitive idea behind the algorithmic approach of Kolmogoroff (1964), who introduced the program complexity, essentially a generalised coding concept for finite and infinite sequences. The need for such a generalisation may be illustrated by the following example: Suppose we have an infinite binary sequence which we want to encode using the noiseless coding theorem. In the k-th order approximation we assign to each block of length k an appropriate code word, for instance by the well known Huffman coding scheme. The optimal choice of the code words depends on the probabilities of the blocks $w \varepsilon X^k$. In a practical application we must substitute relative frequency for probability, which we can calculate for an arbitrary initial segment and then use for the encoding procedure. In the worst case of so called Bernouilli sequences all blocks $w \varepsilon X^k$ have the same relative frequency 2^{-k} for all integers k. The entropy H is one and we can, apart from permuting the symbols, represent the sequence solely by itself. However, there exist Bernouilli sequences which are at the same time computable, a computer is able to calculate as many elements as we wish using a program of finite length.

From this point of view the information content per letter is zero in every computable sequence since, given the finite program, there is no uncertainty about any element. We can also say we have to distribute the finite information content of the finite program to an infinite number of elements (it seems highly probable, but cannot be proven, that the computable transcendental numbers π,e provide examples for Bernouilli sequences).

Therefore, instead of the (computable) correspondence between an initial segment of a binary sequence and its coded representation generated by the purely statistically oriented Huffman coding scheme we consider a general computable correspondence. As a base for computability we allow all mappings $A: X^* \to X^*$ (finite binary sequences into finite binary sequences), which can be realised by Turing machines, each Turing machine representing an effective algorithm or a partial recursive function. We make the following definition:

Definition 1

The program complexity $K_A(x)$ of $x \varepsilon X^*$ with respect to the algorithm $A: X^* \to X^*$ is

$$K_A(x) := \min_{p \varepsilon X^*} \{|p| \; |A(p) = x\} \;, \; \min \emptyset := \infty$$

$|p|$ denotes the length of the program p. In the case of an infinite sequence $z \varepsilon X^\infty$ we consider successively the initial segments z(n) of length n and put

$$\overline{K}_A(z) := \lim_{n \to \infty} \frac{K_A(z(n))}{n}$$

which we call the generalized entropy of z and which represents the information content of z per symbol with respect to A (provided that the limit exists). The fundamental quantity in this theory is the function $n - K_A(z(n))$, which measures the difference between the length of an initial segment and the program necessary to compute it, called the generalized total redundancy of the initial segment z(n) of z with respect to A.

Clearly, these definitions are vacuous if we cannot quantify the program complexity. This problem could not be solved in a satisfactory way before the important results due to C.P. Schnorr (1971,1973), in which he gives a definite answer to the old questions about randomness. His approach to define randomness is an algorithmic one and yields not only an algorithmic foundation of probability theory but also a quite natural foundation of information theory.

Intuitively, randomness and information content are strongly correlated. A recursive binary sequence like the binary representation of the real number π is highly regular and we can represent them by a finite program. On the other hand, a random sequence has by definition no regularities and structure which can be expressed effectively, that is, by an algorithm, and therefore we can represent or encode a random sequence essentially only by itself. One would expect in this case $K_A(z(n))$ to be of the order of n asymptotically for all Turing-computable algorithms $A: X^* \to X^*$. The theory of Schnorr yields not only a distinction between random and regular sequences but also a quantification of the degree of randomness, which is directly correlated with the program complexity.

3. Effective Random Tests and the Program Complexity

An as simple as powerful approach to define randomness is the old concept of a game of chance. Suppose we have an arbitrary infinite sequence $z \varepsilon X^\infty$ and we play a binary roulette on z serving as a random number generator. We request the following rules of the game:

(a) The game begins with a finite, nonnegative amount V of money, for convenience, we put V = 1.

(b) After n trials, the gambler knows the initial segment $z(n)$ of z and bets by means of a computable strategy the amount $B_a(z(n))$ on the event $z_{n+1} = a$, $a = 0,1$. Debts are forbidden, this means $B_0(z(n)) + B_1(z(n)) \leq V(z(n))$, the total capital at step n.

(c) In the case $z_{n+1} = 0$, the gambler gets back the amount $2B_0(z(n))$ and loses $B_1(z(n))$, analogously for $z_{n+1} = 1$.

The pay-off condition (c) ensures that both the gambler and the "casino" have the same chance of winning assuming the equal distribution of zeros and ones. This is natural, otherwise, by the noiseless coding theorem, the sequence z would not be completely irregular because of the possibility of shorter encoding in the case $p(0) \neq \frac{1}{2} \neq p(1)$. (b) contains a causality condition: the gambler cannot look in the future. Debts are forbidden, otherwise, the gambler could compensate any losses. Of course, we require all amounts of capital to be computable real numbers.

We can summarize the conditions (b) and (c) in the following way: From (c) we get

$$\begin{aligned} V(x0) &= V(x) + B_0(x) - B_1(x) \\ V(x1) &= V(x) + B_1(x) - B_0(x) \end{aligned} \qquad x \in X^*$$

and by addition

$$V(x) = \tfrac{1}{2}V(x0) + \tfrac{1}{2}V(x1) \tag{M}$$

(M) is the so called martingale condition. We write $V(\Lambda)$ for the initial capital as the gambler knows no element when he begins, $\Lambda \in X^*$ denotes the empty sequence. The computable strategy $B_a(x)$, $a = 0,1$ generates a mapping $V\colon X^* \to \mathbb{R}^+$ according to (b). We call such a function V a <u>computable martingale</u>.

The intuitive idea is: If the function $V(z(n))$ grows unlimited for $n \to \infty$, then z has regularities in the sense that, knowing $z(n)$, the event z_{n+1} is not completely uncertain and then z is not random. Conversely, if z withstands all computable strategies, i.e. if $V(z(n))$ remains bounded, then we are forced to consider z as a random sequence. Here, given $z(n)$, the event z_{n+1} is completely unpredictable by effective methods.

To give an example, suppose we have an infinite sequence $z \in X^\infty$ and we know the relative frequencies of the zeros and ones. Let $s(x)$ be the number of ones in the segment $x \in X^*$ and assume that the limit

$$\lim_{n\to\infty} \frac{s(z(n))}{n} = p(1) \qquad 0 < p(1) < 1$$

exists. We construct the following martingale V.

To compute the relative frequency of elements we need not take into account any statistical dependence on previous elements. Thus we choose the constant strategy

$$B_o(x) = V(x)\left(\frac{1-q}{2}\right) , \quad B_1(x) = V(x)\left(\frac{1+q}{2}\right) , \quad -1 < q < 1 ,$$

q a computable real number. Then we get

$$V(z(n)) = (1+q)^{s(z(n))}(1-q)^{1-s(z(n))}$$

$$\text{ld } V(z(n)) = s(z(n))\text{ld }(1+q) + (1-s(z(n)))\text{ld}(1-q)$$

$$\lim_{n\to\infty} \frac{\text{ld } V(z(n))}{n} = p(1)\text{ld }(1+q) + p(0)\text{ld }(1-q) := c$$

This leads to

$$\lim_{n\to\infty} \frac{V(z(n))}{2^{cn+o(n)}} = 1$$

Maximizing c as a function of q we get

$$c_{max} = 1 + p(1)\text{ld } p(1) + p(0)\text{ld } p(0) = 1 - H(X)$$

This is a very interesting result as c_{max} is simply the classical first order redundancy $R(X) = 1 - H(X)$, which is defined by the relative frequencies of zeros and ones too. $H(X)$ gives the amount of "compressibility" of z by an optimal Huffman coding scheme.

The above example is a special case of the following theorem:

<u>Theorem 1</u>

Suppose there is a sequence $z \in X^\infty$ and a computable martingale $V: X^* \to \mathbb{R}^+$ with the property

$$\lim_{n\to\infty} \frac{V(z(n))}{f(n)} = c , \quad 0 < c < \infty \tag{1}$$

Then an algorithm $A: X^* \to X^*$ can effectively be constructed such that the following holds: There exists a constant $k \in \mathbb{N}$ with

$$|n - K_A(z(n)) - (\text{ld } f(n) - 2\text{ld ld } f(n))| < k \tag{2}$$

If only the limsup is assumed in (1), then (2) holds for infinitely many integers $n \in \mathbb{N}$.

$f: \mathbb{N} \to \mathbb{R}$ is a everywhere defined, computable and monotonically increasing function, called <u>growth function.</u>

An outline of the proof can be found in the appendix.

In the above example with $f(n) = 2^{(1 - H(X))n + o(n)}$ we get

$$|n - K_A(z(n)) - (n - H(X)n)| = o(n)$$

or

$$\lim_{n\to\infty} \frac{K_A(z(n))}{n} = \overline{K}_A(z) = H(X)$$

A converse of theorem 1 holds too (see appendix) and we can thus give two equivalent definitions of a random sequence:

Definition 2

A sequence $z \in X^\infty$ is random, if (a) or (b) holds:

(a) There exists no algorithm $A: X^* \to X^*$ with effective program complexity K_A and no growth function $g: \mathbb{N} \to \mathbb{R}$ with

$$\liminf_{n\to\infty} (n - K_A(z(n)) - g(n)) > 0 \tag{3}$$

(b) There exists no computable martingale $V: X^* \to \mathbb{R}^+$ and no growth function $f: \mathbb{N} \to \mathbb{R}$ with

$$\limsup_{n\to\infty} \frac{V(z(n))}{f(n)} > 0 \tag{4}$$

The growth functions in (3) and (4) yield a valuation of randomness. The speed of growth tells us something about the degree of randomness resp. of information content of the sequence z.

The absolute character of this definition of randomness and information content cannot be proven. But just as there are various approaches for defining computability leading to the same class of computable functions, there exist further definitions of randomness which yield the same set of random sequences.

The most important one is the so called total recursive sequential test, first introduced in a weaker form by P. Martin-Löf (1966). It is easy to prove that all binary sequences which are not random in the above sense form a set of measure zero (the product measure on X^∞ induced by the measure $p(0) = p(1) = \frac{1}{2}$ on X). On the other hand, if we approximate a set $N \subset X^\infty$ of measure zero by a total recursive sequential test, we can effectively find a martingale V or an algorithm A such that (3) resp. (4) holds for all $z \in N$.

This equivalence is the link between the measure-oriented classical theory and the algorithmic approach. The probability laws which hold with probability one or "almost everywhere", such as the strong law of large numbers or the law of the iterated logarithm, and which can be defined and tested by effective methods are the same as those defined by martingales, program complexity or the total recursive sequential tests.

4. Some Aspects of the New Concept of Information

There are very remarkable aspects in this general algorithmic foundation of information content. Note that we have made no assumptions at all about stationarity, ergodicity or existence of limits which are essential in the classical theory. Probabilities (that is relative frequencies) and statistical correlations are only one possibility among others to describe regularities. Since a random test is based on a given algorithm we obtain a subjective definition: Information content is not only a property of the information source but also depends on the ability of the observer to construct an optimal random test, especially to find the shortest possible description of the source output. We can describe in a natural way the phenomenon that an observer suddenly detects a regularity inherent in the source output after having seen a certain number of elements. Thus the information content decreases as he can find a shorter description of the source output than before.

Yet another main topic of information theory can be handled in an algorithmic way. The transmission of information through a distorted binary channel is described classically by a probability distribution on input-output sequences. The algorithmic theory yields a more general concept of distortions as we will discuss briefly in the case of a symmetric binary channel without memory. Instead of transition probabilities we describe the channel by a noise sequence $s \varepsilon X^{\infty}$ generating the following input-output mapping: If $x \varepsilon X^{\infty}$ is the input sequence and $y \varepsilon X^{\infty}$ the output sequence, then $y_i = x_i \oplus s_i$, $\oplus$ denoting the ordinary modulo 2 addition on $\{0,1\}$. If s is a regular sequence, a generalized channel coding theorem can be proven (Heim (1974)). As a measure for regularity we have the martingale growing speed f(n) and from the martingale strategy and the growth order we can effectively construct a sequential blockwise coding scheme $S: X^k \times \mathbb{N} \to X^*$ which guarantees that the relative frequency of errors estimating the transmitted code word from the channel output tends to zero as k goes to infinity. S assigns to the i-th block $b_i \varepsilon X^k$ in $x = b_1 b_2 \ldots \varepsilon X^{\infty}$ a code word $c_i \varepsilon X^*$ and induces a mapping $S_k^*: X^* \to X^*$ as to $x(ik) = b_1 b_2 \ldots b_i$ is assigned the encoded sequence $\tilde{x}(h(ik)) = c_1 c_2 \ldots c_i$ of length h(ik). The essential result is that any such growth function $h: \mathbb{N} \to \mathbb{N}$ must be bounded below by a growth function $h_{min}: \mathbb{N} \to \mathbb{N}$ depending logarithmically on f(n) in case the statement of asymptotically vanishing error frequency holds. In the special case of a binary symmetric channel with transition probability $\delta, 1 - \delta$, the channel capacity is $C = 1 - H(\delta, 1-\delta) = 1 - \delta \mathrm{ld}\, 1/\delta - (1-\delta)\mathrm{ld}\, 1/(1-\delta)$.

This means that any classical coding scheme satisfying the channel theorem involves a redundancy generating "expansion" $S_k^*(x(ik)) = \tilde{x}(h(ik))$ with $h(n)/n \geq C$. On the other hand, if we simulate the probability distribution $\delta, 1-\delta$ by a noise sequence $s \in X^\infty$ with a relative frequency δ of the ones we can construct a martingale growing on s with order $f(n) = 2^{1 - H(\delta, 1-\delta)} = 2^{Cn}$. In general we have to use a nonstationary coding scheme $S: X^k \times \mathbb{N} \to X^*$ instead of the simple fixed coding table $S: X^k \to X^l$ used in the classical theory. It is not difficult to extend these methods to the general binary channel. Here we would have to consider various noise sequences substituting the various probability distributions. In this way, a description of deterministic and probabilistic models is made possible exclusively in the language of algorithms.

The transmission of information can still be described from another point of view. Namely, we may ask how much information is lost in the channel or, equivalently, how much information must be added to the channel output to be able to reconstruct the input exactly. At this point, we make the following definition:

Definition 3

Let $x, y \in X^*$, $A: X^* \times X^* \to X^*$ be an algorithm. The conditional program complexity $K_A(y|x)$ of y with respect to x (and A) is

$$K_A(y|x) := \min_{p \in X^*}\{|p| \; | A(p,x) = y\} \; , \; \min \emptyset := \infty$$

The conditional program complexity is the generalized algorithmic measure of the dependence between two sequences x and y. In the case of a binary symmetric channel without memory we find that we can construct from the martingale $V(s(n))$ on the noise sequence s an algorithm A with the property that $K_A(y(n)|\Lambda) - K_A(y(n)|x(n))$ is a logarithmic function of the martingale growing speed $f(n)$.

The algorithmic approach to define the notion of information content involves another aspect of complexity. As we have seen, there exist sequences which allow only a martingale growing speed of at most $f(n)$, no matter what our strategies are. But even in the case of a computable sequence, which is trivially of order 2^n, we must take into account the possibility that the computation of the strategy is so complex that no physically realizable machine can do the job. It may happen that for this reason the sequence is random for all real observers because of their physically limited computational resources. We may thus call such sequences pseudorandom with respect to a class of computationally restricted observers. Some basic investigations of pseudorandom sequences may be found in Schnorr (1971).

Information content with respect to a restricted computational resource can be defined in a natural way. The most general possibility is to use an abstract complexity hierarchy in the sense of M. Blum (1967). Informally, an abstract complexity measure of computation is an effective listening $\{A_i, S_i\}$ of all algorithms A_i together with the corresponding "step counting" function S_i of same domain as in A_i, $i \in \mathbb{N}$. For instance, S_i counts the number of steps or the number of tape cells needed in a computation on a multitape Turing machine. A complexity class is then defined by the set of all algorithms whose associated step counting functions $S_i(.)$ are bounded above by a growth function $t(.)$. An excellent overview on this area is given by Hartmanis and Hopcroft (1971).

Definition 4

The t-bounded program complexity $K^t_{\{A\}}(x)$ of $x \in X^*$ with respect to a complexity measure $\{A_i, S_i\}$ is

$$K^t_{\{A\}}(x) := \min_{p, A_i} \{|p| \ |A_i(p) = x,\ S_i(p) \leq t(|x|)\},\ \min\emptyset := \infty$$

Bounded program complexity measures have been studied in detail for a special class of algorithms by R.P. Daley (1973a,1973b). Of basic interest are certain trade-off phenomena: For instance it's obvious that $K^{t_1}_{\{A\}}(x) \leq K^{t_2}_{\{A\}}(x)$ if $t_1(|x|) \geq t_2(|x|)$. The exchange of information content and computational complexity is of great practical importance. If we restrict our available algorithms in computational complexity we may wonder about the regularities which are still visible and those which are not. Conversely, if we can determine the complexity of an object, we may ask for a class of algorithms or machines able to describe the latter in a nontrivial manner.

Clearly, an abstract complexity measure or even a measure related to Turing machines is not very useful in practice, for those measures are applied to infinite objects, e.g. to mappings from $\mathbb{N}$ into $\mathbb{N}$, and this is a class too large for practical purposes. For instance, finite state automata which are without doubt quite rich in structure are solely represented as algorithms contained in every complexity class which means that algorithms realized by finite state automata are of lowest complexity. This reflects itself in the fact that random tests which are computed by finite automata can only detect deviations from the Bernouilli property, that is, test whether or not the occurrence of a given word $w \in X^*$ in an infinite sequence $z \in X^\infty$ is in accordance with the strong law of large numbers which is rather trivial (Schnorr and Stimm (1972)). On the other hand, if we are concerned with finite objects a finite state automaton can recognize structures which we would consider intuitively as complex. Therefore, we need a complexity theory of finite machines resp.

of functions with finite domain as well as a complexity theory of numerical algorithms. Some interesting results have been found but a detailed discussion would go beyond the scope of this informal contribution. An overview on this topic is given by H. Bremermann (1974).

The algorithmic theory outlined above shows how to bridge the gap between the unsharp but intuitively quite meaningful and widely applicable notion of "information" and the narrow, rather abstract but precise definition in the classical theory. This is of special interest if we wish to describe the information processing behaviour in biological systems. Recognizing and learning are only possible in an environment which has definite regularities. The structures of the environment determine how complex a learning or recognizing algorithm must be at least to detect the inherent order. If a message is transmitted through a neuronal channel from the sensory input to the cortex the question arises as to what a reasonable concept of channel capacity would be. Our idea is that the storage of objects and events is not photographically exact in the brain, but that there are programs in the above sense representing the essential properties. Certainly, there are trade-offs between the complexity of processing and storage requirements. For all this we have an exact mathematical language and it is our conviction that further research about regularity, structure and complexity contributes to a deeper understanding of the capabilities of living systems.

Appendix

In this appendix we try to give a more detailed idea of how the equivalence between the definitions of randomness is established. First, we give the exact definition of the recursive and total recursive sequential test. In the following we consider only random sequences $z \in X^\infty$ with respect to the distribution $p(0) = p(1) = \frac{1}{2}$. $\bar{p}$ denotes the induced product measure on X^∞ with the property $\bar{p}[w] = 2^{-|w|}$, $[w] := wX^\infty$ for all $w \in X^*$.

Definition 5

A recursive sequential test (rst) is a recursively enumerable set $Y \subset X^* \times \mathbb{N}$ with

$$Y_i = \{x \mid (x,i) \in Y\}$$

$$\bar{p}[Y_i] \leq 2^{-i}$$

$$\bigcap_{i \in \mathbb{N}} [Y_i] := N_Y \text{ is a recursive null set with respect to } \bar{p}$$

The sequence $\{Y_i\}_{i \in \mathbb{N}}$ approximates constructively a set $N_Y \subset X^\infty$ of $\bar{p}$-measure zero uniquely associated with a probability law which holds with probability one or almost everywhere. As mentioned above we consider a sequence $z \in X^{\infty}$ as random if $z \notin N_Y$ for all rst's Y. Obviously, we cannot decide in a finite number of steps whether or not an infinite sequence z belongs to a set N_Y which has in general the cardinality of the continuum. But similar as in the martingale or program complexity approach (definition 2) we can construct a valuation of an initial segment z(n) indicating the degree of randomness with respect to Y. Consider $[Y_i]$ as the i-th order approximation of N_Y. Then in the case

$$\bar{p}([Y_i] \cap [z_1(n)]) > \bar{p}([Y_i] \cap [z_2(n)])$$

we say $z_2(n)$ withstands the rst Y in the i-th order approximation better than $z_1(n)$. In order to obtain an effective valuation we must require additionally the computability of $\bar{p}[Y_i]$ which is not assumed in definition 5.

Definition 6

A recursive sequential test Y is a <u>total recursive sequential test</u> (trst) if $\bar{p}[Y_i]$ is computable for all $i \in \mathbb{N}$.

Lemma 1

Let $V: X^* \to \mathbb{R}^+$ be a computable martingale. Define

$$V^{(k)} := \{x|\ V(x) \geq k\}$$

Then the following holds:

$$\bar{p}[V^{(k)} \cap xX^*] \leq \frac{V(x)}{k2^{|x|}} \qquad k > 0,\ x \in X^*$$

$$\bar{p}[V^{(k)}] \leq k^{-1} \quad \text{assuming } V(\Lambda) = 1$$

Proof

The martingale condition (M) is $V(x) = \frac{1}{2}V(x0) + \frac{1}{2}V(x1)$, $x \in X^*$. By induction, we obtain

$$V(x) = \sum_{y \in B} 2^{-|y|} V(xy) \qquad x,y \in X^*$$

where $B \subset X^*$ is <u>prefix free</u>, that is $B \cap BXX^* = \emptyset$, and maximal in the sense that $B \cup \{a\}$ is not prefix free if $a \notin B$. Let $C \subset X^*$ be an arbitrary, prefix free, not necessary maximal set. Then

$$V(x) \geq \sum_{z \in C \cap xX^*} 2^{|x| - |z|} V(z)$$

Let $\overline{V^{(k)}}$ be the least prefix free set with $V^{(k)} \subset \overline{V^{(k)}}X^*$. Then

$$V(x) \geq \sum_{z \in \overline{V^{(k)}} \cap xX^*} 2^{|x| - |z|} k$$

or

$$\sum_{z \in \overline{V^{(k)}} \cap xX^*} 2^{-|z|} = \bar{p}[V^{(k)} \cap xX^*] \leq \frac{V(x)}{k2^{|x|}} \qquad \text{q.e.d.}$$

By lemma 1, we see immediatly that the sequence $\{V^{(2^i)}\}_{i \in \mathbb{N}}$ forms a rst. We have defined a sequence $z \in X^\infty$ as random if there exists no computable martingale V and no growth function f with the property that $\limsup_{n\to\infty} V(z(n))/f(n) > 0$. It can be shown that

$$N_{V,f} := \{z \in X^\infty | \limsup_{n\to\infty} V(z(n))/f(n) > 0\} \subset N_Y$$

where Y is a trst. The converse holds too: From a trst Y one can effectively construct a martingale and a growth function with $N_Y \subset N_{V,f}$. The construction is similar to those in theorem 3 and may be found in Schnorr (1971).

In order to establish a relation between the program complexity, the martingale growing order, and trst's we begin with a special class of algorithms. The classical coding and decoding procedure of the noiseless coding theorem is a bijective mapping $H: X^k \to C$ and $H^{-1}: C \to X^k$, where $C \subset X^*$ is a prefix free set containing the code words of the blocks $w \in X^k$. H as well as H^{-1} can be extended uniquely to homomorphisms $H^*: X^* \to X^*$ and $(H^{-1})^*: X^* \to X^*$ by the requirement $H^*(vw) = H^*(v)H^*(w)$ and $(H^{-1})^*(pq) = (H^{-1})^*(p)(H^{-1})^*(q)$, $v,w \in X^k$, $p,q \in C$. We drop the homomorphism, keep the monotony property and define a generalized class of computable mappings:

Definition 7

A partial recursive algorithm $P: X^* \to X^*$ is monotonic or a process, if $P(xy) \in P(x)X^*$, $\forall\, x,xy \in \text{domain}(P)$.

Lemma 2

Let $P: X^* \to X^*$ be a process. Define

$$Y_i = \{x \in X^* |\; |x| - K_P(x) \geq i\} \qquad i \in \mathbb{N}$$

Then $\bar{p}[Y_i] \leq 2^{-i}$.

Proof

For $B \subset X^*$ let $\bar{B}$ denote the least prefix free set with the property $B \subset \bar{B}X^*$. Let $U_i = P^{-1}(Y_i)$. Then the process property guarantees that $\|\overline{Y_i}\| \leq \|\overline{U_i}\|$ and we have

$$1 \geq \bar{p}[U_i] = \sum_{y \in \overline{U_i}} 2^{-|y|} \geq \sum_{x \in \overline{Y_i}} 2^{-|x|+i} = 2^i \bar{p}[Y_i] \text{ q.e.d.}$$

The sequence $\{Y_i\}_{i \in \mathbb{N}}$, Y_i defined in lemma 2, forms a rst. If we require additionally that $K_P(x)$ is computable for all $x \in X^*$ or, weaker, $|x| - K_P(x) \geq f(x)$ where $f: X^* \to \mathbb{N}$ is a computable function, a recursive lower bound, then we obtain a trst.

Lemma 3

Let $Y_i \subset X^*$ be a recursive set with $\bar{p}[Y_i] \leq 2^{-i}$, $i \in \mathbb{N}$. We can effectively construct a process $P_i: X^* \to X^*$ with the property

$$\forall x \in Y_i X^*: \quad |x| - K_{P_i}(x) = i$$

Proof

We construct a bijective mapping $\sigma_j: X^k \to X^{k-j}$, $j,k \in \mathbb{N}$, $j \leq k$, with the property that $\sigma_j(xw) = \sigma_j(x)w$ and domain$(\sigma_j) = Y_i X^*$. We must have

$$1 \geq \bar{p}[\sigma_j(Y_i)] = \sum_{y \in \sigma_j(\overline{Y}_i)} 2^{-|y|} = \sum_{x \in \overline{Y}_i} 2^{-|x| + j} \leq 2^{-i + j}$$

and thus we can choose $j = i$. By assumption, there exists a recursive function $\phi: \mathbb{N} \to X^*$ which enumerates the set $\overline{Y}_i$ in lexicographical order without repetitions. We define σ_i on $\overline{Y}_i$ as follows. Compute $\phi(1)$ and choose $\sigma_i(\phi(1)) \in X^{|\phi(1)| - i}$ arbitrarily. Compute $\phi(k)$, $k = 2,3,\ldots$ and choose $\sigma_i(\phi(k)) \in X^{|\phi(k)| - i}$ so that $\sigma_i(\phi(k)) \neq \sigma_i(\phi(m))$, $m < k$. An element $x \in Y_i X^*$ is uniquely decomposable in $\phi(k)w$, $k \in \mathbb{N}$, $w \in X^*$, and we extend the mapping σ_i on $Y_i X^*$ by the definition $\sigma_i(x) = \sigma_i(\phi(k))w$. The process P_i is given by

$$\text{domain}(P_i) = \sigma_i(Y_i X^*), \qquad P_i(\sigma_i(x)w) = xw \qquad w \in X^*$$

q.e.d.

Next, we outline the

Proof of theorem 1

In general, the set $V^{(k)}$, $k \in \mathbb{N}$, is only recursively enumerable. But it is obvious that we can construct from a computable martingale $V: X^* \to \mathbb{R}^+$ a recursive martingale $\tilde{V}: X^* \to \mathbb{Q}^+$ which differs from V by an arbitrarily small amount and has the same growth order f. The relation $\tilde{V}(x) \geq k$, $k \in \mathbb{N}$, is recursively decidable and we have

$$\bar{p}[\tilde{V}^{(2^i)}] \leq 2^{-i}, \quad \tilde{V}^{(2^i)} \text{ is a recursive set.}$$

Then lemma 3 is applicable and we construct a process $P_i: X^* \to X^*$ with

$$|x| - K_{P_i}(x) = i \qquad x \in \tilde{V}^{(2^i)}$$

This shows the logarithmic dependence of the general redundancy on the martingale growing speed f(n) in theorem 1. Then we combine all processes to a common process P. Let $i = b_r b_{r-1} \ldots b_o$ be the binary expansion of the integer i. Then $\tau(i)x := b_r b_r b_{r-1} b_{r-1} \ldots b_o b_o 01x$ is uniquely decipherable as the pair (i,x), $i \in \mathbb{N}$, $x \in X^*$ and we define

$$P(\tau(i)x) = P_i(x), \qquad \text{domain}(P) = \bigcup_{i=0}^{\infty} \tau(i)\,\text{domain}(P_i)$$

$$P_o(x) = x, \qquad \text{domain}(P_o) = \{x \in X^* \mid V(x) < 2\}$$

We have $|\tau(i)| \leq 2\text{ld } i + 2$ and thus we obtain the double-logarithmic term in (2). The constant k depends on the limit c. q.e.d.

Theorem 2

Suppose that there exist a sequence $z \in X^{\infty}$, a process $P: X^* \to X^*$, and a growth function $g: \mathbb{N} \to \mathbb{N}$ with

$$n - K_P(z(n)) \geq g(n)$$

for infinitely many integers n and the relation "$\geq$" is decidable. Then we can effectively construct a computable martingale $V: X^* \to \mathbb{R}^+$ with the property

$$\limsup_{n\to\infty} \frac{V(z(n))}{2^{g(n)} - \text{ld } g(n)} > 0$$

Proof

Define the recursive sets

$$Y = \{x \in X^* |\ |x| - K_P(x) \geq g(|x|)\ , \quad Y_n = Y \cap X^n X^*$$

By assumption, there are infinitely many integers i so that $z(i) \in \overline{Y}_i$. Because of the monotony of g, we can construct a $h: \mathbb{N} \to \mathbb{N}$ so that

$$g(h(n)) \geq n^{1+\delta}, \delta > 0, \text{ and } z(h(n)) \in \overline{Y}_{h(n)} \quad \forall n \in \mathbb{N}$$

Define

$$V(x) = \sum_{n \in \mathbb{N}} \frac{1}{g(h(n))} \Big(\sum_{xy \in \overline{Y}_{h(n)}} 2^{g(|xy|) - |y|} + \sum_{k < |x|} \chi_{\overline{Y}_{h(n)}}(x(k))\ 2^{g(|x|)} \Big)$$

q.e.d.

Lemma 4

Let $A: X^* \to X^*$ be an algorithm, $z \in X^{\infty}$ a sequence, and $g: \mathbb{N} \to \mathbb{N}$ a growth function so that

$$n - K_A(z(n)) \geq g(n)$$

holds infinitely many times and the relation "$\geq$" is decidable. Then we can construct a process P with the property that

$$n - K_P(z(n)) \geq g(n) - 2\text{ld } n - 3$$

infinitely often.

Proof

Define the recursive set

$$Y = \{x \in X^* |\ |x| - K_P(x) \geq g(|x|)$$

The set $X^* - X^{k+1}X^*$ contains $2^{k+1} - 1$ elements. Therefore, there exist in $Y \cap X^n$ maximal $2^{n - g(n) + 1} - 1$ elements and this implies

$$\bar{p}[Y \cap X^n] \leq 2^{-n}(2^{n - g(n) + 1} - 1) < 2^{1 - g(n)}$$

By lemma 3 and by the construction in the proof of theorem 1, lemma 4 follows immediately.

Corollary

If we assume in lemma 4 "$\geq$" for all integers n, then the statement holds for all integers, too. If

$$\underline{\lim}_{n\to\infty} \frac{K_A(z(n))}{n} = c$$

for an arbitrary algorithm A: $X^* \to X^*$ then we can construct a process P: $X^* \to X^*$ with

$$\underline{\lim}_{n\to\infty} \frac{K_P(z(n))}{n} = c$$

The logarithmic difference between $K_A(z(n))$ and $K_P(z(n))$ in the lemma 4 reflects itself in the oscillating behaviour of the general program complexity (Martin-Löf (1971)).

Lemma 5

For each sequence $z \in X^\infty$ we can construct an algorithm A: $X^* \to X^*$ with the property

$$\underline{\lim}_{n\to\infty}\sup\ (n - K_A(z(n)) - \text{ld } n) > -\infty$$

Proof

Let $\tau: \mathbb{N} \to X^*$ be an enumeration without repetitions of X^* in lexicographical order. We consider successively the prefixes $z(k)$, $k = 1,2...$ and make the following construction. Determine the number i with $\tau(i) = z(k)$. Then we represent the prefix $z(i+k)$ by $z_{k+1}\dots z_{k+i}$. Such a representation is possible infinitely many times and we define

$$A(x) = \begin{cases} \tau(i)x \text{ if there exists } i \in \mathbb{N} \text{ with } \tau(i)x = z(i + |x|) \\ x \text{ otherwise} \end{cases}$$

Because of $|\tau(i)| \leq \text{ld } i$, the lemma follows. q.e.d.

Lemma 5 tells us that even in the case of a random sequence z we can find for infinitely many integers n relatively short programs of the initial segments $z(n)$. On the other hand, we can show that we cannot represent all initial segments by relatively short programs if z is random.

Theorem 3

Let A: $X^* \to X^*$ be an algorithm, $z \in X^\infty$ a sequence, and $g: \mathbb{N} \to \mathbb{N}$ a growth function with

$$n - K_A(z(n)) \geq g(n)$$

for all except finitely many integers n, "$\geq$" decidable. Then we can construct a computable martingale $V: X^* \to \mathbb{R}^+$ with

$$\limsup_{n\to\infty} \frac{V(z(n))}{2^{g(n) - (1+\delta)\,\mathrm{ld}\, g(n)}} > 0 \qquad \delta > 0$$

Proof

Define $Y = \{x \in X^* \mid |x| - K_A(x) \geq g(|x|)\}$, $Y_n = Y \cap X^n$. Without restricting generality we assume that there exists a growth function $h: \mathbb{N} \to \mathbb{N}$ so that the following holds

$$n + 1 + (1+\delta)\,\mathrm{ld}\, n \geq g(h(n)) \geq n + 1 + (1+\tfrac{\delta}{2})\,\mathrm{ld}\, n\ , \ \delta > 0$$

$$z(h(n)) \in Y_{h(n)}$$

for infinitely many integers n. We define the following martingale:

$$V_n(x) = \begin{cases} 2^n \text{ if } x(h(n)) \in Y_{h(n)} \\ 0 \text{ otherwise} \end{cases} \qquad |x| > h(n)$$

We have $\| Y_{h(n)} \| < 2^{h(n) - g(h(n)) + 1}$ and therefore

$$V_n(\Lambda) = \sum_{x \in X^{h(n)}} 2^{-h(n)} V_n(x) < 2^{1 - g(h(n)) + n} \leq \frac{1}{n^{(1+\delta/2)}}$$

Because of $V(\Lambda) = \sum_{n \in \mathbb{N}} V_n(\Lambda) < \infty$, $V(x) = \sum_{n \in \mathbb{N}} V_n(x)$ is a computable martingale.

For infinitely many integers n we have

$$V(z(h(n))) \geq 2^n,\ g(h(n)) - (1+\delta)\,\mathrm{ld}\, g(h(n)) \leq n \qquad \text{q.e.d.}$$

This proves the equivalence of (3) and (4) in definition 2. Note that the assumptions of theorem 3 are stronger than those in theorem 2. By lemma 4 we see that only in the case of $g(n) = O(n^\delta)$, $\delta > 0$, it is sufficient to require "$\geq$" infinitely often. The reason for this is the monotony of the processes and the logarithmically oscillating behaviour which may occur in the general program complexity K_A.

The main ideas of this appendix are contained in Schnorr(1971), Schnorr (1973) where further results about randomness, program and process complexity may be found.

Acknowledgement

I am grateful to Prof. V. Braitenberg, Prof. E. Pfaffelhuber and Mr. H. Sommer for many helpful discussions.

References

Ash, R.: Information theory. J. Wiley & Sons, New York, London, Sydney (1965)

Blum, M.: A machine-independent theory of complexity of recursive functions, J. Assoc. Comp. Machin. 14, 322 - 336 (1967)

Bremermann, H.: Complexity of automata, brains and behaviour, this volume

Daley, R.P.: Minimal-program complexity of sequences with restricted resources, Information and Control 23, 301 - 312 (1973a)

Daley, R.P.: An example of information and computation resource trade-off J. Assoc. Comp. Machin. 20, 667 - 695 (1973b)

Gallager, R.G.: Information theory and reliable communication. John Wiley and Sons Inc., New York, London, Sydney, Toronto (1968)

Hartmanis, J. and Hopcroft, J.E.: An overwiev of the theory of computational complexity, J. Assoc. Comp. Machin. 18, 444 - 475 (1971)

Heim, R.: The algorithmic foundation of information theory, to appear

Kolmogoroff, A.: Three approaches for defining the concept of information quantity, Information Transmission 1, 3 - 11 (1964)

Martin-Löf, P.: The definition of random sequences, Inform. Control 6, 602 - 619 (1966)

Martin-Löf, P.: Complexity oscillations in binary sequences, Z. Wahrscheinlichkeitstheorie verw. Geb. 19, 225 - 230 (1971)

Scnorr, C.P.: Zufälligkeit und Wahrscheinlichkeit, Lecture Notes in Mathematics 218, Springer-Verlag Berlin, Heidelberg, New York (1971)

Schnorr, C.P.: Process complexity and effective random tests, J. of Comp. and Syst. Sc., 4, 376 - 388 (1973)

Schnorr, C.P. and Stimm, H.: Endliche Automaten und Zufallsfolgen, Acta Informatica 1, 345 - 359 (1972)

INFORMATION THEORY AND LEARNING IN BIOLOGY
SUMMARY OF THE WORKGROUP SESSIONS

E. Pfaffelhuber
Institute for Information Sciences
University of Tübingen, Germany

1. General Remarks

Naively speaking, the brain could be said to handle and store information. Thus, information theory seems a most natural tool for understanding how the brain works. However, one must not forget that, whenever one uses a certain mathematical framework as an approach to a sufficiently complex biological (or other) phenomenon, one is bound to describe and formulate only certain aspects (which are essentially predetermined by the mathematical framework chosen) of the problem of interest, and it seems, therefore, that the very general question of 'understanding how the brain works' is itself rather ill defined. Information theory, e.g., is certainly relevant to the problem of efficient neuronal coding of sensory inputs, but has not much to say, at least at present, about the microscopic dynamics of nerve impulse generation.

The work group on "Information Theory and Learning in Biology" dealt with three (interrelated and not precisely separable) aspects of brain functioning which can be roughly described as coding, storing, and learning. Before we sketch the main topics it seems worthwhile to recall some basic notions of classical information theory.

2. Classical Information Theory

Classical information theory deals with information generators (sources), information receivers (sinks), and information transmitters (channels) (see e.g. [1,2]). A source emits a (random) symbol X (and can be identified with the latter) which takes values x (we restrict ourselves to the discrete case here) with (a priori) probabilities p(x). The average information content of the source X is the entropy

$$H(X) = \sum_{x} p(x) \log(1/p(x)) \tag{1}$$

which provides a measure for the uncertainty about which of the possible symbols is emitted. If Y, with possible values y and corresponding probabilities p(y), is the received symbol, and p(x,y) is the common probability distribution of X and Y, then the conditional (a posteriori) probability of x being emitted given that y is received is

$$p(x|y) = p(x,y)/p(y) \tag{2}$$

and the conditional entropy

$$H(X|Y) = \sum_{x,y} p(x,y) \log(1/p(x|y)) \tag{3}$$

provides a measure for the uncertainty about the emitted symbol given the received symbol is known. As a consequence, the difference

$$T(X,Y) = H(X) - H(X|Y) \tag{4}$$

represents the difference in uncertainty about the emitted symbol X, without, and with, knowledge of the received signal Y, respectively, and hence is to be interpreted as the average amount of information contained in Y about X. (Note that, while classical communication theory assumes that X and Y are input and output, respectively, of a communication channel like a telephone line, e.g., the idea of an information transmitting channel is much more general and requires solely that X and Y be somehow stochastically connected, that is to say, they are defined on a common probability space.)

3. Coding

The work group session on 'coding' dealt with the problem of understanding the manner in which environmental constellations E, producing inputs to an organism's sensory pathways, are coded in terms of activities of neurons. If we consider the codeword C, caused by the environmental constellation E, as a description of the activities of neurons at a certain depth along the sensory pathway from receptor cells to the central brain areas, then there are two intuitively obvious requirements for the code:

(a) The information T(E,C) contained in the codewords about the environmental constellations should be large.

(b) The code should be simple and economical.

These requirements are contradictory in some sense, as is usually true for reasonable optimization problems. While the first requirement can be fairly understood and put into a mathematically equivalent form, the second one has not yet been completely clarified. Barlow suggested the principle of 'economy of impulses' as a reasonable criterion for the simplicity (or redundancy reducing capabilities) of a code (see also [3-6]), and this principle is, in fact, often used in the coding of messages along sensory pathways. Well known examples are lateral inhibition and adaptation where near-by nerve impulses (spatially or temporally, respectively) inhibit each other, thus economizing the total number of spikes while (almost) no information is lost. The principle of economy of impulses may very well have its cause not simply in the need for a small energy consumption but rather in the possibility that decision processes going on in more central parts of the brain may be carried out faster and more easily if the decision mechanisms have to listen to only a few neurons instead of to a whole lot.

If the principle of economy of impulses is translated into a mathematical form (see [6]), it gives rise to a reasonable definition of code redundancy (generalizing the corresponding concept of classical information theory) and generates an optimality criterion for codes. In the case of more parallel neural channels, e.g., the code is optimal if different fibers fire independently and as infrequently as possible, without producing any information loss. The existence of rather specific feature detectors (see also [7,8]) seems to be in agreement with this result, although it appears that the principle of economy of impulses alone is not sufficient to deal with the feature detector problem.

4. Storing

In the work group session on 'storing', associative nets (see [9,10]) and ϕ-machines [11] were briefly discussed and compared with some recent theories of cerebellar function [12,13]. Associative storage devices in general accept, during the storing period, pairs of input patterns, thereby changing the system's internal state, such that if later on, during the retrieval period, the first member of a previously presented pair is again presented, the second member will be evoked and generated as output. Again, two intuitively rather obvious requirements can be written down:

(a) The stored information should be large.

(b) Retrieval should be good and reliable.

The associative net which is a special case of an associative storage device and essentially identical to Steinbuch's learning matrix[14], can be visualized as a crosswork of vertical and horizontal wires, with simple on-off switches at each cross connection. The first and second member, respectively, of each presented input pair is characterized by a pattern of activity and non-activity on each of the horizontal and vertical wires, respectively. A switch at a certain cross connection is turned on whenever activity occurs simultaneously in the corresponding horizontal and vertical wire.

The ϕ-machine, on the other hand, has continuously adaptable connectivities which may be positive (excitatory) as well as negative (inhibitory), instead of simple on-off switches. This results in a larger number of reliably storable pattern pairs.

To apply these theoretical devices to cerebellar function, the horizontal wires are to be identified with the parallel fibers, and the vertical wires with the climbing fiber - Purkinje cell - Purkinje axon - lines. Marr's theory [12] can then be roughly characterized as considering the cerebellum as an associative net, the on-off switches playing the role of two-stage modifiable excitatory parallel fiber - Purkinje cell synapses, while Albus' theory [13] considers the cerebellum essentially as a ϕ-machine, with continuously adjustable excitatory and inhibitory synapses. While both theories are neither proved nor disproved at present, they would explain the large mossy fiber - parallel fiber divergence, since the mathematics of associative nets and ϕ-machines predicts that the number of storable patterns is proportional to the number of input (parallel) fibers.

A series of interesting problems arise here, like that of digital ϕ-machines (to make the associative net theory rigorous), of ϕ-machines with bounded synaptic weights which are always either positive or negative (to describe the properties of synapses more properly), of time pattern storage via an associative net or similar device (to join Braitenberg's theory of the cerebellum as a timing device (see [15,16]) with Marr's or Albus' theory), and finally the problem of obtaining direct data on the number of an organism's possible elemental movements (to be compared with the number of storable movement patterns as predicted by theories).

5. Learning

While the sessions on 'coding' and 'storing' used solely concepts of classical information and probability theory, the session on 'learning' dealt with true extensions thereof, first by applying the notion of subjective probabilities in order to arrive at a general information theoretic definition of learning (see [17]), and secondly by introducing algorithmic characterizations of regularities and redundancies (inherent in observed sequences of environmental events, say) to describe how such regularities are detectable and usable (see [18] and the literature there).

The entropy (1) describes the 'objective' uncertainty about the occurrence of X, i.e. the uncertainty resulting from the true, objective probabilities p(x). As an individual biological organism will in general not know the latter, they have to be replaced by the system's own subjective probabilities q(x) in order to define the system's subjective entropy

$$H_s(X) = \sum_x p(x) \log(1/q(x)) \tag{5}$$

representing its own uncertainty about the occurrence of X. Similarly the conditional subjective entropy $H_s(X|Y)$ is defined. It is particularly useful to characterize an individual system's knowledge about statistical correlations between X and Y.

Since

$$H_s(X) \geqslant H(X), \quad H_s(X|Y) \geqslant H(X|Y) \tag{6}$$

with the equality sign holding if and only if subjective and objective probabilities coincide, we can define learning, on a general information theoretic level, as a temporal decrease of H_s, implying that the learning system's subjective probabilities come closer to the true, objective ones.

This approach bears also on the problem of coding sensory inputs by means of feature detectors, because a feature detector is efficient only if the corresponding feature is sufficiently frequent and biologically relevant, so that the true probabilities of features have to be learned (either in the course of evolution or by the individual -- both possibilities being apparently realized (see [7])) before an efficient code can be generated.

The information theoretic definition of learning is related to statistical decision theory by the fact that, once learning has decreased the subjective entropy to a sufficiently small level, subjectively optimal decisions are necessarily also nearly optimal objectively. Furthermore, the time course of the subjective entropy, when evaluated in the framework of mathematical learning models (see [19-23]), yields quantitative and qualitative statements as to the performance of such models in stationary and non-stationary environments.

The second extension of classical information theory concerns the types of regularities inherent in a (spatial or temporal) sequence of events which may represent, e.g., the observations made by a system about its environment. These regularities are not simply characterized by the information theoretic redundancy contained in the sequence but rather by the more powerful and general concepts of order of growth and (computational or program) complexity. The rough idea is, first, that the more regular a sequence is, the easier it is -- given the right algorithm -- to predict the events to come and to choose the correct actions to be taken leading to an increase (as measured by the order of growth) in the system's profit (assuming that the interplay between the environment, generating the observed sequence, and the system itself is something like a lottery); and secondly, that the more 'transparent' a given type of regularity is the less complex need the device be to detect it.

The theory, at present, is far from being at a reasonable biological level, one particular problem being that the predictions of the events to come and the correct actions to be taken are, in general, determined by an appropriate algorithm, and it is hard to believe, at this stage, in any reasonable similarity between the capabilities of Turing machines and biological organisms. While this problem can in part be circumvented by the fact that rather simple (and hence biologically more relevant) types of regularities can also be detected by appropriate automata, which appear to be closer to biological organisms, a second problem still remains, namely the fact that so far the mathematical theory only guarantees the existence of such automata capable of detecting regularities, while constructive and adaptive procedures are still lacking to a large extent.

6. Conclusions

As was pointed out in the beginning, information theory is certainly not relevant to all possible aspects of brain functioning. In those cases, however, where it can be used, its application does not only lead to a common point of view of technical and biological communication, storing and coding problems, but may also induce extensions of classical concepts which are specifically useful at a biological level.

References

1. Gallager, R.G.: Information Theory and Reliable Communication, Wiley, New York, 1968 (Chapter 2).
2. Pfaffelhuber, E. and Güttinger, W.: Kybernetik (Informations-, System- und Nachrichtentheorie) I,II, University of Munich Lecture Notes, 1967/68.
3. Barlow, H.B.: Trigger features, adaptation and economy of impulses. In: Information Processing in the Nervous System, ed. by K.N. Leibovic. Springer-Verlag, New York, 1969.
4. Barlow, H.B.: The coding of sensory messages. In: Current Problems in Animal Behavior, ed. by W.H. Thorpe and O.L. Zangwill. Cambridge University Press, Cambridge, 1961.
5. Barlow, H.B.: Redundancy and perception, this volume.
6. Pfaffelhuber, E.: Sensory coding and economy of nerve impulses, this volume.
7. Barlow, H.B.: Detection of form and pattern by retinal neurones, this volume.
8. Gross, C.G., Rocha-Miranda, C.E. and Bender, D.B.: Visual properties of neurons in inferotemporal cortex of Macaque, J. Neurophysiol. 35, 96 (1972).
9. Willshaw, D.J., Buneman, O.P. and Longuet-Higgins, H.C.: Non-holographic associative memory, Nature 222, 960 (1969).
10. Willshaw, D.J. and Longuet-Higgins, H.C.: Associative memory models, Machine Intelligence 5, 351 (1970).
11. Nilsson, N.J.: Learning Machines, McGraw-Hill, New York, 1965 (Chapter 2).
12. Marr, D.: A theory of the cerebellar cortex, J. Physiol. 202, 437 (1969).
13. Albus, J.S.: A theory of cerebellar function, Math. Biosciences 10 25 (1971).
14. Steinbuch, K.: Die Lernmatrix, Kybernetik 1, 36 (1961).
15. Braitenberg, V.: Functional interpretation of cerebellar histology, Nature 190, 539 (1961).
16. Braitenberg, V.: Is the cerebellar cortex a biological clock in the millisecond range?, Progress in Brain Research 25, 334 (1967).
17. Pfaffelhuber, E.: Learning and information theory, Intern. J. Neuroscience 3, 83 (1972), and in: Physical Principles of Neuronal and Organismic Behavior, ed. by M. Conrad and M. Magar. Gordon and Breach, New York, 1973.
18. Heim, R.: An algorithmic approach to information theory, this volume.
19. Pfaffelhuber, E.: A model for learning and imprinting with finite and infinite memory range, Kybernetik 12, 229 (1973).
20. Pfaffelhuber, E. and Damle, P.S.: Learning and imprinting in stationary and non-stationary environment, Kybernetik 13, 229 (1973).
21. Pfaffelhuber, E.: Finite and infinite memory range learning processes in stationary and nonstationary environments, Proc. 12th IEEE Symp. Adaptive Processes, San Diego, Paper No. TA5-8 (1973).

22. Pfaffelhuber, E.: Mathematical learning models and neuronal networks, J. theor. Biol. 40, 63 (1973).

23. Pfaffelhuber, E. and Damle, P.S.: Mathematical learning models and modifiable synapses, Intern. J. Neuroscience 6, 35 (1973).

PART VIII

Dynamical and Chemical Systems

MATHEMATICAL MODELS AND BIFURCATION THEORY IN BIOLOGY*

J. George
Department of Mathematics
University of Wyoming
Laramie, Wyoming

1. Introduction

The mathematical description of any physical process begins with a repeatable experiment. A mathematical theory is then sought which duplicates the experimental results. Usually physical laws or principles are applied to the experimental system to obtain the form and structure of the mathematical model. In many applications, the parameters in the mathematical model are not precisely known, and must also be obtained from the experiment.

A biological example of this type of modeling is the Hodgkin-Huxley model of the squid giant axon. The model consists of a partial differential equation for the current in the axon, and a number of chemical reactions, all coupled together (see FitzHugh[1] for a detailed description of this model.) The constants in the Hodgkin-Huxley model were then empirically determined from the experimental results. The model of Hodgkin-Huxley has been shown to be lacking in some respects[1], and criticisms have been made; however, no equally detailed model has been proposed as an alternative.

The physical laws and principles in biological systems are not so well understood as, e.g., the corresponding theories in fluid mechanics and celestial mechanics. For example, in chemical-enzyme reactions the exact reaction chain must be known, otherwise drastic changes in the concentrations can be observed. If the resulting mathematical model is linear, the solutions are of entirely different behavior and structure than if the chemical kinetic equations were quadratic. The trouble is that the exact chain of reactions is usually <u>not</u> known.

As another example of the difficulties occurring in biological mathematical modeling, consider morphogenesis, in which an organism develops a more complicated structure as it evolves in time. Explanations could be that a) the most complicated structure appearing in the organism was created at the "birth" of the organism, and only needed to be

* This research was supported by the Alexander von Humboldt Senior Scientist Program and carried out at the Institute for Information Sciences, University of Tübingen.

"turned on", b) new structures are created in the organism as it develops. To explain these ideas mathematically, consider the following system

$$\frac{dx}{dt} = f(x,\lambda), \quad x(0) = x_o \qquad (1)$$

where x is an n-dimensional "state" vector, λ is an m-dimensional parameter vector, and f is a sufficiently smooth mapping from $R^n \times R^m$ to R^n.

A morphogenetic structure could be obtained from (1) by "creating" a new differential system

$$\begin{aligned} \frac{dx}{dt} &= f(x,\lambda) \qquad x(0) = x_o \\ \frac{d\lambda}{dt} &= g(x,\lambda) \qquad \lambda(0) = \lambda_o \end{aligned} \qquad (2)$$

where g is also sufficiently smooth. Using singular perturbation theory as discussed by H. Hahn at this conference, the solution of (2) behaves under certain technical conditions like the solution of (1). By changing g we can evolve to a more "developed" organism.

If biological structures exhibit a more complex behavior than described above, what chance does one have of analyzing such problems? We must begin by applying the methods and ideas which have been successful in the more rigid world of classical macroscopic physics.

Let us begin by separating the idea of the local and global behavior of a system. By local behavior we mean the behavior in a neighborhood of a known periodic or steady state solution. The global behavior refers to the ways local behavior around every known periodic or steady state solution interact. The determination of local behavior is already a difficult problem for ordinary differential equations of order n, since no necessary and sufficient conditions exist for the determination of periodic solutions for large n, and usually a complicated nonlinear algebraic equation must be solved for the steady state solutions.

Poincaré worked long on the global problem, developing the familiar concepts of center, focus, etc. However, many of his techniques are restricted to second order ordinary differential equations.

René Thom has conceived "catastrophe theory"[2] to visualize the global theory for gradient systems. Since a physical system can only be considered gradient by applying some approximation arguments, the usefulness of Thom's theory must still be demonstrated.

Thom shows that if solutions bifurcate, only seven types of singularities of the (gradient) field are possible, under certain smoothness restrictions, and a limit on the number of physical parameters is imposed. It is possible that nongradient systems can have more complicated solution behavior. Since W. Güttinger has concentrated on this theory I will refer the reader to his paper and emphasize more the classical bifurcation theory.

2. Classical Bifurcation Theory

To see better what is meant by classical bifurcation theory let us consider the following example. Consider the equation

$$\frac{dx}{dt} = -x(\lambda - x) \qquad t \geqslant 0 \tag{3}$$

where x, λ are scalars. Here, clearly

$$x \equiv 0$$

and

$$x \equiv \lambda$$

are two solutions of (3) for all $t \geqslant 0$ and are represented pictorially in Figure 1. It can be easily seen that branch I and branch II are asymptotically stable, i.e. small perturbations in the initial conditions of (3) produce a solution which approaches the sheets as $t \to \infty$. Branches III and IV are unstable, i.e. small changes in the initial conditions produce a diverging solution from the sheets. The steady state ($t \to \infty$) behavior is given in Figure 1b.

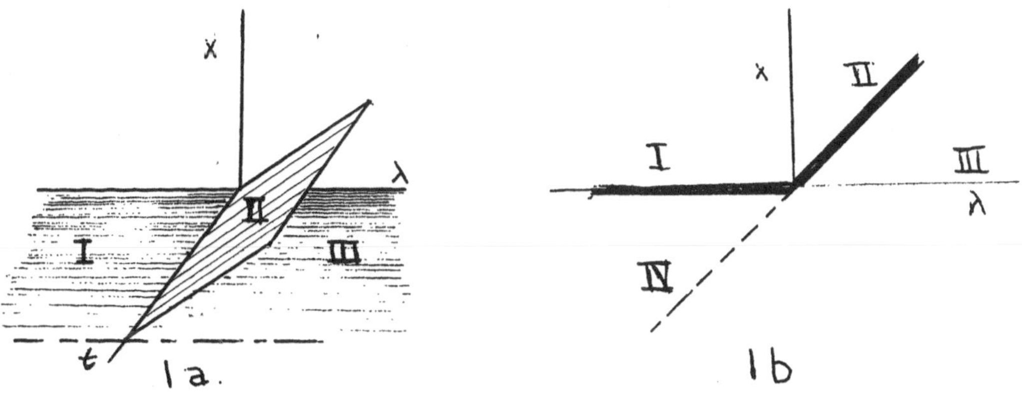

Figure 1

To formulate precisely the idea of a bifurcating solution, consider a nonlinear operator A in a Banach space B, with A(0) = 0. Assume the Frechet derivative A' exists and A'(0) = L, L = a constant linear operator. In R^n the Frechet derivative can be computed as the Jacobian matrix. Consider

$$Au = \lambda u \tag{4}$$

<u>Definition</u>. A value of $\lambda = \lambda_o$ is a branch point of A if in every neighborhood of λ_o there exists a nonzero solution of (4).

For example, in (3), $x \equiv \lambda$ is a nonzero solution, so $\lambda_o=0$ is a branch point. It is relatively easy to add another stable branch to Figure 1b. For example,

$$\frac{dx}{dt} = -\lambda x + x^2 \tag{5}$$

has three solutions, $x \equiv 0$

$x = \pm\lambda^{\frac{1}{2}}$.

The steady state (i.e., $\dot{x} = 0$) branching can be described as:

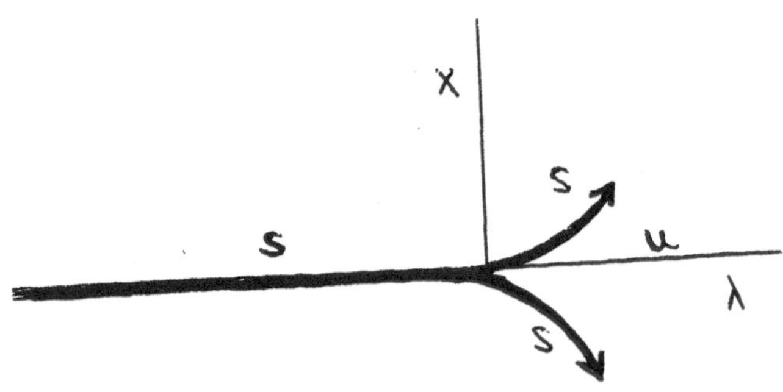

Figure 2

where $\lambda_o = 0$ is still a branch point, but now with two stable branches. More stable branches can be added in a similar fashion as we did in going from (3) to (5). Our definition says that there is <u>a</u> solution which is nonzero. There may be many.

The following theorems are taken from (3) and are classic results in bifurcation theory.

<u>Theorem 1</u>. In equation (4) $\lambda = \lambda_o$ is a branch point of A only if λ_o is an eigenvalue of $A'(0) = L$.

<u>Theorem 2</u>. (Leray-Schauder) If A is a finite dimensional operator, and

$\lambda = \lambda_o \neq 0$ is an eigenvalue of L with odd multiplicity, then λ_o is a branch point.

Theorem 3. (Krasnoselskii) Let A be a finite dimensional operator which is the gradient of a uniformly differentiable functional and let A have a second derivative at the origin. Then any nonzero eigenvalue of L is a branch point of A.

Remarks. 1. Both theorems 2 and 3 can be extended to a Banach space, which is needed to apply this theory to partial differential equations, using the concept of complete continuity[3].

2. Theorem 1 gives a necessary condition for bifurcation, while theorems 2 and 3 give sufficient conditions.

3. Many important physical problems, such as the formation of the Benard cell patterns in fluid dynamical convection can be explained by using bifurcation theory[3].

4. P. Rabinowitz[15] has obtained global results from the local bifurcation theory using topological degree arguments.

3. Applications of Bifurcation Theory in Biology

The ideas of bifurcation theory are presently being vigorously applied to many biological phenomena. An interesting application is in the paper of Gurel[10]. He applies bifurcation theory ideas to the Hodgkin Huxley model of nerve membrane dynamics, and indicates the existence of an unstable bifurcating solution, suspected earlier by numerical methods. He does not use the more general theory[3,6], but restricts himself to two dimensional projections of the four dimensional space. Other biological applications by Gurel are cited[10].

The coupling between biological oscillators such as chicadian clocks and periodic patterns occurring in chemical enzyme synthesis[11], and the examples of H. Hahn can be described in terms of bifurcation theory.

Cronin[12] considers the equation

$$\ddot{V} + \alpha V - \beta V^2 + \gamma V^3 = F \cos wt \qquad (6)$$

where V = velocity of blood flow in an aneurysm, F = constant, α, β, γ are positive constants. In earlier work, it is established that any abrupt change in V may cause turbulence, or an increase of turbulence already present, and this turbulence may cause a rupture in the aneurysm. If w is such that a resonance condition in (6) occurs, this is one such change. Cronin then uses bifurcation theory to predict periodic solutions if w is in a neighborhood of $\frac{\sqrt{2}}{n}$ or $n\sqrt{2}$, $n=1,2,\ldots$. This is called the phenomenon of synchronization, or frequency entrainment.

Another interesting application by Cronin[13] of bifurcation theory deals with the problem of periodic catatonic schizophrenia. It is assumed that the physiological phenomenon of catatonic schizophrenia is directly caused by a chemical mechanism, which is modeled by the following equations.

$$\begin{aligned}
\frac{dP}{dt} &= c - h\theta - gP \qquad & \theta < \frac{c}{h} \\
\frac{dP}{dt} &= -gP \qquad & \theta > \frac{c}{h} \\
\frac{dE}{dt} &= mP - kE & \\
\frac{d\theta}{dt} &= aE - b\theta &
\end{aligned} \tag{7}$$

P = concentration of thyrotropin

E = enzyme

θ = thyroid

c, h, g, m, k, a, b are positive constants.

Cronin shows that it is possible to have bifurcating solutions of this equation when the linearized equation (7) has a pair of pure imaginary eigenvalues. Because of the existing work which indicates a high correlation between equation (7) and the symptoms of catatonic schizophrenia, the ideas of bifurcating solutions can be used to treat the disease. For example, amytal produces a temporary remission of catatonic stupor, which corresponds to a change in the coefficients in (7).

Pimpley[14] has used bifurcation theory to study an antigen-antibody model. He shows the existence of stable limit cycles, which represent a situation where the antigen invasion is under control, and the organism is healthy.

This is only a sample of how bifurcation theory is being used in biology.

4. New Vistas

Many areas of biological application should be explored along these lines. For example:

1. Classical combustion theory[1] has reached a high degree of sophistication. Methods used successfully in combustion theory combined with bifurcation theory should produce new results in biological systems. Here, combustable material is converted to heat plus waste products, a

familiar cycle in mammals. Photosynthesis is another related phenomenon which has only recently been studied in terms of the enzyme kinetics[16].
2. The concept of phase transition, the change of a solid to a liquid or both to a gas and converse changes, is now being studied in detail. The Stefan formulation of the free boundary problem model of phase transition[9] is essentially a singularity occurring at the free boundary. While this causes some mathematical troubles[9,17] the phase transition problem can be considered from a macroscopic point of view. It is possible that the phenomenon of photophosphorylation, the illumination ability of fireflies[16], has an explanation using the free boundary formulation. An interesting application of bifurcation theory is to the crystal structure in heavy methane.[8] The boundary between two interacting substances such as ink and water can also be formulated as a free boundary problem.
3. Concepts relating the behavioral responses of an organism to some chemical system[13] should be studied in more detail. The ideas of chemotherapy would greatly profit here.

References

1. FritzHugh, R.: Mathematical models of excitation and propagation in nerve. In: Biological Engineering, ed. by H.P. Schwan, pp. 1-84. McGraw-Hill, New York, 1969.
2. Thom, R.: Stabilitê Structural et Morphogenêsê, Benjamin, Reading, Mass., 1972.
3. Stakgold, I.: Branching of solutions of nonlinear equations, S.I.A.M. Rev. 13, 289-332 (1971).
4. Keller, J. and Antman, S.: Bifurcation Theory and Nonlinear Eigenvalue Problems, Benjamin, Reading, Mass., 1969.
5. Zeeman, E.: Differential equations for the heart beat and nerve impulse. In: Towards a Theoretical Biology, vol. 4, ed. by C.H. Waddington, pp. 8-67. Edinburgh University Press, Edinburgh, 1973.
6. Sattinger, D.: Topics in Stability and Bifurcation, Lecture Notes in Mathematics No. 309, Springer Verlag, New York, 1973.
7. Frank-Kamenetzky, D.: Diffusion and Heart Exchange in Chemical Kinetics, Princeton University Press, Princeton, N.J., 1955.
8. Lemberg, H.L. and Rice, S.A.: A re-examination of the theory of phase transitions in crystalline heavy methane, Physica 63, 48-64 (1973).
9. Rubinstein, H.: The Stefan problem, Translations of Mathematical Monographs 27 (1971).
10. Gurel, O.: Bifurcations in nerve membrane dynamics, Intern. J. Neuroscience 5, 281-286 (1973).

11. Landahl, H.D. and Licko, V.: On coupling between oscillators which model biological systems, Int. J. of Chronobiology 1, 245-252 (1973).

12. Cronin, J.: Biomathematical model of aneurysm of the circle of Willis: a qualitative analysis of the differential equation of Austin, Mathematical Biosciences 16, 209-225 (1973).

13. Cronin, J.: The Danziger-Elmergreen theory of periodic catatonic schizophrenia, Bull. Math. Biophysics 35, 689-707 (1973).

14. Pimpley, G.: On Predator-Prey Equations Simulating an Immune Response, Lecture Notes in Mathematics No. 322, Springer Verlag, New York, 1973.

15. Rabinowitz, P.H.: Some global results for nonlinear eigenvalue problems, J. Functional Anal. 7, 487-513 (1971).

16. Mahler, H. and Cordes, E.: Biological Chemistry, Harper and Row, New York, 1966.

17. George, J. and Damle, P.S.: On the numerical solution of free boundary problems, preprint.

BIOCHEMICAL KINETICS

E.G. Meyer
University of Wyoming
Laramie, Wyoming

There are two basic systems for controlling biochemical reactions. One prevents the synthesis of the various enzymes throughout the reaction chain, and the other inhibits the activity of the first enzyme in the chain. In this review we describe the basic mechanisms of control and their relation to energy and entropy transformations in biological systems.

1. Biochemical Control

A. Repression and induction: preventing (or inducing) enzyme production

E. coli cells in tryptophan medium stop producing tryptophan synthetase (repression). Yeast cells not grown in lactose lack lactase (the enzyme needed to cause lactose breakdown), but put in a lactose medium they start producing lactase in about 14 hours and then proceed to ferment the sugar (induction). Mutated cells produce lactase in the presence or absence of lactose, and the lactase is fully effective in causing fermentation (lactose breakdown). Hence the production of the fermentation enzyme is apparently due to one yeast cell enzyme, while its structure is due to another. This dictates the existence of a regulatory plus operator or triggering gene system and a separate structural gene.

The regulatory/operator gene system must be able to either repress or induce the structural gene to produce the essential enzyme. It is postulated to work by first producing a repressor molecule which is capable of reacting with the regulatory substrate. In a repressive system the repressor molecule, when reacted with the substrate, inhibits the operator and hence the enzyme synthesis, while in the absence of the substrate this inhibition does not occur and the enzyme is synthesized. In the inductive system the repressor molecule plus the regulatory substrate form an inactive system which does not inhibit the operator gene and thus does not control enzyme synthesis.

It is possible to induce enzyme generation by "false" regulatory substrates. Thus molecules similar to lactose can trigger the generation of lactase. This indicates some lack of specificity in the repressor molecule. However, no such lack exists in the action of the lactase; it will not act on the false lactose. Similarly, an analog of tryptophan, 5-methyl tryptophan can repress the production of tryptophan synthetase and hence tryptophan. Since the cell cannot incorporate the 5-methyl tryptophan in proteins and has produced no tryptophan, it dies.

If two molecules are needed in controlled amounts for synthesizing a cellular unit, then the same repressor/inductor system acts on the production of both molecules (e.g., in the synthesis of purine and pyrimidine nucleotides for use in the production of nucleic acids). Pyrimidine synthesis is controlled not only by the pyrimidine bases produced but also by the purine bases from the purine synthesis chain, the former acting as a repressor and the latter as an inductor.

In DNA synthesis itself there is a signaling molecule that acts like the repressor molecule in enzyme synthesis; however, it activates rather than deactivates the sequence. This activator molecule is synthesized by a regulatory gene and triggers DNA replication after interacting with the cell membrane.

B. Control of enzyme activity

The presence of L-isoleucine in E. coli not only stops synthesis of the whole series of enzymes needed for its production but also inhibits the activity of the L-threonine deaminase, the first enzyme in the synthetic chain. This control by inhibiting enzyme activity occurs in the cells' production of amino acids and polypeptides, vitamins, purine and pyrimidine bases, and other molecules of prime importance. In these cases the effect is inhibitory; however, like the control of enzyme synthesis which can be either repressive or inductive, control of enzyme activity can be negative or positive.

Reserve energy in animal cells is stored in the form of glycogen. It is synthesized from glucose-6-phosphate to glucose-1-phosphate to uridine diphosphate D-glucose to glycogen (each step being enzymatically controlled). If there is an adequate supply of energy, in the form of adenosine triphosphate ATP, then the cell produces glucose-6-phosphate which triggers the synthesis of glycogen. When the energy supply drops so that glycogen must be utilized, the enzyme glycogen phosphorylase, which degrades the glycogen, is activated by adenosine monophosphate AMP, the "denergized" product of ATP.

Hemoglobin and myoglobin behave like enzymes in that they will bind a small molecule (oxygen); however, the former has four cooperative heme groups each of which acts as an oxygen binding site. Thus the initial binding in hemoglobin promotes further binding and a sigmoid relationship between the oxygenated form and oxygen partial pressure exists. The relationship for myoglobin is hyperbolic.

Explanation of the reaction kinetics of "ordinary" and "regulatory" enzymes shows an interesting difference. A plot of enzyme activity vs. substrate concentrations gives a hyperbolic curve for "ordinary" enzymes and a sigmoid curve for "regulatory" enzymes. The former is due to the fact that as substrate concentration is increased more enzyme binding sites are occupied. So at high concentrations of substrate when nearly all of the binding sites of the enzymes are occupied, the reaction rate levels off and the hyperbolic relationship results. With regulatory enzymes it appears that there is a cooperative effect during the initial stages of substrate addition that may involve not only the substrate itself, but a group of molecules that act as regulatory signals. Thus in the initial stages of binding the cooperation gives greater than expected rates (hence the sigmoid relationship) and further provides a mechanism whereby signal molecules can play a role. The geometry of regulatory enzymes apparently makes possible the recognition of both inhibitor and activator signals as well as substrate molecules. Further, the enzyme apparently is capable of responding only if substrate and signal molecules are above a certain threshold concentration.

A model proposed by Monod and co-workers postulates an enzyme geometry that will accommodate the substrate and signal in separate binding sites. Moreover in those cases where there can be both activation and inhibition there are three sites. One plus that for the substrate is available in one enzyme conformation, while the other plus that for the substrate is available in another enzyme conformation. Thus the regulatory enzyme could bind activators in one geometry and inhibitors in another.

2. Entropy and Information

All of the foregoing processes are dependent on the system's ability to transfer information, hence let us consider briefly information theory and its relationship to entropy.

If we consider a question, Q, which is well-defined (that is, it can be answered) and about which we have some knowledge, X, then according to the Shannon formulation the entropy $S(Q/X) = -K\Sigma p_i \ln p_i$, where K is a scale factor and p_i is the probability that the ith answer is correct. And the information, I, in a message is the difference between two entropies or uncertainties in information before and after transmission of the message. Therefore

$$I = S(Q/X) - S(Q/X').$$

The analogy to the Clausius definition is

$$S' - S = \int_{state}^{state'} \frac{dq_r}{T}$$

In fact, Shannon's (really Boltzmann's) and Clausius' functions are equivalent. From the relationship we can draw the following statements:

Entropy $\propto$ Disorder or disorganization $\propto$ Ignorance or missing Information

Information $\propto$ Order or organization $\propto$ Negative of entropy or negentropy

$$\Delta I = -\Delta S$$

Now the scale factor, K, can be chosen. If one lets K = 1/ln2, then S can be measured in bits of information. If K = the gas constant, as is common in thermodynamics, then entropy will be expressed in joules per degree.

The simplest thermodynamic system to which the information concept can be applied is that of an ensemble of molecules, each of which can be in either of two equally probable states. Here, since $p_1 = p_2 = 1/2$, S = kln2 or approximately 10^{-23} joule/o_K. From this we get the important information that the smallest thermodynamic entropy change than can be associated with a measuremtn yielding one bit of information is 10^{-23} joule/o_K. If we were to mix a half mole each of two isotopes of an element, the resulting entropy change would be $N_o$kln 2 (where N_o is Avogadro's number, 6×10^{23}) or about 6 joules/o_K or about 6×10^{23} bits. Thus we would need to make 6×10^{23} decisions to sort the mixture out.

Applying these concepts to DNA, one calculation for a DNA molecule with one million molecular weight and triplet coding indicates the possession of about 8000 bits of information. Hence a typical cell would contain 10^{12} bits and the human body with 10^{14} cells would contain 10^{26} bits of information. One must, of course, be cautious in making such extrapolations as they assume binary choices, and, further, the idea of information in DNA is still not well defined. There may be several different types or levels of information in DNA. A good summary of the applications of information theory to biology is given in the article by H.A. Johnson (1970).

Finally we can formalize the relationship between classical thermodynamics and information:

$$S_o = \frac{E + P_o U - \sum_i \mu_{io} N_i}{T_o}$$

where S_o is the uncertainty when energy E, volume V, and the number of moles of various chemical species N_i are unrecognizable because they are distributed in an environment at a temperature T_o, a pressure P_o, and chemical potentials μ_{io}. And since I = So - S where S is the entropy of a system of energy volume and composition when it is diffused into (i.e. indistinguishable in) a reference environment

$$I = \frac{E + P_o V - \sum_i \mu_{io} N_i - T_o S}{T_o}$$

(These relationships were derived by Robert B. Evans of the Georgia Institute of Technology, and give an excellent measure of disequilibrium or potential work).

It should be kept clearly in mind that just as the entropy of information has meaning only in relation to a well-defined question, the entropy of thermodynamics has meaning only in relation to a well-defined system. In our present understanding of science that system is defined by quantum theory. Thus the question, "In what quantum state is this system?" is answered "In some statistical combination of states defined by the quantum mechanical solution of the wave equation." In fact, these solutions define the possibilities alluded to in discussing information and uncertainty. The probabilities encode our knowledge about the occupancy of the possible states; possible, that is, for a given state of knowledge.

3. The Applicability of the Second Law

Living organisms maintain their steady state within relatively narrow physiological limits in spite of wide fluctuations in environmental conditions. This is accomplished by a continual struggle to maintain a small island of order in an ocean of disorder through information and negative feedback which alter the nature and rate of the necessary chemical reactions. Death can occur by excessive ordering with limited adaptability to environmental stress or by excessive disorder where any critical part of the life process fails. The steady state represents life's way of temporarily escaping the increasing entropy imposed by the second law. If living systems deviate from the steady state they become inefficient and produce entropy at a higher rate. They survive by producing entropy at a minimal rate in the steady state. Thus living systems per se must be described in terms of nonequilibrium, or irreversible or steady state thermodynamics. There are numerous reference to this subject including Morowitz (1968) and Prigogine and Babloyantz (1972a,b).

There have been questions about the possible or "apparent" violation of the second law by living organisms. There are two good and sufficient reasons why these are false. As Schroedinger (1969) first suggested, living organisms feed on the negentropy of their surroundings. Hence there is an over-all increase in entropy or disorder and decrease in information when both system and surroundings are considered. Secondly the second law applies to the equilibrium or reversible thermodynamics of isolated or closed systems. However, life involves an open or flowing system maintained by a balanced exchange of matter and energy with the environment. This system cannot attain a state of dynamic equilibrium, instead it attains a steady state which is maintained by a continual flow of energy and matter into, and out of, the system.

System	Attains	G	S	Exchange with Surroundings
closed	Equilibrium	low	high	energy only
open	Steady State	high	low	energy and matter

The basic principles of nonequilibrium thermodynamics still involve energy and entropy, but the emphasis is on energy flow through the steady state system. Such an open system is constantly subjected to forces tending to move it toward thermodynamic equilibrium. To prevent this drift, which means death for a living system, the organism must constantly perform work to maintain its steady state. The system must therefore be connected to an outside energy source providing for an energy flow into it, and so free energy flow rather than entropy becomes the focus of steady state thermodynamics.

4. Oxidation of Glucose

Bioenergetically the cell may be considered to be fueled (normally) by glucose which provides the free energy for it to perform work.

$$C_6H_{12}O_6 \ (S) + 6O_2 \ (g) = 6CO_2 \ (g) + 6H_2O \quad (1)$$

$\Delta G^{o}_{298} = -675.9$ kcal

$\Delta H^{o}_{298} = -667.0$ kcal

$\Delta S^{o}_{298} = 63$ e. u.

This energy, if released at one time, would be too large for the cell to handle and much would be wasted or lost as heat. This would produce a temperature so high that the enzymes that control reaction kinetics would be denatured. Thus the living cell needs (1) a means for capturing and storing free energy from its "fuel" molecules, (2) a means for releasing this energy in the appropriate amount when and where it is needed and (3) a means for bringing about various nonspontaneous reactions that are essential to the cellular processes. These tasks are accomplished by the capture and storage of energy in ATP and ADP, high (free) energy molecules.

The oxidation of glucose is a series of reactions involving at least thirty steps. During this process small increments of free energy are released, but of course the total cannot exceed 675.9 kcal. The adenosine phosphates are the most important compounds for the energy absorption and release.

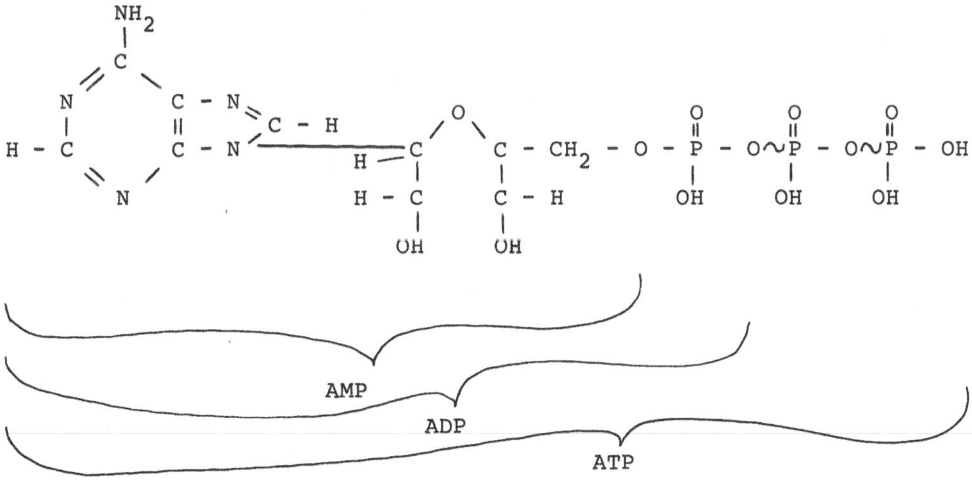

```
          NH2
          |
          C
        //  \
     N        C - N
     |        ||     \\ C - H           O                          O          O          O
H  - C        C - N /              /        \                      ||         ||         ||
      \\     /    ------------ H == C        C - CH2 - O - P - O~P - O~P - OH
         N                         |         |                 |          |          |
                              H  - C         C - H             OH         OH         OH
                                   |         |
                                   OH        OH

                       AMP
                                  ADP
                                                 ATP
```

Thermodynamically we can write

$$ATP + H_2O = ADP + H_3PO_4 \qquad \Delta G^{o} = -7300 \text{ cal}$$

$$ADP + H_2O = AMP + H_3PO_4 \qquad \Delta G^{o} = -7300 \text{ cal}$$

$$AMP + H_2O = A + H_3PO_4 \qquad \Delta G^{o} = -3400 \text{ cal.}$$

Since the reaction is reversible, free energy can be both released (by hydrolysis) or stored (by phosphorylation). Since the cellular conditions are not standard the free energy of hydrolysis is actually about 12,500 cal.

Three types of oxidation occur in the cell: direct oxidation with oxygen, hydrogen removal, and electron transfer (actually all oxidations ultimately involve electron removal). Most cellular oxidations occur by hydrogen removal or electron transfer. The purpose of these reaction sequences is to release small energy packets which couple with the oxidative phosphorylation of ADP. The efficiency of this process can be approximated as follows. In the aerobic respiratory process one mole of glucose produces a net of 38 moles of ATP. Since the standard free energy of formation of ATP from ADP is 7300 cal, 38 moles would trap (38)(7300) = 277.4 kcal. Since the maximum yield is 675.9 kcal, the efficiency is $\frac{277.4}{675.9}$ (100) or 41%. If the yield is considered to be 12,500 cal under the real cellular conditions then the efficiency is increased to $\frac{(38)(12.5)(100)}{675.9}$ or 70%. This is certainly a good conversion of available energy to used energy. The remainder, of course, is lost as heat or is unrecovered.

References

Morowitz, H.J.: Energy Flow in Biology, Academic Press, New York, 1968.

Prigogine, I., Nicolis, G. and Babloyantz, A.: Thermodynamics of evolution, Part I, Physics Today 25, No. 11, 23-28 (1972a).

Prigogine, I., Nicolis, G. and Babloyantz, A.: Thermodynamics of evolution, Part II, Physics Today 25, No. 12, 38-44 (1972b).

Schrödinger, E.: What is Life?, Cambridge University Press, Cambridge, England, 1969 (reprinted)

GEOMETRICAL ASPECTS OF THE PSEUDO STEADY STATE HYPOTHESIS IN ENZYME REACTIONS

H. Hahn
Institute for Information Sciences
University of Tübingen, Germany

1. A Simple Reaction Scheme and its Generalization

The simplest enzymatic reaction is the irreversible conversion of a substrate A into a product P by means of an enzyme E.

$$A \xrightarrow{E} P$$

Even in case of this simple reaction there exist a lot of different reaction schemes which may furnish a convenient model for the underlying mechanisms (Czerlinski, 1971; Henri, 1902; Michaelis Menten, 1913). In what follows, let us choose the irreversible Michaelis-scheme, which assumes that the reaction $A \xrightarrow{E} P$ proceeds in two steps:

$$A + E \underset{k_2}{\overset{k_1}{\rightleftharpoons}} X_1 \xrightarrow{k_3} P + E. \tag{1}$$

Here X_1 denotes the complex formed between the subrate A and the enzyme E; P is the product of the reaction. The rate constants will be denoted by k with the appropriate suffix. Let c_1, c_2, c_3, c_4 denote the concentrations of A, P, X_1, E, respectively. Assuming mass action law kinetics (which has been extremely successful, even when applied to quite complex chemical mechanisms), the reaction scheme (1) leads to a system of differential equations which describe the dynamic behaviour of the reaction; it is as follows.

$$\begin{aligned}
&a) \quad \frac{dc_1}{dt} = -k_1c_1c_4 + k_2c_3 \\
&b) \quad \frac{dc_2}{dt} = k_3c_3 \\
&c) \quad \frac{dc_3}{dt} = k_1c_1c_4 - (k_2+k_3)c_3 \\
&d) \quad \frac{dc_4}{dt} = -k_1c_1c_4 + (k_2+k_3)c_3
\end{aligned} \tag{2}$$

As is easily seen from equations 2c and 2d, the following relation will hold:

$$\frac{dc_3}{dt} + \frac{dc_4}{dt} = 0 \tag{3a}$$

This equation reflects a conservation constraint on the enzyme E and may be written in the form

$$E_t = c_3 + c_4 \tag{3b}$$

where E_t denotes the total amount of enzyme in the reaction.

Choosing the initial conditions

$$\begin{aligned} c_1(t=0) &= c_{10} \\ c_4(t=0) &= c_{40} \\ c_2(t=0) &= c_{20} = 0 \\ c_3(t=0) &= c_{30} = 0 \end{aligned} \tag{3c}$$

we get from equations 3b and 3c : $c_{40} = E_t$.

Taking into account equations (2) and (3) and eliminating the variable c_4, we get

$$\begin{aligned} &\text{a)} \quad \frac{dc_1}{dt} = -(k_1E_t)c_1 + k_2c_3 + k_1c_1c_3 \\ &\text{b)} \quad \frac{dc_2}{dt} = k_3c_3 \\ &\text{c)} \quad \frac{dc_3}{dt} = (k_1E_t)c_1 - (k_2+k_3)c_3 - k_1c_1c_3 \; ; \end{aligned} \tag{4}$$

or in form of a vector equation

$$\begin{aligned} &\dot{\tilde{c}} = \tilde{A}\tilde{c} + [\tilde{B}_1^\tau\tilde{c}, \tilde{B}_2^\tau\tilde{c}, \tilde{B}_3^\tau\tilde{c}]^\tau \, \tilde{c} \\ &\tilde{A} := \begin{pmatrix} -k_1E_t & 0 & k_2 \\ 0 & 0 & k_3 \\ k_1E_t & 0 & -(k_2+k_3) \end{pmatrix}, \tilde{B}_1^\tau := \begin{pmatrix} 0 & 0 & +\frac{1}{2}k_1 \\ 0 & 0 & 0 \\ +\frac{1}{2}k_1 & 0 & 0 \end{pmatrix} \\ &\tilde{B}_2^\tau := \underline{0} \quad ; \quad \tilde{B}_3^\tau := -\tilde{B}_1^\tau \quad ; \quad \tilde{c} := \begin{pmatrix} c_1 \\ c_2 \\ c_3 \end{pmatrix} \quad ; \end{aligned} \tag{5}$$

As is easily seen, the equations (4) are not independent. Therefore it is sufficient to treat the following system

$$\dot{c} = Ac + [B_1^{\tau}c, B_2^{\tau}c]^{\tau}c$$

$$A := \begin{pmatrix} -k_1E_t & k_2 \\ k_1E_t & -(k_2+k_3) \end{pmatrix} , \quad B_1^{\tau} := \begin{pmatrix} 0 & -\frac{1}{2}k_1 \\ -\frac{1}{2}k_1 & 0 \end{pmatrix} = -B_2^{\tau} \tag{6a}$$

$$c := \begin{pmatrix} c_1 \\ c_3 \end{pmatrix}$$

In general the kinetic equations of enzyme reactions based on mass action law kinetics and written in tensor notation take the form

$$\dot{x}_i = a_{ij}x_j + b_{i\lambda}u_\lambda + c_{ijk}x_jx_k + d_{ij\lambda}x_ju_\lambda$$
$$i,j,k = 1(1)n \; ; \quad \lambda = 1(1)m \tag{6b}$$

which is equivalent to the following system of differential equations written in vector notation.

$$\dot{x} = Ax + Bu + [C_1^{\tau}x_1 \ldots, C_n^{\tau}x]^{\tau}x + [D_1^{\tau}x_1 \ldots, D_n^{\tau}x]^{\tau}u$$

$$x \in \mathbb{R}^n \; ; \quad A = (a_{ij})_{\substack{i=1,\ldots,n \\ j=1,\ldots,n}} \in \mathbb{R}^{n,n} \; ; \quad C_i = (C_{ijk})_{\substack{j=1,\ldots,n \\ k=1,\ldots,n}} \in \mathbb{R}^{n,n} \; ; \; i=1(1)n \tag{6c}$$

$$u \in \mathbb{R}^m \; ; \; B = (b_{i\lambda})_{\substack{i=1,\ldots,n \\ \lambda=1,\ldots,m}} \in \mathbb{R}^{n,m} \; ; \quad D_i = (d_{ij\lambda})_{\substack{j=1,\ldots,n \\ \lambda-1,\ldots,m}} \in R^{n,m}, \; i=1(1)n$$

x is the state vector and the input vector of the system; A^{τ} means A transposed.

It is worth remarking that chemical kinetics focus on the internal state of the system whereas in thermodynamics input-output models are preferred.

It should be noted that equations (6) may formally include besides chemical kinetics different types of couplings such as diffusion coupling.

2. The Pseudo Steady State Hypothesis (P.S.S.H.)

Even though in the case of simple reaction scheme (1) the corresponding differential equations (6a) can be treated directly, biochemists usually simplify these equations by means of the so-called p.s.s.h. in order

(i) to simplify the theoretical considerations,

(ii) to facilitate numerical computations,

(iii) to obtain a direct relationship between at least some of the rate constants and other constants accessible to experimental measurement and

(iv) to make it possible to treat more complex chemical situations far away from equilibrium.

Assuming the classical laboratory situation, the p.s.s.h. postulates that the inequality

$$E_t / c_{10} \ll 1 \tag{7a}$$

implies

$$dc_3 / dt = dc_4 / dt = 0. \tag{7b}$$

This hypothesis asserts that if the enzyme is present in small quantity relative to the substrate (at time t = 0) then, for a considerable portion of the time occupied by the reaction, we may take $dc_3 / dt = 0$.

It should be mentioned that there exist other <u>versions</u> of a p.s.s.h., some of which will be discussed later. Besides providing a powerful method for simplifying kinetic equations, this hypothesis is justified from the <u>chemical</u> point of view by excellent agreement with experiments in a large number of chemical situations.

But there are also important chemical situations where inequality (7a) will not hold and where the p.s.s.h. is chemically not justified. Because of the relatively high enzyme concentration certain elementary transient processes may become detectable in this case. This implies that the reaction can no longer be represented by one single <u>input-output</u> rate equation - which, as will be shown later, is a consequence of the p.s.s.h. - but by a convenient <u>state</u> equation which takes into account the intermediates or the internal state of the reaction (M. Eigen 1968). The latter case will be excluded in what follows.

From the <u>mathematical</u> point of view, the p.s.s.h. cannot in general be justified. The assumption $E_t / c_1 \ll 1$ (7a) does not in general assert anything about the smallness of dc_3 / dt (7b).

But as is well-known from the literature (Heineken et al., 1967;

Bowen et al., 1963; Alberty et al., 1958; Young, 1954) the p.s.s.h. is closely related to the problem of the degenerate system in singular perturbation theory of differential equations. On certain assumptions this hypothesis may be mathematically justified by a fundamental theorem due to Tikhonov (Tikhonov, 1952; Vasil'eva, 1963).

In order to show this, we introduce the following transformations.

$$y_1 := c_1/c_{10} \qquad y_2 := c_2/c_{10} \qquad z := c_3/E_t$$

$$\mu := E_t/c_{10} \qquad \tau := (k_1 E_t)\cdot t \qquad \sigma := \tau/\mu \;;\; \mu \neq 0 \tag{8}$$

$$\lambda_2 := \frac{k_2}{k_1 c_{10}} \qquad \lambda_3 := \frac{k_3}{k_1 \cdot c_{10}}$$

The definitions (8) together with equations (6a) yield

$$\text{a)} \quad \frac{dy_1}{d\tau} = - y_1(1 - z) + \lambda_2 z$$

$$\text{b)} \quad \mu \frac{dz}{d\tau} = y_1(1 - z) - (\lambda_2 + \lambda_3)z \tag{9}$$

with initial conditions

$$y_{1_o} = 1 \;;\; y_{2_o} = 0 \;;\; z_o = 0 \quad \text{at } \tau = 0$$

or, introducing a new time scale $\sigma := \tau/\mu$, $\mu \neq 0$,

$$\text{a)} \quad \frac{dy_1}{d\sigma} = \mu \, \{\lambda_2 z - y_1(1-z)\}$$

$$\text{b)} \quad \frac{dz}{d\sigma} = \{y_1(1-z) - (\lambda_2+\lambda_3)z\} \tag{10}$$

3. Some Geometrical Aspects of the P.S.S.H.

As will be shown in the sequel, the central aspects of the above problem can be formulated in the language of the theory of differential dynamical systems with the advantage of yielding conceptual clarity and a clear geometrical insight into the problem. Moreover by means of this language the problem may be fitted (on further assumption) into the framework of Thom's catastrophe theory.

In order to illustrate further considerations and motivate the notations that follow, let us consider the simple example of equation

$$\mu \frac{d\tilde{z}}{d\tau} = -\tilde{z} \qquad (11a)$$

$$; \quad \tilde{z},\tilde{y} \in \mathbb{R}^1$$

$$\frac{d\tilde{y}}{d\tau} = -\tilde{y} \qquad (11b)$$

(due to Zeeman, 1973).

Assuming that $\mu << 1$ (7a) and that the system (11) is asymptotically stable, and comparing the solutions of equations (11a,b)

$$\tilde{z} = \tilde{z}_o e^{-\tau/\mu} \qquad (11c)$$

$$\tilde{y} = \tilde{y}_o e^{-\tau} \qquad (11d)$$

it is easily seen that the variable $\tilde{z}(\tau)$ decays much more rapidly than the variable $\tilde{y}(\tau)$. Therefore equation (11a) and the variable $\tilde{z}(\tau)$ are called <u>fast equation</u> and <u>fast variable</u>, respectively, whereas equation (11b) and the variable $\tilde{y}(\tau)$ are called <u>slow equation</u> and <u>slow variable</u>.

The orbits of equation (11) are schematically sketched in Figure 1. Figure 2 illustrates the analogous situation for $y \in \mathbb{R}^2$, $z \in \mathbb{R}$.

With this simple example in mind let us return to equations (9) and write down these equations in a more general form.

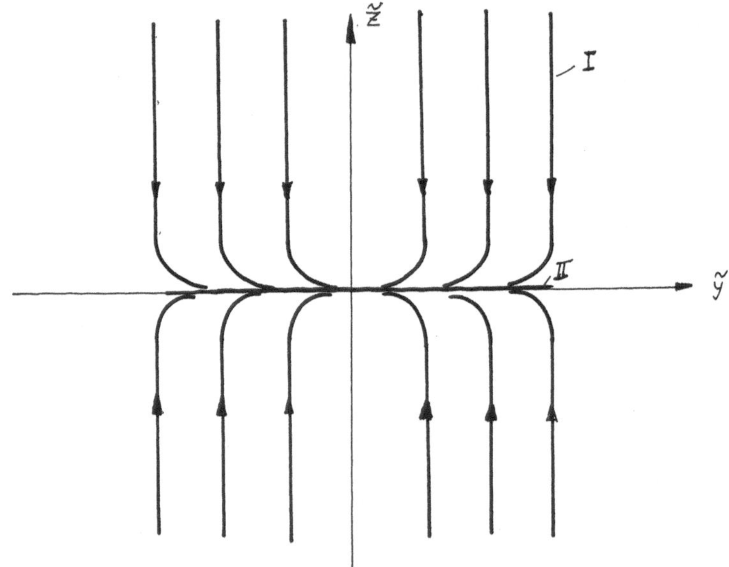

Figure 1

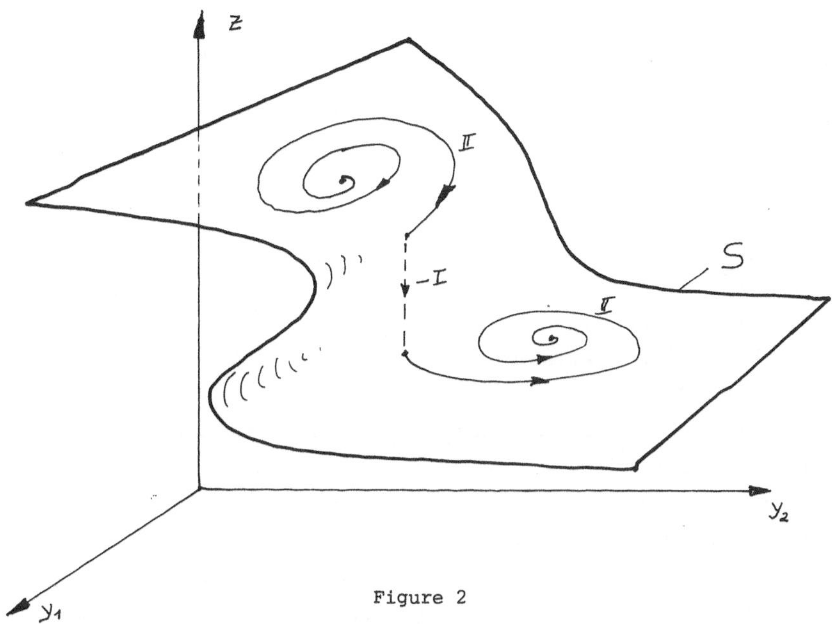

Figure 2

Let $y \in \mathbb{R}^m$, $z \in \mathbb{R}^p$, m+p = :n ; $\mathbb{R}^m$= m-dimensional Euclidian space. Instead of equation (9) let us write

$$\text{a)}\quad \mu \frac{dz}{d\tau} = F(y,z) \quad , \quad z(\tau=0) = z_o$$

$$\text{b)}\quad \frac{dy}{d\tau} = f(y,z) \quad , \quad y(\tau=0) = y_o \qquad (12)$$

where the functions $F : \mathbb{R}^m \times \mathbb{R}^p \to \mathbb{R}^p$
$f : \mathbb{R}^m \times \mathbb{R}^p \to \mathbb{R}^m$

satisfy a Lipschitz-condition.

Rewriting equation (12) in the time scale of $\sigma(\sigma := \tau/\mu \ ; \ \mu << 1; \mu \neq 0)$ we get

$$\text{a)}\quad \frac{dz}{d\sigma} = F(y,z)$$

$$\text{b)}\quad \frac{dy}{d\sigma} = \mu f(y,z) \qquad (13)$$

which shows explicity that two different time scales (τ,σ) are involved. Equations (12) and (13) are the basic equations in what follows. Clearly, on condition (7a) $(\mu << 1)$, equations (12a) and (13a) are the fast equations and equations (12b) and (13b) are the slow equations.

Let us introduce some further useful definitions which are well known in the calculus of differential dynamical systems, thereby following some of the notations of Zeeman (Zeeman, 1972).

In the limit as $\mu \rightarrow \theta$ the fast differential equation (12a) degenerates into the following set of p algebraic equations

$$F(y,z) = 0 \tag{14a}$$

which in general defines an m-dimensional hyperplane S in the n-dimensional space $\mathbb{R}^m \times \mathbb{R}^p$. S is called <u>slow manifold</u> in what follows. This means that the fast equation (12a) determines the slow manifold.

In the above simple example (eq. 11) the $\tilde{y}$-axis represents the slow manifold in $\mathbb{R}^2$.

Let us further define the <u>fast foliation</u> to be the family of p-dimensional hyperplanes which are orthogonal to $\mathbb{R}^m$.

In the example of equation (11) the fast foliation is the family of lines parallel to the $\tilde{z}$-axis.

Let $z = \phi(y(\tau))$ a root of equation (14a), i.e. a singular point of the fast equation for fixed y. Then the slow equation (12b) takes the form

$$\frac{dy}{d\tau} = f(y,\phi(y)) \tag{14b}$$

which is called <u>degenerate system</u> of (12).

As all singular points of the fast equation (12a) lie on the slow manifold the degenerate system (14b) determines the behaviour of the system (12) on the slow manifold. (It should be mentioned that the system (14b) can only fulfill free initial conditions $y(\tau=0) = y_o$ and no longer free initial conditions z_o).

Let in what follows the solution of the degenerate system (14b) be denoted as $\bar{y}(\tau)$ and $\bar{z}(\tau) = \phi(\bar{y}(\tau))$

Urging the two different time scales in equations (12) and (13) and taking into account that these systems change their character abruptly in time, Table 1 summarizes the main steps of the above discussions.

The appearance of two extremely different time scales in equations (12) and (13) motivates the construction of solutions of these equations in two steps, well-known in singular perturbation theory, where the concept of inner and outer solutions corresponds to the idea of a system changing its character abruptly in time ($\phi \leftrightarrow \tau$).

Choosing $\mu (\mu << 1)$ as a perturbation parameter, the inner solution is a perturbation solution of equation 13 in the time scale of σ, taken

TABLE I

Equation		Result	Range	Sketch
$\tau = \mu \cdot \sigma$		$\sigma \longrightarrow \tau$ time contraction		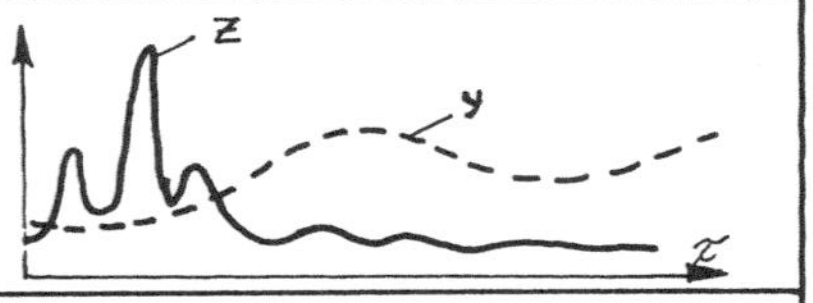 z, y, τ
(12a) $\mu \frac{dz}{d\tau} = F(y,z)$ fast equation	$\mu \to o$	$F(y,z) = o$ or $z = \phi(y)$ slow manifold	for large τ	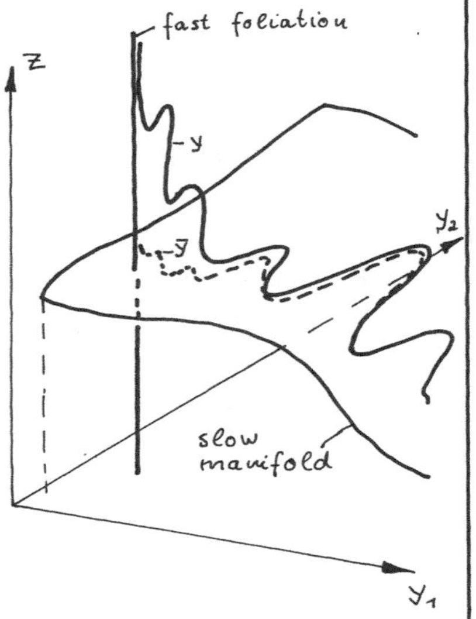 fast foliation, z, y, $\bar{y}$, y_2, y_1, slow manifold
(12b) $\frac{dy}{d\tau} = f(y,z)$ slow equation	$\mu \to o$	$\frac{d\bar{y}}{d\tau} = f(\bar{y}, \phi(\bar{y}))$ degenerate system; orbits on the slow manifold	for large τ	
(13a) $\frac{dz}{d\sigma} = F(y,z)$ fast equation	$\mu \to o$	jumps; orbits on the fast foliation	for small σ	
(13b) $\frac{dy}{d\sigma} = \mu f(y,z)$ slow equation	$\mu \to o$	$\frac{dy}{d\sigma} \approx o$ or $y(\sigma) = \text{const.}$ adiabatic behaviour of the slow variable y	for small σ	
$\sigma = \frac{1}{\mu} \cdot \tau$		$\tau \longrightarrow \sigma$ time dilatation		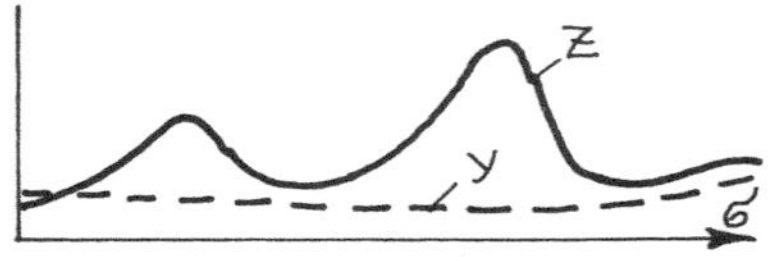 z, y, σ

for times close to $\tau=0$, whereas the outer solution is a perturbation solution of equation 12 in the time scale of τ, taken for large τ.

On constructing these solutions one has to be careful that the initial conditions for the outer solution are obtained by interpreting the inner and the outer solutions as belonging to two extremely different time scales so that the inner solution as σ becomes very large should correspond to the outer solution as τ becomes small (Heineken et al., 1967).

The inner and the outer solutions of the systems (11) and (12) are labeled by I and II, respectively in Figures 1 and 2.

4. The Mathematical Justification of the P.S.S.H.

In this context the following question is of basic importance: On what conditions will the solutions $z = z(\mu,\tau)$ and $y = y(\mu,\tau)$ of the overall system (12) tend towards the solutions $\bar{y}(\tau)$ and $\bar{z}(\tau) = \phi(\bar{y}(\tau))$ of the degenerate system (14b) as $\mu \to o$?

This means on what conditions can we to a large extent forget about the dynamics of the fast variables z if we are interested in the asymptotic behaviour of the system for large τ, or on what conditions can the asymptotic behavior of the overall system (12) or (13) be discussed by studying only the degenerate system (14b), i.e. by investigating only the orbits on the slow manifold and forgetting about the fast foliation. Clearly the answer to this question delivers a first step towards a rigorous mathematical justification of the process of simplifying models and of getting rid of information which is without any relevance with respect to the questions one is interested in (Hahn, George, 1974).

In case m=1, p=1 (eq. 11) this question can partly be answered by a lemma of Zeeman (Zeeman, 1972).

A more general answer to the above question is given by Tikhonov's Theorem which in this context may be stated in a somewhat weaker form (for autonomous systems) as follows:

Theorem of Tikhonov

a) Let $z = \phi(y)$ an isolated positively stable root of the algebraic equation (14a) in a certain domain $D \in \mathbb{R}^m$.

b) Let the initial point $(y(\tau=0), z(\tau=0))$ lie within the domain of influence of this root.

c) Let the solution $\bar{y}(\tau)$ of the degenerate system (14b) belong to D for $0 \leq \tau$.

Then the solution $y = y(\mu,\tau)$

$z = z(\mu,\tau)$

of the overall system (12) tends to the degenerate solution $\overline{y}(\tau)$ of the small system (14b) in the sense that

$$\lim_{\mu \to 0} y(\tau,\mu) = \overline{y}(\tau) \qquad \text{for all } 0 \leq \tau;$$

$$\lim_{\mu \to 0} z(\tau,\mu) = \phi(\overline{y}(\tau))$$

Remarks:

(1) Z is called positive stable in D for a fixed point $y^* \in D$ if the solution $z(\sigma)$ of the fast equation $\frac{dz}{d\sigma} = F(y^*,z)$ (equation 13a) tends to $\phi(y^*)$ as σ tends to infinity. (i.e. $\lim_{\sigma\to\infty} z(\sigma) \to \phi(y^*)$). In other words, z is positive stable if it is an attractive singularity of the fast equation (13a) for fixed $y=y^* \in D$.

(2) The domain of influence of an isolated positive stable root $z = \phi(y)$ is the set of all initial points (y^*,z^*) such that if $z(\sigma=0) = z^*$ is the initial condition of the solution of the fast equation $\frac{dz}{d\sigma} = F(y^*,z)$ this solution will fulfill the condition $\lim_{\sigma\to\alpha} z(\sigma) \to \phi(y^*)$ (compare Figure 3).

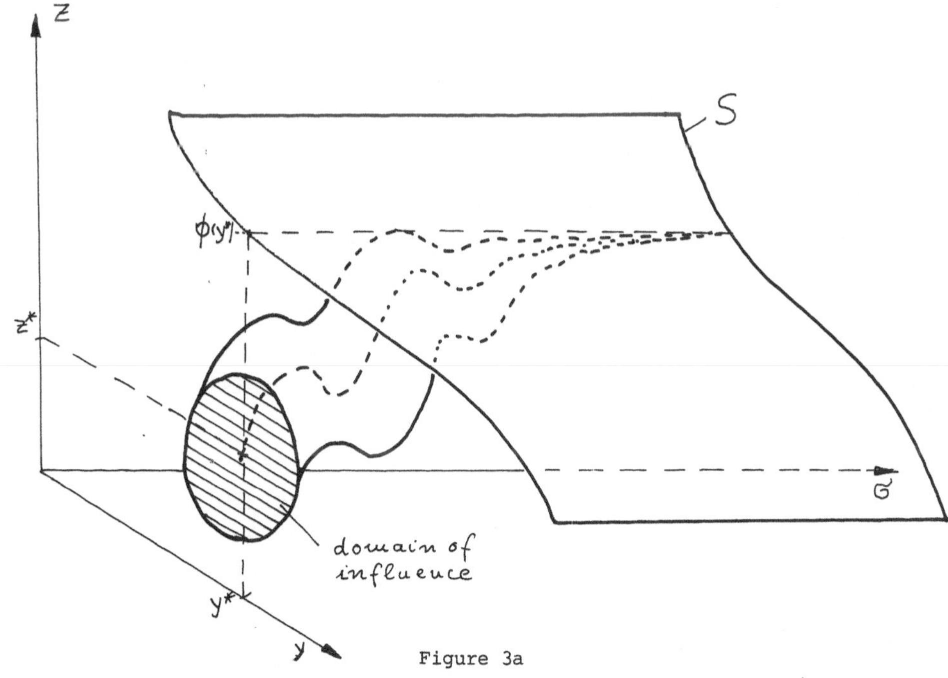

Figure 3a

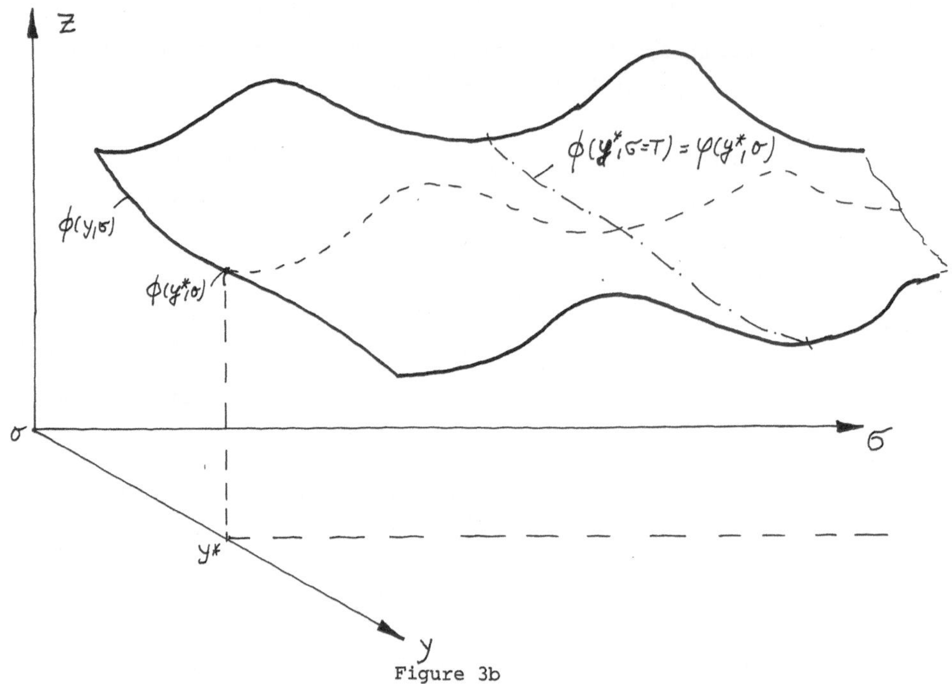

Figure 3b

5. Some Consequences of the Main Theorem for Chemical Situations

Returning to chemical systems the question arises whether the application of the p.s.s.h. to chemical situations is in general mathematically justified, i.e. whether chemical systems in general satisfy the assumptions of Tikhonov's Theorem.

Clearly, all thermodynamic processes which fit into the framework of linear nonequilibrium thermodynamics possess for a large variety of boundary conditions a Liapunov-function which is closely related to the entropy production of the system. If this Liapunov-function is radially unbounded, the assumptions of the above theorem are satisfied.

On the other hand, enzyme reactions usually don't fit globally into this framework. This can be proved by many well-known counterexamples.

Let us define now two different classes of enzyme reactions:

> Reactions of class SR (scale reducible systems) shall satisfy the assumptions of Tikhonov's Theorem, whereas
> reactions of class SI (scale irreducible systems) shall not fulfill these assumptions.

Only for class SR chemical processes the pseudo-steady-state-hypothesis can be (at the present time) mathematically justified.

To avoid mathematical subtlety let us assume in what follows that the roots of equation 14a are isolated.

Usually it is extremely complicated or even impossible to verify for a special situation each of the assumptions a,b,c of the theorem. For instance, in general it is rather cumbersome if not impossible to compute the whole domain of influence (b) of a root of equation (14a) or to verify whether the solution $\bar{y}(\tau)$ of the degenerate system (14b) belongs to a certain domain D for all $\tau \geq 0$ (c). On the other hand, if the domain of influence of a root is unknown, it may be so small, that though this root may be locally attractive it is practically not attractive (problem of practical stability).

Instead of studying all these questions separately by means of lengthy computations let us look for some more simple criteria which satisfy all of these assumptions at once. This leads us, as a first step, towards the concept of <u>global</u> asymptotical stability of enzyme reactions. Clearly systems of this type cover only a <u>subclass</u> of SR. One theorem which assures global asymptotic stability of a general chemical system of the type of equation (6c) may be stated as follows.

<u>Theorem</u>

Let $x_1 \in \mathbb{R}^p$; $P_1, A_1 \in \mathbb{R}^{p,p}$; $B_1 \in \mathbb{R}^{p,r}$

$x_2 \in \mathbb{R}^m$; $P_2, A_2 \in \mathbb{R}^{m,m}$; $B_2 \in \mathbb{R}^{m,r}$

$F_i \in \mathbb{R}^{m,p}$, $i = 1(1)m$; $C_§ \in \mathbb{R}^{p,m}$, $§ = 1(1)p$

$u \in \mathbb{R}^r$; $Q_§ \in \mathbb{R}^{r,m}$, $§ = 1(1)p$

Let A_1, A_2 stable matrices (whose eigenvalues possess negative real parts);

Let $V_j = x_j^T P x_j$ Liapunov-functions of the linear systems $\dot{x}_j = A_j x_j$

$j = (1,2)$

then the system

$$\dot{x}_1 = A_1x_1 + B_1u + [C_1^Tx_1,\dots,C_p^T \cdot x_1]^T \cdot x_2 - P_1^{-1} \cdot [F_1^Tx_2,\dots,F_m^Tx_2] \cdot P_2x_2 + [Q_1^Tu,\dots,Q_p^Tu]^T \cdot x_2$$

$$\dot{x}_2 = A_2x_2 + B_2u - P_2^{-1} \cdot [C_1^Tx_1,\dots,C_p^Tx_1] \cdot P_1x_1 + [F_1^Tx_2,\dots,F_m^Tx_2]^T \cdot x_1 - P_2^{-1} \cdot [Q_1^Tu,\dots,Q_p^Tu] \cdot P_1x_1 \quad (15)$$

is global asymptotically stable for bounded inputs u and for all matrices C_i, F_i, Q_i.

Remark:

It should be mentioned that the variables x_1 and x_2 in equation (15) are linear decoupled though there exist strong nonlinear couplings among them. The specific form of equation (15) may be interpreted as describing the overall behaviour of an adaptive chemical process or with some small modifications a chemical identification scheme which gives rise to interesting biochemical considerations (Hahn, 1974).

The above theorem guarantees that the chemical process described by equation (15) is of class SR. On the other hand a system which is globally asymptotically stable is unable to produce phenomena which are called exotic and can be produced in experiments (smooth oscillations, relaxation oscillations, hysteresis phenomena, frequency entrainment, bifurcations....).

When performing initial velocity studies of enzyme reactions biochemists very often use the p.s.s.h. In simple cases they call the degenerate system 14b Michaelis-Menten kinetic law. It determines the system behaviour on the slow manifold.

In case of the irreversible Uni-Uni-reaction of equations 1 and 2 equation (14b) takes the simple form

$$\frac{dy_1}{dt} = -\frac{dy_2}{dt} = -\frac{\lambda_3 y_1}{(\lambda_2+\lambda_3)+y_1} \quad \text{or with} \quad V_1 := k_3 E_t$$

$$K_1 := (k_2+k_3)/k_1 \tag{16a}$$

$$\frac{dc_1}{dt} = -\frac{dc_2}{dt} = \frac{-(k_3E_t)c_1}{(k_2+k_3)/k_1 + c_1} = \frac{-V_1\, c_1}{K_1+c_1}$$

In case of a reversible Uni-Uni-reaction

$$A + E \underset{k_2}{\overset{k_1}{\rightleftharpoons}} X_1 \underset{k_4}{\overset{k_3}{\rightleftharpoons}} P + E$$

the degenerate system takes the form

$$\frac{dc_2}{dt} = \frac{V_1 c_1 - K_1(V_2/K_2)c_2}{K_1(1 + c_2/K_2) + c_1} \tag{16b}$$

$$V_2 := k_3E_t\ , \quad K_2 = (k_2+k_3)/k_4$$

which indicates the phenomena of linear competitive product or substrate inhibition.

In both of these cases the fast foliation has dimension 1 whereas the slow manifold is a two dimensional hyperplane in $\mathbb{R}^3$ and the trajectories on this manifold are governed by equation 16a respectively 16b, the degenerate system.

Going to more complex reaction schemes, for instance to the cases of

c) ordered Ter-Bi-Reactions or
d) random Bi-Bi-Reactions or
e) Ping-Pong-B-Bi-Reactions

the slow manifold has dimension

c) 5
d) 4
e) 4

whereas the fast foliation is of dimension

c) 4
d) 5
e) 3

The appropriate equations (16c,d,e) which can easily be derived by the method of King and Altman show a similar though a somewhat more complicated form than equations 16a and 16b.

In even more complicated cases, for instance dealing with multienzyme-complexes or with enzymes consisting of a number of subunits, the corresponding degenerate equations (16) may lead, depending on additional hypotheses for instance to the equation of Monod-Wyman-Changeux or to Hill's equations, both of which may describe allosteric transitions. In certain cases they reflect sigmoid control behaviour of the enzyme which may give rise to all or none behaviour of the control function, a behaviour which seems to be of great importance in metabolic control processes. It is interesting to notice that all of these different types of equations (16) which form the basis for the theory of enzyme kinetics are restricted to the system behaviour on the slow manifold, forgetting about its dynamic behaviour on the fast foliation.

Before weakening the severe assumption of global asymptotic stability of the overall system which excludes most of the exotic phenomena in chemical systems let us classify them as follows:

Those phenomena which are produced mainly by the fast equation, i.e. at least in part outside the slow manifold are of the fast type (F) whereas those produced by movements inside the slow manifold are of the slow type (S).

In agreement with the definition of class SR systems the severe assumption of global asymptotic stability of the overall system (12) can be weakened by postulating that only the fast equation (12a) has to be global asymptotically stable. This implies that class SR systems can only produce exotica of type S, whereas exotica of class F are excluded.

Geometrically this means that the slow manifold is a global attractor for all solutions of the fast equations (12a).

As a consequence of the p.s.s.h., i.e. the assumptions of Tikhonov's Theorem, the slow manifold has to be transversal to the fast foliation in the whole space $\mathbb{R}^u$ or at least if local tangency between these two manifolds exists, the system cannot be generic, which implies a structurally unstable situation as small parameter changes in the system may turn it from class SR to class SI type.

The above transversality condition excludes for instance the cases that the slow manifold of Figure 1 or 2 is S- or C-shaped. These implications exclude e.g. oscillations not completely inside the slow manifold or trigger phenomena, threshold behaviour or jumps that start on the slow manifold and will leave it for some time, i.e. exotica of the fast type (type F phenomena). On the other hand, oscillations or jumps within the slow manifold, i.e. exotica of the slow type (type S phenomena) are still allowed. But it must be kept in mind that e.g. S-type jumps are far from being as rapid as jumps of type F, as is easily seen from equations (12) or (13).

As a consequence, if biochemists want to attack problems concerned with chemical exotica on the basis of the laws of equation (16) and in agreement with the underlying mathematical theory, they have to restrict their investigations to type S phenomena.

If we restrict our considerations even more and admit only gradient type degenerate systems (16), it is possible--following the lines of Thom's catastrophe theory--to classify the admissible S-type exotica. It is worthwhile to note that even though the overall system (12)--in the language of René Thom--can at most produce the two types of catastrophes, the elliptic and the hyperbolic umbilic, the degenerate system (16) is able to produce not only all types of catastrophes of Thom's table, but also a lot of other types of catastrophes which are not listed in Thom's table (Hahn, 1974).

6. Other Types of P.S.S.H.

The pseudo steady state hypothesis as discussed until now (7a,b) is due to Briggs and Haldane. There exist some other versions of a steady state hypothesis which, though often related to equation 16, lead to degenerate systems of a completely different structure compared to the structure of equation 16. Heineken et al. (1967) have shown that assuming $\frac{k_3}{k_1} << 1$ instead of (7a) in the reaction scheme (1) implies a degenerate system of a structure quite different from the structure of equations (16).

An interesting situation arises if one leaves in vitro experiments and treats situations which are normally true in vivo, for instance in a living cell. In this case the chemical situation is described by an open system, where substrates are continually supplied and products are continually removed and where the concentrations of the enzyme and the concentrations of the substrates and products may be of the same order of magnitude.

For instance the substrate concentration may be maintained within certain limits by other reactions producing the substrate as a product and the product concentration may be kept constant within certain limits by subsequent reactions using the product as a substrate.

But taking into account the additional assumption

$$\left|\frac{dy}{d\tau}\right|\Bigg|_{\text{without control}} =: |f(y,z)| >> \left|\frac{dy}{d\tau}\right|\Bigg|_{\text{controlled}} =: |\tilde{f}(y,z)| \qquad (17)$$

or

$$\frac{dy}{d\sigma} = \tilde{f}(y,z) := \mu f(y,z) \;, \quad o < \mu << 1$$

where μ represents the influence of cell regulation on the velocity $\frac{dy}{d\tau}$ of y (and not on the accuracy of the controlled variable y), equations 12 and 13 result again.

Therefore when investigating in vivo experiments a formalism completely analogous to the procedure of the p.s.s.h. can be applied and leads to the same type of system equation (16) if condition (17) is fulfilled.

Concluding Remarks

The mathematical justification of the p.s.s.h. delivers an instructive example for a first step rigorous procedure to simplify models by forgetting information which is of no relevance for the understanding of certain limited aspects of a dynamic system. Especially it provides a method to separate the continuous (slow) from the discontinuous (fast) processes of the system. There is some hope that a refinement of this method may lead to a rigorous procedure separating the switching- and decision making part of the system from its continuous transient behaviour.

On the other hand there exist many situations where we are forced by new experimental observations to refine existing models and to increase their dimension. This is just a situation inverse to the problems which are treated on the p.s.s.h. Instead of deriving a degenerate system we now have to construct a "regenerate" system. Mathematical ideas analogous to those of the singular perturbation theory should allow for a vigorous treatment of this problem.

Summary: Starting from a simple reaction scheme a set of equations is set up which delivers a compact formulation of the dynamic equations of mono- and bimolecular reactions. The mathematical and chemical consequences of the pseudo steady state hypothesis (p.s.s.h.) are discussed. The geometrical aspects of this hypothesis give rise to a rough classification of enzyme reactions and show that the different basic equations in enzyme chemistry may be interpreted from a common geometrical point of view. A theorem is stated which guarantees the global asymptotic stability of a class of reactions, justifies the p.s.s.h. and allows for an interpretation of the reaction as an adaptive system. A criterion is formulated which justifies the applicability of the p.s.s.h. even to in vivo reactions. A comment is made concerning the consequences of the p.s.s.h. with respect to the catastrophe theory of R. Thom.

References

Bowen, J.R. et al.: Chem. Eng. Sci. 18, 177 (1963).

Czerlinski, G.H.: J. theor. Biol. 32, 373-383 (1971).

Eigen, M.: Q. Rev. of Biophys. I, 3-33 (1968).

Hahn, H.: (1974) to appear.

Hahn, H. and George, J.: (1974) to appear.

Heineken, F.G., Isuchiya, H.M. and Aris, R.: Math. Biosc. 1, 95-113 (1967).

Henri, V.: C.r. hebd. Séanc. Acad. Sci., Paris 135, 916 (1902).

Michaelis, L. and Menten, M.: Biochem. Z. 49, 333 (1913).

Tikhanov, A.N.: Math. Sb. 31(73), 575 (1952).

Vasileva, A.B.: Russ. Math. Surveys 18, 13 (1963).

Young, Ch. Ch.: Arch. Biochem. Biophys. 51, 419 (1954).

Zeeman, C.: In: Towards a Theoretical Biology, ed. by C.H. Waddington, 1972.

A SYNTHETIC APPROACH TO EXOTIC KINETICS (WITH EXAMPLES)

O.E. Rössler
Division of Theoretical Chemistry, University of Tübingen
Tübingen, Federal Republic of Germany

Introduction

The approach to be outlined in the following is "classical" in the sense that only concentrations of individual substances are considered, rather than potentials of chemical reactions; furthermore, use is made of the "nonexplicit" convention for catalyzed reactions, so that the mass action law is violated in the equations; finally, the existence of "pool" substances, i.e. non-exhaustible sources, is presupposed. Hence neither the elegant tools of the recent potential-theoretic "network" approach of Oster et al. (1973), nor the recent methods of algebraic redundance-reduction in closed mass-action systems and their generalizations (cf. Feinberg, 1972; Horn & Jackson, 1972; and Horn, 1973) can be exploited. However, the reader should be offered sufficient methods in order to be able to invent new, non-trivial exotic chemical circuits for himself.

As an aid for intuition, three sources for the generation of new chemical circuits with non-trivial, i.e. exotic, dynamical behavior are offered at the outset: (a) the analogy to electronic circuits; (b) the analogy to neural elements; and (c) the analogy to already existing chemical circuits, invented by mathematical biophysicists. Accordingly, the presentation will be divided into three parts, followed by a concluding part devoted to generalizing applications. In this last part, fluid analogs to neural structures will be considered briefly, whereby both Thom's (1972) abstract catastrophy theory and the actual chemical reaction system of Belousov (1959) and Zhabotinsky (1964) will find an application.

Part I: An Analogy between Certain Building-elements of Electronic Circuits and Certain Building-elements of Chemical Circuits, with Applications

We start with a rather trivial analogy, namely that between a chain of RC elements in electronics, and a series of consecutive irreversible first-order reactions in chemistry (Fig. 1). Isothermous, homogeneous conditions and an appropriate concentration range are presupposed as usual.

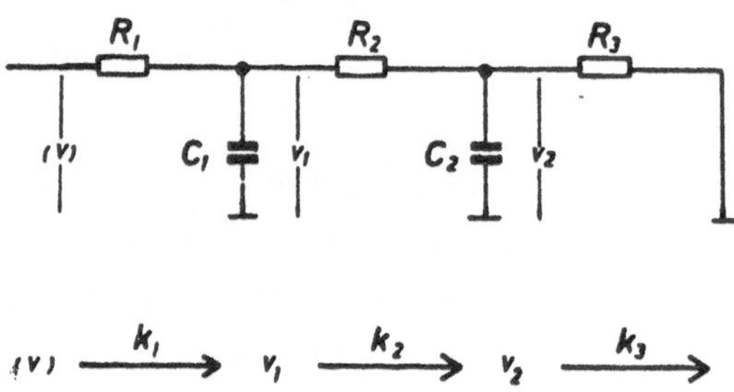

Fig. 1. A Chemical RC-chain and electronic equivalent.

Here and in the following, brackets denote sources (with low internal resistance); all voltages refer to ground, unless indicated otherwise.

The differential equations in the electronic case are (using Ohm's law: $v(t) = Ri(t)$, and the characteristic of a linear, time-invariant capacitor: $dv/dt = \dot{v} = C^{-1}\,i(t)$):

$$\dot{v}_1 = \frac{1}{R_1C_1}\,v - \frac{1}{R_2C_1}\,v_1 - \frac{1}{R_1C_1}\,v_1 + \frac{1}{R_2C_1}\,v_2$$

$$\dot{v}_2 = \frac{1}{R_2C_2}\,v_1 - \frac{1}{R_3C_2}\,v_2 - \frac{1}{R_2C_2}\,v_2\,, \tag{1}$$

reducing, for $R_1 \gg R_2 \gg R_3$, to

$$\dot{v}_1 = k_1v - k_2v_1$$

$$\dot{v}_2 = k_2v_1 - k_3v_2, \tag{2}$$

where $k_1 = (R_1C_1)^{-1}$, $k_2 = (R_2C_1)^{-1} = (R_2C_2)^{-1}$, $k_3 = (R_3C_2)^{-1}$.

The latter equation is just identical with that describing the behavior of the chemical circuit. Hence it is evident that <u>concentrations of chemical substances may be analogous to voltages across capacitors</u>. It is seen further that a chain of first-order reactions or pseudo-first-order reactions (including the participation of additional pool substances), respectively, may be analogous to a simple passive linear electronic circuit.

For the sake of completeness it is mentioned that the voltage-reduction occurring from step to step in the idealized electronic circuit can be avoided, in the chemical case, by an appropriate re-scaling of both the variables and the constants. Finally it is observed that the above-made requirement of irreversibility for all chemical reactions may also be dropped. It had been introduced only for the sake of the existence of an immediate electronic analog (according to the here exploited type of analogy). Abstract chemical circuits are in general more flexible for abstract design purposes than electronic ones, due to the (relative) independence of forward and reverse rate constants. It is only when the question of realization by actual chemical substances occurs that the problems begin. The specifically chemical restriction that not every chemical capacitor (substance) can be wired via a chemical resistor (reaction) to every other one constitutes an almost unsurmountable barrier at the present. It should be overcome soon, however, when the techniques of computer-aided wiring will have been adapted to this constraint.

We now come to a second major element which is analogous in both domains: the three-terminal (or, synonymously, controllable) resistor (Fig. 2).

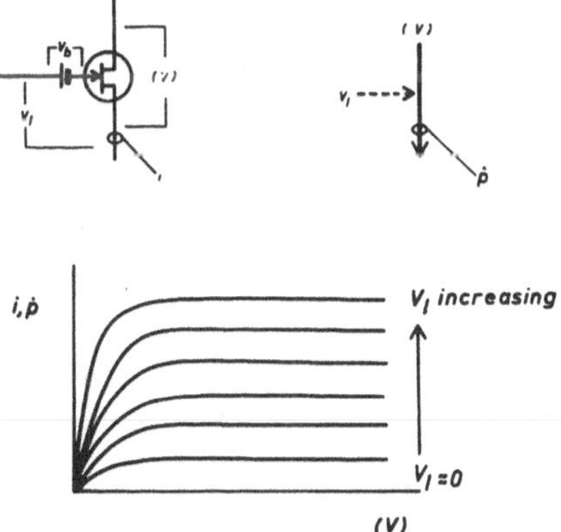

Fig. 2. A chemical "FET-type" three-terminal resistor and electronic equivalent.

As the electronic example, the N-channel field-effect transistor (FET) is depicted, but a vacuum-tube pentode possesses just the same output-characteristics (lower graph). On the chemical side, the frequent case of a so-called Michaelis-Menten type kinetics is distinguished by corresponding output characteristics (lower graph). The input characteristic is also the same in both cases (practically infinite resistance for nonnegative values of the input variable). Hence the large set of catalyzed reactions obeying this type of kinetics can be classified as "three-terminal resistors of the FET-type". (It is recalled that this type of a "chemical resistor" is based on the voltage-concentration as well as current-reaction rate correspondence. It should not be confounded, therefore, with the three-terminal (=two-port) chemical resistors which appear in the more physical approach of Oster et al., 1973. A single three-terminal resistor of the present type corresponds to a whole network, involving 2 resistive and 6 transducing two-ports, of the thermodynamic type. The exact relation between both approaches has not yet been worked out.)

A simple RCR circuit using the new resistive element is the following non-inverting chemical amplifier, depicted in Fig. 3 together with its electronic analog.

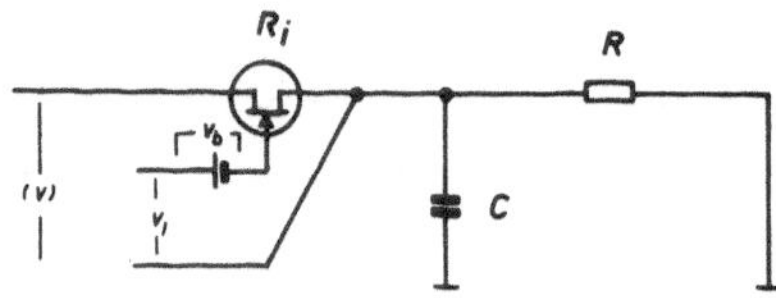

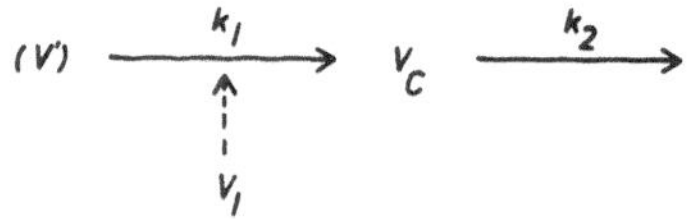

Fig. 3. A chemical noninverting amplifier and electronic equivalent.

The latter obeys the differential equation

$$\dot{v}_C = \frac{1}{R_i C} v - \left(\frac{1}{R_i C} + \frac{1}{RC} \right) v_C ,$$

where

$$R_i = \frac{v - v_C}{i_{R_i}}$$

and (with minor idealization)

$$i_{R_i} = \alpha v_i \frac{(v-v_C)}{(v-v_C)+K} ;$$

α being the gain factor of the FET and K its phenomenological saturation constant (cf. Chirlian, 1971). Introduction of the restriction $R_i >> R$ (such that also $v >> v_C$) yields the desired "ideal" equation

$$\dot{v}_C = \frac{\alpha}{C} v_1 \frac{v}{v+K} - \frac{1}{RC} v_C$$

which can also be written as

$$\dot{v}_C = k_1 v_1 \frac{v'}{v'+K'} - k_2 v_C , \qquad (3)$$

where $k_1 = C^{-1}\alpha$, $k_2 = (RC)^{-1}$, $a = v/v' = K/K' >> 1$, and K' = Michaelis-constant of the catalyzed reaction. Eq. (3) implies that $v_C \to v_1 k_1/k_2$ when $v' >> K'$ and $\dot{v}_C >> \dot{v}_1$. Hence it describes a linear amplifier under the named conditions, with amplification factor k_1/k_2. Note that, once more, the electronic circuit is more restrictive, due to the requirement $R_i >> R$ which implies $v_C >> v$. This is avoided in the chemical case, so that v_C may be of the same order of magnitude as v' or even of greater magnitude. (It is mentioned that the so-called source-follower of electronics, where v_1 is replaced by v_1-v_C, is not a basic circuit in the chemical case.)

The corresponding inverting chemical amplifier is given in Fig. 4.

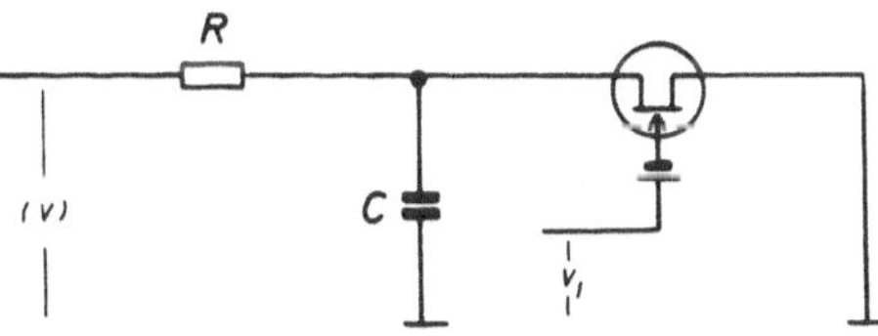

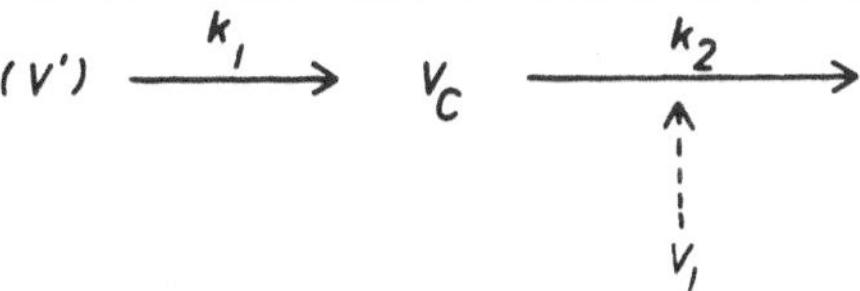

Fig. 4. A chemical inverting amplifier and electronic equivalent.

Its differential equation, obtained in an analogous way (but for $R >> R_i$), is

$$\dot{v}_C = k_1 v' - k_2 v_1 \frac{v_C}{v_C + K'} , \tag{4}$$

where $k_1 = (RC)^{-1}a$, $k_2 = C^{-1}\alpha$, $a = v/v' >> 1$, $K = K'$.

On the basis of these few elementary circuits, we are now able to translate two simple, but famous electronic circuits into chemistry: the linear RC-oscillator, and the Eccles-Jordan trigger. The former system was first outlined by Seelig (1970): Fig. 5.

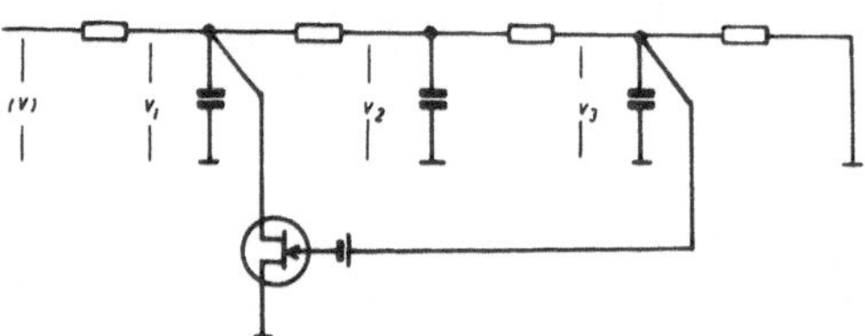

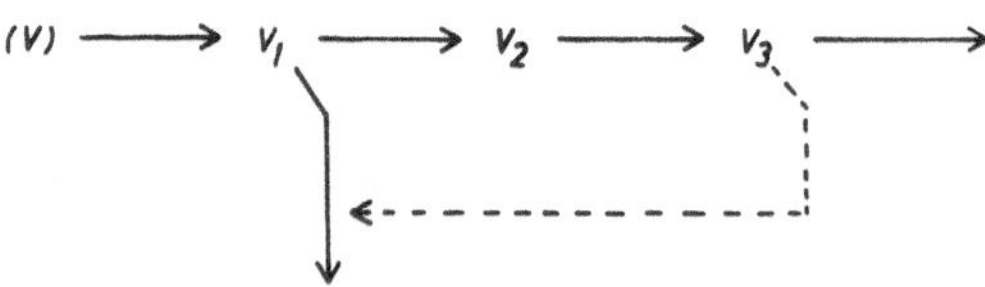

Fig. 5. A chemical RC-oscillator and electronic equivalent.

The equations are, after appropriate normalization,

$$\begin{pmatrix} \dot{v}_1 \\ \dot{v}_2 \\ \dot{v}_3 \end{pmatrix} = \begin{pmatrix} -1 & 0 & -8 \\ 1 & -1 & 0 \\ 0 & 1 & -1 \end{pmatrix} \begin{pmatrix} v_1 \\ v_2 \\ v_3 \end{pmatrix} + \begin{pmatrix} 1 \\ 0 \\ 0 \end{pmatrix} \tag{5}$$

(if $v_1 >> K$), i.e. those of a simple three-component linear oscillator (Seelig & Göbber, 1971). Of course, linear oscillators are structurally unstable and cannot be realized by physical means (cf. Rosen, 1970). The essential nonlinearity of the term $-8 \frac{v_1}{v_1+K}$ in the first line of Eq. (5), even for $v_1 >> K$, is the reason for a slight intrinsic damping of the system, increasing with the amplitude, which has to be compensated by a constant, small "negative damping" of the matrix (in the form of a small positive real part of the complex pair of eigenvalues). The system, therefore, is a self-excited limit-cycle oscillator producing almost sinusoidal oscillations. In the chemical case, the two additional problems of (a) resolving the non-explicit catalytic reaction into explicit elementary reactions of the mass-action type, and of (b) admitting reversible reactions in the chain, had to be solved. A successful explicit version could indeed be offered (Seelig, 1971a).

The announced "chemical Eccles-Jordan trigger" is displayed in Fig. 6.

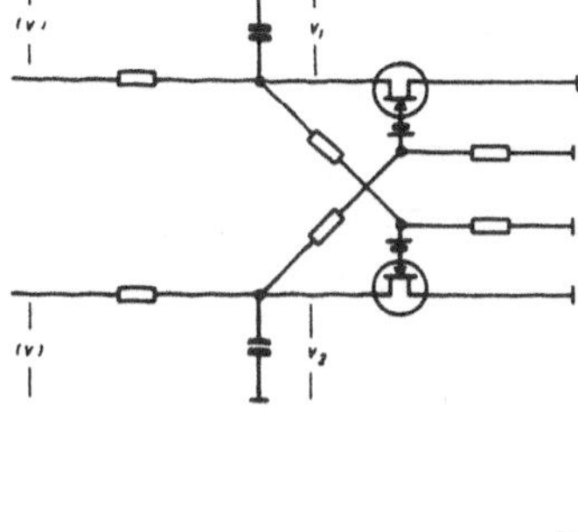

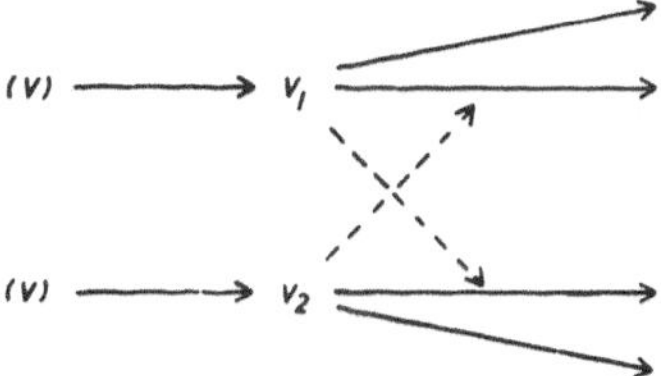

Fig. 6. A chemical Eccles-Jordan trigger and electronic equivalent.

It is described by the equations

$$\dot{v}_1 = k_1' - k_2 v_1 - k_3 v_2 \frac{v_1}{v_1+K}$$
$$\dot{v}_2 = k_1' - k_2 v_2 - k_3 v_1 \frac{v_2}{v_2+K} , \qquad (6)$$

where $k_1' = k_1 v$ (Rössler, 1974b).

The system can be considered as a linear system for all values of v_1 and $v_2 \gg K$. In this region it possesses a saddle-point for sufficiently large K_3. Two further stable steady states (nodes) are present near the positions $(\frac{k_1'}{k_2}, 0)$ and $(0, \frac{k_1'}{k_2})$, as is readily seen by setting one variable equal to zero. A graphical solution is obtained by computing both nullcines ($\dot{v}_1 = 0$ and $\dot{v}_2 = 0$) which are hyperbolas (cf. Fig. 7 below) and determining their intersection points which are the steady states. The presence of further limit sets is excluded with the aid of Bendixson's negative criterion: the sum

$$\frac{\partial \dot{v}_1}{\partial v_1} + \frac{\partial \dot{v}_2}{\partial v_2}$$

is less than zero, and hence does not change sign,in the positive quadrant (see Rosen, 1970).

The characteristic feature of the electronic trigger (Eccles & Jordan, 1919) is that it can be pushed, by exogeneous manipulation, from one stable steady state toward the other and vice versa, as a resettable flip-flop ("RS flip-flop"; Hurley, 1961). The same applies to the chemical system if, for example, a sufficiently strong additional input is applied to one side (for instance, by increasing k_1'). Then the two hyperbolas mentioned above intersect in either three points or one, two points disappearing simultaneously at a distinct bifurcation value of k_1', and vice versa: Fig. 7.

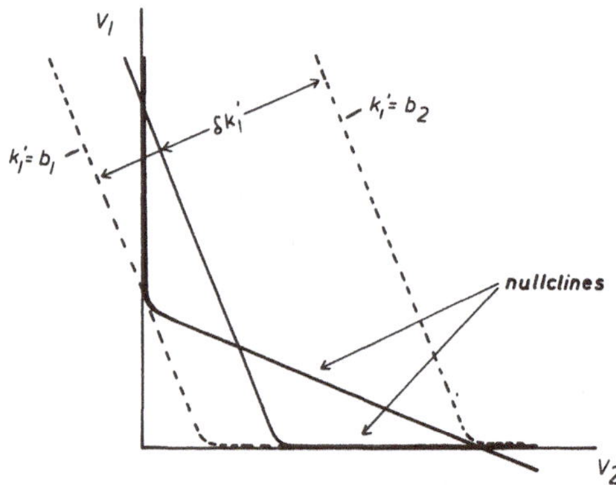

Fig. 7. Shifting of steady states in the system of Fig. 6, Eq. (6).

(A similar picture may be found in the book of Andronov et al., 1966.) The whole system, therefore, acts as a "hysteresis system" possessing an almost Z-shaped steady state curve: Fig. 8.

Speaking more distinctly, a whole family of systems is being parameterized by k_1', whereby adjacent systems possess the same qualitative behavior at all values of k_1' except for the bifurcation values b_1 and b_2. Due to the "structural stability" shown by the different systems under parametric perturbation at almost all values of k_1', the whole family of systems can be considered as a single system which preserves its qualitative structure under a sufficiently slow parametric change at all values of k_1' except those lying in the immediate neighborhood of b_1 and b_2. At these points, a "catastrophic" change of the system's

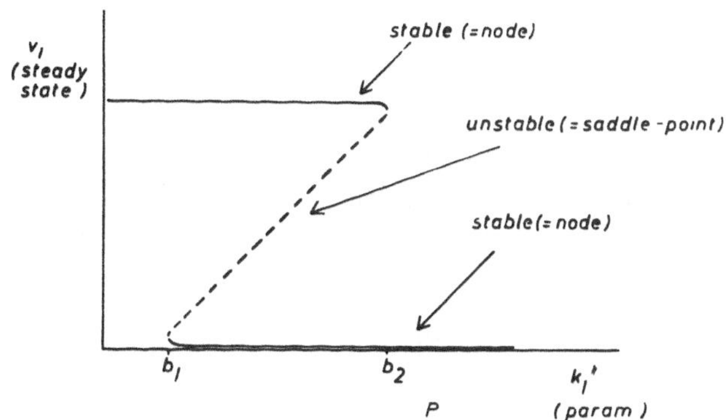

Fig. 8. Steady states of the system of Fig. 6 in dependence on the parameter k_1' (Eq. 6).

qualitative structure (and of the position of the next stable steady state which can be approached) occurs. Hence a hysteresis cycle (Fig. 9) is formed under a recurrent up- and down-variation of k_1'.

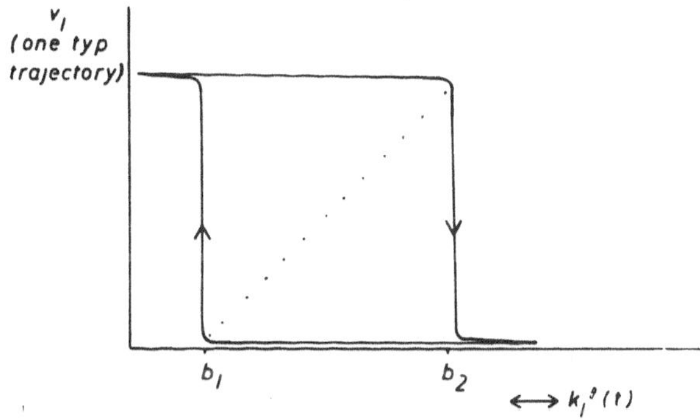

Fig. 9. Hysteresis loop shown by the system of Fig. 6 under a recurrent slow variation of the "parameter" k_1' (Eq. 6).

There are two events of "hard" bifurcation involved (see Part IV).

There exists a variety of time-courses of k_1' which all yield the same behavior of v_1 (quasi-steady state) as long as certain regions (being hatched in the Figure) are avoided: Fig. 10.

Among these inputs there may be very slowly changing ones, (so-called DC-inputs), and very fast-changing, reversible ones (i.e. so-called AC-inputs). Hence the trigger of Fig. 6 is both an AC- and a DC-input trigger (see Rössler, 1973).

Before proceeding to the electronic multivibrator as the next circuit to be realized by homogeneous chemical means, it is appropriate to ask whether the RS flip-flop of Fig. 6 cannot immediately be transformed

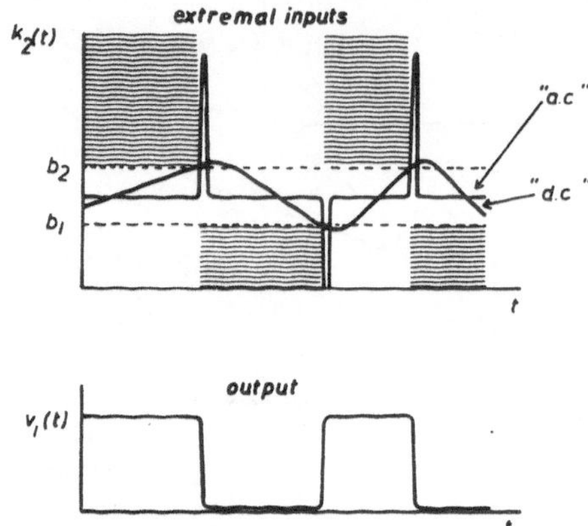

Fig. 10. Extremal time-courses of the "input-parameter" $k_1'(t)$ determining the same "output-behavior" $v_1(t)$ in the system of Fig. 6, Eq. (6).

into an astable flip-flop, i.e. an autonomously flipping "multivibrator". This is indeed trivially possible on the basis of the DC principle named: it is only necessary to connect the system to its own flipping-state via a slowly charging and discharging capacitor (substance): Fig. 11.

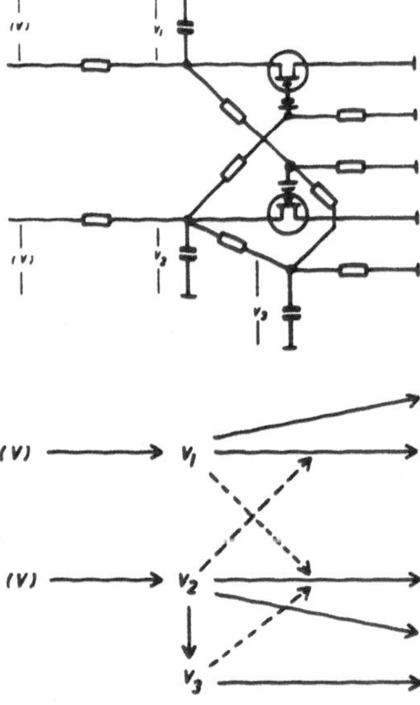

Fig. 11. An astable modification of the chemical Eccles-Jordan trigger and electronic equivalent.

The dynamical equations are

$$\dot{v}_1 = k_1' - k_2'v_1 - k_3v_2 \frac{v_1}{v_1+K}$$

$$\dot{v}_2 = k_1' - k_2v_2 - k_3(v_1+v_3) \frac{v_2}{v_2+K} \qquad (7)$$

$$\dot{v}_3 = k_4 v_2 - k_5 v_3,$$

where $k_1' = k_1 v$; $k_2' > k_2$; and k_4 and k_5 are relatively small.

Curiously enough, this multivibrating circuit has never been described in electronics (see below). If one considers instead the famous Abraham-Bloch multivibrator of electronics, which takes four elements less than this circuit (one capacitor more, 5 resistors less), one finds that the chemical analog is much more complicated than that of the former circuit, as evidenced by Fig. 12.

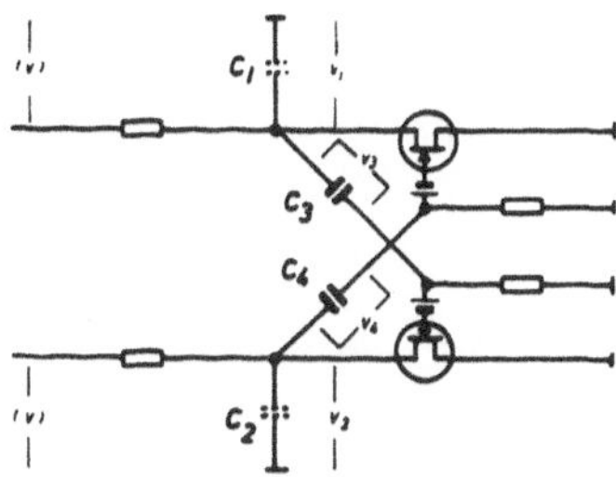

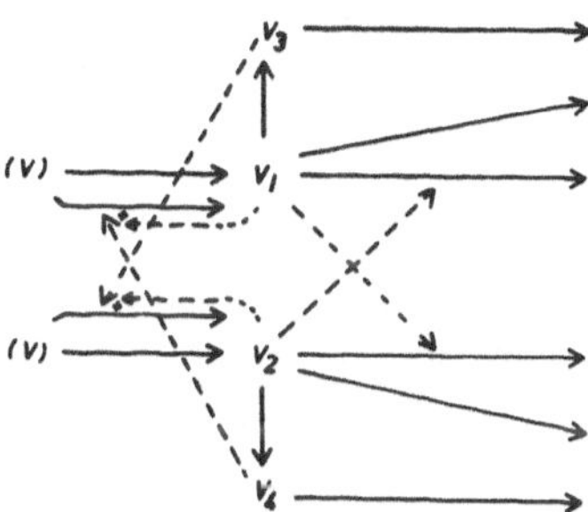

Fig. 12. A chemical Abraham-Bloch multivibrator and electronic equivalent.

It has the equations

$$\dot{v}_3 = k_5 (v_1 - v_3)$$

$$\dot{v}_1 = k_1' - k_2 v_1 - k_3 (v_2 - v_4) \frac{v_1}{v_1 + K}$$

$$\dot{v}_2 = k_1' - k_2 v_2 - k_3 (v_1 - v_3) \frac{v_2}{v_2 + K} \qquad (8)$$

$$\dot{v}_4 = k_5 (v_2 - v_4).$$

From the equations it is evident that the action of a coupling-capacitor, being one of the characteristic building elements of electronics, can be simulated by chemical means only in an indirect manner.

Due to the fact that the voltage across R on the other side of a coupling-capacitor C is defined by $v_R = v - v_C$, whereby v_C is approaching v by virtue of $\dot{v}_C = (RC)^{-1}(v-v_C)$, the connection can be replaced by a direct one. It is only necessary that v_C, the voltage of C being loaded via R, be subtracted from v (in the differential equation for v_1 where $v_1 >> K > 0$). The resulting two dynamically equivalent electronic circuits are shown in Fig. 13. The second circuit possesses a chemical analog, as depicted.

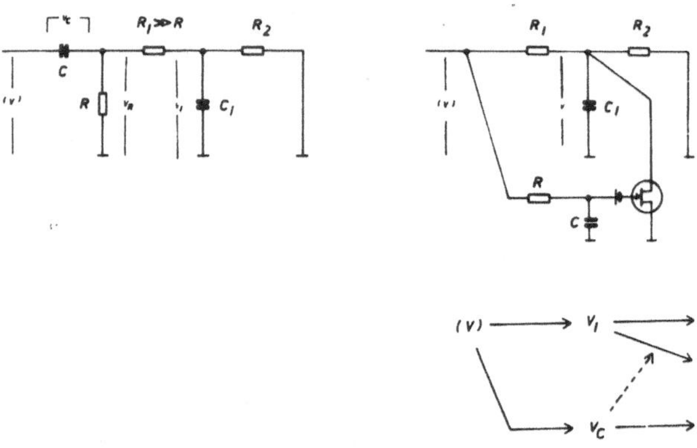

Fig. 13. A chemical circuit simulating an electronic circuit which contains a coupling capacitor.

Speaking more in general, the action of a coupling-capacitor can always be simulated by adding a delayed-acting compensation to a direct connection. This can be seen also from Eq. (8), where a direct cross-inhibition is provided between v_1 and v_2 by means of reciprocal outflux-catalysis, just like in the chemical Eccles-Jordan trigger (Eq. 6). The only difference is that, in the present circuit (Eq. 8), this cross-coupling is going to be suspended spontaneously through a delayed-acting compensation <u>via</u> v_3 and v_4, respectively. The activation of the catalyzed influxes by the respective products v_1 and v_2 up to saturation (see Eq. 8) guarantees that the input becomes effective only after the product has been raised above the value of K. Without this restriction, the circuit would tend to be globally unstable.

It is now evident how this chemical scheme can be simplified. Assuming both a relatively weak cross-inhibition and a small K, the self-activations can be omitted without appreciable error (supposing that global stability can be insured). In addition, one of the two "wings" of the circuit can be omitted now (because of the just additive effect of both wings in this mode of operation), hereby dispensing completely with the problem of global stability. The resulting one-wing scheme is, however, equivalent to the preceding circuit of Fig. 11, because an asymmetry of influx means the same as a reversed asymmetry of outflux: see Fig. 14.

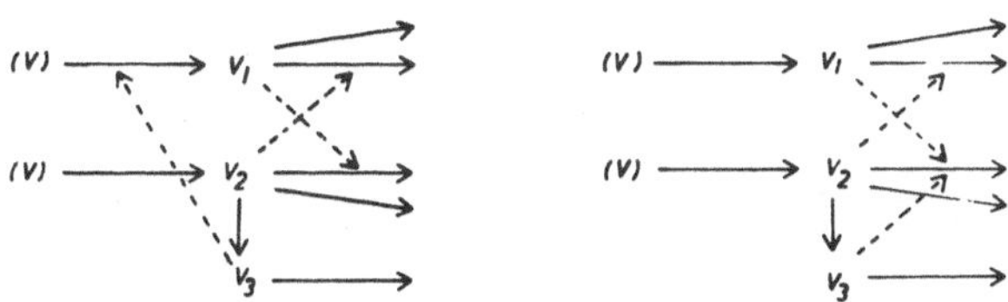

Fig. 14. "Modified chemical Abraham-Bloch oscillator" versus "self-triggering chemical Eccles-Jordan flip-flop": two equivalent systems.

By means of numerical simulation it can be demonstrated easily that the thus obtained scheme (left-hand side of Fig. 14) indeed shows the whole set of possible behaviors of the Abraham-Bloch multivibrator: astable running, monostable activity (as a so-called mono-flop), synchronizability, and T flip-flop behavior: Fig. 15.

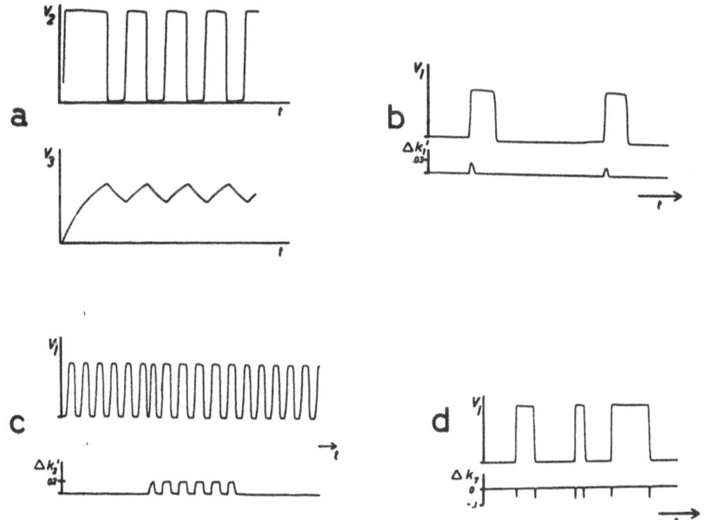

Fig. 15. Four behavioral modes of the modified chemical Abraham-Bloch multivibrator.

a = astable flip-flop
b = monostable flip-flop
c = synchronization
d = single-input (T) flip-flop

Bistability with enhanced sensitivity to triggering inputs can also be obtained (see Rössler, 1974b, for further details).

The same trick which has been used above in the modeling of the

Abraham-Bloch multivibrator (namely to substitute one electronic circuit for another, before beginning with the "translation") can be employed in other instances, too. As an example, a "translation" of the two functions of limitation and rectification (usually realized by a diode in electronic circuits) is to be mentioned. First the cutting-off of the lower portion of a voltage (or concentration, respectively) is considered: Fig. 16.

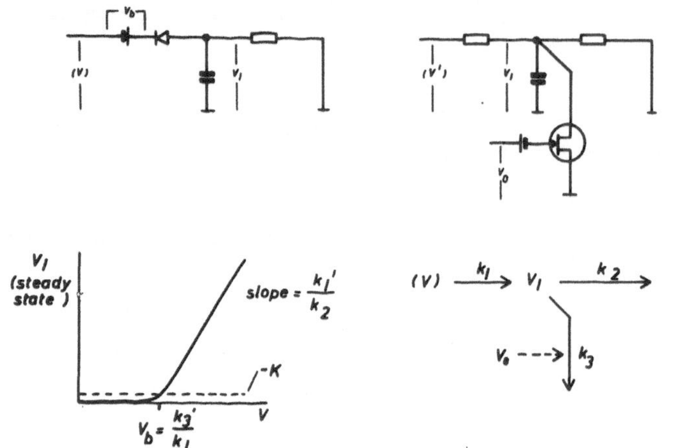

Fig. 16. A biased chemical rectifier and two electronic equivalents.

The corresponding differential equation is

$$\dot{v}_1 = k_1 v - k_2 v_1 - k_3' \frac{v_1}{v_1 + K'} \tag{9}$$

where $k_3' = k_3 v_o$. The changes of v have to be sufficiently slow when compared with the relaxation times of v_1, i.e. max $|\dot{v}|$ << max $|\dot{v}_1|$. It is evident that the two parameters of the depicted steady state curve, k_3'/k_1 and K, are mutually independent, so that an arbitrarily sharp "knee" of the hyperbola can be obtained by simply increasing the ratio between both values. An analogous result has been described be Licko (1972) without using the electronic analogy.

The dual case of upper limitation is shown in Fig. 17.

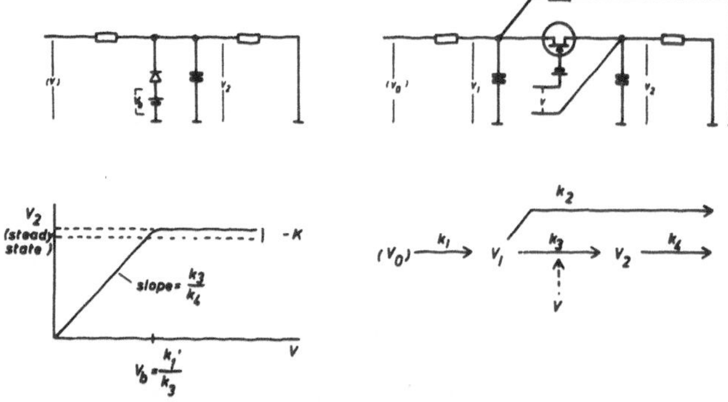

Fig. 17. A chemical limiter and two electronic equivalents.

Two differential equations apply in this case,

$$\dot{v}_1 = k_1' - k_2 v - k_3 v \frac{v_1}{v_1+K}$$
$$\dot{v}_2 = k_3 v \frac{v_1}{v_1+K} - k_4 v_2, \quad (10)$$

where $k_1' = k_1 v_o$. Again, an arbitrarily sharp knee can be obtained for the steady state curve of v_2 in dependence on the parameter v, by simply increasing the ratio of k_1'/k_3 to K.

The just-described two simple circuits (Fig. 16 and 17) may be of interest in the generation of impulses out of sinusoidal chemical oscillations, or even in the rectification of an amplitude-modulated chemical oscillator.

Before concluding this part, two further analogies between an electronic and a chemical building-element may be pointed out. First an analog to a tunnel diode (or a vacuum-tube tetrode, respectively) which possesses a so-called N-shaped output characteristic is considered (cf. Franck, 1965): Fig. 18.

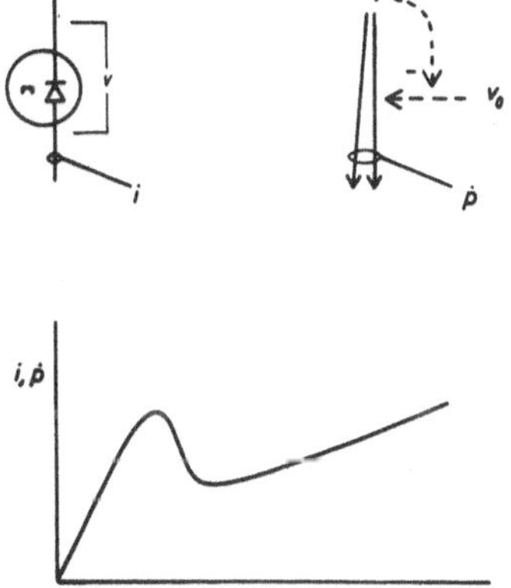

Fig. 18. A chemical nonlinear resistor possessing an N-shaped characteristic and electronic analog (tunnel diode).

An immediate application is the bistable tunnel diode-circuit which is an asymmetrically-built hysteresis system (RS flip-flop), as shown in Fig. 19. The corresponding differential equation has the form

$$\dot{v}_1 = k_1' - k_2 v_1 - k_3' \frac{v_1}{v_1+K+K'v_1^2}, \quad (11)$$

where $k_1' = k_1 v$ and $k_3' = k_3 v_o$.

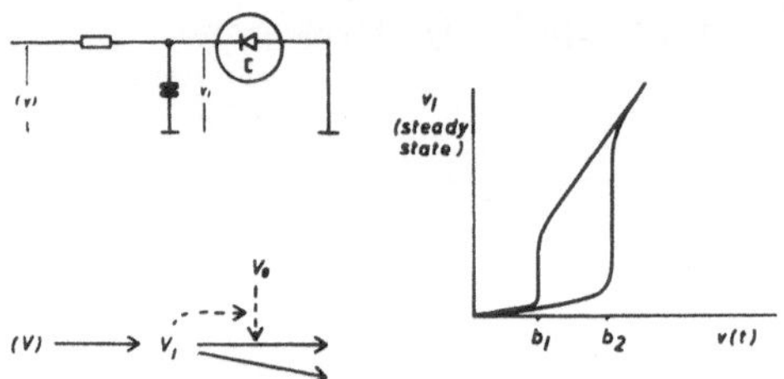

Fig. 19. A chemical system analogous to a tunnel diode bistable system.

This abstract chemical system has been detected independently by several authors (Kettner, 1969; O'Neill, 1971; Rössler, 1972; Seelig & Denzel, 1972). Most recently, all possible explicit versions of this system, based on mass-action type elementary reactions, have been treated in an exhaustive way (Denzel, 1973).

As a final example in this section, a chemical analog to a glow tube (e.g. neon bulb) is considered: Fig. 20.

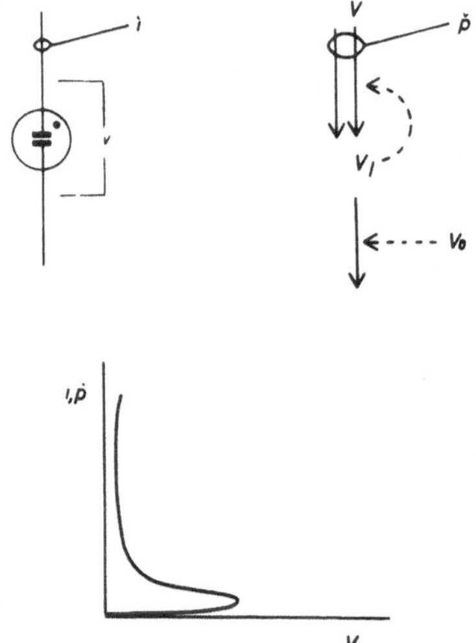

Fig. 20. A chemical system simulating the dynamical behavior of a glow-tube.

Here a whole circuit replaces the single electronic element. (Nonetheless a similar dynamics may apply internally to the glow tube, too.) The corresponding equation reads:

$$\dot{v}_1 = k_1' + k_2'v_1 - k_3' \frac{v_1}{v_1+K} , \tag{12}$$

where $k_1' = k_1v$, $k_2' = k_2v$, and $k_3' = k_3v_o$.

This differential equation describes a chemical "switch", as the system has been called (Rössler, 1972a). Any triggerable chain reaction, defined by a balance between chain propagation and chain termination, is described qualitatively by this equation. (The first right-hand term in Eq. (12) refers to "chain initiation", the second to "chain propagation", and the third to "chain termination".)

An immediate application of the last-outlined correspondence can be seen in the chemical analog of the well-known <u>neon bulb sawtooth oscillator</u> (Fig. 21),

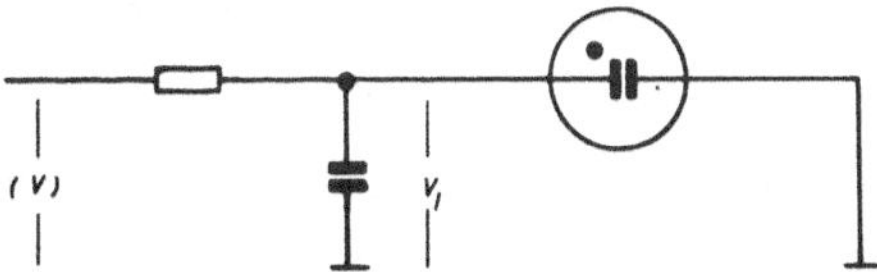

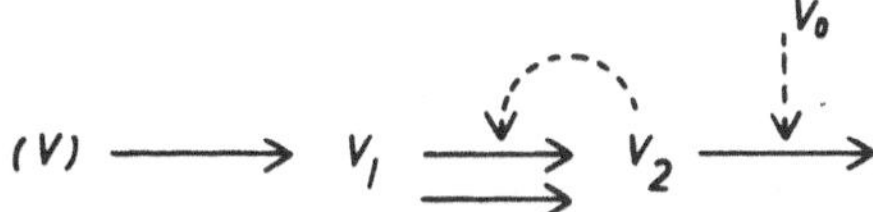

Fig. 21. A chemical sawtooth oscillator and electronic equivalent (glow tube oscillator).

possessing the equations

$$\dot{v}_1 = k_1' - k_2 v_1 - k_3 v_1 v_2$$
$$\dot{v}_2 = k_2 v_1 + k_3 v_1 v_2 - k_4' \frac{v_2}{v_2 + K} \quad , \tag{13}$$

where $k_1' = k_1 v$, $k_4' = k_4 v_o$.

For a sufficiently small k_1', v_2 shows the well-known sawtooth behavior, whereas v_1 produces spikes at the moments of quick relaxation: Fig. 22.

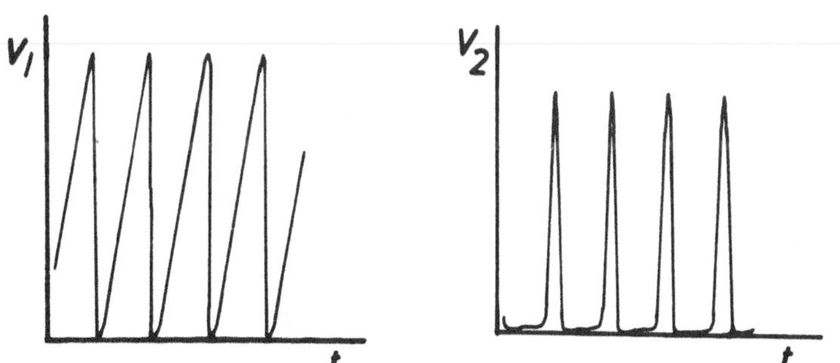

Fig. 22. Simultaneous sawtooth and spike behavior in the system of Fig. 21.

The first real chemical oscillator possessing this mode of action was described in 1834 by Rosenschöld: In the gas phase of a bottle which contains yellow phosphorus kept under water, and into which atmospheric oxygen is allowed to diffuse in very slowly, periodic light flashes of chemoluminescence are produced. The involved chain reaction in the gas phase has not yet been elucidated completely (cf. Degn, 1972).

Part II: Homogeneous Chemical Analogs to Neural Behavior

First some preliminary considerations. A nerve membrane is an "excitable system", that is one which, following the application of a sufficiently strong appropriate stimulus, shows a stereotypic, autonomously proceeding response. This property is typical for a mono-flop (cf. Fig. 15b).

The simplest mathematical system showing the qualitative behavior of a mono-flop is Liênard's (1928) equation, modified by the addition of a constant non-zero term α:

$$\dot{x} = -x^3/3 + x + y$$

$$\dot{y} = -\mu(x - \alpha), \qquad (14)$$

where $\mu \to 0$ and $\alpha = 1.01$ (e.g.).
The behavior of this system is readily understood: The first equation obviously describes a hysteresis system (i.e. RS flip-flop) whereby y serves as the hysteresis-eliciting parameter (see the cubic-shaped nullcline in Fig. 23).

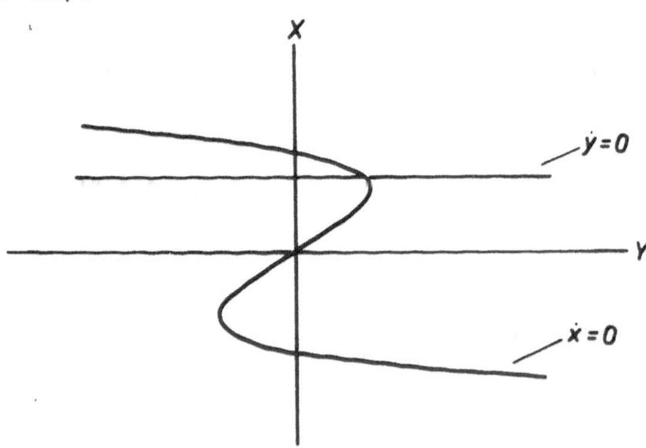

Fig. 23. Nullclines of Eq. (14).

The second equation determines the motions of the "slowly changing parameter y" (so to speak) in dependence on the actual state of x. The position and hence stability properties of the steady state (intersection point of nullclines in Fig. 23) determines the overall behavior just as in the previously discussed 3-dimensional case (Figs. 14 and 15). For $|\alpha| < 1$, an astable multivibrator is obtained (whose mode of action is completely analogous to that of the above-described electronic and chemical multivibrators). For $|\alpha| > 1$ (but close to 1), a monostable multivibrator is obtained (again in complete analogy to the three-component paradigm; Fig. 15b). Eq. (14) has the advantage that it is simpler to handle analytically, and indeed there does not exist any other nonlinear differential system better known than Eq. (14). See, for example, the textbook by Reissig, Sanzone, and Conti (1963). However, relatively little mathematical work has been done on the monostable version of this system (α being greater than one, but close to one; see also Zeeman, 1972). The equation seems to have been invented by Karl Friedrich Bonhoeffer (1943) who proposed it as a nerve model, but did not write down the equation explicitly. Later on, FitzHugh (1961) arrived at practically the same equation in his attempt to reduce, by mathematical means, the number of variables in the well-known phenomenological Hodgkin-Huxley (1952) equations of nerve action from four to two. In FitzHugh's equation, an additional term containing $\underline{y}$ is introduced into the second equation, without any resulting change of qualitative behavior. FitzHugh proposed the name Bonhoeffer-van der Pol (BVP) model for the equation; indeed Liénard's equation can be transformed into van der Pol's equation by means of a nonlinear transformation.

Due to the fact that any chemical hysteresis system, when complemented by a slowly changing triggering variable, gives rise to a chemical multivibrator of the same type as Eq. (14), there exist at least so many chemical nerve analogs as there are chemical hysteresis systems. In the preceding electronic section, two chemical hysteresis systems were offered: the homogeneous Eccles-Jordan trigger, and the homogeneous tunnel diode circuit. We add here that a third specimen is obtained by means of a modest modification of the chemical glow tube analog described above: Fig. 24. It was first described in an explicit, less general version by Edelstein (1971).
The corresponding differential equation is

$$\dot{b} = k_1' + k_2'b - k_3' \frac{b}{b + K} - k_4 b^2, \qquad (15)$$

whereby $k_1' = k_1 a$, $k_2' = k_2 a$, and $k_3' = k_3 c$.

The introduction of the third right-hand term in the equation, limiting the possible growth of $\underline{B}$, changes the nullcline into a cubic. (As a

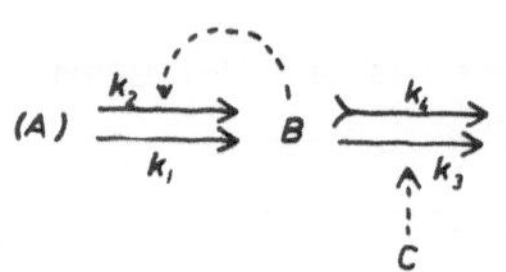

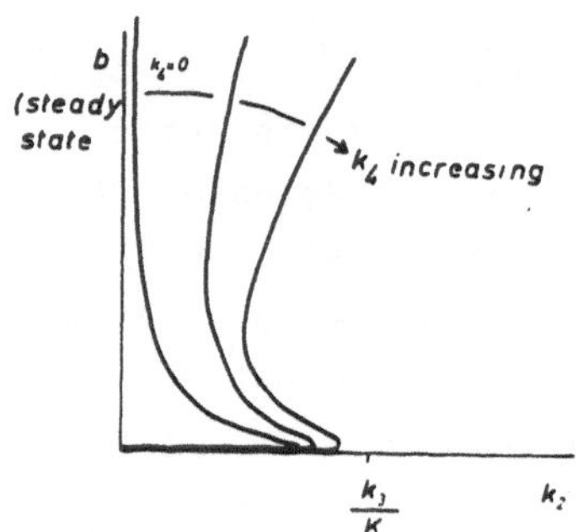

Fig. 24. Another chemical hysteresis system (modification of the system of Fig. 20).

matter of fact, the characteristics of the glow tube have a similar form, if sufficiently large values of B are taken into consideration. However, this has been of no importance in the previous context.) Further examples of more or less complicated chemical RS flip-flops could be added (indeed another two will be encountered below in the context of self-differentiating chemical systems).

The reaction schemes of those three mono-flops which can be obtained on the basis of the so far described three hysteresis systems are shown in Fig. 25. The thick arrows mark relatively slow reactions (in contrast to the usual convention).

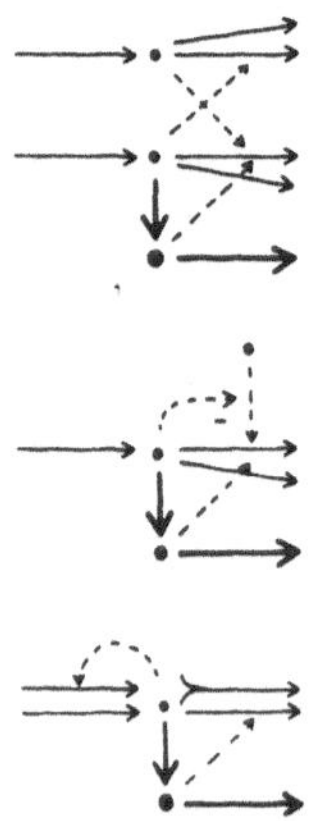

Fig. 25. Three chemical monoflop circuits.

We now come to a second class of chemical nerve analogs which are not based on the hysteresis principle, so that Eq. (14) no longer constitutes the mathematical prototype. This is possible because mono-flops can in general be devised not only on the basis of multivibrators, but equally well on the basis of sawtooth oscillators. (Even more generally speaking, they can be derived from any self-excited oscillator whose limit cycle involves the slow transgression of at least one bifurcation-threshold of a sub-system.) As an example, the glow tube oscillator is to be reconsidered (Fig. 26).

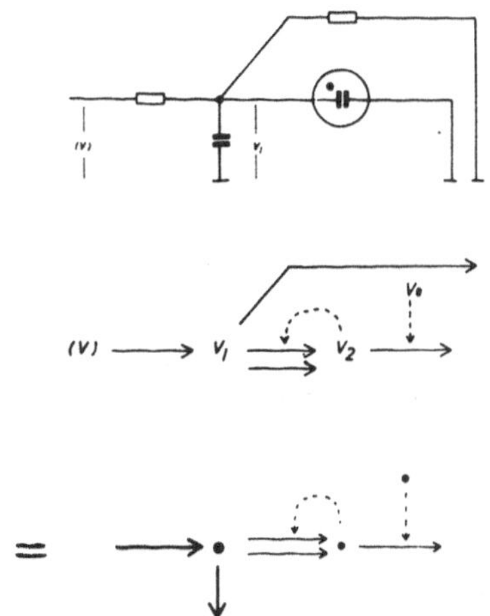

Fig. 26. A single-threshold chemical monoflop and electronic equivalent (cf. Fig. 21).

The additional resistive load to the capacitor C may cause the re-charging process to stop short just before the bifurcation threshold of the consecutive element is being reached. This modification renders the sawtooth oscillator (or, synonymously, "time-base generator") a so-called single-sweep time-base generator (Chirlian, 1971). Of course, any other "non-resettable chemical switch" (Rössler, 1972a) might replace the one employed in Fig. 26. This yields the following series of circuits (Fig. 27).

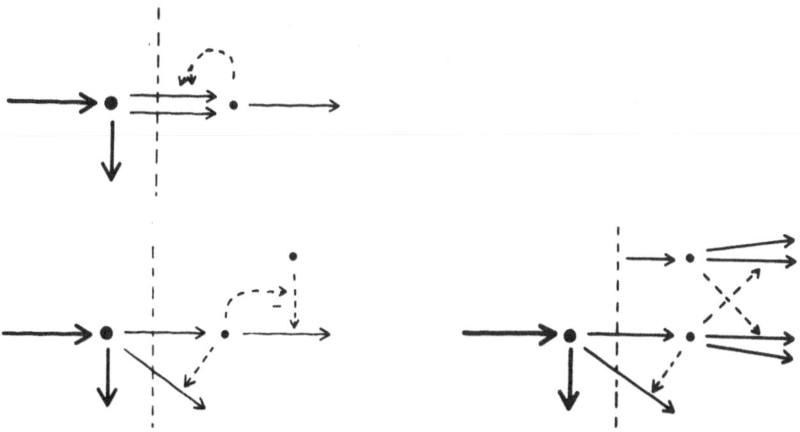

Fig. 27. Three additional single-threshold chemical monoflops.

The switch employed in the first example of the series is especially simple from the mathematical point of view (although its chemical realizability is not straightforward, since it involves, when translated into an explicit scheme, a whole network of second-order reactions standing for the nonchemical third-order reaction assumed). The switch has the equation

$$\dot{x} = x^2 - x ,$$

so that the whole system is described by:

$$\begin{aligned} \dot{x} &= x^2 y - x \\ \dot{y} &= \mu - x^2 y - \alpha y \end{aligned} \tag{16}$$

(if the "leak-reaction" is omitted for the sake of simplicity). This equation is the mathematical prototype of the presently considered class of systems. (An explicit version has been considered by Karfunkel & Seelig, 1972). Unlike Eq. (14), the exotic behavior of this system is confined to the positive quadrant of state space even in the mathematically simplest case.

Since any resettable chemical switch (i.e. simple hysteresis system) also involves an "explosion phase", these systems are equally appropriate for the generation of a sawtooth type mono-flop, when employed as a load to a slowly charged chemical capacitor. (An example is the already discussed glow-tube oscillator when a second-order outflux term is added to the second right-hand side of Eq. 13). And of course, any resettable chemical switch can, through the deletion of some parts, also be changed into a non-resettable switch. (But this change is unnecessary, as just outlined.) Hence in general at least <u>four</u> different mono-flops can be derived from any single chemical hysteresis system: two of square-wave type, and two of sawtooth (or, equivalently, spike) type. The number can be raised up to six, if the possibility to generate "negative sawtooth oscillations" (by deleting the other threshold) is also taken into account.

After having shown the existence of many abstract possibilites of how to arrive at a homogeneous chemical nerve analog, it is worthwhile to ask whether <u>real</u> chemical systems falling into the described classes also exist. Indeed two examples can be mentioned. The first is the already described sawtooth oscillator in the gas phase of Rosenschöld: there is little doubt that this system can be rendered a mono-flop by adding a mild oxygen-absorbent (corresponding to the additional load to the first variable in the neon-bulb oscillator of Fig. 26). The second example is the well-known fluid oscillator of Belousov (1959) and Zhabotinsky (1964) for which an excitable (i.e. triggerable) condition has been found experimentally (Winfree, 1972; 1974). Despite the fact that the involved reaction network has not yet been elucidated into the last detail (although Noyes et al., 1972 and Field et al., 1972, have done impressive work), it is well established that the system contains an autocatalytic system (Vavilin & Zhabotinsky, 1969) which is potentially bistable as a subsystem (Rössler & Hoffman, 1972; Rössler, 1973). This subsystem can be switched on and off by the addition of bromide or bromide-releasing substances (cf. the experiments done by Vavilin & Zhabotinsky, 1969). In addition, bromide seems to be part of a regenerative cycle in the overall reaction (Rössler & Hoffman, 1972), so that it acts as a "generalized catalyst" (see Rössler, 1971, for this notion). This leads to the following proposed reaction chart (Fig. 28; Rössler & Hoffman, 1972).

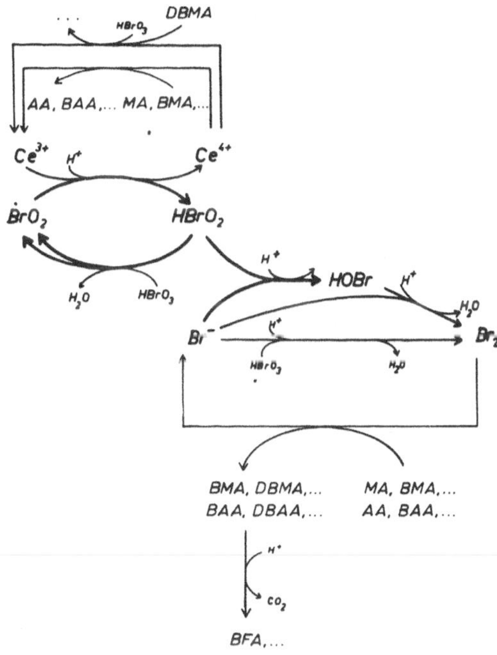

Fig. 28. A possible representation of the Belousov-Zhabotinsky reaction network.

Even if not all details of this half-schematic Figure will prove correct, it seems permissable to conclude that the system is a hysteresis type mono-flop of the same class as the prototypical system, shown in Fig. 25, lowest part (Rössler, 1972b). By historical reasons, systems of this structure deserve the name "Bonhoeffer multivibrators" (Rössler & Hoffmann, 1972), so that the Belousov-Zhabotinsky reaction becomes a "Bonhoeffer fluid".

We now proceed to chemical analogs of whole neural circuits. The following homogeneous chemical circuit (Fig. 29) shows "classical conditioning" in the sense that after a series of (almost) coincident events of exogeneous stimulation (mono-flop I being stimulated by a shortlived rise of S_1, and II by S_2), stimulation of mono-flop II alone will also result in the elicitation of a response of monoflop I (conditional response). The duration of the "conditioned" state depends on the outflux rate constant of C and the number of prior simultaneous stimulations.

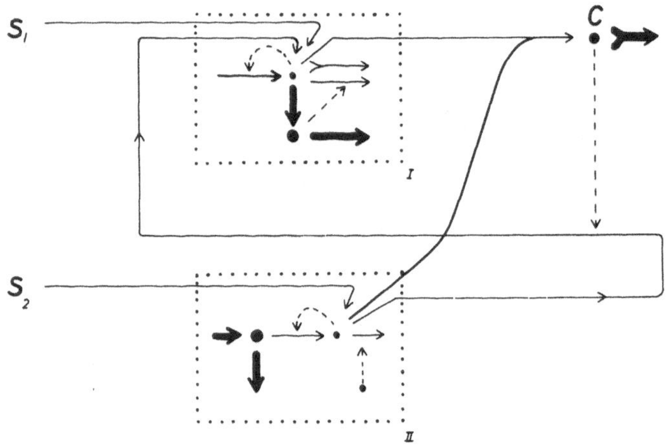

Fig. 29. A "conditionable" homogeneous abstract reaction system.

Abstract neural circuits showing the same behavior, as for example the one suggested by Walter (1953), usually involve more than 2 mono-flops (neurons) for two reasons: (i) the analog-storing device (substance C) of the chemical scheme is assumed to be realized by another neuron (in the form of a reverberator) being in either of two discrete dynamical states, rather than by a (two-input) modifiable synapse which would correspond directly to C; (ii) the incorporation of two additional neurons allows for an adjustment of the relative durations of the specific delays of both the nonconditioned and the conditioned response to realistic values. The "nonrealistic" chemical scheme

(Fig. 29) thus corresponds to the abstractly simplest situation.

Considering for a moment the problem of actual realization, evidently any homogeneous "neuron" (monoflop) may be inserted for I and II. Apart from the excitable Belousov-Zhabotinsky reaction (Winfree, 1972) which might realize box I, an analogous modification of Bray's (1921) oscillator (which in fact has not been looked for) may be appropriate for box II. However, the problem of restricting and canalizing the interaction of both excitable systems in the indicated (or some equivalent) way will impose additional and, probably, unsurmountable difficulties. Thus, the paper-and-pencil model (Fig. 29) will remain unmatched by chemical reality in the near future.

This notwithstanding, the conclusion that real chemical systems simulating neural networks are in general nonfeasible would be misleading. For there exists another type of chemical "wiring" (and "resistors"), based on diffusion-coupling between identical substances, rather than on reaction-coupling between different substances. Since one and the same reaction scheme applies throughout in the whole reaction volume , the realizability is straightforward and almost trivial, once the corresponding homogeneous reaction is found.

Indeed the excitable Belousov-Zhabotinsky-Winfree reaction provides a physico-chemical nerve analog which can be manipulated experimentally not only in the homogeneous state. By allowing for diffusion, models of excitation-propagation in one, two, and three dimensions are obtained. The two-dimensional model has a certain medical interest because it can be used to test theories on normal and abnormal excitation propagation in the heart-muscle. The theory of a so-called excitable medium was detected as an area of mathematical research by van der Pol and van der Mark (1928); it was later resumed by Wiener and Rosenblueth (1946). Analytical techniques are hardly applicable to the resulting set of parabolic differential equations. The following one forms a sort of prototype system:

$$\frac{\partial x}{\partial t} = (1 - \mu)x - y\,\frac{x}{x + K} + D_1\nabla^2 x$$

$$\frac{\partial y}{\partial t} = \mu(x - ky) + D_2\nabla^2 y, \qquad (17)$$

where, e.g., $K = 0.1$, $\mu = 0.01$, $D_1 = D_2 = 0.003$, $k \approx 0.03$.

The equation refers to the nonstirred version of the reaction system of Fig. 25, lowest part (Rössler, 1972b) and hence corresponds to the simplest "Bonhoeffer medium".

It can be expected that with the rising effectivity of numerical simulation methods, a better understanding of the spatio-temporal pattern observed in the Belousov-Zhabotinsky reaction will become possible. In the numerical simulation of arrays of excitable systems, so far (with one exception -- hybrid computer; Gul'ko and Petrov, 1972 --) only discrete models of excitable systems have been investigated (Krinsky, 1966). Hereby the famous two-dimensional reverberator, consisting of a turning spiral which (like the grooves on a record disk) is the involute of a circle, could be found again. A generalization of the reverberator to 3 dimensions, the so-called scroll, was recently detected by Winfree (1973) in the Belousov-Zhabotinsky reaction, when examining the 3-dimensional excitable reaction in a stack of Millipore filter sheets.

A theory on the qualitative behavioral differences between discrete and continuous monoflops, and on the influence of different types of continuous monoflops and of different degrees of coupling, etc., could not yet be attempted, due to the lack of flexible computer programs for reaction-diffusion systems. Most recently, however, Karfunkel (1974) and Winfree (personal communication) have obtained such programs.

Apart from the heart muscle, there is another biological system whose activity is simulated by the two-dimensional Zhabotinsky reaction: the spatial patterns observed during the morphogenetical aggregation of cellular slime molds (Gerisch, 1971). Here an intracellular monostable (or astable) chemical system seems to be coupled by extracellular diffusion. The resulting wave-patterns are used as locomotive clues, whereby a Doppler-effect comes into operation which is absent in the Belousov-Zhabotinsky reaction (Gerisch, 1971). A first simulation result was recently obtained by Novak and Seelig (1974).

The possible role of three-dimensional excitable media as analogs of neural networks will be discussed in Part IV.

Part III: Specifically Chemical Exotic Systems

We now come to the consideration of two prototypical reaction systems which have no immediate electronic or neural counterpart. Both deal with a sort of differentiation, one being homogeneous and the other inhomogeneous.

The first system has been proposed as theoretical evidence for the possibility of spontaneous generation of strong optical asymmetry in a reaction mixture of racemic substances. It is presupposed that the optical antipodes already possess (at least moderately) a selective autocatalytic activity: Fig. 30.

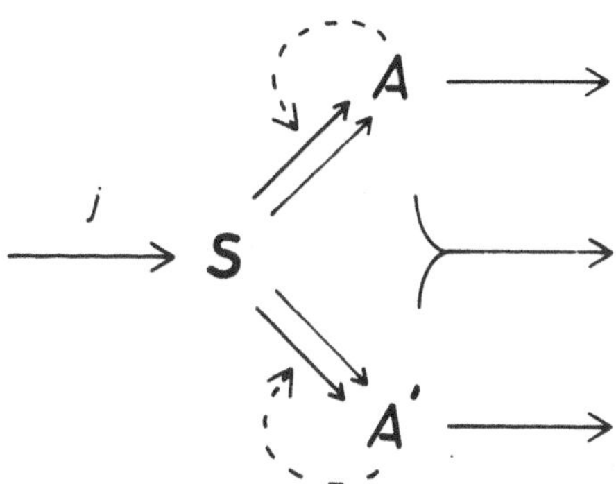

Fig. 30. A "self-differentiating" homogeneous reaction system (conditional flip-flop, see text).

The system has been invented in different versions independently by several authors (Frank, 1953; Kacser, 1960; Seelig, 1971; Decker, 1973). In the present form, which is due to Seelig (1971; slightly simplified), the system is a conditionally bistable system, depending on the influx $\underline{j}$ to the common substrate. Beyond a certain value of $\underline{j}$, the symmetrical steady state becomes unstable (a saddle-point), being flanked by two mirror-inverted stable steady states (nodes). This behavior follows easily from a computation of the eigenvalues of the Jacobian matrix of the system at the symmetrical steady state, if the steady state value of $\underline{S}$ is considered as a parameter, so that the system becomes two-dimensional. Any symmetrical two-dimensional system showing both cross-inhibition (i.e. $\frac{\partial \dot{A}}{\partial A'} = \frac{\partial \dot{A}'}{\partial A} < 0$) and self-inhibition ($\frac{\partial \dot{A}'}{\partial A'} = \frac{\partial \dot{A}}{\partial A} < 0$) at a symmetrical steady state exhibits the transition from a stable node to a saddle-point (i.e. from two negative eigenvalues to a negative and positive one), whenever the numerical value of cross-inhibition exceeds that of self-inhibition (in ecology the same principle is known as Gause's principle; Gause, 1934). This is exactly what occurs at a certain bifurcation value of $\underline{j}$. The resulting picture (Fig. 31) corresponds to the typical case of "soft" bifurcation in dynamical systems (Rössler, 1973). This type of bifurcation is distinguished by the preservation of continuity between stable steady states on this side and beyond the bifurcation value. In terms of Thom's (1972) catastrophe notation, the same behavior amounts to a "silent catastrophe", in the sense that, for a sufficiently slow change of $\underline{j}$, no "jump" is observed. In terms of Zeeman's (1972) "cusp" notation, on the other hand, the present bifurcation parameter $\underline{j}$ points

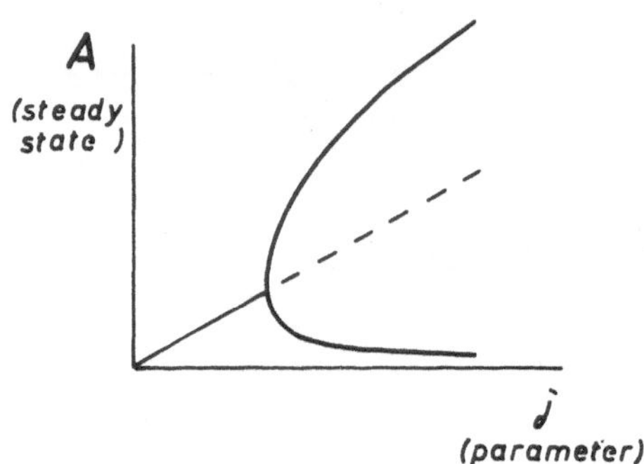

Fig. 31. Bifurcation of steady states in dependence on the parameter j in the system of Fig. 30. (Cf. Fig. 8).

into the direction of, and goes through, the edge of the cusp (where the overhanging "cliff" disappears). This cusp visualization is particularly natural since it shows at the same time that the same system must with respect to another, perpendicular parameter (namely bias of influx), show the behavior of an RS flip-flop (supposing that j is held fixed at a supra-threshold value), involving two hard bifurcations rather than a single soft one. Indeed any RS flip-flop (see also Figs. 6, 19, 24, and 32) possesses a cusp-type manifold of steady states and is, therefore, dynamically equivalent to the system of Fig. 30.

The second system to be considered here is the Rashevsky-Turing model of morphogenesis. It consists, in the simplest case, of the following reaction scheme (Fig. 32). The indicated two-compartment version is interesting in its own right, if interpreted as a model of cellular differentiation, but can be considered equally well as the simplest approximation of a more general case which is described by partial differential equations.

It is first observed that the system consists of two potentially oscillating systems coupled by selective diffusion. Indeed either partial system can be run as a two-component, "linear" chemical oscillator, for sufficiently small K, as is easily seen from the linear differential system which results for all values of a and b >> K. (This oscillator, too, could be regarded as a typically chemical exotic circuit, because a two-component RC-oscillator is unknown in electronics.) However, the potential oscillatory activity of the system plays no role in the present context.

The composed system (Fig. 32) is a soft-bifurcation system, again,

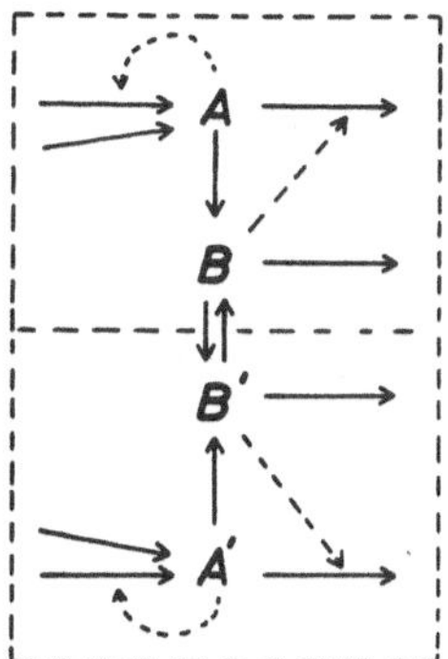

Fig. 32. The simplest Rashevsky-Turing type morphogenetical two-cellular system.

i.e. a conditionally bistable system, in dependence on the selective diffusion (or permeability) coefficient for B acting as a parameter. Therefore, the system constitutes a possible model for differentiation in chemical and cellular systems, since either of the two stable steady states generated refers to a distinct state of spatial differentiation.

This statement can be verified by computing the eigenvalues of the four-dimensional Jacobian matrix in the symmetrical steady state (Turing, 1952; Rössler & Seelig, 1972). A more intuitive explanation can be found in the following way: since only steady-state considerations are at stake the variables B may be conceptually contracted, so that A is considered to catalyze its own outflux by itself (via B). Then B just functions to separate A's autocatalytic activity from its own auto-inhibitory activity. Thus the latter activity is mediated in part to the other side, as cross-inhibition (via the diffusion between B and B'), whereas the former activity is insulated against mediation. Hence the above-mentioned two-dimensional bifurcation law for the determination of a cusp in a symmetrical system ("numerical value of cross-inhibition exceeding that of self-inhibition") can be fulfilled once more, at a certain value of the selective diffusion coefficient.

Historically speaking, an abstract morphogenetical system possessing very similar equations and the same mode of action was proposed first by Rashevsky (1940) as a model of polar differentiation within cells. In 1952, Turing independently devised the indicated system as a prototype model in his attempt to elucidate "the chemical basis of morphogenesis" by means of a dynamical metaphor (cf. Rosen, 1970). (The prototype system -- Fig. 32 -- is described only verbally on

p. 43 of Turing's paper. This fact and the ensuing discussion of the system's behavior in the linear neighborhood of the steady state has often caused a misapprehension of the system and the whole theory as being "linear". However, a variety of other, even "more" nonlinear systems showing the same behavior are also mentioned in the same paper.) Most recently, very similar equations have been proposed in an attempt to model hydra morphogenesis (Gierer & Meinhardt, 1972, their Eqs. 12-16). Finally, the general morphogenetical significance of bifurcations of $\underline{V}^4$-type potentials has been redetected by Thom (1970, 1972). A $\underline{V}^4$-type catastrophe refers to the "silent" generation of bistability out of global stability (as schematically illustrated in Fig. 33), leading to the steady-state picture of Fig. 31 ("soft bifurcation" or, to use Turing's term, "evocation").

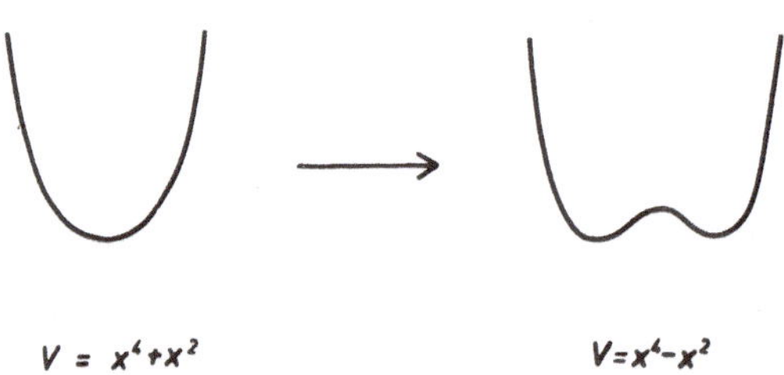

Fig. 33. Potential-theoretic description of soft bifurcation.

It is hereby understood that the dynamical system in question can be interpreted as a gradient field generated by a scalar function (like the one shown). This technique was first introduced into the theory of physico-chemical systems in a little-known paper by Rashevsky (1929; see also Rashevsky, 1960). Thom succeeded in applying the same technique to gradient-type partial differential systems, by defining boundaries (catastrophic sets) between those spatial domains showing the same local qualitative behavior (cf. Rosen, 1970).

The above-discussed two-cellular cusp-type system (Fig. 32) can, like any other chemical RS flip-flop, easily be complemented to a two-cellular "morphogenetical multi-vibrator" (Fig. 34) in accordance with the formerly described rules. (It corresponds to a symmetric version of the system of Fig. 11.) An existence proof for an at least eight- (rather than six-) component morphogenetical oscillator has been given by Smale (1972). The morphogenetical outcome in the more than two-cellular (and partial) case has not yet been investigated. One could expect an ordered "pattern reversal" to occur from time to time.

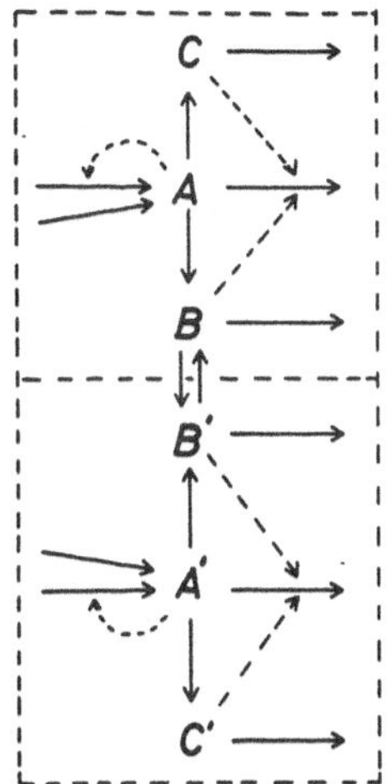

Fig. 34. An astable morphogenetical flip-flop ("traffic light").

As a curiosity it may be mentioned in this context that the Rashevsky-Turing system itself acts not only as a conditional flip-flop and as an RS flip-flop, but also as a T flip-flop, i.e. as a passive morphogenetical oscillator which drops from one differentiation state into the other every time an appropriate input is exogeneously applied. The input consists of a shortlived reduction of selective permeability for B (Rössler & Seelig, 1972). This is an example where a dynamical system designed for one purpose reveals, in addition, a rather unexpected and almost opposite behavioral capability.

Part IV: Two Possible Applications

Two sorts of applications will be considered: the possible design of homogeneous reaction systems showing a complicated prescribed behavior and the possible significance of "fluid neurons" for the theory of neural nets.

The existence of chemical systems which are triggerable by means of slowly varying substances enables the formulation of a building-block principle applying to the design of complicated reaction systems (Rössler, 1973). Just as the properties of a multivibrator can be predicted on the basis of knowing its components, namely a hysteresis system and a slowly charging and discharging variable (Figs. 14, 25), it is possible to determine the behavior of larger or even very large systems, *a priori*, when these systems are also composed of triggerable subsystems and slowly changing connecting variables. (The behavior of the whole system then can be interpreted as a concatenation of cusp type catastrophes; cf. Güttinger, 1974.)

Part of the larger systems obtained with the aid of this building-block principle even admit a much simpler description, without loss of any "essential" information, in terms of automata theory (see Rössler, 1974a).

The very existence of this class of "decomposible" chemical (and dynamical) systems is highly astonishing in view of the well-known fact that the majority of dynamical systems react in a counter-intuitive way to the addition or deletion of elements (Rosen, 1970).

A second possible application involves the question of how the study of simple reaction systems showing "exotic" behavior may contribute to the understanding of the central nervous system or some simply organized parts of it.

It is, of course, the class of fluid nerve analogs whose consideration appears to be most promising. As we have seen, the nonstirred two-dimensional Zhabotinsky reaction is a fluid analog to the heart-muscle. In the same manner, the nonstirred three-dimensional reaction does, in the form of Winfree's excitable modification, provide a functional analog ("analog computer") for a primitive neural net, being (like the heart-muscle) characterized by next-neighbor connections. A considerably more complicated model of a neural net would be obtained if a mixture of two (or more) weakly coupled triggerable liquids could be formed. The hypothetical neural analog of such an _n_-mixture would not possess a trivial connection scheme any longer. It is tempting to speculate about a possible triggerable modification of Bray's (1921) iodine-hydrogen peroxide reaction as a possible second constituent.

Presently it is very hard to predict the possible behavioral outcomes of such a 2-mixture. Even the spatio-temporal behavior of the simple Belousov-Zhabotinsky-Winfree reaction alone has not yet been classified according to the types of possible attractors (cf. Winfree, 1974). The reason lies in the fact that a qualitative theory of this class of reaction-diffusion systems is still lacking. Thom's (1970) theory is not immediately applicable, because only slowly changing bifurcation parameters are admitted, whereas a fast triggering, occurring between adjacent mono-flops, is the characteristic of the present class of systems. The preceding remark does not mean, however, that an attempted application of Thom's theory to neural nets made up of continuous formal neurons (Rössler & Hoffmann, 1972) will not prove rewarding. Cowan (1972) postulated a "neural field theory", being based on Bonhoeffer-type partial differential equations (in the present terminology), which also allows for catastrophical shifts between different regimens of qualitative behavior. An analogous simulation result ("hysteresis behavior") has been obtained in discrete neural nets by Anninos (1972). Thus a "topological theory of the brain" (Zeeman, 1965) begins to emerge in operational and even in chemical terms.

Concluding Remarks

The competence of chemical kinetics for the generation of dynamical systems showing nontrivial ("exotic") dynamical behavior has been found to be at least comparable to that of electronics. It must be added however that this result is, so far, strictly confined to paper. An additional characteristic shortcoming of exotic reaction systems is their being relatively slow-acting. On the other hand, chemical systems possess a constitutive asset too, at least in their nonstirred modification: they appear to be "pre-adapted" to a parallel processing activity. This fact seems to provide the reason for the (otherwise quite astonishing) heuristic model value of chemical broths for the understanding of neural computers.

Summary: On the basis of non-physical analogy between homogeneous reaction systems and electronic systems (concentrations corresponding to voltages), spatially homogeneous chemical analogs to a number of well-known electronic circuits can be devised *in abstracto* (amplifier, rectifier, RC oscillator, Eccles-Jordan trigger, multivibrator, time-base generator, monoflop, single-sweep time-base generator), some of which are identical to already existing abstract reaction systems. The fact that for the majority of these systems, several equivalent chemical circuits exist increases the chance for their eventual concrete realization.

Another analogy connects homogeneous chemical monoflops with neural elements. A homogeneous chemical system showing "classical conditioning" is indicated. Apart from "chemical wiring", diffusion-type wiring leads to analogs of simply organized neural nets. Here even a concrete example exists (Winfree's modified Belousov-Zhabotinsky reaction). Electronic computers of the same functional type ("extreme parallel processing") have not yet been realized. Also, a mathematical theory on the qualitative behavior of these dynamical systems is still lacking.

As a third group of "exotic" chemical systems, some abstract examples of "differentiating" and "morphogenetic" reaction systems are considered, both types being based on "soft" bifurcation. A two-cellular morphogenetic oscillator, involving hard bifurcation in addition, is also indicated.

References

Abakumov, A.S., A.A. Gulyayev, and G.A. Gulyayev: Propagation of excitation in an active medium, Biophysics 15, 1113-1120 (1970).

Abraham, N. and F. Bloch: Mesures en valeur absolue des périodes des oscillations électriques de haute fréquence, Ann. der Phys. (Ser. 9) 12, 237 (1919).

Andronov, A.A., A.A. Vitt, and S.E. Khaikin: Theory of Oscillators, Pergamon Press, Oxford and New York, 1966, pp. 309-312.

Anninos, P.A.: Mathematical model of memory trace and forgetfulness, Kybernetik 10, 165-170 (1972).

Balakhovsky, I.S.: Several modes of excitation movement in an ideal excitable tissue, Biophysics 10, 1175-1179 (1965).

Belousov, B.P.: Sborn. Referat. Radiats. Meditsin. za 1958 (1958 Collection of Abstracts on Radiation Medicine), Medgiz. Publ. House, Moscow, 1959, p. 145.

Bonhoeffer, K.F.: Zur Theorie des elektrischen Reizes, Naturwissenschaften 31, 270-275 (1943).

Bray, W.C.: J. Amer. Chem. Soc. 43, 1262 (1921).

Chirlian, P.M.: Electronic Circuits: Physical Principles, Analysis and Design, McGraw-Hill, New York, 1971.

Cowan, J.: Stochastic models of neuroelectric activity. In: Towards a Theoretical Biology, 4, C.H. Waddington, ed., University Press, Edinburgh, 1972, pp. 169-187.

Decker, P.: Evolution in open systems: bistability and the origin of molecular asymmetry, Nature New Biology, 241, 72-74 (1973).

Degn, H.: Oscillating chemical reactions in homogeneous phase, J. Chem. Educ., 49, 302-307 (1972).

Denzel, B.: Berechnungen zum Hystereseverhalten von Enzymsystemen mit Substrathemmung. Diploma Work, University of Tübingen (1973).

Eccles, W.H. and F.W. Jordan: A trigger relay utilizing three electrode thermionic vacuum tubes, Radio Review, 1, 143 (1919).

Edelstein, B.B.: Biochemical model with multiple steady states and hysteresis, J. theor. Biol., 29, 57-62 (1970).

Feinberg, M.: On chemical kinetics of a certain class, Arch. Rational Mech. Anal., 46, 1-41 (1972).

Field, R.J., E. Körös, and R.M. Noyes: Oscillations in chemical systems. II. Thorough analysis of temporal oscillation in the bromate-cerium-malonic-acid system, J. Amer. Chem. Soc., 94, 8649-8664 (1972).

Fitzhugh, R.: Impulses and physiological states in theoretical models of nerve membrane, Biophys. J., 1, 445-466 (1961).

Franck, U.F.: Modelle zur biologischen Erregung, Studium Generale, 18, 313-329 (1965).

Franck, U.F.: Oscillatory behavior, excitability and propagation phenomena on membranes and membrane-like interfaces. In: Biological and Biochemical Oscillators, ed. by B. Chance, A.K. Ghosh, E.K. Pye, and B. Hess, pp. 7-30. Academic Press, New York, 1973.

Frank, F.C.: On spontaneous asymmetric synthesis, Biochim. et Biophysica Acta 11, 459-463 (1953).

Gause, G.I.: The Struggle for Existence, Williams and Wilkins, Baltimore, 1934.

Gerisch, G.: Periodische Signale steuern die Musterbildung in Zellverbänden, Naturwissenschaften 58, 430-438 (1971).

Gierer, A. and H. Meinhardt: A theory of biological pattern formation, Kybernetik 12, 30-39 (1972).

Gul'ko, F.B. and A.A. Petrov: Mechanism of formation of closed pathways of conduction in excitable media, Biophysics 17, 271-281 (1972).

Güttinger, W.: Catastrophe geometry in physics and biology, These Proceedings (1974).

Horn, F.: Stability and complex balancing in mass-action systems with three short complexes, Proc. R. Soc. Lond. A, 334, 331-342 (1973).

Horn, F. and R. Jackson: General mass action kinetics, Arch. Rational Mech. Anal. 47, 81-116 (1972).

Hodgkin, A.L. and A.F. Huxley: A qualitative description of membrane current and its application to conduction and excitation in nerve, J. Physiol. 117, 500-544 (1952).

Hurley, R.B.: Transistor Logic Circuits, Wiley, New York, 1961.

Kacser, H.: Kinetic models in development and heredity, Symp. Soc. Exptl. Biol. 14, 13-27 (1960).

Karfunkel, H.R.: A computer program for the simulation of the temporal and spatial behavior of multi-component reaction-diffusion systems, Comput. Biol. Med. (in press) (1974).

Karfunkel, H.R. and F.F. Seelig: Reversal of inhibition of enzymes and the model of a spike oscillator, J. theor. Biol. 36, 237-253 (1972).

Kettner, F.: Zur Theorie und Phänomenologie elektrischer, chemischer, hydraulischer und mathematischer Analogmodelle biologischer Oszillatoren, Doctoral Thesis, Technical University of Aachen (1969).

Krinsky, V.I.: Spread of excitation in an inhomogeneous medium (state similar to cardiac fibrillation), Biophysics 11, 776-784 (1966).

Licko, V.: Some biochemical threshold mechanisms, Bull. math. Biophysics 34, 103-112 (1972).

Liénard, A.: Etude des oscillations entretenues (2 parts), Revue Génerale de l'Electricité 23, 901-912; 946-954 (1929).

Novak, B. and F.F. Seelig: Phase-shift model for the aggregation of cellular slime mold, a computer study, J. theor. Biol. (submitted) (1974).

Noyes, R.M., R.J. Field, and E. Körös: Oscillations in chemical systems. I. Detailed mechanism in a system showing temporal oscillations, J. Amer. Chem. Soc. 94, 1394-1395 (1972).

O'Neill, S.P., M.D. Lilly, and P.N. Rowe: Multiple steady states in continuous flow stirred tank enzyme reactors, Chem. Enging. Sci. 26, 173-175 (1971).

Oster, G., A.S. Perelson, and A. Katchalsky: Network thermodynamics: dynamic modelling of biophysical systems, Quarterly Rev. Biophys. 6, 1-134 (1973).

Rashevsky, N.: Über Hysterese-Erscheinungen in physikalischen und chemischen Systemen, Z. Physik 3, 102-106 (1929).

Rashevsky, N.: An approach to the mathematical biophysics of biological self-organization and of cell polarity; Further comments to the theory of cell polarity and self-regulation; and Physicomathematical aspects of some problems of organic form, Bull. math. Biophysics, 2, 15-25; 65-67; and 109-121 (1940).

Rashevsky, N.: Mathematical Biophysics, 3rd ed., Vol. II, Dover Publications, New York, 1960, pp. 69-73.

Reissig, G., G. Sansone, and R. Conti: Qualitative Theorie nichtlinearer Differentialgleichungen, Edizione Cremonese, Rome, 1963.

Rosen, R.: Dynamical System Theory in Biology I, Wiley-Interscience, New York, 1970.

Rosenschöld, P.S. Munck af: Poggendorf's Annalen der Physikalischen Chemie, 32, 216 (1834).

Rössler, O.E.: A system-theoretic model of biogenesis (in German), Z. Naturforsch. 26b, 741-746 (1971).

Rössler, O.E.: Grundschaltungen von flüssigen Automaten und Reaktionssystemen (Basic circuits of fluid automata and relaxation systems), Z. Naturforsch. 27b, 333-343 (1972a).

Rössler, O.E.: A principle for chemical multivibration, J. theor. Biol. 36, 413-417 (1972b).

Rössler, O.E.: Existenz eines Bausteinprinzips beim Entwurf von komplizierten Reaktionssystemen, Habilitationsschrift, University of Tübingen (1973).

Rössler, O.E.: Chemical automata in homogeneous and reaction diffusion kinetics, These Proceedings (1974a).

Rössler, O.E.: A multivibrating switching network in homogeneous kinetics, Bull. math. Biol. (in press) (1974b).

Rössler, O.E. and D. Hoffmann: Repetitive hard bifurcation in a homogeneous reaction system. In: Analysis and Simulation of Biochemical Systems, H.C. Hemker & B. Hess, eds. North Holland and Elsevier, Amsterdam and New York, 1972, pp. 91-101.

Rössler, O.E. and F.F. Seelig: A Rashevsky-Turing system as a two-cellular flip-flop, Z. Naturforsch. 27b, 1445-1448 (1972).

Seelig, F.F.: Undamped sinusoidal oscillations in linear chemical reaction systems, J. theor. Biol. 27, 197-206 (1970).

Seelig, F.F.: Activated enzyme catalysis as a possible realization of the stable linear chemical oscillator model, J. theor. Biol. 30, 497-514 (1971a).

Seelig, F.F.: Systems-theoretic model for the spontaneous formation of optical antipodes in strongly asymmetric yield, J. theor. Biol. 31, 355-361 (1971b).

Seelig, F.F. and B. Denzel: Hysteresis without autocatalysis: simple enzyme systems as possible binary memory elements, FEBS Letters 24, 283-286 (1972).

Seelig, F.F. and F. Göbber: Stable linear reaction oscillator, J. theor. Biol. 30, 485-496 (1971).

Smale, S.: On periodic attractors in cell biology, Lecture held at the University of Tübingen, November, 1972.

Thom, R.: Topological models in biology. In: Towards a Theoretical Biology, Vol. 3, C.H. Waddington, ed., University Press, Edinburgh, 1970, pp. 89-116.

Thom, R.: Stabilité Structurelle et Morphogênese, Essai d'une Théorie Générale des Modeles, Benjamin, Reading, Mass., 1972.

Turing, A.M.: The chemical basis of morphogenesis, Phil. Trans. Roy. Soc., London, Ser. B, 237, 37-72 (1952).

Van der Pol, B. and J. van der Mark: The heartbeat considered as relaxation oscillation and an electrical model of the heart, Phil. Mag. 6, 763 (1928).

Vavilin, V.A. and A.M. Zhabotinsky: Autocatalytic oxidation of trivalent cerium by bromate ion, I, Kinetika Kataliz 10, 83-88 (in Russian) (1969).

Walter, W.G.: The Living Brain, Duckworth, London, 1953.

Wiener, N. and A. Rosenblueth: The mathematical formulation of the problem of conduction of impulses on a network of excitable elements, specifically in cardiac muscle, Arch. Inst. Cardiologia de Mexico 16, 105 (1953) (re-evaluated in Abakumov et al, 1970).

Winfree, A.T.: Spiral waves of chemical activity, Science 175, 634-636 (1972).

Winfree, A.T.: Spatial and temporal organization in the Zhabotinsky reaction. In: Aharon Katchalsky Memorial Symposium, Berkeley, California, March 21, 1973 (Preprint).

Winfree, A.T.: Wave-like activity in biological and chemical media. In: Lecture Notes in Biomathematics, P. van den Driessche, ed., Springer Verlag, Berlin-Heidelberg-New York (forthcoming) (1974).

Zeeman, E.C.: Topology of the Brain (Mathematics and Computer Sciences in Biology and Medicine), Medical Research Council, 1965.

Zeeman, E.C.: Differential equations for the heartbeat and nerve impulse. In: Towards a Theoretical Biology, Vol. 4, C.H. Waddington, ed., Edinburgh University Press, Edinburgh, 1972, pp. 8-67.

Zhabotinsky, A.M.: Periodic oxidation reactions in the liquid phase, Dokl. Akad. Nauk. SSSR 157, 392-395 (1964) (Cover-to-cover translation, pp. 701-704).

CONCLUDING REMARKS

J.G. Taylor
Department of Mathematics
King's College, London, England

There are over a million species of living neural nets. Yet we are still far from understanding a single one of them. The talks given at this summer school indicated that a great deal is known about such systems, and progress is certainly being made. The mathematics and physics of nervous systems is based upon models of these systems. There is the most basic question of all: what sort of model is the most appropriate? It is not clear that a single model is satisfactory; possibly various ones are needed to understand different aspects of brain activity. When we turn to more general properties of nervous systems then a multiplicity of models is to be expected.

There are at present different types of models being discussed, and representatives of each type were presented during the summer school. We can classify these types as:

a) descriptive, e.g. the cerebellum as giving a loop control of movement, or the cerebral cortex as containing feature detectors.
b) wiring diagrams, e.g. the connectivity producing lateral inhibition in the retina, or the retinal system of the fly to produce the optomotor response.
c) information-theoretic, e.g. the removal of redundancy of the feature detectors in the cerebral cortex.
d) small neural nets
e) large neural nets
f) abstract models, e.g. the theory of automata.

There is overlap between these categories, especially (b), (d) and (f), but this is reduced if the approach in small or large neural nets includes information processing more completely than in (b).

These various levels of modelling require different types of mathematics -- system theory, information theory, non-linear equations, matrix theory, group theory, etc. One crucial level that was represented much less than might be expected is that of artificial intelligence, both in practice -- pattern recognition, automatic problem solvers -- as well as at the theoretical level. The mathematical theory of artificial intelligence is not at all well developed. It is clear that this must be remedied; without it the problem of the intelligent brain will remain unsolved.

One clear pitfall which seems to have claimed its due toll of victims, is the neglect of the data. This is partly due to the fact that the most suitable blend of mathematical disciplines relevant to brain research has not yet been discovered. To theorists this is the crucial challenge. But it can only be resolved by turning to the living neural nets and attempting to understand them directly. We ignore the data at our peril!